Methods in Cell Biology

VOLUME 110

Computational Methods in Cell Biology

Series Editors

Leslie Wilson

Department of Molecular, Cellular and Developmental Biology

University of California

Santa Barbara, California

Paul Matsudaira

Department of Biological Sciences

National University of Singapore

Singapore

Methods in Cell Biology

VOLUME 110

Computational Methods in Cell Biology

Edited by

Anand R. Asthagiri

Department of Chemical Engineering, Northeastern University, Boston, MA, USA

Adam P. Arkin

Department of Bioengineering, University of California, Berkeley, CA, USA

ELSEVIER

AMSTERDAM • BOSTON • HEIDELBERG • LONDON
NEW YORK • OXFORD • PARIS • SAN DIEGO
SAN FRANCISCO • SINGAPORE • SYDNEY • TOKYO

Academic Press is an imprint of Elsevier

Academic Press is an imprint of Elsevier
225 Wyman Street, Waltham, MA 02451, USA
525 B Street, Suite 1900, San Diego, CA 92101-4495, USA
32, Jamestown Road, London NW1 7BY, UK
Linacre House, Jordan Hill, Oxford OX2 8DP, UK

First edition 2012

ISBN: 978-0-12-388403-9
ISSN: 0091-679X

For information on all Academic Press publications visit our website at elsevierdirect.com

Printed and bound by CPI Group (UK) Ltd, Croydon, CR0 4YY

Transferred to Digital Print 2012

CONTENTS

CONTRIBUTORS

Numbers in parentheses indicate the pages on which the author's contributions begin.

Mark Alber (367), Department of Applied and Computational Mathematics, University of Notre Dame, Notre Dame, Indiana, USA; Department of Medicine, Indiana University School of Medicine, Indianapolis, Indiana, USA

Alphan Altinok (285), Division of Biology, California Institute of Technology, Pasadena, California, USA

Anil Aswani (243), Department of Electrical Engineering and Computer Sciences, University of California, Berkeley, California, USA

Julio M. Belmonte (325), Department of Physics, Biocomplexity Institute, Indiana University, Bloomington, Indiana, USA

Peter Bickel (243), Department of Statistics, University of California, Berkeley, California, USA

Mark D. Biggin (243, 263), Genomics Division, Lawrence Berkeley National Laboratory, Berkeley, California, USA

Richard Bonneau (19), Department of Biology, Center for Genomics and Systems Biology, New York University, New York, NY, USA; Computer Science Department, Courant Institute of Mathematical Sciences, New York, NY, USA

Joseph P. Califano (139), Department of Biomedical Engineering, Cornell University, Ithaca, New York, USA

Shawn P. Carey (139), Department of Biomedical Engineering, Cornell University, Ithaca, New York, USA

Scott Christley (367), Department of Surgery, University of Chicago, Chicago, Illinois, USA

Ann E. Cowan (195), R. D. Berlin Center for Cell Analysis and Modeling, University of Connecticut Heath Center, Farmington, CT, USA

Alexandre Cunha (285), Center for Advanced Computing Research, Division of Engineering and Applied Science, California Institute of Technology, Pasadena, California, USA; Center for Integrative Study of Cell Regulation, California Institute of Technology, Pasadena, California, USA

Hana El-Samad (111), Department of Biochemistry and Biophysics, California Institute for Quantitative Biosciences, University of California, San Francisco, CA

Ernest Fraenkel (57), Department of Biological Engineering, Massachusetts Institute of Technology, Cambridge, Massachusetts, USA; Computer Science and Artificial Intelligence Laboratory, Massachusetts Institute of Technology, Cambridge, Massachusetts, USA

Ambhighainath Ganesan (1), Department of Biomedical Engineering, Johns Hopkins University, Baltimore, Maryland, USA
James A. Glazier (325), Department of Physics, Biocomplexity Institute, Indiana University, Bloomington, Indiana, USA
Alex Greenfield (19), Computational Biology Program, New York University School of Medicine, New York, NY, USA
Cameron Harvey (367), Department of Physics, University of Notre Dame, Notre Dame, Indiana, USA
Jason M. Haugh (223), Department of Chemical and Biomolecular Engineering, North Carolina State University, Raleigh, North Carolina, USA
Dimitrij Hmeljak (325), Department of Physics, Biocomplexity Institute, Indiana University, Bloomington, Indiana, USA
Alexander Hoffmann (81), Signaling Systems Laboratory, San Diego Center for Systems Biology of Cellular Stress Responses, Program in Bioinformatics and Systems Biology, University of California San Diego, La Jolla, California, USA
Shao-shan Carol Huang (57), Department of Biological Engineering, Massachusetts Institute of Technology, Cambridge, Massachusetts, USA; Current address: Plant Molecular and Cellular Biology Laboratory, Salk Institute for Biological Studies, La Jolla, California, USA
Tommy Kaplan (263), Department of Molecular and Cell Biology, California Institute of Quantitative Biosciences, University of California, Berkeley, California, USA; School of Computer Science and Engineering, The Hebrew University, Jerusalem, Israel
Oleg Kim (367), Department of Applied and Computational Mathematics, University of Notre Dame, Notre Dame, Indiana, USA
Casey M. Kraning-Rush (139), Department of Biomedical Engineering, Cornell University, Ithaca, New York, USA
Andre Levchenko (1), Department of Biomedical Engineering, Johns Hopkins University, Baltimore, Maryland, USA
Joshua Lioi (367), Department of Applied and Computational Mathematics and Statistics, University of Notre Dame, Notre Dame, Indiana, USA
Leslie M. Loew (195), R. D. Berlin Center for Cell Analysis and Modeling, University of Connecticut Heath Center, Farmington, CT, USA
Paul M. Loriaux (81), Signaling Systems Laboratory, San Diego Center for Systems Biology of Cellular Stress Responses, Program in Bioinformatics and Systems Biology, University of California San Diego, La Jolla, California, USA
Elliot M. Meyerowitz (285), Division of Biology, California Institute of Technology, Pasadena, California, USA
Eric Mjolsness (285), Department of Computer Science, University of California, Irvine, California, USA
Ion I. Moraru (195), R. D. Berlin Center for Cell Analysis and Modeling, University of Connecticut Heath Center, Farmington, CT, USA

Robert F. Murphy (179), Lane Center for Computational Biology, Carnegie Mellon University, Pittsburgh, Pennsylvania, USA; Freiburg Institute for Advanced Studies, University of Freiburg, Freiburg, Germany

Christopher S. Poultney (19), Department of Biology, Center for Genomics and Systems Biology, New York University, New York, NY, USA

Cynthia A. Reinhart-King (139), Department of Biomedical Engineering, Cornell University, Ithaca, New York, USA

Adrienne H.K. Roeder (285), Center for Integrative Study of Cell Regulation, California Institute of Technology, Pasadena, California, USA; Division of Biology, California Institute of Technology, Pasadena, California, USA

Elliot D. Rosen (367), Department of Medical and Molecular Genetics, Indiana University School of Medicine, Indianapolis, Indiana, USA

James C. Schaff (195), R. D. Berlin Center for Cell Analysis and Modeling, University of Connecticut Heath Center, Farmington, CT, USA

Abbas Shirinifard (325), Department of Physics, Biocomplexity Institute, Indiana University, Bloomington, Indiana, USA

Boris M. Slepchenko (195), R. D. Berlin Center for Cell Analysis and Modeling, University of Connecticut Heath Center, Farmington, CT, USA

Jacob Stewart-Ornstein (111), Department of Biochemistry and Biophysics, California Institute for Quantitative Biosciences, University of California, San Francisco, CA

Wenzhao Sun (367), Department of Applied and Computational Mathematics, University of Notre Dame, Notre Dame, Indiana, USA

Maciej H. Swat (325), Department of Physics, Biocomplexity Institute, Indiana University, Bloomington, Indiana, USA

Paul T. Tarr (285), Division of Biology, California Institute of Technology, Pasadena, California, USA

Gilberto L. Thomas (325), Department of Physics, Biocomplexity Institute, Indiana University, Bloomington, Indiana, USA; Instituto de Física, Universidade Federal do Rio Grande do Sul, C.P. 15051, Porto Alegre, Brazil

Claire Tomlin (243), Department of Electrical Engineering and Computer Sciences, University of California, Berkeley, California, USA

Erik S. Welf (223), Department of Chemical and Biomolecular Engineering, North Carolina State University, Raleigh, North Carolina, USA

Zhiliang Xu (367), Department of Applied and Computational Mathematics, University of Notre Dame, Notre Dame, Indiana, USA

PREFACE

Computation is an essential part of the cell biologist's toolbox. The value of computation in analyzing systems involving numerous, interconnected mechanisms has long been appreciated. Computational models provide a framework not only to formally represent and simulate the mechanisms, but also predict the response of an integrated system to new perturbations and thereby lead to testable hypotheses. In this way, computational modeling and analysis can suggest new experiments that challenge and help revise our mechanistic understanding of the cell system.

Prediction and hypothesis-generation, however, tells only part of the story. The need for computation is now far more pervasive in cell biology. Cell biological data is increasingly gathered with high bandwidth, often exploiting heterotypic measurement modalities. This flood of data includes changes in gene expression, post-translational modifications, and the subcellular location of key regulatory events. The '-omic' scale *in vivo* imaging of spatiotemporal patterns in gene expression during the development of model organisms is a compelling example. Extracting meaningful data from such images is a key challenge and involves reliable segmentation, annotation, storage and data management, bioinformatics, and data mining.

Having acquired the data, one seeks to infer salient mechanistic relationships and models. Deriving a model of how a system works based on experimental data is, of course, not new. The challenge now is that the volume, the spatiotemporal resolution, and the heterotypic nature of the data make such inferences difficult to execute by intuition alone. Computational algorithms to sift through the data and extract models consistent with the data are essential. Furthermore, model schematics, whether derived by computation or intuition, are conceptual until they are used to generate concrete, testable predictions. Making such predictions, however, is encumbered by a dearth of information regarding parameter values and by the fact that cellular mechanisms often operate over multiple time and spatial scales, in many cases combining biochemical and mechanical elements. Thus, inferring *computable* models that are amenable to simulation requires inference not only of the mechanistic connections, but also the parameters that describe the strength of those connections and interactions.

This remarkable breadth of applications of computation in cell biology impresses the fact that computation is more than a module in a multi-step process that involves iterative feedback between model and experiment. It is also increasingly integral to how data is gathered and interpreted, how mechanistic models are inferred, and how new mechanisms are hypothesized and uncovered. This volume captures this broad integration of computation in experimental cell biology. The volume covers the role of computation in the extraction of quantitative information from raw data; inference of mechanistic computable (i.e., parameterized) models from large, heterotypic datasets; and prediction and hypothesis-generation to drive new experiments.

The contributors to this volume were presented with a difficult challenge: to tailor each chapter in a way that provides both high-level and in-depth tutorials of key computational methods, while also communicating the biological question that inspired the computational approach and the biological insights that were uncovered. The contributors have, in our opinion, succeeded admirably in tackling this challenge. The chapters are organized into three parts that focus on (1) molecular regulatory networks, (2) spatial and biophysical aspects of cell regulation, and finally (3) multicellular systems. Each part of the volume contains chapters that deal with the different applications of computation in cell biology: measurements and data extraction, model development and inference, and prediction and hypothesis generation.

With acknowledgment and deepest gratitude to the tremendous efforts of the contributors and to the many anonymous peer reviewers, we are pleased to present this volume and trust that it will provide inspiration and instructive tutorial in your search for the right computational tool for your cell biology quest.

Anand R. Asthagiri

Department of Chemical Engineering,
Northeastern University, Boston,
Massachusetts, USA

Adam P. Arkin

Department of Bioengineering,
University of California, Berkeley,
California, USA

September 30, 2011

CHAPTER 1

Principles of Model Building: An Experimentation-Aided Approach to Development of Models for Signaling Networks

Ambhighainath Ganesan and Andre Levchenko

Department of Biomedical Engineering, Johns Hopkins University, Baltimore, Maryland, USA

Abstract

Living cells continuously probe their environment and respond to a multitude of external cues. The information about the environment is carried by signaling cascades that act as "internal transducing and computing modules," coupled into complex and interconnected networks. A comprehensive understanding of how cells make decisions therefore necessitates a sound theoretical framework, which can be achieved through mathematical modeling of the signaling networks. In this chapter, we conceptually describe the typical workflow involved in building mathematical models that are motivated by and are developed in a tight integration with experimental analysis. In particular, we delineate the steps involved in a generic, iterative experimentation-driven model-building process, both through informal discussion

0091-679X/10 $35.00
DOI 10.1016/B978-0-12-388403-9.00001-1

and using a recently published study as an example. Experiments guide the initial development of mathematical models, including choice of appropriate template model and parameter revision. The model can then be used to generate and test hypotheses quickly and inexpensively, aiding in judicious design of future experiments. These experiments, in turn, are used to update the model. The model developed at the end of this exercise not only predicts functional behavior of the system under study but also provides insight into the biophysical underpinnings of signaling networks.

I. Introduction

Models represent useful abstractions of reality, and are clearly a part of how we learn about and understand various aspects of the world around us. These models we all seem to have are frequently conceptual, but they can also be quite quantitative, for instance in developing intuitive predictive abilities in applying just enough force in lifting a full glass of water, in applying brakes while driving on a busy road, or in catching a ball during a windy afternoon game. Many initial models, formed in the early childhood, turn out to be wrong or overly simplistic when faced with increasingly complex realities of testing them in a real world. Arguably following a very similar tendency, we try to build models while engaged in scientific research. These models also frequently start out as simplistic and largely incorrect during the infancy of a scientific discipline, being gradually refined and honed as they face the reality checks provided by experimental analysis. As experiments become more precise, they provide more stringent tests of related models, enabling model development in more quantitative, mathematical fashion. This gradual refinement of our understanding of a particular phenomenon through iteration of modeling and experiment is at the heart of the scientific method itself, and as models are never complete, nor are they meant to be, they provide us with the best hope for making continuous progress in furthering our understanding of complex processes.

Computational models of biological phenomena, particularly at the level of description of subcellular, molecular processes, are still very much in their infancy. And it is still very much unclear what metaphors and what mathematical and computational concepts and tools would provide a useful platform for developing these models. Arguably, the ultimate test here is again provided through a tight linkage between the model and experiment. If a model, whatever its mathematical embodiment or degree of sophistication, is able to provide a hitherto unavailable insight, useful generalization or abstraction, or make unanticipated prediction, its value becomes increasingly high, a kind of justification of its use *a posteriori*. In this regard, many models, based on and using the concepts and mathematics for applications in engineering and physics, turn out to be still quite predictive and thus justified in their use, when applied to biological processes. In this chapter, we provide some examples of such models.

Recently, the "classical" approaches to model development described above have gradually become challenged due to the exceedingly rapid progress in how many biological variables can be measured, and how fast it can be done. In many ways, there is now a requirement for building models that need to be multidimensional from the start, dealing with hundreds or even thousands of simultaneously measured entities, accompanying complex biological events. In some ways, this rapid technological development heralded the "age of Kepler" in biology, the age of finding statistical relations that can capture many aspects of the biological processes, while also being predictive of the ultimate outcomes. However, the understanding of the primary causes of relations between the underlying ultimate controlling processes may still be awaiting the "age of Newton," the age of important conceptual breakthroughs. Arguably, these breakthroughs in understanding the processes described from the large-scale, multivariable, "bottom up" perspective can and will emerge from a more of a classical, iterative "top-down" description, aimed at accounting for processes, however ostensibly complex, using the simplest models possible.

II. Signaling Systems and Mathematical Models

An area of biological research that has been extremely amenable to and hence benefited from mathematical modeling is the study of signaling systems, largely facilitated by the fact that signaling cascades are, in their basest form, nothing but elementary chemical reactions. What started as an attempt to quantitatively describe the action of a single enzyme by Leonor Michaelis, Maud Menten, and others has now blossomed into a full-fledged undertaking to model the workings of large signaling cascades involving not one but scores of enzymes, their substrates, and myriad other biomolecules. The fact that steps in a signaling cascade can be construed as chemical reactions lends itself easily to the development of mathematical models comprising simple Ordinary Differential Equations (ODEs), each of which describes a particular reaction. The real power of these models, however, arises from the fact that signaling systems are built not unlike many man-made control systems, replete with nonlinear connections between different components. The presence of such nonlinear connections, which give rise to interesting dynamics, makes it difficult to intuitively predict the response of a system under different conditions. Mathematical models are immensely useful in not only helping us quantitatively predict system responses but also allowing us to generate additional hypotheses for experimental testing. More and more frequently, models of signaling networks provide insights into the abstract principles that have guided Nature in the evolutionary "design" of signaling networks, facilitating efforts in understanding of the existing and (re-)designing of novel networks of desired properties (Lim, 2010; Antunes *et al.*, 2009; Toettcher *et al.*, 2010).

III. Experimentation-aided Model Development

Nonlinear connections in signaling cascades inherently give rise to dynamically interesting behavior such as oscillations. Over the course of evolution, life forms seem to have exploited such temporal dynamics to their advantage. Crucial information about external stimuli can be embedded in the parameters of oscillations such as frequency and amplitude (Cheong and Levchenko, 2010). Oscillations are exhibited by multiple signaling systems and are thought to underlie the rhythmic beating of the heart, insulin secretion, and memory formation. Central to many such oscillating systems is the ubiquitous second messenger, calcium. This, combined with a long history of observations of intracellular calcium dynamics, has led to many mathematical models describing putative mechanisms of how calcium oscillations can arise in diverse settings (see reviews: (Schuster *et al.*, 2002; Dupont *et al.*, 2011). However, experimental validation of such models is frequently not undertaken. More recently, we and many other groups have used experimental monitoring of calcium oscillations in conjunction with computational models to address basic questions of signaling. In this chapter, we will illustrate the experimentation-aided development, refinement, and implementation of computational models in general, using the specific example of a recent modeling–experimental analysis project (Ni *et al.*, 2011), which we believe captures many archetypal features of an integrative effort relying on both modeling and experimental research in equal measure. The principles involved in such a model-building process can be generalized as follows:

1. Template identification: Mathematical models have already been built for many signaling cascades. A good starting point therefore in the development of a computational model may therefore be to identify an existing model that is suitable for the system of choice.
2. Module development: No models are ever complete. Experimental results could identify new components or links in a signaling pathway, which need to be incorporated in the model. Newly identified components and links often occur as subsystems, which could be modeled semi-independently as individual modules.
3. Architectural revision: Although the modules within a complex model could be relatively independent of each other, the modules have to be integrated in a seamless manner in the final model to reproduce the experimental results. In many cases, all the links between different components may not be known, in which the most likely configuration of a signaling network has to be selected.
4. Model simulations: The complete model is then simulated to replicate experimental results and to generate additional hypotheses, which are subsequently verified by experiments.

Although, these principles have been enumerated as a defined list, it is not uncommon to employ them in a combined and iterative fashion.

A. Template Identification

As the efforts to employ modeling in the analysis of biological processes accelerate, it is becoming more and more common to initiate the analysis, with existing prior art in the form of published models capturing a certain aspect of the biological process of interest. Models for calcium oscillations, for instance, abound in literature (Schuster *et al.*, 2002; Dupont *et al.*, 2011). Choosing the right template to start with depends mainly on the system under study and the purpose of the intended final model. For example, stochastic models are immensely helpful in addressing questions of dynamics pertaining to single molecules, such as ion channels (Dupont *et al.*, 2008; Cannon *et al.*, 2010), or in the exploration of how noise affects transcriptional control and signaling dynamics (Roberts *et al.*, 2011; Ko *et al.*, 2010). Multiscale models may be necessary to explain tissue-level functionalities using molecular mechanisms (See reviews: (Greenstein and Winslow, 2011; Du *et al.*, 2010). In most cases, however, simple deterministic models can serve the purpose of explaining how signaling cascades regulate cellular functions. When setting out to address potential mechanisms of cross-talk between calcium, cyclic adenosine monophosphate (cAMP), and protein kinase A (PKA) signaling, our choice of the template model was based on our initial experimental results and the particular set of questions we sought to address. The fact that intracellular levels of calcium oscillate in pancreatic β-cells upon membrane depolarization has been well documented for more than two decades (Bertram *et al.*, 2010; Grapengiesser *et al.*, 1988; Corkey *et al.*, 1988). Using genetically encoded biosensors based on Förster Resonance Energy Transfer (FRET) (Zhang *et al.*, 2001; DiPilato *et al.*, 2004), we observed that cAMP and the associated kinase, PKA, also exhibited temporal oscillatory dynamics, matching the calcium oscillations in a calcium-dependent manner. Further, our experiments indicated that calcium oscillations could be regulated by PKA through a putative feedback loop. These experiments presented us with the need and the opportunity to expand current models describing the generation and regulation of calcium oscillations. The main purpose of our modeling analysis was then to develop a minimal yet sufficiently detailed model accounting for and constrained by the experimented observations.

We chose the original Chay–Keizer model (Chay and Keizer, 1983) and its later version detailed by Sherman, Li, and Keizer (Sherman *et al.*, 2002) as our initial modeling templates and thus built a "voltage module" to describe the membrane potential dynamics and its connection to calcium concentration dynamics. When the parameters present in the original model were used, we observed that the frequency of oscillations was much lower than the frequency of oscillations experimentally observed. Hence, some of the parameter values needed to be refined in order to corroborate the experimental data. Certain parameters are likely cell-type specific and hence are most likely to vary when distinct experimental systems are analyzed. The net conductance of a type of channels in a cell, for instance, is a function of the number of available channels, and so conductance values vary among different cell types. In our model, the conductance of voltage-dependent calcium channels (VDCCs) was altered, as has been similarly done in a model published by Fridlyand *et al.* (Fridlyand *et al.*, 2007). To match the change in VDCC activity owing to this reduced conductance and to match the experimentally observed frequency of oscillations, the conductance of the delayed rectifier K^+ channels also was reduced. The value of another parameter that reflects the fraction of free calcium in a cell was chosen by Chay and Keizer on the basis of the time scale of the oscillations in their model, in the absence of any concrete experimental measurements. Because the frequency of oscillations experimentally observed in our system was much lower than that in the study by Chay and Keizer, this parameter was also reduced accordingly. Apart from these frequency-related parameters, the reversal potential for VDCCs, E_{Ca}, was modified to 100 mV, as previous reports have consistently used this value (Chay and Keizer, 1983; Bertram *et al.*, 2000). Following these modifications, the calcium module (see "B. Module Development") was developed to simulate the calcium dynamics. Simulations of the modified template model with the new parameter values matched the experimental results quite well (Fig. 1).

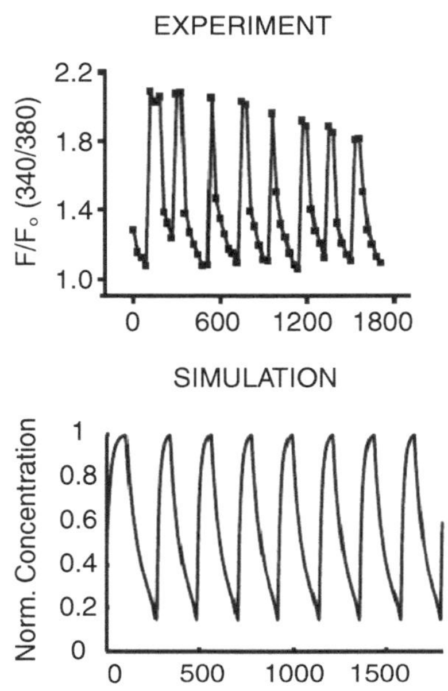

Fig. 1 (A) Experimental measurements of intracellular calcium in a single MIN6 pancreatic β-cell. MIN6 cells were loaded with Fura2-AM (2 μM), and the ratio of excitation at 340 nm to that at 380 nm was recorded. (B) Simulations of the modified template model with new parameters. F, fluorescence intensity at 340 nm; Fo, fluorescence intensity at 380 nm.

B. Module Development

Regulatory biochemical networks and the corresponding mathematical and computational models can be frequently decomposed into constituent "modules," subsets of reactions that always occur in particular combination, whenever they are encountered. Modules can have variable "linkages," or cross-activation reactions, which may be cell- or tissue-specific, or display other types of specificity. One can therefore first attempt to model each of the modules in detail, and then attempt to understand how they may be "linked" in a particular system of interest. Below we describe examples of modules modeled within the context of the pancreatic β-cell signaling.

1. Calcium Module

Apart from calcium influx across the plasma membrane, calcium release from the internal stores is also assumed to play a major role in many processes that involve calcium oscillations, as evidenced by the glucose-regulated expression of IP_3 receptors (IP_3Rs) in rat pancreatic islets (Bezprozvanny *et al.*, 1991). Hence, we derived equations to describe the net calcium release from and uptake by internal stores mediated by IP_3Rs and Smooth Endoplasmic Reticulum Calcium ATPases (SERCAs), respectively. As in the case of the voltage module, we made initial use of the models by Gorbunova and Spitzer (Gorbunova and Spitzer, 2002) and Tang and Othmer (Tang and Othmer, 1995) as templates for modeling calcium release from internal stores. Although the equations of the voltage module employ parameters that are specifically suited for the pancreatic β-cell system, the equations for flux across the internal stores employed by Gorbunova and Spitzer and by Tang and Othmer were developed in the context of aplysia neurons and cardiac myocytes, respectively. Hence, using the equations with the values derived from these models could lead to oscillations with frequencies different from that observed in our experiments. The parameters defining IP_3R density, for instance, or others governing the flux through internal stores would be expected to differ when used in the pancreatic β-cell system. In the absence of any published values for the number or density of IP_3Rs in these or other cell types, we again resorted to matching of experimental results and model simulations. Similarly, the parameters pertaining to SERCA activity were also refined so as to ensure robust oscillations of frequencies matching those in the initial experimental results shown in Fig. 1. This process is an integral part of model "training." The trained model was then used to make further predictions.

2. cAMP and PKA Modules

We used kinetic parameters published earlier (see Supplementary material of the study by Bhalla and Iyengar (Bhalla and Iyengar, 1999) to develop the cAMP and PKA modules. Certain parameters pertaining to kinetics of adenylyl cyclases (ACs, enzymes that synthesize cAMP) were also derived from other studies with appropriate assumptions (see Supplementary methods of (Ni *et al.*, 2011) for complete

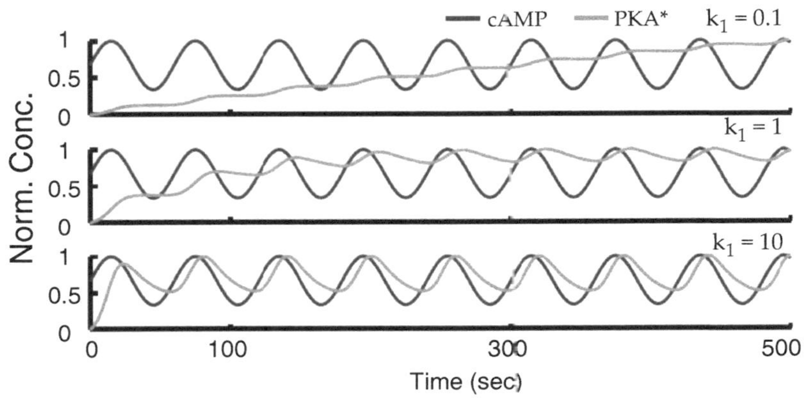

Fig. 2 Simulation of PKA activity (PKA*) in the presence of oscillatory cAMP, showing different activity patterns depending on the characteristics of the oscillations and parameters of PKA activation and deactivation. The parameter, κ_1, reflecting the binding of cAMP to PKA homodimer was varied in this simulation. The parameter κ_1 is the ratio of the new value to the nominal value of κ_1. cAMP, cyclic adenosine monophosphate; PKA, protein kinase A. (For color version of this figure, the reader is referred to the web version of this book.)

derivation of the equations). The advantage of developing such semi-independent modules is that they could be used to test certain hypotheses even at the early stages of model development. These results may further guide the development of appropriate experiments and in refining of the model itself. For instance, in the calcium–cAMP–PKA system that we were investigating, we had experimentally observed that cAMP could oscillate in tandem with calcium oscillations. Given that cAMP oscillates, we wondered if it automatically translates into PKA oscillations. In order to test this hypothesis, we used the PKA module independently and simulated PKA dynamics in response to a sinusoidal cAMP input signal. In theory, as the model results suggested, cAMP input can lead to diverse PKA dynamics (Fig. 2) depending on the value of a parameter that reflects the binding constant of cAMP to the PKA homodimer. Based on these modeling results, we were able to conclude that cAMP oscillations do not always necessarily translate into PKA oscillations. Based on these modeling results, we decided to monitor PKA activity dynamics concurrently with calcium dynamics. Using a FRET-based biosensor, the A Kinase Activity Reporter (AKAR), to monitor changes in PKA activity in live cells, we observed that PKA activity does in fact oscillate in tandem with calcium oscillations – an observation that formed the central basis of our whole study.

C. Architectural Revision

Modules constituting complex regulatory systems may be linked in a variety of ways, displaying cell, tissue, and condition specificity. The number of such linkages in fact can be combinatorially large for an increasing number of modules. Thus, if one was to include all possible regulatory interactions and feedbacks regulating individual modules, the resulting model can be too complex to be of predictive value, and more importantly, potentially irrelevant to the particular cell type and regulatory situation considered in the model. If, therefore, there is a way to restrict the type and number of putative regulatory links in the modeled system, the resulting model can be more powerful, relevant, and predictive. Again, contrasting model predictions with experimental observations can be of considerable help. An example of such a process is described below.

Several possible simplified versions of how the calcium and cAMP modules in the signaling network of pancreatic β-cells might interlink were initially considered for consistency with the experimental data (Fig. 3). In particular, we focused on the need by the model to account for the experimental observation that calcium and cAMP oscillations are out of phase. This could be explained by the activation of calcium-inhibited ACs (enzymes that synthesize cAMP) and/or calcium-activated phosphodiesterases (PDEs, enzymes that degrade cAMP). We therefore explored different possible combinations in which the components of the signaling circuit could be connected as shown in Fig. 3. Furthermore, the experiments suggested that calcium rise phase and cAMP decay phase in each peak are coincident and very sharp, suggesting another criterion for the model "pruning."

ACs can be activated by calcium-calmodulin ($CaM.Ca_4$) or inhibited by calcium. There are therefore two ways to represent the calcium-AC link in the model in terms of kinetic parameters (which are equivalent to exponents in S-system models (Voit, 2000): inactivation by calcium is denoted by −1, whereas no dependence on calcium

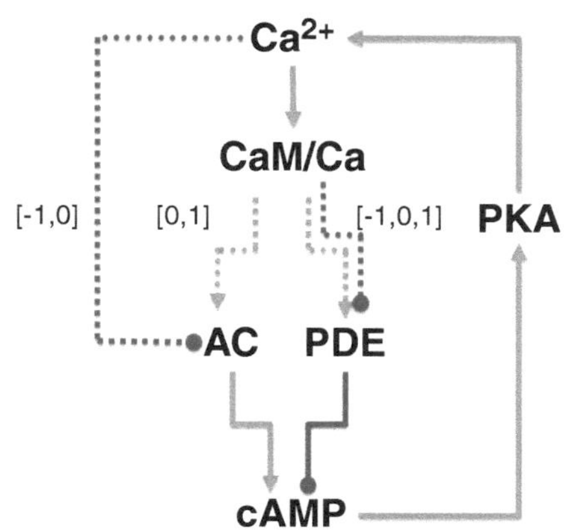

Fig. 3 Possible topologies for the circuit with the components shown. Solid lines indicate fixed connections. Dotted lines indicate variable connections. The numbers −1, 0, and 1 indicate if the link is inhibitory, absent, or activating, respectively. AC, adenylyl cyclase; cAMP, cyclic adenosine monophosphate; CaM, calmodulin; PDE, phosphodiesterases; PKA, protein kinase A. (For color version of this figure, the reader is referred to the web version of this book.)

is denoted by 0. Similarly, the effect of $CaM.Ca_4$ on AC can be represented in two ways: activation by $CaM.Ca_4$ is denoted by 1, whereas no dependence on $CaM.Ca_4$ is denoted by 0. PDEs likewise can be activated or inactivated by $CaM.Ca_4$ or can be independent of $CaM.Ca_4$ activity. Following the same logic, the parameter for this link can be represented as 1, 1, or 0, respectively. Accordingly, we have $2 \times 2 \times 3 = 12$ possible circuits.

As the experimental criteria used in the analysis are essentially dynamic, we developed a "Time Delay Metric," whose value quantifies the phase delay between the calcium and cAMP oscillations, using a circular cross-correlation function as a function of time of oscillations. To quantify the "sharpness" of the calcium rise and cAMP decay phases, we determined the time taken to reach half maximum or minimum ($t_{1/2}$) by these species. Finally, a metric to quantify the coincidence of sharp rises in calcium and decays in cAMP was also evaluated. The model was a simplified version of the full ODE model, primarily designed to capture the overall positive or negative effect of one variable on another, without accounting for precise temporal kinetics.

We also defined four sensitivity parameters, which were varied one-by-one to describe different possible circuits. In particular, three different values for the parameter under investigation were chosen: "−1," to represent a "low" value; "0," to represent a nominal value; or "1," to represent a "high" value, respectively. An instance of one such simulation is presented in Fig. 4. The ordered set of numbers in the plots should be read as: [sensitivity parameter value, Calcium-AC link, calmodulin (CaM)-AC link, CaM-PDE link]. The sensitivity parameter value changes across the rows. Therefore, each row corresponds to the full set of 12 circuits at a fixed sensitivity parameter value. Conversely, each column corresponds to a particular circuit with the sensitivity parameter value spanning the complete range. In the plots, the "warmer" (red being the warmest) colors indicate a higher value for the corresponding metric and "cooler" (blue being the coolest) colors correspond to lower values. White patches indicate that the corresponding circuits did not produce oscillations or produced complex oscillations with varying amplitude, a feature that we do not observe in our experimental results. As mentioned above, we looked for circuits that satisfied certain key criteria in accordance with our experimental results. So, we identified circuits that could produce antiphasic oscillations (which correspond to warmer patches in the Time Delay Metric plots) and had a simultaneously fast calcium rise phase and cAMP decay phase (which correspond to cooler patches in the Rise–Decay Metric plots).

The results in Fig. 4 correspond to variation in the parameter that relates to the PKA feedback to calcium. Analyzing all such plots, we identified the five following circuits that repeatedly satisfied the required criteria:

(−1, 1, 1), (−1, 0, 0), (−1, 0, 1), (0, 1, 1), and (0, 0, 1).

Of these, the three following circuits appeared to be relatively robust to parameter variation in each case:

(−1, 1, 1), (−1, 0, 1), and (0, 0, 1).

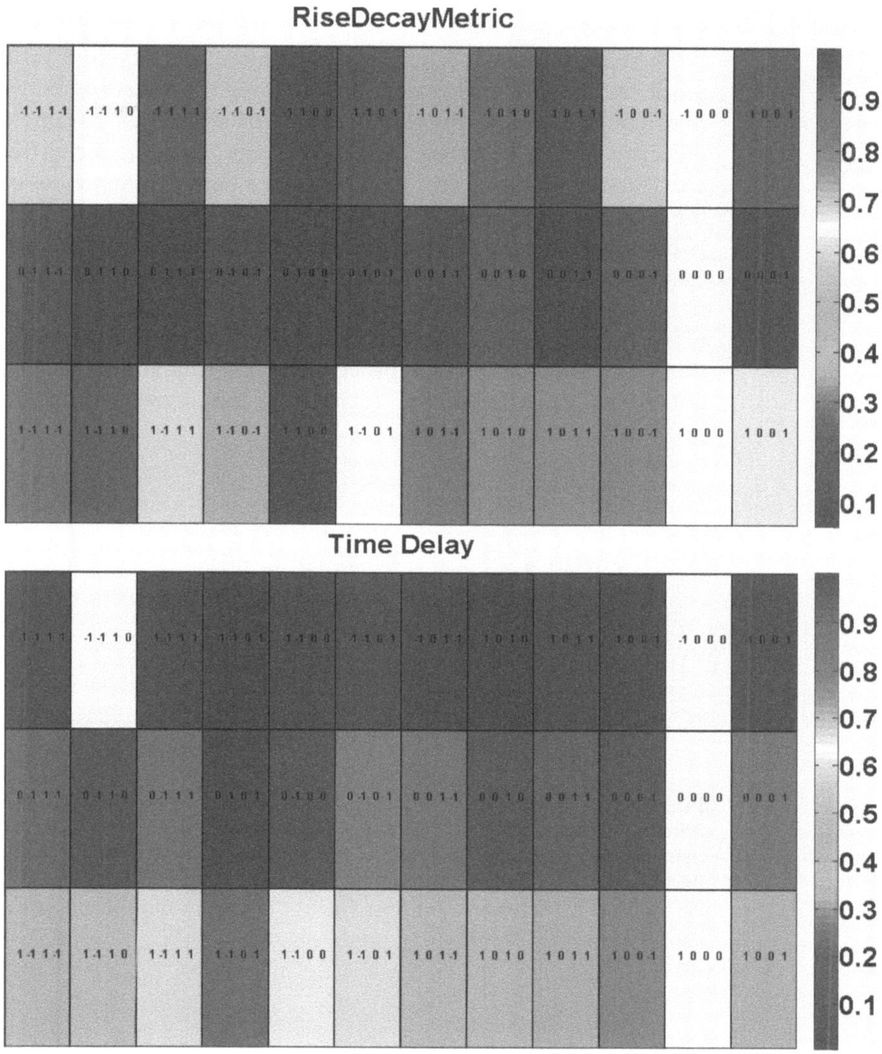

Fig. 4 Changes in (A) the time delay between Ca^{2+} and cAMP oscillations and (B) the coincidence of rapid Ca^{2+} rise phase and a rapid cAMP decay phase due to change in one of the sensitivity parameters, KPKAvar. KPKAvar takes one of the values (−1, 0, 1) to represent a "low," "nominal," or "high" value, respectively. cAMP, cyclic adenosine monophosphate. (See color plate)

In all of the three circuits above, we found a common feature, namely that the third number in the set was always 1. In other words, PDE activation by CaM is sufficient to produce calcium and cAMP oscillations out of phase with each other, with calcium having a sharp rise and cAMP having a sharp decay phase. We used this result in formulating our model for the calcium–cAMP–PKA circuit as detailed below.

D. Model Simulations

The results obtained during the architectural revision process can be used to guide the model development in finalizing the signaling network architecture. At this juncture, a few simple equations or additional parameters may need to be incorporated to describe the actual links between the different modules. This complete model may still need to have a few parameter revisions so as to match the experimental results. In our study of the signaling network in pancreatic β-cells, we were led to assume that cAMP dynamics in the final model is regulated by CaM.Ca_4-dependent PDE and PKA feeds back to calcium. The model simulations were able to capture most of the salient features in the experimental results (Figs. 5 and 6). This complete model was then used to generate hypotheses and test them experimentally. This process has led to a variety of interesting predictions confirmed experimentally, increasing the level of confidence in the model precision.

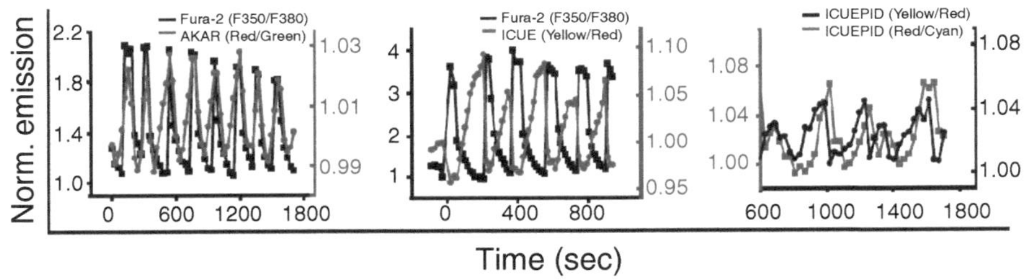

Fig. 5 Coordinated oscillatory changes, observed upon membrane depolarization, in (A) $[Ca^{2+}]_i$ (Fura-2 traces, in black) and PKA (monitored by PKA-specific biosensor, AKAR-GR, in red). (B) $[Ca^{2+}]_i$ (monitored by Fura-2, in black) and cAMP (monitored by cAMP biosensor, ICUE, in red). (C) cAMP (black) and PKA (red), monitored simultaneously using the dual-specificity FRET-based biosensor ICUEPID. (For interpretation of the references to color in this figure legend, the reader is referred to the web version of this book.)

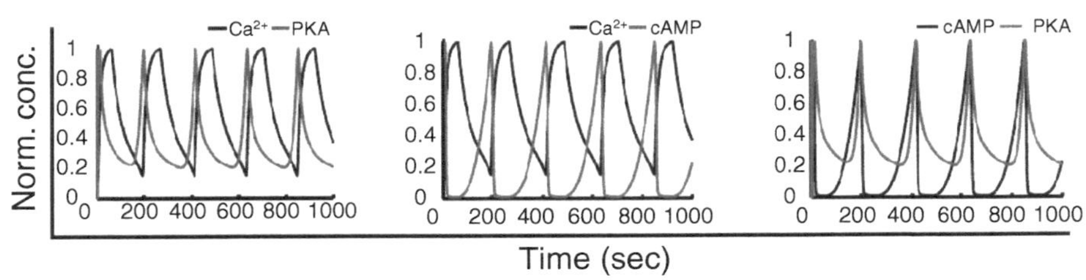

Fig. 6 Simulations of mathematical model showing oscillatory changes in (A) Ca^{2+} and PKA, (B) Ca^{2+} and cAMP, and (C) cAMP and PKA. cAMP, cyclic adenosine monophosphate; PKA, protein kinase A. (For color version of this figure, the reader is referred to the web version of this book.)

One important aspect of having a working model is that it can be used to address certain questions that would require infeasible experiments. Thus, in addition to generation of experimentally testable hypotheses, a model can be used to theoretically test the importance of certain responses. For instance, although the experimental results showed that PKA oscillated in tandem with calcium and that PKA feedback is necessary for the calcium oscillations, it was not clear if oscillations of elevated PKA were essential for oscillations of calcium and cAMP. In other words, it was of interest to examine whether a constant elevated level of PKA activity would be sufficient to trigger and sustain calcium oscillations. As experimental "pegging" of the PKA activity to a constant level is not easily achievable, we explored this question by model simulation. The results (Fig. 7) revealed that calcium and cAMP entered into an oscillatory regime even when a constant level of PKA activity was maintained. We modeled this by eliminating the ODEs in the PKA module and fixing the concentration of active PKA as a parameter in the model. This modeling result indicated that PKA activity oscillations might not be required for Ca^{2+} oscillation *per se*, but rather have other regulatory roles. Indeed, based on other results, we noted that PKA activity oscillations can help make this molecule a frequency modulator, and that this frequency modulation can enable PKA to switch from acting locally (restricted to a certain intracellular domain) to acting globally (controlling gene expression).

Among particularly useful model analyses that can lead to experimental validation are the tests of perturbation of signaling and other networks through the use of genetic and pharmacological inhibitors of molecular function. One can investigate whether such perturbations can lead to considerable disruption of particular signaling functions or specific other network components. In the calcium oscillatory circuit, one can explore, for instance, the role of PKA feedback to calcium. If the feedback is "abolished" computationally by setting the concentration of active PKA to 0 in simulations, one observes complete termination of the calcium oscillations

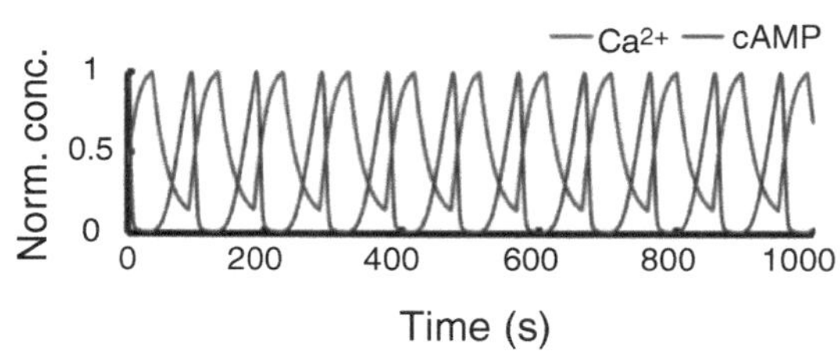

Fig. 7 A hypothetical circuit with a constant normalized PKA activity of 0.4 produces oscillations at a much higher frequency than when PKA is in feedback. The concentration of each species was normalized with respect to its maximal level during the course of the oscillations. The concentration of active PKA was normalized with respect to the maximal level of [PKA*] achieved when the PKA is in feedback in the nominal system. cAMP, cyclic adenosine monophosphate; PKA, protein kinase A. (For color version of this figure, the reader is referred to the web version of this book.)

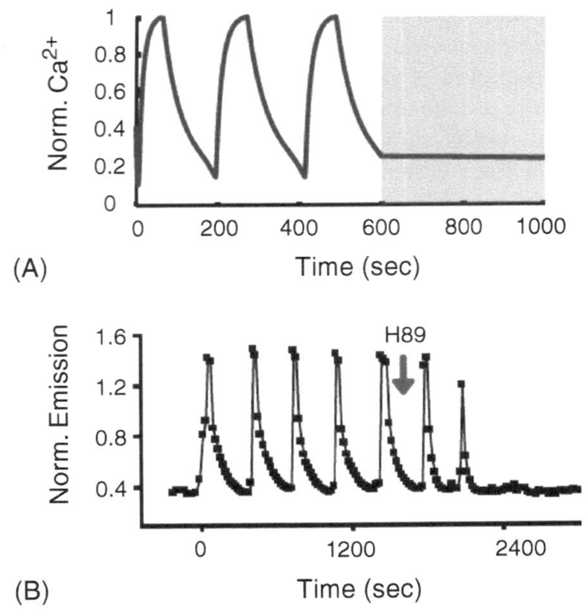

Fig. 8 (A) Simulation of the model in the presence or absence of PKA (shaded region). (B) The effect of inhibiting PKA by H89 (10 μM) on calcium oscillations. PKA, protein kinase A. (For color version of this figure, the reader is referred to the web version of this book.)

(Fig. 8). This computationally generated hypothesis was then validated using the PKA inhibitor H89, which indeed completely abolished calcium oscillations.

The ultimate test and benefit of a model lies not just in testing interactions but in providing novel insights into the workings of the system – explaining why a system behaves the way it does. Again, taking the example of the calcium-PKA cross-talk, the model helped us address a long-standing paradox of localization and control of PKA activation. The molecular mechanisms of PKA activation indicate that increasing input signal to PKA would result in continued diffusion of its catalytic subunits away from the regulatory subunits, ultimately losing its ability to reset itself and allowing all the catalytic subunits to translocate to the nucleus, in response to the naturally present nuclear localization signal. Oscillatory PKA activity might help address this potential problem. At low frequency of oscillations, the number of the catalytic subunits escaping a local signaling domain (proportional to the time-average of the PKA activity) would be relatively low, allowing the local PKA activation to transiently exceed a threshold needed for spatially localized substrate activation, while avoiding escape into the cytosol and the nucleus. An increase in PKA activation can, however, change the oscillation frequency, increasing the average PKA activity and thus the escape of the catalytic subunits from the local signaling domain, allowing them to have global cellular activity, including activation of nuclear targets and regulation of gene expression (Fig. 9).

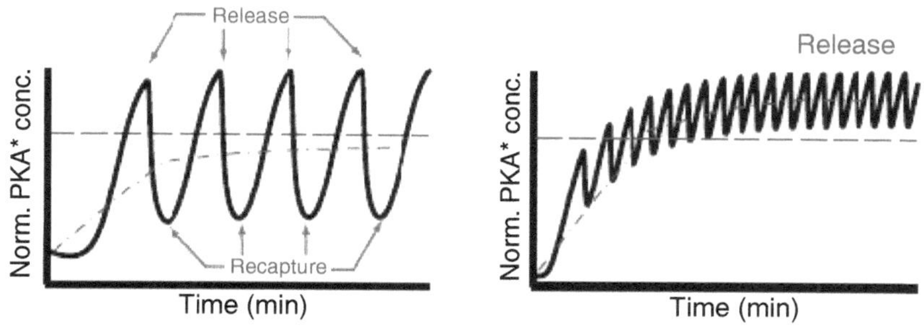

Fig. 9 Simulations of the model indicate that at low-frequency conditions (left panel), catalytic subunits of PKA would be periodically released and captured for "local" target phosphorylation. However, at high-frequency conditions (right panel), the mean PKA activity (red line) may cross a threshold (black dotted line) leading to continued release of catalytic subunits resulting in phosphorylation of "global" targets. PKA, protein kinase A. (For interpretation of the references to color in this figure legend, the reader is referred to the web version of this book.)

The oscillation frequency can, therefore, control the switch between the local PKA activity controlled by anchor proteins and global PKA activity in pancreatic β-cells and, potentially, many other cell types. We tested this hypothesis by monitoring the nuclear activity of PKA using AKAR-NLS (a nucleus-targeted version of the PKA activity biosensor). At a "low dose" of cAMP input, which was expected to correspond to low-frequency oscillations, we noticed that the nuclear activity was low. However, a "high dose" of cAMP input, expected to correspond to high-frequency oscillations, resulted in a dramatic increase in nuclear PKA activity (Fig. 10).

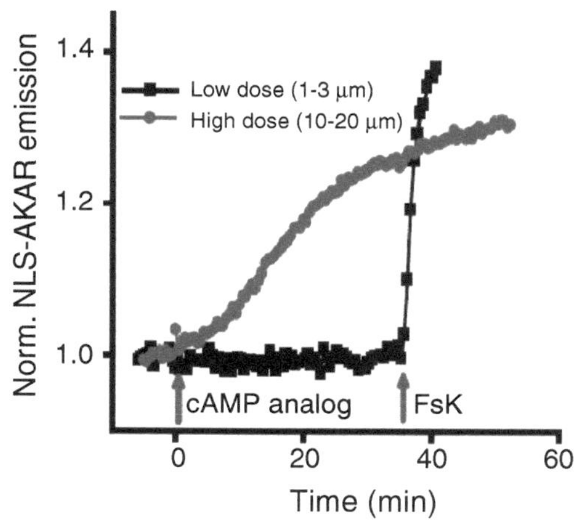

Fig. 10 Representative time courses of nuclear localized AKAR (NLS-AKAR) showing the absence and presence of nuclear PKA activity upon stimulation with low (1–3 μM) and high (10–20 μM) doses of a PKA-specific cAMP analogue, respectively (n = 7 and 4, respectively). cAMP, cyclic adenosine monophosphate; PKA, protein kinase A. (For color version of this figure, the reader is referred to the web version of this book.)

IV. Conclusion

Modeling can be a very powerful tool in defining the framework for the analysis and understanding of a variety of biological systems, including the biomolecular systems responsible for signal transduction. Modeling can be used both to generate progressively sophisticated testable hypotheses about the workings of the systems under investigation and to gain a better understanding of the systems' design and properties. Whatever its use, modeling analysis can be at its most effective if tightly coupled to experimental validation. Only through this coupling, it can be more evident whether model assumptions are fit for a specific system, from the standpoint of both specific molecular interactions and the parameter values defining them. Only through a very tight coupling between model and experiment can one hope to see instances of model invalidations, usually reflective of unanticipated, novel connections between the constituent components or suggestive of the presence of novel mechanistic details. These are the most exciting points of scientific discovery, which will depend more and more on our ability to recognize the necessity for making a breakthrough due to an essential conceptual missing link, as expressed in a model. As systems biology is providing a rapidly increasing and detailed information about the complexity of a variety of regulatory processes, experimental and computational biology will have to be intimately interlinked, as it has happened in many other areas of human knowledge and endeavor. Some lessons discussed here will be hopefully useful in guiding this process and making it more effective and enjoyable.

Acknowledgments

This work was supported by NIH grants: DK073368 and RR020839.

References

Antunes, M. S., Morey, K. J., Tewari-Singh, N., Bowen, T. A., Smith, J. J., Webb, C. T., Hellinga, H. W., and Medford, J. I. (2009). Engineering key components in a synthetic eukaryotic signal transduction pathway. *Mol. Syst. Biol.* **5**, 270.

Bertram, R., Previte, J., Sherman, A., Kinard, T. A., and Satin, L. S. (2000). The phantom burster model for pancreatic beta-cells. *Biophys. J.* **79**, 2880–2892.

Bertram, R., Sherman, A., and Satin, L. S. (2010). Electrical bursting, calcium oscillations, and synchronization of pancreatic islets. *Adv. Exp. Med. Biol.* **654**, 261–279.

Bezprozvanny, I., Watras, J., and Ehrlich, B. E. (1991). Bell-shaped calcium-response curves of Ins(1,4,5) P3- and calcium-gated channels from endoplasmic reticulum of cerebellum. *Nature.* **351**, 751–754.

Bhalla, U. S., and Iyengar, R. (1999). Emergent properties of networks of biological signaling pathways. *Science.* **283**, 381–387.

Cannon, R. C., O'Donnell, C., and Nolan, M. F. (2010). Stochastic ion channel gating in dendritic neurons: morphology dependence and probabilistic synaptic activation of dendritic spikes. *PLoS. Comput. Biol.* **6**, pii: e100088.

Chay, T. R., and Keizer, J. (1983). Minimal model for membrane oscillations in the pancreatic beta-cell. *Biophys. J.* **42**, 181–190.

Cheong, R., and Levchenko, A. (2010). Oscillatory signaling processes: the how, the why and the where. *Curr. Opin. Genet. Dev.* **20**, 665–669.

Corkey, B. E., Tornheim, K., Deeney, J. T., Glennon, M. C., Parker, J. C., Matschinsky, F. M., Ruderman, N. B., and Prentki, M. (1988). Linked oscillations of free Ca^{2+} and the ATP/ADP ratio in permeabilized RINm5F insulinoma cells supplemented with a glycolyzing cell-free muscle extract. *J. Biol. Chem.* **263**, 4254–4258.

DiPilato, L. M., Cheng, X., and Zhang, J. (2004). Fluorescent indicators of cAMP and Epac activation reveal differential dynamics of cAMP signaling within discrete subcellular compartments. *Proc. Natl. Acad. Sci. U S A.* **101**, 16513–16518.

Du, P., O'Grady, G., Davidson, J. B., Cheng, L. K., and Pullan, A. J. (2010). Multiscale modeling of gastrointestinal electrophysiology and experimental validation. *Crit. Rev. Biomed. Eng.* **38**, 225–254.

Dupont, G., Abou-Lovergne, A., and Combettes, L. (2008). Stochastic aspects of oscillatory Ca^{2+} dynamics in hepatocytes. *Biophys. J.* **95**, 2193–2202.

Dupont, G., Combettes, L., Bird, G. S., and Putney, J. W. (2011). Calcium oscillations. *Cold. Spring. Harb. Perspect. Biol.* **3**, pii: a004226.

Fridlyand, L. E., Harbeck, M. C., Roe, M. W., and Philipson, L. H. (2007). Regulation of cAMP dynamics by Ca^{2+} and G protein-coupled receptors in the pancreaticbeta-cell: a computational approach. *Am. J. Physiol. Cel.l Physiol.* **293**, C1924–C1933.

Gorbunova, Y. V., and Spitzer, N. C. (2002). Dynamic interactions of cyclic AMP transients and spontaneous $Ca(^{2+})$ spikes. *Nature.* **418**, 93–96.

Grapengiesser, E., Gylfe, E., and Hellman, B. (1988). Glucose-induced oscillations of cytoplasmic Ca^{2+} in the pancreatic beta-cell. *Biochem. Biophys. Res. Commun.* **151**, 1299–1304.

Greenstein, J. L., and Winslow, R. L. (2011). Integrative systems models of cardiac excitation-contraction coupling. *Circ. Res.* **108**, 70–84.

Ko, C. H., Yamada, Y. R., Welsh, D. K., Buhr, E. D., Liu, A. C., Zhang, E. E., Ralph, M. R., Kay, S. A., Forger, D. B., and Takahashi, J. S. (2010). Emergence of noise-induced oscillations in the central circadian pacemaker. *PLoS. Biol.* **8**, e1000513.

Lim, W. A. (2010). Designing customized cell signaling circuits. *Nat. Rev. Mol. Cell. Biol.* **11**, 393–403.

Ni, Q., Ganesan, A., Aye-Han, N. N., Gao, X., Allen, M. D., Levchenko, A., and Zhang, J. (2011). Signaling diversity of PKA achieved via a Ca^{2+}-cAMP-PKA oscillatory circuit. *Nat. Chem. Biol.* **7**, 34–40.

Roberts, E., Magis, A., Ortiz, J. O., Baumeister, W., and Luthey-Schulten, Z. (2011). Noise contributions in an inducible genetic switch: a whole-cell simulation study. *PLoS. Comput. Biol.* **7**, e1002010.

Schuster, S., Marhl, M., and Hofer, T. (2002). Modelling of simple and complex calcium oscillations. From single-cell responses to intercellular signaling. *Eur. J. Biochem.* **269**, 1333–1355.

Sherman, A. S., Li, Y. -X., and Keizer, J. E. (2002). Whole cell models. In "*Computational Cell Biology*," (C. P. Fall, ed.), pp. 101–139. Springer-Verlag, New York.

Tang, Y., and Othmer, H. G. (1995). Frequency encoding in excitable systems with applications to calcium oscillations. *Proc. Natl. Acad. Sci. U S A.* **92**, 7869–7873.

Toettcher, J. E., Mock, C., Batchelor, E., Loewer, A., and Lahav, G. (2010). A synthetic-natural hybrid oscillator in human cells. *Proc. Natl. Acad. Sci. U S A.* **107**, 17047–17052.

Voit, E. O. (2000). *Computational analysis of biochemical systems: a practical guide for biochemists and molecular biologists.* Cambridge University Press, pp. 49–58.

Zhang, J., Ma, Y., Taylor, S. S., and Tsien, R. Y. (2001). Genetically encoded reporters of protein kinase A activity reveal impact of substrate tethering. *Proc. Natl. Acad. Sci. U S A.* **98**, 14997–15002.

CHAPTER 2

Integrated Inference and Analysis of Regulatory Networks from Multi-Level Measurements

Christopher S. Poultney[*,1], **Alex Greenfield**[†,1] **and Richard Bonneau**[*,‡]

[*]Department of Biology, Center for Genomics and Systems Biology, New York University, New York, NY, USA

[†]Computational Biology Program, New York University School of Medicine, New York, NY, USA

[‡]Computer Science Department, Courant Institute of Mathematical Sciences, New York, NY, USA

[1] These authors contributed equally to this work.

0091-679X/10 $35.00
DOI 10.1016/B978-0-12-388403-9.00002-3

Abstract

Regulatory and signaling networks coordinate the enormously complex interactions and processes that control cellular processes (such as metabolism and cell division), coordinate response to the environment, and carry out multiple cell decisions (such as development and quorum sensing). Regulatory network inference is the process of inferring these networks, traditionally from microarray data but increasingly incorporating other measurement types such as proteomics, ChIP-seq, metabolomics, and mass cytometry. We discuss existing techniques for network inference. We review in detail our pipeline, which consists of an initial biclustering step, designed to estimate co-regulated groups; a network inference step, designed to select and parameterize likely regulatory models for the control of the co-regulated groups from the biclustering step; and a visualization and analysis step, designed to find and communicate key features of the network. Learning biological networks from even the most complete data sets is challenging; we argue that integrating new data types into the inference pipeline produces networks of increased accuracy, validity, and biological relevance.

I. Introduction

Regulatory networks (RNs) can provide global models of complex biological phenomena, such as cell differentiation or disease progression. Knowledge of the underlying RNs has been key to understanding the functioning of diseases such as certain cancers (Carro *et al.*, 2010; Suzuki *et al.*, 2009), the creation of biofuels, and understanding the functioning of newly sequenced organisms (Bonneau *et al.*, 2007). Although some cancers can be traced to a single causative mutation, many cancers are much more functionally complex, requiring simultaneous mutations in multiple genes that result in aberrations in the functions of multiple signaling pathways. Elucidation of the global RN allows for the study of disease-associated mutations in their global context. Biological regulation is a process that inherently occurs on multiple levels, such as transcription, translation, phosphorylation, and metabolism that span varying temporal and physical scales. Effective methods for RN inference must likewise integrate multiple types and scales of data – transcriptomic, proteomic, metabolomic – in order to most accurately recapitulate the complex underlying RNs. Our work, as described in this chapter, focuses on methods that can integrate multi-level data to elucidate an RN-scale view of complex biological processes.

The current explosion in the quantity, quality, and availability of high-throughput, genome-scale measurements provides powerful new tools to understand complex processes. Such measurements are now becoming available at different biological levels (e.g., transcriptomics and proteomics) for the same cell types or disease processes. At present, the most readily available genome-wide data type is microarray data, capturing the "transcriptomic state" of the cell. We first discuss this data type in the context on network inference, then discuss other equally relevant data types. Microarray data provide genome-wide measurements of the abundance of mRNA for every transcript for which there is a probe on the microarray (typically thousands). Online compendia such as the Gene Expression Omnibus (Edgar *et al.*, 2002) and Microbes online (http://www.microbesonline.org/) contain many thousands of such microarrays spanning many species and diseases, making this the most complete data available for the purpose of RN inference. Since these data are collected on the transcript level, they allow for the interrogation of only transcriptional effects. The mediators of these effects are transcription factors (TFs), which are proteins that bind the DNA and modulate the transcript abundance of their downstream targets, and environmental factors (EFs), which are environmental cues that modify the transcriptional program.

Inferring accurate, global RNs from such data remains a challenge for a multitude of reasons. The error (or variance in replicate measurements) in the measurement of transcript abundance is proportional to the expression level that is being measured (more expression means more error in the measurement) and many statistical methods do not properly account for this heteroskedasticity. Additionally, many of the data compendia used contain experiments from different laboratories and can thus contain batch effects (changes in expression that are mostly due to variations in experimental procedure from different labs). Even if a data compendium were to be normalized for batch effects and the other types of noise, the best data set would contain many more variables (genes) than data points that can be used for inferring regulatory interactions (conditions), leading to a computationally underdetermined problem. Finally, as transcriptomics data only capture one level of regulatory interactions, it provides an inherently biased and incomplete view of the underlying RN. Despite these caveats, it has been shown by us and others that novel biological interactions can be elucidated from these data.

A. Experimental Design

No technique or technology can provide a "one size fits all" solution to network inference that is optimal with respect to the completeness or accuracy of the learned network topology or the ability of the model to describe system behavior. Further, careful experimental design is needed to balance the biological goals of any given systems biology effort (what cell processes are of interest to the effort as a whole, what biology is interesting to the graduate person doing the work). In general, microarray experimental designs fall into two broad categories: (1) steady-state experiments, and (2) time-series experiments. Balancing steady-state measurements following perturbations (both genetic and environmental) with time-series experiments that provide measurements of the system in action (capturing key changes post-perturbation and providing a means of characterizing system dynamics) is key to the success of efforts to elucidate biological networks. In steady-state experiments, a perturbation (i.e., drug or genetic perturbation such as knock-down of a gene by RNAi) is introduced for a period of time, presumably until the system has reached a steady state, at which point the state of the system is assayed via microarray. We refer to these experiments as "steady state" even though the system may not have achieved a true steady state when measured. In a time-series experiment, a perturbation is introduced, and the response of the system is measured at multiple time points. Although these time-series experiments are more costly, this type of information can aid in resolving causality, and inferring not only topological structure but also the degree of regulation (i.e., kinetic parameters).

However, these types of data cover only the transcriptional level of regulation, and even state-of-the-art methods (Greenfield *et al.*, 2010; Huynh-Thu *et al.*, 2010; Pinna *et al.*, 2010; Prill *et al.*, 2010) still make a significant number of false predictions. The accuracy of these predictions can be improved with the addition of other data types. Recently, chromatin immunoprecipitation followed by sequencing (ChIP-seq) has become a widely used method for collecting direct TF–DNA binding data. These data can be used to help infer direct binding events. Note that these data typically contain many false positives as many binding events are non-functional. Another approach involves using single nucleotide polymorphism (SNP) data in conjunction with mRNA expression data to learn the extent to which each mutation can have a functional effect (Lee *et al.*, 2009). Such an approach can be used to uncover the TFs that are more likely to have a phenotypically important effect. Current approaches to learning RNs combine (1) binding data (from ChIP-seq and scans using well-characterized binding sites), (2) priors on network structure from known/validated regulatory interactions, (3) perturbation/genetic data, and (4) expression/proteomics data to triangulate regulatory interactions. We discuss our pipeline in the context of expression data (to describe our core model and prior work) and then develop a method that can integrate these four sources of information (Chen *et al.*, 2008; Christley *et al.*, 2009; De Smet and Marchal, 2010; Friedman and Nachman, 1999; Geier *et al.*, 2007; Gevaert *et al.*, 2007; Husmeier and Werhli, 2007; Huynh-Thu *et al.*, 2010; Ideker *et al.*, 2001; Lee *et al.*, 2009).

B. Estimating Co-regulated Genes Prior to Network Inference

If only mRNA expression data are available, as is the case in many processes/diseases of interest, other steps can be taken to improve the quality of the final output network. One means of dealing with the ambiguities of direct inference from microarray data is to reduce the complexity of the problem by estimating co-regulated groups via clustering or biclustering, which we discuss here specifically in the context of RN inference. Automatic learning of genetic RNs from microarray data presents a severely under-constrained problem: even in the most complete data set, the number of genes is greater than the number of experimental conditions. This is traditionally addressed by applying dimensionality reduction techniques to reduce the number of genes, for example, eliminating genes based on signal-to-noise ratio, or clustering genes based on similarity of expression.

Clustering methods, when applied correctly, can reflect the known biological property that many gene products work together in functional modules under identical regulatory control, forming components of tightly conserved pathways or molecular machines. Thus, applying a clustering method prior to network inference serves not only as a dimensionality reduction technique, but also as an additional method to capture the relevant underlying biology. Standard clustering groups together genes that show common expression across all experimental conditions (referred to as co-expression). However, co-expression may not extend across all conditions, particularly as the number of conditions in a data set increases. A subset of genes may be co-expressed over only a subset of conditions; or a gene may participate in multiple processes, and therefore be co-expressed with several different subsets of genes across different subsets of conditions.

Biclustering (Cheng and Church, 2000) refers to simultaneous clustering of both genes and conditions, and can account for these more complex patterns of co-expression (Cheng and Church, 2000; Lazzeroni and Owen, 1999). Both genes and conditions can belong to multiple biclusters, and each bicluster's subset of genes and conditions represents a putative functional module reflecting the organization of known biological networks into modules (Singh *et al.*, 2008). Early works (Morgan and Sonquist, 1963) introduced the idea of biclustering as "direct clustering" (Hartigan, 1972), node deletion problems on graphs (Yannakakis, 1981), and biclustering (Mirkin, 1996). More recently, biclustering has been used in several studies to address the biologically relevant condition dependence of co-expression patterns (Ben-Dor *et al.*, 2003; Bergmann *et al.*, 2003; Cheng and Church, 2000; DiMaggio *et al.*, 2008; Gan *et al.*, 2008; Kluger *et al.*, 2003; Lu *et al.*, 2009; Supper *et al.*, 2007; Tanay *et al.*, 2004). Biclustering also provides another advantage relating to increasing the signal (relative to the noise) of microarray data. These data are noisy due to both random noise (e.g., fluctuations in the scanner's laser) and systematic effects (e.g., sequence-specific differences in performance of probes or PCR amplification), as well as inherent biological noise, all of which occur per-gene. When genes are combined into modules, the average expression of the module is used, and thus the per-gene noise is averaged out, and the expression of the signal (relative to the noise) is increased.

Traditional biclustering is based solely on microarray data. Additional genome-wide data (such as association networks and TF binding sites) greatly improves the performance of these approaches (Elemento and Tavazoie, 2005; Huttenhower *et al.*, 2009; Reiss *et al.*, 2006; Tanay *et al.*, 2004). Examples include the most recent version of SAMBA, which incorporates experimentally validated protein–protein and protein–DNA associations into a Bayesian framework (Tanay *et al.*, 2004), and *cMonkey* (Reiss *et al.*, 2006), an algorithm we recently introduced. Bicluster inference has also been extended to detect conservation of modules across multiple species (Waltman *et al.*, 2010). These integrative biclustering methods provide more accurate and biologically relevant biclusters, and provide a template for the use of integrative methods in network inference.

Biclusters are of particular interest for network inference: inference on these putative functional modules is both more tractable and more easily interpreted than inference on individual genes. The regulation of biclusters (or individual genes) by the relevant TFs and EFs in the system can be learned in a variety of ways. Common difficulties for any network inference method include determining the direction of a regulatory relationship (does gene a regulate gene b or does gene b regulate gene a), and separating direct from indirect regulatory relationships (does gene a regulate gene b directly, or does gene a regulate gene c which then regulates gene b) (Marbach *et al.*, 2010). The ability to resolve the directionality of a regulatory relationship can be improved by using microarray data collected from time-series or genetic-knockout experiments, as such data allow for causal inferences to be made (Chaitankar *et al.*, 2010; Madar *et al.*, 2010; Marbach *et al.*, 2009b; Pinna *et al.*, 2010; Schmitt *et al.*, 2004; Yip *et al.*, 2010). However, it is still difficult to distinguish direct interactions from indirect: again, data such as ChIP-seq and ChIP-chip help resolve this ambiguity, and would ideally be available for the construction of an accurate, global RN. Even in cases where multiple, putatively complementary data types are available (i.e., microarray and ChIP-seq), validation of the output RN and comparison of RNs generated by different methods is a challenging task. For example, many top-performing methods are likely to involve data-integration methods that may integrate data with complex relationships and co-dependencies. Also, as the full integrated data set used for network inference becomes more complex, generating leave-out test sets that are completely separate from the integrated inference data becomes a research problem of its own.

C. Validation of Network Inference Methods is Key to Progress

The plethora of different methodologies available for RN inference makes the comparison of the RNs produced by different algorithms a challenging problem. Until recently, a group developing an RN inference algorithm would generate a long list of hypothesis, experimentally validate their first few predictions, and consider their method successful. This tradition of validating top predictions makes good sense when one considers that biologists may only have the capacity to follow up on a limited number of top predictions. This focus on top predictions, however, is insufficient for comparing network inference methods and assessing their relative strengths and weaknesses, as a typical RN inference method generates thousands of predictions. For such comparisons, a gold-standard RN inference data set is needed in which the topology of the underlying network is unambiguously and completely determined (Marbach *et al.*, 2009c, 2010; Prill *et al.*, 2010; Stolovitzky *et al.*, 2007). Databases for model organisms that collate thousands of validated regulatory interactions (such as Transfac, RedFly, and RegulonDB) are also critical to developing and validating RNs. A fully complete RN gold standard, however, cannot currently be obtained from real biological experiments, as known biological networks, even in the simplest organisms, are both extremely complex and considered to be incomplete.

In an effort to standardize the comparison and assessment of algorithms for RN inference, the Dialogue in Reverse Engineering Assessments and Methods

(DREAM) has posed a set of challenges to the network inference community in a double-blind fashion (Marbach *et al.*, 2009c, 2010; Prill *et al.*, 2010; Stolovitzky et al., 2007). Participating groups only see the microarray data (either synthetically generated, or from real data compendia) and not the underlying topology. Likewise the evaluators only see the predictions from each group (in the form of ranked lists of regulatory interactions), and not the method that was used to generate them. When such gold standards exist, metrics such as area under the precision recall curve (AUPR) and area under the receiver operator curve (AUROC) (Davis and Goadrich, 2006) can be used to assess an algorithm's performance. However, when applying RN inference techniques to mammalian data, relatively little of the true underlying topology is known, and AUPR and AUROC are not nearly as informative as for simpler systems.

D. Visualization

Analysis of inferred RNs for such systems presents a difficult set of problems. RNs have an intuitive visual representation as graphs consisting of nodes connected by directed or undirected edges, and programs such as Cytoscape (Shannon *et al.*, 2003) provide a straightforward means of rendering these graphs and annotating them with manifold types of associated information. This visual representation can be used by a researcher with domain knowledge of the underlying biological problem to extract the most meaningful and interesting parts of the network. Unfortunately, for networks larger than tens of nodes connected by at most hundreds of edges, this straightforward visualization becomes too dense to comprehend as a whole, presenting visually as the familiar network "hairball." While this dense representation contains much valuable information that can be interpreted by a researcher who has spent days or weeks investigating it, to the uninitiated it is essentially meaningless. Dimensionality reduction (e.g., via biclustering) can reduce visual complexity, but imposes other issues: a gene name is unambiguous, but how best to label a collection of genes and conditions? Whether inference is performed directly on all genes or on a reduced set of TFs and biclusters, the challenges are the same: to tease out the meaningful information contained in the network, and to convey this information effectively to other researchers not intimately familiar with the overall network. Thus, a set of visualization and analysis tools is necessary to query the network in an intuitive, meaningful, and easily accessible manner. Below, we describe one coordinated visualization system that allows users to explore biclusters, networks, and annotations.

II. Overview of Model/Algorithm

Our pipeline for network inference (Fig. 1) consists of three main steps: (1) inference of co-regulated modules using cMonkey, (2) RN inference using the

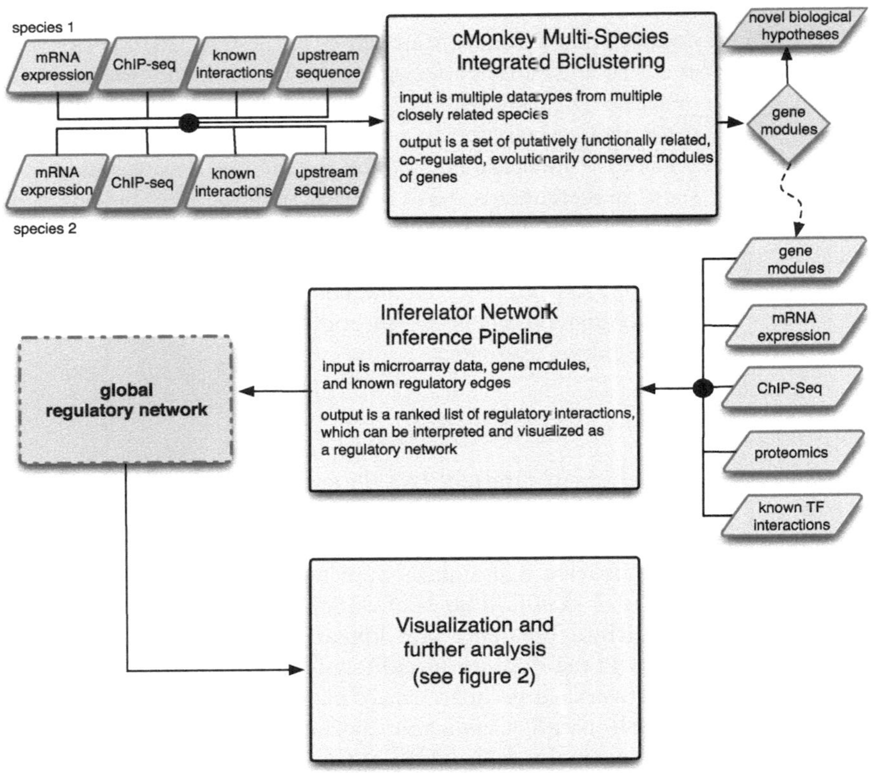

Fig. 1 Overall inference pipeline. Our inference pipeline is composed of three main steps: (1) inference of biclusters, which are putative functionally related, co-regulated modules of genes, by Multi-species cMonkey (MScM); (2) inference of the regulation of these biclusters by transcription factors (TFs) via our Inferelator inference pipeline; and (3) analysis and visualization using a collection of Gaggle-connected tools. The input to cMonkey consists of mRNA expression data, known as interactions (some of which come from ChIP-seq), and upstream sequence information from two or more species. The output of MScM is biclusters that are conserved between multiple species. These biclusters can be used for hypothesis generation, and also serve as the input to the Inferelator network inference pipeline. Along with biclusters, the Inferelator also uses mRNA expression data, known interactions between relevant TFs and their targets, proteomics data, and ChIP-seq data. The output of the Inferelator inference pipeline is a set of regulatory interactions between the biclusters and TFs. This putative regulatory network can be visualized and analyzed by the Gaggle-connected set of tools shown in Fig. 2. (For color version of this figure, the reader is referred to the web version of this book.)

Inferelator (Inf) pipeline, and (3) network analysis and visualization using software tools connected by the Gaggle (Fig. 2). Here, we present an overview of biclustering methods, RN inference methods, and a detailed overview of each step of our pipeline.

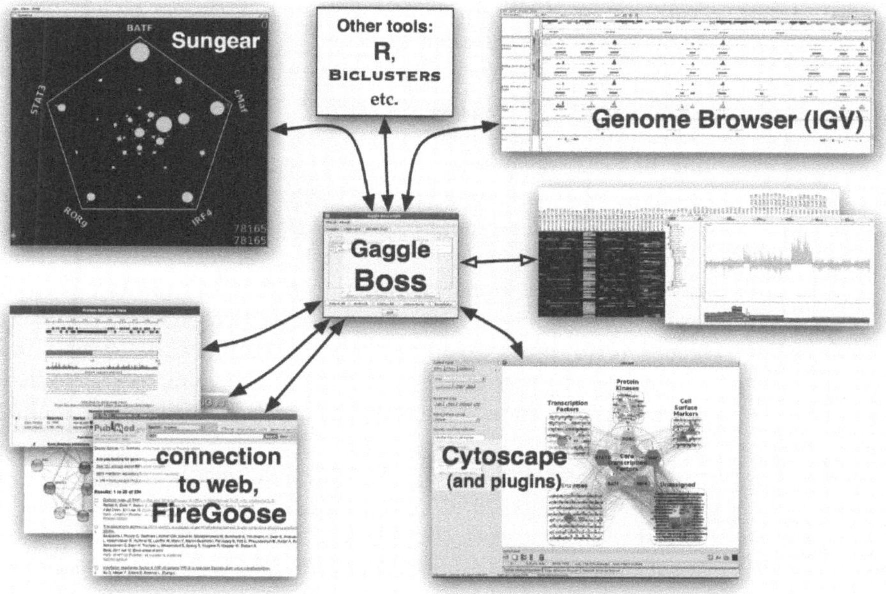

Fig. 2 Gaggle visualization and analysis framework. The Gaggle Boss, shown in the center, coordinates communication among the various member tools (geese), removing the need for file import/export and format translation. Also shown is a subset of geese, including two – Cytoscape and Sungear – that are used as part of the analysis discussed in Biological Insights section and Methods section. Each of the geese can both send to and receive from the Boss, which permits an iterative workflow: for example, a small set of genes from Sungear can be sent to Cytoscape, analyzed to find its 1-hop network, then sent back to Sungear for further analysis. In addition, several geese provide extensible means to connect to a larger set of tools: Cytoscape and Sungear via plug-in frameworks, FireGoose via its connections to other websites, and R via its downloadable packages. (For color version of this figure, the reader is referred to the web version of this book.)

A. Biclustering

Biclustering methods can be broken into three categories, which we will refer to as co-expression, co-regulation, and conserved co-regulation. Some methods, such as that of Cheng and Church (2000), rely solely on gene expression data to find groups of genes that are co-expressed. More recently, algorithms such as cMonkey (Reiss *et al.*, 2006; Waltman *et al.*, 2010), COALESCE (Huttenhower *et al.*, 2009), and the most recent version of SAMBA (Tanay *et al.*, 2004) consider additional types of data such as common binding motifs, protein–DNA binding, and protein–protein interaction networks. These integrative techniques infer modules that are co-regulated rather than simply co-expressed. This distinction is of particular importance for RN inference, as genes in co-regulated biclusters are more likely to exhibit shared transcriptional control. Finally, several techniques (Bergmann *et al.*, 2003;

Waltman *et al.*, 2010) extend the integrative approach by searching for conserved biclusters across different species.

The biclustering method cMonkey was designed to produce putatively co-regulated biclusters that are optimal for network inference. In addition to microarray expression data, cMonkey also incorporates upstream sequences and interaction networks into the biclustering process. Upstream sequences are used to find putative common binding motifs among genes in a bicluster, providing additional evidence for possible co-regulation. Co-regulated genes are also more likely to share other functional couplings, which will be reflected as an above-average number of connections between genes within a bicluster according to databases of known interactions such as BIND (Bader *et al.*, 2003) and DIP (Salwinski *et al.*, 2004) – in other words, these genes form small, highly connected sub-networks within these larger networks. Compared to other methods, cMonkey generates biclusters that are "tighter" (have lower variance across bicluster gene expression values) yet include more experimental conditions.

Multi-species cMonkey (MScM) (Waltman *et al.*, 2010) is an extension of the cMonkey method to allow discovery of modules conserved across multi-species datasets. Recent work (Ihmels *et al.*, 2005; Tirosh and Barkai, 2007) shows significant conservation of co-regulated modules across species. Therefore, biclusters that are highly conserved between organisms are most likely to be biologically relevant. In addition, by pairing a well-studied model organism such as yeast or mouse with a closely related but less well-studied organism, MScM is more likely to find meaningful biclusters in the other organism. Even pairing well-studied organisms may be beneficial as different processes may be better elaborated in each organism. The regulation of these putative conserved functional modules of genes can be inferred using the Inf-based inference pipeline.

B. Regulatory Network Inference

The key question that RN inference aims to answer is which EFs and TFs regulate which genes? In other words, given a set of observations (e.g., expression data), what is the underlying network responsible for observed data? Furthermore, can predictions be made from the output network? In order for quantitative predictions to be made about the response of the system to new perturbations, the dynamics of the system must be learned from time-series data. A multitude of inference methods exist, using varying underlying assumptions and modeling principles. We limit ourselves to the discussion of the following broad groups of methods: (1) Bayesian methods, (2) mutual information (MI)-based methods, and (3) ordinary differential-equation (ODE)-based methods. We briefly describe each grouping, and then proceed with a description of our network inference method. Here, we focus on methods that scale to systems with thousands of interactions.

A Bayesian network is defined as a graphical model that represents a set of random variables and their conditional dependencies. Such a framework naturally applies to RN inference, as RNs can intuitively be though of as directed graphs. The observed data are used to compute the model whose probability of describing the data is the highest, and such methods have resulted in several notable works (Friedman *et al.*, 2000; Friedman and Nachman, 1999; Husmeier and Werhli, 2007; Sachs *et al.*, 2002; Sachs *et al.*, 2005; Segal *et al.*, 2003). Bayesian methods also allow for the incorporation of priors such as sparsity constraints and structured priors (Geier *et al.*, 2007; Gevaert *et al.*, 2007; Mukherjee and Speed, 2008). However, Bayesian methods have difficulty in explicitly handling time-series data. Additionally, many Bayesian methods suffer from the identifiability problem: multiple network topologies produce equally high probabilities. In this situation, it is unclear which topology is best.

Differential-equation-based methods for RN inference attempt to learn not only the topology of the network but also the dynamical parameters of each regulatory interaction. RN models resulting from these methods can be used to predict the system-wide response to previously unseen conditions, future time points, and the effects of removing system components. A drawback of these methods is that they generally require time-series data and more complete datasets than many alternative methods. Typically these methods are based on ordinary differential equations (ODEs) due to several assumptions that improve the computational cost for parameterizing these models. ODE-based methods model the rate of change in the expression of a gene as a function of TFs (and other relevant effectors) in the system. Differential-equation-based methods differ in their underlying functional forms, how the system of equations is solved or parameterized (coupled or uncoupled solution, optimization procedures, etc.), and how structured priors and sparsity constraints are imposed on the overall inference procedure. For example, several methods have been proposed that use complex functional forms (Mazur *et al.*, 2009) and solve a coupled system (Madar *et al.*, 2009; Mazur *et al.*, 2009), while other methods solve a simplified linear system of ODEs (Bansal *et al.*, 2006; Bonneau *et al.*, 2007; Bonneau *et al.*, 2006; di Bernardo *et al.*, 2006; di Bernardo *et al.*, 2005). Several methods have been developed that are able to incorporate structured priors into network inference (Christley *et al.*, 2009; Yong-a-poi *et al.*, 2008).

A number of methods for detecting significant regulatory associations are based on similarity metrics derived from information theory, such as MI (Shannon, 1948). The MI between two signals (in this case the expression of a TF and its target) is calculated by subtracting the joint entropy of each signal from the sum of their entropies. It is similar to correlation (higher values connote stronger relationships), but is more generally applicable as it assumes neither a linear relationship between two signals nor continuity of signal. At their core, methods that rely on MI generally infer undirected interactions, as the MI between two variables is a symmetric quantity (Butte and Kohane, 2000; Faith *et al.*, 2007; Margolin *et al.*, 2006); however, modifications can be made that allow for the inference of direction (Chaitankar *et al.*, 2010; Liang and Wang, 2008; Madar *et al.*, 2010).

Each RN inference method has its own simplifying assumptions, biases, and data requirements. Recently, there has been much interest and progress in combining methods that use multiple different data types and modeling algorithms into RN inference pipelines. For example, it has been demonstrated by us and others (Greenfield *et al.*, 2010; Pinna *et al.*, 2010; Prill *et al.*, 2010; Yip *et al.*, 2010) that the response of a system to a genetic knockout is a very powerful data type for uncovering the topology of the underlying RN. Methods that take this into account performed very well in the DREAM3 and DREAM4 network inference challenges (Greenfield *et al.*, 2010; Pinna *et al.*, 2010; Yip *et al.*, 2010).

It has also been shown that when multiple network inference methods, or ensembles of networks generated by the same method, are combined, the overall performance is better than that of any individual method (Greenfield *et al.*, 2010; Marbach *et al.*, in press, 2009a; Prill *et al.*, 2010). This improvement in performance due to combining multiple methods is an important technique that can be applied to complex biological problems where complete knockout data are not available. In such cases it is also important to supplement microarray data with other available data types. The Encyclopedia of DNA Elements Consortium (ENCODE) has been compiling a vast database of high-sequence data such RNA-seq, ChIP-seq, and genome-wide distribution of histone modifications. These data can be used in many ways to influence the confidence that a network inference algorithm assigns to a regulatory interaction. We have incorporated these ideas into our network reconstruction methods in two forms: (1) topology dominated, where evidence from different data types is combined to rank interactions by converting all regulatory hypothesis derived from each data type into p-values or ranks, then combining them to form an overall p-value or rank for all regulatory interactions (Greenfield *et al.*, 2010; Marbach *et al.*, in press), and (2) model dominated, where information from different data types is used as structure priors during the network inference step (described below).

Our inference pipeline is built on three core principles: (1) combining multiple methods and data types in a mutually reinforcing manner, (2) using time-series information to infer putative causal, directed relationships (as opposed to undirected associations), and (3) inferring sparse models of regulation using model selection. The input to our method is a microarray dataset consisting of multiple types of experiments. All data sets include steady-state data (in response to perturbation), time-series data is often available; and in the best-case scenario, genetic-knockout steady-state data are available as well. The core of our inference pipeline comprises two methods that work in tandem: time-lagged context likelihood of relatedness (tlCLR) and the Inferelator 1.0. tlCLR computes a prediction of the RN that is further refined and optimized by the Inferelator 1.0. The output of tlCLR is the input to Inf, and we refer to the combined method as tlCLR-Inf. tlCLR-Inf uses all available microarray data and treats all steady-state data the same (making no distinction between knockout perturbations and any other perturbations). tlCLR-Inf takes advantage of the time-series data to learn putatively causal, directed edges, and assign dynamical parameters (see Methods).

tlCLR (Greenfield *et al.*, 2010; Madar *et al.*, 2010) is based on the well-known RN inference algorithm context likelihood of relatedness (CLR) (Faith *et al.*, 2007). CLR uses MI followed by background correction to calculate the confidence in the existence of any regulatory interaction. tlCLR uses the same CLR strategy of MI followed by background correction, but takes advantage of the time-series data to learn the direction of the regulatory interaction. This method is described in detail in the Methods section. The output of this method is a set of predicted regulators for each target, and is used by the Inf to remove the least likely regulatory interactions and improve accuracy and computational efficiency.

The Inf models the network as a system of linear ODEs. The rate of change for each gene is modeled as a function of the known regulators in the system. This function can take many different functional forms, and can be easily modified to capture biologically relevant behaviors. For example, it is common in biological systems that two TFs must act in tandem in order to affect their target. The core Inf model allows for these non-linear combinatorial interaction terms. Additionally, it is known that the activation of a target by its regulator follows a hill-type curve (multiple functions with a roughly sigmoidal shape can be used to model biologically relevant activation thresholds, cooperation, and saturation of TF-target response). This can be incorporated into the core Inf model by approximating this behavior via sigmoidal functions compatible with efficient learning methods, such as constrained logistic regression. Once a functional form is chosen, the parameters for each regulatory interaction are calculated using least angle regression (LARS) (Efron *et al.*, 2004) which is a constrained linear-regression approach that imposes an l_1 constraint on the model parameters. This constraint ensures that sparse models are learned (in concordance with the known properties of TF RNs). Importantly, we have modified this core model selection algorithm, LARS, such that we can influence the degree to which a predictor is incorporated into or removed from a model. Using this modification, we can incorporate structured priors (derived from validated interactions, literature search algorithms, or alternate data types) into our network inference approach. We have shown that using a simple linear model with (and also without) interaction terms performs well in recovering the topology of the network.

C. Network Visualization and Analysis

RNs often consist of hundreds or thousands of nodes connected by thousands or more of regulatory edges. Analysis methods for networks of this scale generally fall into two categories that we will refer to as "network-centric" and "gene-centric," with some techniques bridging the two. Network-centric (or "holistic") techniques accumulate statistics about the network as a whole that can provide a sense of the validity of the overall network (e.g., by comparing statistics with those of validated biological networks) or guide further exploration (e.g., by pointing out the existence of highly connected nodes or densely inter-connected sub-networks). The simplest of these network-centric techniques is simply to count each node's in- and out-degrees, that is, its incoming and outgoing regulatory edges, respectively. Analysis of node out-degree will highlight network "hubs": those TFs that regulate many more genes than average. Examination of the distribution of node in- and out-degree also provides valuable information. Biological networks, such as metabolic networks (Jeong *et al.*, 2000), as well as many other types of complex networks, tend to be "scale free" networks: the probability of a node having k in or out edges is described by $P(k) \approx k^{-\gamma}$. This is considerably different from random networks generated according to the classical Erdös-Rényi model, where any two nodes in a graph have an equal probability of being connected: such graphs are characterized by a Poisson distribution that peaks strongly at the average number of connections (Jeong *et al.*, 2000). Average shortest-path length (the "small world" property), average clustering coefficient distribution (Ravasz *et al.*, 2002), and many other have metrics have been shown or theorized to have biological relevance. Zhou *et al.* (2010) provide an example of using general network statistics to characterize and differentiate between ecological networks under different conditions. Cytoscape plug-ins such as NetworkAnalyzer (http://med.bioinf.mpi-inf.mpg.de/netanalyzer/index.php), the R packages *sna* (Butts, 2008), and *igraph* (http://cneurocvs.rmki.kfki.hu/igraph/) are designed to perform these and many other types of network analysis.

In general, and particularly with inferred networks, these network-centric metrics act as a guide to suggest areas of further exploration – such as network hubs – rather than an explicit measure of network plausibility. Gene-centric (or "constructive") analysis techniques tend to follow a "find and connect" approach. They start with a small set of nodes – such as a set of genes of interest, a small sub-network of known function, a bicluster with significant functional annotations, or a set of network hubs identified through network-centric analysis – then gradually add connected nodes to grow the size of the network. The most basic approach is to start with a single gene, then examine its "1-hop" sub-network within the full network: the genes directly connected to it, that is, its direct targets and regulators. One can also make a 1-hop network for multiple genes that is simply the union of the 1-hop networks of the individual genes. These sub-networks can be expanded to an arbitrary number of "hops," with each additional step adding all nodes directly connected to those already in the sub-network. Typically, the hope is that the small 1- or 2-hop networks will include some known regulatory edges (as a "sanity check" of the inference process) as well as some plausible novel edges that bear further investigation.

Another gene-centric approach is to find, for some set of genes or biclusters, the smallest sub-network that includes all these genes or biclusters of interest (the phrase "gene-centric" is used generically to refer to network consisting of genes or biclusters). This sub-network may resemble a known functional module (another "sanity check"); it may connect known genes or biclusters in a novel way; and it may include unknown or unexpected genes in an otherwise well-described functional module. While no Cytoscape plug-in provides this functionality directly, the Subgraph Creator (http://metnet.vrac.iastate.edu/MetNet_fcmodeler.htm) plug-in can be used to find the sub-network with a given number of directed hops for a set of starting genes, and so iteratively find a sub-network containing all the desired genes.

1. Gaggle Tools

We supplement the network viewer-based approaches above by providing a collection of tools that provide analysis of different types of information at varying scales. The Gaggle (Shannon *et al.*, 2006) connects together many independent tools into a cohesive framework where the component tools (geese) can exchange information such as lists of genes directly without the need for intermediate files or format conversion. A key aspect of this type of approach is that it enables iterative exploration across multiple tools, where results are repeatedly sent from one tool to the next and further refined with each step in this process. Gaggle-enabled tools include

- Network viewers Cytoscape (Shannon *et al.*, 2003) and nBrowse (http://www.gnetbrowse.org)
- Firegoose (Bare *et al.*, 2007), a Firefox plug-in that provides data exchange with external web resources such as STRING (Snel *et al.*, 2000; Szklarczyk *et al.*, 2011)
- The Comparative Microbial Module Resource (CMMR) (Kacmarczyk *et al.*, 2011), a comprehensive bicluster visualization and analysis tool
- The Data Matrix Viewer (DMV) (http://gaggle.systemsbiology.net/docs/geese/dmv.php), a data matrix exploration tool
- MultiExperiment Viewer (MeV) (Saeed *et al.*, 2003), a sophisticated analysis tool for microarray data
- Sungear (Poultney *et al.*, 2007), a set analysis and exploration tool
- The Integrative Genomics Viewer (IGV) (Robinson *et al.*, 2011), a browser for associating annotations and other data with chromosomal locations
- The statistical programming language R (http://www.R-project.org).

III. Biological Insights

In this section, we will focus mainly on ways of extracting potential insights or points for further investigation. The networks discussed here were chosen to show a range of inference and analysis techniques across different network scales. For details of the methods used to create these networks, see the Computational Methods section.

Fig. 3 shows a subset of a larger network inferred on biclusters derived from the Immunological Genome Project (IMMGEN) (Painter *et al.*, 2011) mouse immune cell data set and human immune cell experimental data from GEO (see Methods for details). This sub-network has been chosen to show the subset of biclusters from the full network that are most strongly linked to various hallmarks of cancer (Hanahan and Weinberg, 2000, 2011). An immediately striking feature of this network is that different hallmarks separate naturally into sub-regions of the network, joined by the TF MTA2. This is not a deliberate design feature of this sub-network, but rather an intriguing consequence of choosing the set of biclusters with high-confidence connections (via significant GO terms) to hallmarks of cancer. The top region, whose

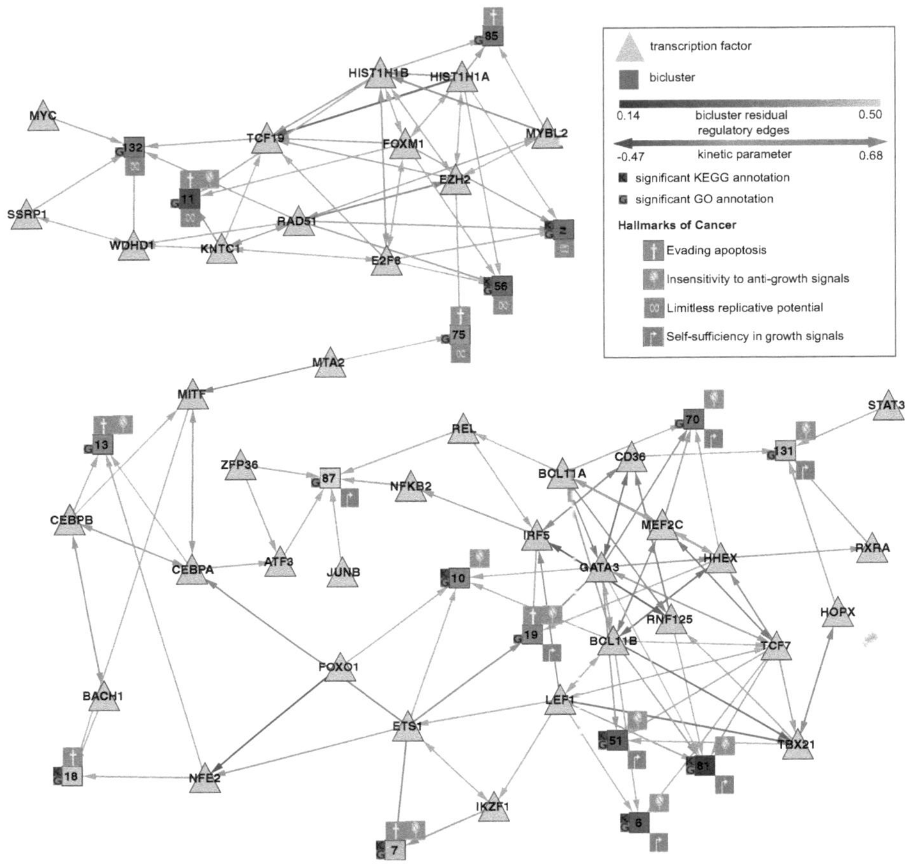

Fig. 3 Hallmarks of cancer shown overlaid on a sub-network of biclusters and transcriptions factors (TFs). Biclusters are shown as squares, with shading indicating the bicluster residual (variance in gene expression values). Surrounding icons indicate the putative hallmarks of cancer. A small K or G to the bicluster left indicates particularly significant enrichment for one or more KEGG or GO terms, respectively. TFs are shown as triangles, with regulatory edges to biclusters and other TFs. Green edges indicate upregulation, and red edges downregulation. Four of the six original hallmarks are represented in the network: biclusters associated with self-sufficiency in growth signals and insensitivity to anti-growth signals are clustered together, as are those associated with limitless replicative potential; biclusters inferred to be involved in evading apoptosis are spread through the network. (See color plate)

regulators include FOXM1 and MYC, includes all biclusters annotated with the hallmark "limitless replicative potential" (blue icon). The bottom region includes all biclusters annotated with the hallmark "self-sufficiency in growth signals" and all but one bicluster annotated with the hallmark "insensitivity to anti-growth signals." The few biclusters annotated with "evading apoptosis" are spread evenly between the network clusters.

Figs. 4–6 illustrate a comparative analysis of two different cell lines: normal human breast epithelial tissue (MCF-10) and invasive, metastatic breast cancer (MDA-MB-231) (see Methods for details). In all three figures, a blue-to-yellow continuum is used to indicate relative specificity of a gene, gene product, or regulatory edge to MDA-MB-231 (blue) or MCF-10A (yellow), with the intermediate gray denoting neutrality. Fig. 4 illustrates a typical network "hairball": with 1866

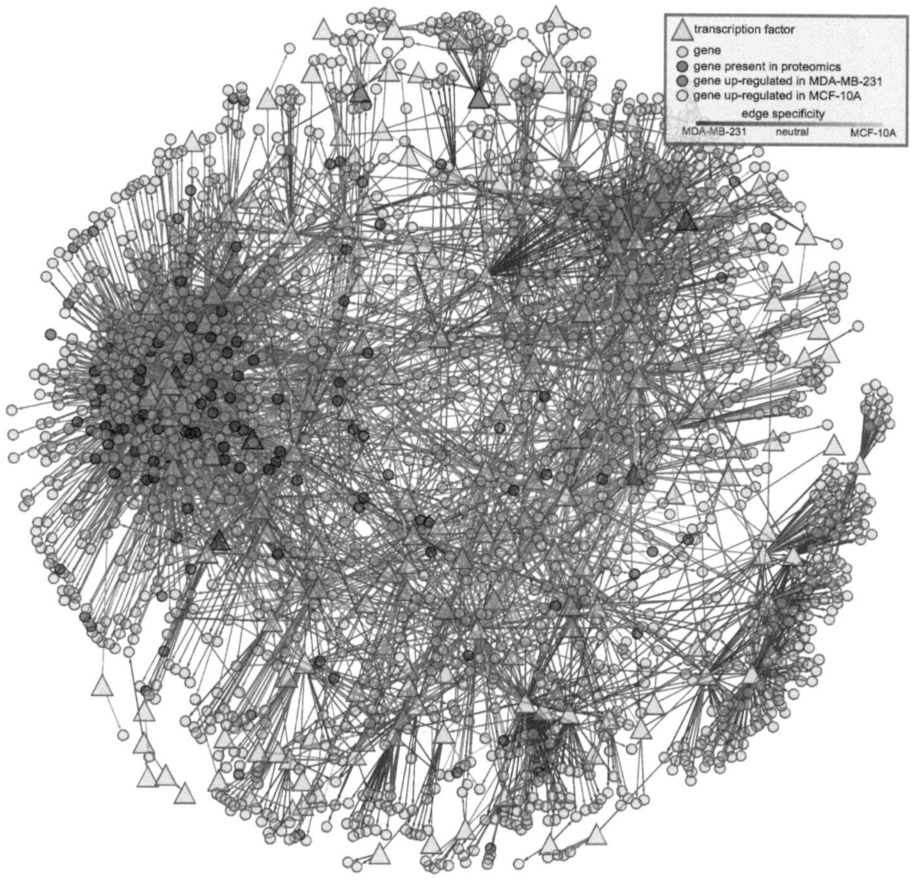

Fig. 4 Breast cancer network with the top 4822 edges ranked by combined confidence from the two cell line inference runs. Edge color denotes differential inferred regulation on a yellow-to-blue gradient from MCF-10A (yellow) to MDA-MB-231 (blue). Nodes are rendered semi-transparent so that the distribution of cell-line-specific regulatory edges can be clearly seen. Proteomics data from MCF-10A/MDA-MB-231 comparison are also shown using node colors: differential expression in MCF-10A is shown in yellow, and MDA-MB-231 in blue. Genes present but not differentially expressed are shown in darker gray. (See color plate)

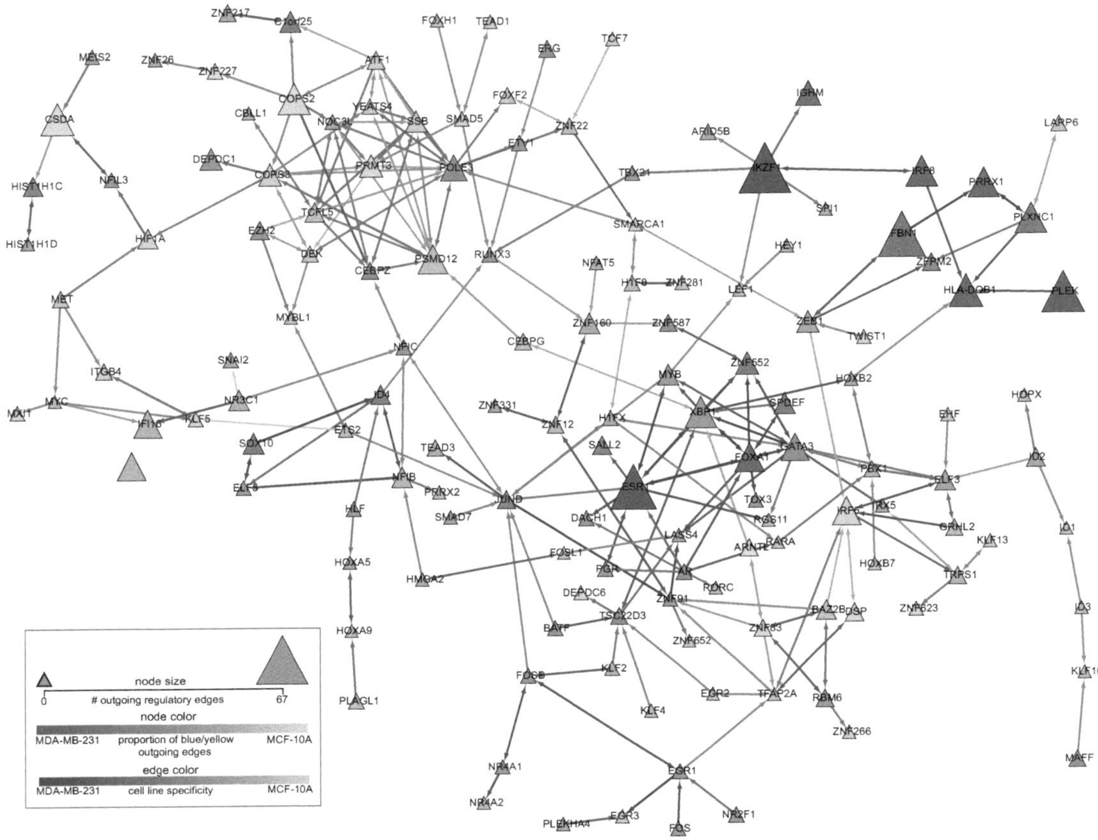

Fig. 5 Largest connected sub-network of transcription factors (TFs) from the overall cell line comparison network. A "summary" of the entire network is provided by (a) hiding all targets of the shown TFs that are not themselves TFs, and (b) setting the size and color of each remaining TF node to reflect its number and proportion of cell-line-specific edges. Node size shows the number of edges in the master network that were above a cutoff for specificity to either cell line. Larger nodes have more cell-line-specific edges; the largest, IKZF1, has 67 edges above the threshold. Node color is determined by the ratio of above-cutoff edges specific to MCF-10A versus MDA-MB-231, with yellow denoting more MCF-10A edges and blue more MDA-MB-231 edges. Nodes with many edges specific to one cell line or the other are therefore large and brightly colored, such as IKZF1 or COPS2. Edges are colored on a yellow-to-blue gradient based on the inferred confidence of the edge in the MCF-10A cell line (yellow) or MDA-MB-231 cell line (blue). (See color plate)

nodes and 4822 regulatory edges, it is useful mostly for giving a general sense of the proportion of edges more active in MDA-MB-231 (blue) and MCF-10 (yellow), as well as the abundance of proteomics data (nodes colored yellow, blue, or dark gray).

Fig. 5 is designed to present a summary of Fig. 4 that allows much more intuitive identification of features of interest. It represents the largest connected sub-network of TFs (142 of 220 total TFs in the original network). The number of regulatory targets of each TF is represented by the size of the node, while the node color denotes the ratio of regulatory edges strongly active in one cell type or another on a gradient from blue (MDA-MB-231) to yellow (MCF-10A). This "summary" representation of targets and regulatory edges allows the removal of all non-TF targets and their corresponding regulatory edges so that hubs such as CSDA, COPS2, IKZF1, and FBN1 are easily spotted: they are large and brightly colored. This constitutes a powerful use of simple network-centric techniques to simplify network visualization and analysis.

Fig. 6 shows a putative sub-network involved in cell motility. Our data set includes differential proteomics data for two conditions, shown in this network using node

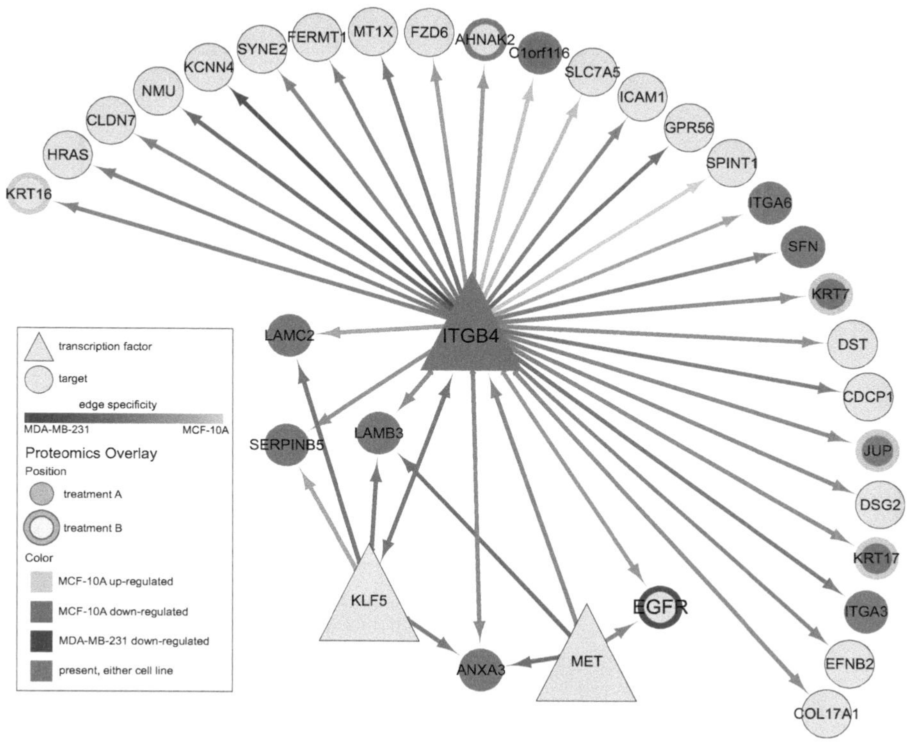

Fig. 6 A sub-network extracted from the cell line comparison network illustrating all interactions with ITGB4 along with overlays of experimental proteomics (SILAC) data. Shown is the 1-hop network from gene ITGB4 along with differential expression in two experimental conditions, referred to as treatment A and treatment B. ITGB4 was identified *a priori* as a gene of interest, and is inferred to regulate gene of interest EGFR and several Laminins. Differential expression in treatment A is shown using node center, and in treatment B using node border, as follows: bright yellow denotes upregulation in MCF-10A, bronze denotes downregulation in MCF-10A, and blue denotes downregulation in MDA-MB-231. Gray denotes proteins that were present in either cell line but that did not meet the differential expression cutoff. Therefore, KRT17 (bottom right) is downregulated in MCF-10A with treatment A but upregulated in MCF-10A with treatment B, while EGFR is downregulated in MDA-MB-231 with treatment B. Edge colors are as in Fig. 5. (See color plate)

center and border colors. An analysis of sub-networks containing all differentially expressed proteins in both conditions found a sub-network centered on ITGB4 – identified *a priori* as a protein of interest involved in cell matrix, cell–cell adhesion, and motility – that contained an unusual number of differentially expressed proteins given the relatively small number of differentially expressed proteins in the network ($\rho = 5 \times 10^{-7}$ via hypergeometric distribution). Among the genes inferred to be regulated by ITGB4 are two members of the laminin family also thought to be involved in motility, providing a degree of "sanity check" as mentioned earlier. The presence of JUP in this sub-network is particularly interesting because of (a) its differential expression in one of the proteomics conditions, and (b) its known participation in c-MET and EGFR signaling cascades (Guo *et al.*, 2008).

IV. Open Challenges

Combining multiple data types in the inference of RNs is still in its beginning stages, and many questions remain to be answered. Among these are the integration of additional data types into both the biclustering and inference processes, integrating across multiple temporal and physical scales, validation of inferred networks, using multiple-species datasets, and visualization of networks that are multi-scale and change across time and conditions.

A. Integrating New Data Types

New types of experimental data are becoming available that will be informative to the network inference process. Metabolomic data can provide detailed measurements of changes in hundreds of metabolite levels in response to changing cell state or environment. Techniques such as surface plasmon resonance imaging (SPRi) (Smith and Corn, 2003) can provide additional high-throughput data on protein-binding constants via measurements of association and dissociation rates, potentially providing small but high-accuracy interaction networks. Mass cytometry can provide single-cell measurements of phosphorylation on a very fine time scale (Bendall *et al.*, 2011). New data types can be added to cMonkey fairly easily since its basic model is already integrative (see Methods). The network inference pipeline can accommodate some of these data, such as SPRi-derived interaction networks, by using them to influence the likelihood that a regulatory interaction is incorporated into the model. However, other data types – particularly those that are on different time scales, like mass cytometry – pose a more difficult challenge for network inference. Even integrating proteomics data – which may superficially resemble microarray data – into the inference pipeline, rather than simply overlaying it on inferred microarray-derived networks, poses new challenges. Proteomics measurements still produce sparser data sets than microarrays, and techniques such as SILAC (Ong *et al.*, 2002) will be systematically biased against certain proteins. A more serious issue is that TFs tend to have low expression values, and proteomics techniques do a poor job of capturing proteins expressed at low levels.

B. Validation

Validation of inferred networks of biclusters and genes is a key issue that we address explicitly in Methods. It should be emphasized, however, that new data types as discussed above will not only improve the quality of the inferred biclusters and networks, but will aid validation as well. Bicluster enrichment analysis already provides an example of using independent data types (KEGG and GO pathways) for validation as the annotations used for enrichment analysis are independent from those used in bicluster inference: because of this independence, significant enrichment provides one indicator of bicluster quality. After such enrichment analysis, it is crucial for experts with domain knowledge to highlight their most interesting genes and pathways. With the thousands of predictions that are made in a single run of our pipeline and the lack of a true gold-standard data set, such biological expertise is crucial to fully realize the hypothesis-generating potential of our methods.

C. Visualization

One issue that needs to be addressed with current visualization tools concerns displaying per-gene measurements, like the proteomics overlays in Fig. 6, in networks consisting mostly of biclusters like the hallmarks network in Fig. 3 – in other words, what is the best way to indicate differential expression of a small subset of genes contained in one or more biclusters? This may only be relevant until overlays of data from other sources are replaced by integration of these data into the inference pipeline, but for now the issue of overlaying single-gene data on biclustered networks remains open.

A larger issue is that network visualizations such as those produced by Cytoscape show a single view of a network as it might exist at one point in time. This network view may also represent a superset of the RNs that produced the data: any regulatory interaction with enough support across the various conditions is reproduced in the final network. But networks change over time, as is shown in many cancers; and different parts of any network will be active under different conditions. In other words, what is currently shown might be called a union or average of many potentially valid inferred networks. As inference tools and data availability improve, what is really desired is a tool (or set of tools) that can be used to explore this multiplicity of possible networks. This will probably require tools that can display changes in networks, in real time and in interpretable fashion, extending the "network-centric/gene-centric" metaphor introduced earlier: network-centric techniques would summarize the possible network changes over time and/or condition with the goal of steering the user to interesting features of the data, gene-centric techniques would create network sets from one or more networks of interest, and hybrid techniques might answer questions posed by the user about specific alterations in the network.

V. Computational Methods

A. cMonkey Integrative Biclustering

The steps below describe the cMonkey algorithm. Examples of data sources used are those for *Escherichia coli* in a multi-species biclustering by Kacmarczyk *et al.* (2011). For further details on cMonkey and MScM see Reiss *et al.* (2006) and Waltman *et al.* (2010).

1. Data Preparation

cMonkey uses three main data types: microarray expression, upstream sequences, and networks of associations or interactions. Data preparation and translation into a format that cMonkey can use is a key and non-trivial part of running cMonkey. Specifics of this process will be addressed in the section for each data type. The overall data preparation process involves (a) finding appropriate expression data, (b) determining upstream sequence information for the relevant organism(s), (c) downloading the association and interaction network data to be used, and (d) processing network data as necessary to reduce it to a list of interacting pairs of genes. A crucial issue across all these steps is determining a single-gene naming convention across all input data types and converting as necessary. cMonkey uses the Global Translator goose for this (http://err.bio.nyu.edu/cytoscape/bionetbuilder/translator.php).

2. Expression Data

Expression data for cMonkey is given in matrix form, where rows represent genes and columns represent experimental conditions. Expression data are row-normalized to have mean = 0, SD = 1. *E. coli* expression data for the multi-species biclustering by Kacmarczyk *et al.* were comprised of 507 conditions covering 16 projects from the Many Microbe Microarrays Database (M3D) (Faith *et al.*, 2008).

We denote the expression levels of the genes by $x = (x_1, \ldots, x_{N_g})^T$. We store the C observations of these N_g genes in an $N_g \times C$ matrix, X, where the columns correspond to the experimental observations. For a given bicluster k, if $p(x_{ij})$ is defined as the normally distributed likelihood of the expression value x_{ij} within bicluster k, then the co-expression p-value r_{ik} for gene i relative to bicluster k is $r_{ik} = \sum_{j \in J_k} p(x_{ij})$ where J_k indexes the conditions in bicluster k. The co-expression p-value r_{jk} for condition j is defined similarly.

3. Sequence Data

Methods for obtaining and processing upstream sequence data depend on the organism. Generally the regulatory sequence analysis tools (RSAT) (van Helden, 2003) are used to extract upstream cis-regulatory sequences: sequence length depends on whether the organisms are archaea, bacteria, or eukaryotes, and additional processing may be required to account for the presence of operons (see Reiss *et al.*, 2006, for details). *E. coli* data for the Kacmarczyk *et al.* biclustering were obtained using RSAT as above and adding network edges between genes known to share operons to the network data (see below).

For a given bicluster k, MEME (Bailey and Elkan, 1994) is used to determine a set of motifs common to some or all of the upstream sequences of the genes in that cluster. MAST (Bailey and Gribskov, 1998) is then used to calculate a motif value s_{ik} for each gene i relative to bicluster k (motif values for conditions are set to zero).

4. Network Data

Network data are the most varied, generally comprising multiple network types for a given biclustering analysis. These data break down into two types: association and metabolic networks, such as Prolinks (Bowers *et al.*, 2004) and Predictome (Mellor *et al.*, 2002); and interaction networks, such as DIP (Salwinski *et al.*, 2004) and BIND (Bader *et al.*, 2003). While data sources such as DIP provide pairs of interacting proteins directly, others must be processed to generate these lists of interacting pairs. For example, KEGG (Kanehisa and Goto, 2000) metabolic pathways are examined for pairs of genes that participate in a reaction sharing one or more ligands (excluding water and ATP). Network data are also the most species-dependent as different network data types are available for different organisms. This is reflected in the number and diversity of network data types in the Kacmarczyk *et al. E. coli* biclustering: operon edges between genes known to lie on the same operon; metabolic edges from KEGG as described above; gene neighbor, phylogenetic profile, and gene cluster edges from Prolinks; and COG-code edges from COG (Tatusov *et al.*, 2000).

For a given bicluster k, gene i, and network N, the network association p-value q_{ik}^{N} is computed using a hypergeometric distribution based on the number of connections between gene i and bicluster k, connections between gene i and genes not in bicluster k, and connections within and between genes in k and not in k. This metric assigns better p-values to densely connected sub-networks of genes that are likely to participate in common functional modules.

5. cMonkey Bicluster Model

cMonkey determines biclusters by iteratively (a) updating the conditional probability of each bicluster based on its previous state, and (b) further optimizing the bicluster by adding or dropping genes and/or conditions. This constitutes a Markov chain process where the probabilities in the optimization step depend only on the previous state of the bicluster. Additions and deletions are made by sampling from the conditional probability distribution using a Monte Carlo procedure. The component contributions to the conditional probability come from the expression, sequence, and network p-values described above, which are combined into a regression model.

Denoting an arbitrary gene or condition by i, we define the vector g_{ik} as the projection into one dimension of the space defined by r_{ik}, s_{ik}, and q_{ik}, as follows:

$$g_{ik} = r_0 \log(\tilde{r}_{ik}) + s_0 \log(\tilde{s}_{ik}) + \sum_N q_0^N \log(\tilde{q}_{ik}^N) \qquad (1)$$

The use of $\tilde{r}_{ik}$ instead of r_{ik} denotes that the $\log(r_{ik})$ values have been normalized, for each bicluster, to have mean = 0, SD = 1; the same applies for $\tilde{s}_{ik}$ and $\tilde{q}_{ik}^N$, placing all three on the same scale for each bicluster. The likelihood of any gene or condition i belonging to bicluster k is then

$$\pi_{ik} \propto \exp(\beta_0 + \beta_1 g_{ik}) \qquad (2)$$

The parameters β_0, β_1 determine the conditional probability of membership of gene or condition i in bicluster k. The importance of each evidence type can be adjusted using the "mixing parameters" r_0, s_0, q_0^N.

A cMonkey run starts with "seeding" of initial biclusters, with each bicluster randomly seeded according to one of several algorithms. After seeding, each iteration (a) updates the bicluster motifs, (b) recalculates the probabilities π_{ik} described above for each gene or condition i, and (c) preferentially adds or drops genes or conditions according to their probability of membership using a simulated annealing protocol. Unlikely moves (additions or deletions) are permitted according to an annealing temperature T that is decreased over time. Mixing parameters r_0, s_0, q_0^N are also varied according to a set schedule: s_0 starts small early in the process, when biclusters are unlikely to have coherent motifs, and is gradually ramped up until its influence is equivalent to that of r_0. Values of q_0^N follow a schedule that depends on the networks involved.

6. Multi-Species cMonkey

MScM is similar to single-species cMonkey as described above, with a few additional steps. The overall MScM process, assuming a two-species run, is to (1) find orthologous genes between the two species; (2) perform the cMonkey Markov Chain Monte Carlo procedure, using orthologous gene pairs identified in step 1 instead of individual genes, to produce biclusters of "orthologous core" genes; (3) for each organism, elaborate these orthologous core biclusters by adding and dropping individual genes (instead of orthologous gene pairs) using the normal single-species cMonkey process, with the restriction that no orthologous core genes are dropped; and optionally (4) perform separate single-species cMonkey runs on the remaining genes for each species. Orthologous genes are identified using existing tools, such as the Mouse Genomics Informatics database (Bult *et al.*, 2008) or the InParanoid algorithm (Remm *et al.*, 2001). Determination of biclusters in step 2 begins by calculating values for each species U and V, g_{ik}^U and g_{ik}^V as in Eq. (1). These are combined to produce a likelihood of an orthologous pair i belonging to cluster k similarly to Eq. (2)

$$\pi_{ik} \propto \exp(\beta_0 + \beta_1 (g_{ik}^U + g_{ik}^V)) \qquad (3)$$

7. Enrichment Analysis

Analysis and validation of biclusters is a key component of the cMonkey design. As a post-biclustering step, biclusters are analyzed for significant enrichment according to standard annotations such as GO (Ashburner *et al.*, 2000), KEGG, and COG (Tatusov *et al.*, 2000). These annotations provide a standard way to assign putative functions to biclusters, somewhat resolving the issue of giving meaningful names to biclusters in inferred networks. With the exception of the shared-ligand network derived from KEGG, these annotations are separate from the data used to infer the biclusters, so enrichment analysis also provides a means of assessing bicluster quality (see the Validation section below).

8. Integrating New Data Types

Integration of new data types into cMonkey is relatively straightforward. Additional network types are easily added as additional q_{ik}^N terms. New data, such as relative expression levels from proteomics experiments, could be incorporated as a fourth major data type (in addition to microarray expression, sequence, and networks) and added to the calculation of g_{ik}. In both cases, an appropriate annealing schedule for the weight given to the new network or data type would have to be determined.

B. Inferelator Pipeline

We have applied our network inference pipeline to a variety of different data sets (synthetic, prokaryotic, yeast, human white blood cells). We have developed several closely related variants of the core pipeline, which is composed of two core methods: (1) tlCLR, and (2) the Inferelator 1.0. A coarse prediction of the topology is made using tlCLR, which is further refined by the Inf. This pipeline of tlCLR followed by Inf is repeated for multiple permutations of the data set (resampling), resulting in an ensemble of predicted RNs, which is then combined into one final network. Here we present a brief description of tlCLR (for a more detailed description we refer to the reader to Greenfield *et al.*, 2010 and Madar *et al.*, 2010). Additionally, we present a modification to the core Inf method that allows for the incorporation of *a priori* known regulatory edges.

1. Problem Setup

As in the description of cMonkey, we denote the expression levels of the genes by $x = (x_1, \ldots, x_{N_g})^T$. We store the C observations of these N_g genes in an $N_g \times C$ matrix, where the columns correspond to the experimental observations. These observations can be of two types: time-series data (X^{ts}), and steady-state data (X^{ss}). Since we make explicit use of the time-series data in the description of our inference procedure, we denote time-series conditions by $t_1, t_2, \ldots, t_k$, (i.e., $x(t_1), x(t_2), \ldots, x(t_k)$ are the k

time-series observations that constitute the columns of X^{ts}). Our inference procedure produces a network in the form of a ranked list of regulatory interactions, ranked according to confidence. We refer to the final list of confidences as an $N_g \times N_p$ matrix Z^{final}, where N_p denotes the possible predictors. Element i,j of Z^{final} represents our confidence in the existence of a regulatory interaction between x_i and x_j.

2. Core Method 1: Time-Lagged Context Likelihood of Relatedness

tlCLR (Greenfield *et al.*, 2010; Madar *et al.*, 2010) is a MI-based method that extends the original CLR algorithm (Faith *et al.*, 2008) to take advantage of time-series data. MI (Shannon, 1948) is an information theory metric of mutual dependence between any two random variables. The original formulation of CLR was unable to learn directionality of regulatory edges as MI is a symmetric measure. In the tlCLR algorithm, we make explicit use of the time-series data to learn directed regulatory edges. We describe, in brief, three main steps: (1) model the temporal changes in expression as an ODE, (2) calculate the MI between every pair of genes, and (3) apply a background correction (filtering) step to remove least likely interactions. We refer the reader to Greenfield *et al.* (2010) and Madar *et al.* (2010) for a thorough description of this method.

We assume that the temporal changes in expression of each gene can be approximated by the linear ODE:

$$\frac{dx_i(t)}{dt} = -\alpha_i x_i + \sum_{j=1}^{N} \beta_{i,j} x_j(t), \quad i = 1, \ldots, N \tag{4}$$

where α_i is the first-order degradation rate of x_i and the β_{ij} s are a set of dynamical parameters to be estimated. Note that the functional form presented above treats the rate of change of the response (x_i) as linear function of the predictors (x_js). Here, we describe only this linear form for simplicity, but in several applications we employ more complex functional forms. The value of β_{ij} describes the extent and sign of the regulation of target gene x_i by regulator x_j. We store the dynamical parameters in a $N \times P$ matrix β, where N is the number of genes, and P is the number of possible regulators. Note that β is typically sparse, that is, most entries are 0 (reflecting the sparsity of transcriptional RNs). Later, we describe how to calculate the values β_{ij} by a constrained linear-regression scheme. First, we briefly describe how to use the time-series data in the context of improving the calculation of MI values between a gene x_i and its potential regulator x_j.

We first apply a finite approximation to the left-hand side of Eq. (4), for each x_i, $i = 1, \ldots, N_g$ and rewrite it as a response vector y_i, which captures the rate of change of expression in x_j. We pair the response y_i with a corresponding explanatory variable x_j, $j = 1 \ldots N_p$. Note each x_j is time-lagged with respect to the response y_i, that is, $x_j(t_k)$ is used to predict $y_j(t_{k+1})$. For more details of this transformation, we refer the reader to Greenfield *et al.* (2010). As a measure of confidence for a directed regulatory interaction between a pair of genes ($x_j \rightarrow x_i$), we use MI, $I(x_i, x_j)$, where a pair that shows a high MI score (relative to other pairs) is more likely to represent a true regulatory interaction. Note that $I(y_i, x_j) \neq I(y_j, x_i)$, making one regulatory direction more likely than the other. We refer to the MI calculated from $I(y_i, x_j)$ as dynamic MI, as it takes advantage of the temporal information available from time-series data (distinguishing time-series data from steady-state data). As described above, we calculate $I(x_i, x_j)$ and $I(y_i, x_j)$ for every pair of genes and store the values in the form of two $N_g \times N_p$ matrices M^{stat} and M^{dyn}, respectively. Note that M^{stat} is symmetric, while M^{dyn} is not. We now briefly describe how tlCLR integrates both static and dynamic MI to produce a final confidence score for each regulatory interaction. For a more detailed explanation, we refer the reader to Greenfield *et al.* (2010) and Madar *et al.* (2010).

For each regulatory interaction $x_j \rightarrow x_i$, we compute two positive Z-scores (by setting all negative Z-scores to zero): one for the regulation of x_i by x_j based on dynamic-MI (i.e., based on M^{dyn}), $Z_1\left(x_i, x_j\right) = \max\left(0, \frac{M_{i,j}^{\{dyn\}} - \frac{\sum_{j'} M_{i,j'}^{\{dyn\}}}{N}}{\sigma_i}\right)$, where σ_i is the standard deviation of the entries in row i of M^{dyn}; and one for the regulation of x_i by x_j based on both static and dynamic MI, $Z_1\left(x_i, x_j\right) = \max\left(0, \frac{M_{i,j}^{\{dyn\}} - \frac{\sum_{i'} M_{i',j}^{\{stat\}}}{N}}{\sigma_j}\right)$ where σ_j is the standard deviation of the entries in column j of M^{stat}. We combine the two scores into a final tlCLR score, $Z_{i,j}^{tlCLR} = \sqrt{(Z_1^2 + Z_2^2)}$. Note that some entries in Z^{tlCLR} are zero, that is, Z^{tlCLR} is somewhat sparse. The output of tlCLR, Z^{tlCLR}, is used as the input to Inf, as only the highest ranked predictors from row i of Z^{tlCLR} are considered as possible predictors for gene i

3. Core Method 2: Inferelator 1.0

We use Inf to learn a sparse dynamical model of regulation for each gene x_i. As potential regulators of x_i, we consider only the P highest confidence (non-zero) regulators (i.e., the P^i most-highly ranked regulators from row i of Z^{tlCLR}). Accordingly, for each gene, x_i, we denote this subset of potential regulators as x^i. We then learn a sparse dynamical model of regulation for each x_i as a function of the potential regulators x^i using Inf. We assume that the time evolution in the x_is is governed by $\frac{dx_i(t)}{dt} = -\alpha_i x_i + \sum_{j=1}^{P^i} \beta_{i,j} x_j(t), \quad i = 1, \ldots, N$ which is exactly Eq. (4) with our constraint on the number of regulators. LARS (Efron *et al.*, 2004) is used to efficiently implement l_1 constrained regression to determine a sparse set of the parameters β. This is done by minimizing the following objective function, amounting to a least-square estimate based on the ODE in Eq. (4) under an l_1-norm penalty on regression coefficients,

$$\sum_{j=1}^{P^i} |\beta_{i,j}| < s_i \sum_{j=1}^{P^i} |\beta_{i,j}^{\text{ols}}| \tag{5}$$

where β^{ols} are the values of β determined by ordinary least squares regression (ols), and s_i, the shrinkage parameter. This parameter is in the range [0,1], and controls the sparsity of the model, with $s_i = 0$ amounting to a null model, and $s_i = 1$ amounting the full ols model. We select the optimal values of s_i by 10-fold cross validation. After applying this l_1 regression, we have β, an $N_g \times N_p$ matrix of dynamic parameters β_{ij} for each regulatory interaction $x_j \rightarrow x_i$. We use the percentage of explained variance of each parameter β_{ij} as confidences in these regulatory interactions, as described in Greenfield *et al.* (2010). We store these confidences in Z^{Inf}. We combine these confidences in a rank-based way such that each method is weighted equally, as described in Greenfield *et al.* (2010), to generate $Z^{\text{tlCLR-Inf}}$, which represents our confidence in each regulatory interaction after running our pipeline one time. We now describe how we resample our network inference pipeline to generate an ensemble of predicted networks (i.e., lists of confidences for each regulatory interaction).

4. Using Resampling to Improve Network Inference

To further improve the quality of our ranked list, we apply a resampling approach to the pipeline described above to generate an ensemble of putative RNs. We denote the matrix of response variables y_i, $i = 1, \ldots, N_g$ by Y. Similarly we denote the matrix of predictor variables x_j, $j = 1, \ldots, N_p$ by X. We sample with replacement from the indices of the columns of Y, generating a permutation of the indices, c^*. We use this permutation c^* to permute the columns of Y and X, generating Y^* and X^*, respectively. Note that (1) c^* is typically picked to be the number of conditions in the dataset (i.e., we sample from all experimental conditions), and (2) the columns of Y match the columns of X in the sense that the time-lagged relationship between the response and the predictors is preserved. We generated Z^{tlCLR}, Z^{Inf}, and $Z^{\text{tlCLR-Inf}}$ as described before, with the only difference being that we use the response and explanatory vectors from Y^* and X^* instead of Y and X. We repeat this procedure B times. This generates an ensemble of B predicted RNs. The final weight we assign to each regulatory interaction is the median weight assigned to that interaction from each of the B networks. Thus, the final weight can be considered an "ensemble vote" of the confidence the ensemble of networks has in that edge: $Z_{i,j}^{\text{final}} = \text{median}(Z_{i,j}^{\text{tlCLR-Inf}}(1),\ Z_{i,j}^{\text{tlCLR-Inf}}(2),\ \ldots, Z_{i,j}^{\text{tlCLR-Inf}}(B))$.

5. Incorporating Prior Information Directly into Network Inference

Our tlCLR–Inf pipeline is capable of inferring not only topology but also dynamical parameters, which can be used to predict the response of the system to new perturbations (Greenfield *et al.*, 2010). Our predictions, like those of any network inference method, contain false-positive interactions. One way to improve the performance of network inference is to constrain the model selection procedure to incorporate regulatory interactions that are known *a priori*, as many databases of known regulatory interactions exist (Aranda *et al.*, 2010; Bader *et al.*, 2003; Ceol *et al.*, 2010; Chautard *et al.*, 2011; Croft *et al.*, 2011; Goll *et al.*, 2008; Knox *et al.*, 2011; Lynn *et al.*, 2008; Michaut *et al.*, 2008; Prieto and Rivas, 2006; Razick *et al.*, 2008; Stark *et al.*, 2011). However, if one is studying a particular process (e.g., lymphoma) not all of the known interactions will be relevant in lymphoma. Thus, a method is needed that incorporates a known edge only if it is supported by the given data. We do so by solving Eq. (4) subject to the following constraint:

$$\sum_{j=1}^{P^i} \theta_{ij}\left|\beta_{i,j}\right| < s_i \sum_{j=1}^{P^i} \left|\beta_{i,j}^{\text{ols}}\right| \qquad (6)$$

which is exactly Eq. (5) with the parameter θ_{ij} (Yong-a-poi *et al.*, 2008; Zou, 2006). This parameter is referred to as the adaptive weight, and regulates the degree to which β_{ij} is shrunk out of the model. If it is known from an external data type (e.g., literature mining, ChIP-seq, etc.) that x_j regulates x_i, then this knowledge can be incorporated by setting $\theta_{ij} < 1$, which will make it less likely that β_{ij} will be shrunk (removed from the model) by LARS. If there exists negative prior knowledge (i.e., knowledge that x_j does not regulate x_i), this can be incorporated by setting $\theta_{ij} > 1$. The exact values of θ_{ij} that are needed to incorporate an *a priori* known interaction vary from dataset to dataset and must be chosen heuristically. This behavior is similar to that of many other methods for incorporating priors, including Bayesian methods, which require a heuristically chosen hyper-parameter to determine the shape of the prior (Mukherjee and Speed, 2008). In our method, once an informed choice of θ_{ij} is made, an edge is incorporated only if it is supported by the data. Even if θ_{ij} is set to a very low value (approaching zero, reflecting strong belief in the existence of this edge), the corresponding parameter, β_{ij}, will be non-zero only if there is support from the data set. This is exactly the desired behavior when we are given *a priori* knowledge that may or may not be completely relevant for our data sets.

C. Analysis and Visualization

Given the wide range of network properties, features of interest, and intended audiences, there is no "silver bullet" approach to visualizing biological networks. The most effective visualizations come from detailed analysis of the network, followed by a careful linking of important network properties to visual features such that interesting properties are immediately and intuitively obvious. The steps below show how Figs. 3–6 were created, and are intended to provide an arsenal of examples and tools to arrive at an effective combination of analysis and representation.

Fig. 3 uses publicly available mouse and human microarray data from GEO. The mouse data consisted of 508 conditions from the IMMGEN (Painter *et al.*, 2011) data set of experiments on characterized mouse immune cell lineages (GEO accession number: GSE15907). Human microarray data were gathered from 23 different experiment sets measuring the response of human immune cells to different stimuli. In an attempt to mirror the conditions of the IMMGEN data set, only the control conditions from the different experiment sets were used, yielding a total of 140 conditions. The network was generated from a full run of the MScM and Inf pipelines on the data described above as follows:

1. Run MScM to generate a collection of 176 mouse and human biclusters.
2. Perform enrichment analysis over all biclusters using generic GO slim, GO, and KEGG.
3. Run the Inf pipeline using the mouse biclusters and known TFs for mouse alone to produce a preliminary mouse-specific network. Although the human biclusters are not used directly, their presence in the MScM run should improve the biological relevance of the mouse biclusters as discussed above.
4. Remove low-confidence edges (Inf z-score < 3.5, or $|\beta| < 0.1$) to produce a refined preliminary network.
5. Find biclusters with significantly enriched GO slim terms and label them with hallmarks of cancer associated with these terms. This results in 17 biclusters with hallmark annotations.
6. Reduce the network to the smallest possible network containing all 17 biclusters identified above along with their regulators, giving the final Fig. 3 network.

Further analysis of this network would begin with further investigation of the biclusters to obtain a better sense of the function represented by each bicluster. The Annotation Viewer (http://gaggle.systemsbiology.net/docs/geese/anno.php) is a Gaggled tool that allows browsing of arbitrary gene or bicluster annotations – in this case, bicluster GO and KEGG annotations. The CMMR provides more detailed examination of all facets of the biclusters: genes, conditions, residuals, etc. When bicluster functions are better understood, one can then ask whether the inferred regulatory interactions make sense, and investigate the significance of the observed separation of cancer hallmarks into two different clusters.

The next three examples (Figs. 4–6) use breast cell lines from normal breast epithelial tissue (MCF-10A) (Soule *et al.*, 1990) and invasive, metastatic breast cancer tumor tissue (MDA-MB-231) (Cailleau *et al.*, 1978). Data for each cell line were gathered from a total of eight GEO data sets, giving a total of 103 MCF-10A conditions and 121 MDA-MB-231 conditions covering roughly 12,000 genes. Proteomics data consisting of genes up- and downregulated in each cell line under two treatments were also provided. Here, we infer regulation of individual genes directly, instead of regulation of biclusters, so that we can overlay the proteomics data on the corresponding genes in the resulting network. As a result, we are unable to take advantage of the dimensionality reduction and noise reduction provided by cMonkey, and used the following heuristic approach instead. From the initial 12,000 genes, we selected those genes whose standard deviation across experiments was at the 75th percentile or better, then added the 2000 genes with the most differential expression according to significance analysis of microarrays (SAM) (Tusher *et al.*, 2001). The final input set for inference on each cell line consisted of 4619 genes, 289 of which were TFs. Network inference proceeded as follows:

1. Perform separate Inf pipeline runs on each cell line to produce a ranked list of putative regulatory edges for each cell line.
2. Combine separate network edges from the individual runs into a single "differential network" where edge color shows the likelihood of each regulatory edge being active in one cell type or another. The rank of each edge in this differential network is determined by using Stouffer's method to combine the Inf scores from the individual networks. Specificity to one cell line or the other is calculated as the log ratio of individual edge ranks.
3. Retain the top-ranked 5000 edges of the differential network; remove those with $|\beta| < 0.05$ for a final total of 4822 edges and 1866 nodes. Overlay proteomics data from two experimental conditions to produce the network are shown in Fig. 4.
4. Starting with the network in Fig. 4, find the largest connected sub-network of TFs using the Subgraph Creator Cytoscape plug-in. Map node out-degree to node size. Map node color to fraction of edges specific to one cell line or the other, counting only those edges with absolute value rank ratio above 4. This provides the network "summary" shown in Fig. 5.
5. Given lists of genes up- and downregulated in each cell line in the two proteomics experiments, load these lists into Sungear. Send each gene list to Cytoscape using the Gaggle and annotate genes according to cell line, experimental condition, and up- or downregulation.
6. Find the smallest sub-network for each condition that includes all differentially expressed genes. This identifies ITGB4 as a likely key gene involved in motility as discussed in Biological Insights.
7. Find the 1-hop network around ITGB4. Distinguish up- and downregulated genes for each treatment using the annotations assigned earlier to the Sungear-derived gene lists, producing Fig. 6.

Further analysis of the network in Fig. 6 might proceed as follows: send the set of differentially regulated genes back to Sungear to look for interesting intersections, such as over-representation within a particular intersection of conditions; or broadcast ITGB4 and some of its targets to the Firegoose, then from there to EMBL STRING to look for additional evidence for the inferred edges or grow the network further.

D. Validation

Validation of inferred networks of genes or biclusters (i.e., of predicted regulatory topology and kinetic models) is a critical challenge that has not been well resolved. In all cases, the best validation is of course follow-up experimentation to verify the computational results, but this approach is inherently limited by available time and resources.

Bicluster validation is currently a more tractable problem than network validation. Several metrics used to compare biclustering methods (Waltman *et al.*, 2010) can also be used to assess the quality of individual biclusters and biclustering runs. Each bicluster has a residual score that shows the variance in expression data within the bicluster; lower residuals mean higher coherence in expression values. Significant bicluster enrichment implies that a cluster contains co-functional genes. High coverage of the input expression matrix, in terms of fraction of overall genes and conditions included in the overall set of biclusters, is favorable, as is a low degree of overlap between biclusters. For multi-species biclustering, the degree of conservation between species in a bicluster is also important.

Inferred networks present a more significant challenge, especially when no gold standard is available. A simple approach is to calculate some of the network statistics mentioned earlier: for example, compare the distribution of node in- and out-degrees to the expected power-law curves. However, the inference technique itself, as well as any means used to filter results or select sub-networks of interest, may skew these distributions: for example, the Inf limits the in-degree of any gene to a user-defined threshold. Another means of network validation is to determine the degree to which the inferred network recapitulates known network edges. However, this can become circular when known edges are used to provide priors for inference. Recent work by the DREAM consortium incorporated the prediction of multiple methods by different research groups into one "community" prediction, and experimentally validated the top-ranked predictions. Ideally, such integrative methods will continue to be developed and shed light on previously unknown biology. Although the notion of "community predictions" is novel and exciting, such vast resources not always exist. Even in such cases, methods that are integrative in terms both the algorithms and data types used show great promise in building global, predictive RNs of complex biological phenomena.

References

Aranda, B., Achuthan, P., Alam-Faruque, Y., Armean, IBridge, A., Derow, CFeuermann, M., Ghanbarian, A. T., Kerrien, S., Khadake, J., Kerssemakers, J., Leroy, C., Menden, M., Michaut, M., Montecchi-Palazzi, L., Neuhauser, S. N., Orchard, S., Perreau, V., Roechert, B., van Eijk, K., and Hermjakob, H. (2010). The IntAct molecular interaction database in 2010. *Nucleic Acids Res.* **38**, D525–D530.

Ashburner, M., Ball, C. A., Blake, J. ABotstein, D., Butler, H., Cherry, J. M., Davis, A. P., Dolinski, K., Dwight, S. S., Eppig, J. T., Harris, M. A., Hill, D. P., Issel-Tarver, L., Kasarskis, A., Lewis, S., Matese, J. C., Richardson, J. E., Ringwald, M., Rubin, G. M., and Sherlock, G. (2000). Gene ontology: tool for the unification of biology. The Gene Ontology Consortium. *Nat. Genet.* **25**, 25–29.

Bader, G. D., Betel, D., and Hogue, C. W. (2003). BIND: the Biomolecular Interaction Network Database. *Nucleic Acids Res.* **31**, 248–250.

Bailey, T. L., and Elkan, C. (1994). Fitting a mixture model by expectation maximization to discover motifs in biopolymers. Proceedings /.. International Conference on Intelligent Systems for Molecular Biology; ISMB. *Int. Conf. Intell. Syst. Mol. Biol.* **2**, 28–36.

Bailey, T. L., and Gribskov, M. (1998). Combining evidence using p-values: application to sequence homology searches. *Bioinformatics* **14**, 48–54.

Bansal, M., Gatta, G. D., and di Bernardo, D. (2006). Inference of gene regulatory networks and compound mode of action from time course gene expression profiles. *Bioinformatics* **22**, 815–822.

Bare, J. C., Shannon, P. T., Schmid, A. K., and Baliga, N. S. (2007). The Firegoose: two-way integration of diverse data from different bioinformatics web resources with desktop applications. *BMC Bioinformat.* **8**, 456.

Ben-Dor, A., Chor, B., Karp, R., and Yakhini, Z. (2003). Discovering local structure in gene expression data: the order-preserving submatrix problem. *J. Comput. Biol.* **10**, 373–384.

Bendall, S. C., Simonds, E. F., Qiu, P., Amir el, A. D., Krutzik, P. O., Finck, R., Bruggner, R. V., Melamed, R., Trejo, A., Ornatsky, O. I., Balderas, R. S., Plevritis, S. K., Sachs, K., Pe'er, D., Tanner, S. D., and Nolan, G. P. (2011). Single-cell mass cytometry of differential immune and drug responses across a human hematopoietic continuum. *Science* **332**, 687–696.

Bergmann, S., Ihmels, J., and Barkai, N. (2003). Iterative signature algorithm for the analysis of large-scale gene expression data. *Phys. Rev. E Stat. Nonlin. Soft Matter Phys.* **67**, 031902.

Bonneau, R., Facciotti, M. T., Reiss, D. J., Schmid, A. K., Pan, M., Kaur, A., Thorsson, V., Shannon, P., Johnson, M. H., Bare, J. C., Longabaugh, W., Vuthoori, M., Whitehead, K., Madar, A., Suzuki, L., Mori, T., Chang, D. -E., Diruggiero, J., Johnson, C. H., Hood, L., and Baliga, N. S. (2007). A predictive model for transcriptional control of physiology in a free living cell. *Cell* **131**, 1354–1365.

Bonneau, R., Reiss, D. J., Shannon, P., Facciotti, M., Hood, L., Baliga, N. S., and Thorsson, V. (2006). The Inferelator: an algorithm for learning parsimonious regulatory networks from systems-biology data sets de novo. *Genome Biol.* **7**, R36.

Bowers, P. M., Pellegrini, M., Thompson, M. J., Fierro, J., Yeates, T. O., and Eisenberg, D. (2004). Prolinks: a database of protein functional linkages derived from coevolution. *Genome Biol.* **5**, R35.

Bult, C. J., Eppig, J. T., Kadin, J. A., Richardson, J. E., and Blake, J. A. (2008). The Mouse Genome Database (MGD): mouse biology and model systems. *Nucleic Acids Res.* **36**, D724–D728.

Butte, A. J., and Kohane, I. S. (2000). Mutual information relevance networks: functional genomic clustering using pairwise entropy measurements. 418–429.

Butts, C. T. (2008). Social network analysis with sna. *J. Stat. Softw.* **24**, 1–51.

Cailleau, R., Olive, M., and Cruciger, Q. V. (1978). Long-term human breast carcinoma cell lines of metastatic origin: preliminary characterization. *in vitro* **14**, 911–915.

Carro, M. S., Lim, W. K., Alvarez, M. J., Bollo, R. J., Zhao, X., Snyder, E. Y., Sulman, E. P., Anne, S. L., Doetsch, F., Colman, H., Lasorella, A., Aldape, K., Califano, A., and Iavarone, A. (2010). The transcriptional network for mesenchymal transformation of brain tumours. *Nature* **463**, 318–325.

Ceol, A., Aryamontri, A. C., Licata, L., Peluso, D., Briganti, L., Perfetto, L., Castagnoli, L., and Cesareni, G. (2010). MINT, the molecular interaction database: 2009 update. *Nucleic Acids Res.* **38**, D532–D540.

Chaitankar, V., Ghosh, P., Perkins, E. J., Gong, P., and Zhang, C. (2010). Time lagged information theoretic approaches to the reverse engineering of gene regulatory networks. *BMC Bioinformat.* **11**(Suppl 6), S19.

Chautard, E., Fatoux-Ardore, M., Ballut, L., Thierry-Mieg, N., and Ricard-Blum, S. (2011). MatrixDB, the extracellular matrix interaction database. *Nucleic Acids Res.* **39**, D235–D240.

Chen, X., Xu, H., Yuan, P., Fang, F., Huss, M., Vega, V. B., Wong, E., Orlov, Y. L., Zhang, W., Jiang, J., Loh, Y. -H., Yeo, H. C., Yeo, Z. X., Narang, V., Govindarajan, K. R., Leong, B., Shahab, A., Ruan, Y., Bourque, G., Sung, W. -K., Clarke, N. D., Wei, C. -L., and Ng, H. -H. (2008). Integration of external signaling pathways with the core transcriptional network in embryonic stem cells. *Cell* **133**, 1106–1117.

Cheng, Y., and Church, G. M. (2000). Biclustering of expression data. *Proc. Int. Conf. Intell. Syst. Mol. Biol.* **8**, 93–103.

Christley, S., Nie, Q., and Xie, X. (2009). Incorporating existing network information into gene network inference. *PloS One* **4**, e6799.

Croft, D., O'Kelly, G., Wu, G., Haw, R., Gillespie, M., Matthews, L., Caudy, M., Garapati, P., Gopinath, G., Jassal, B., Jupe, S., Kalatskaya, I., Mahajan, S., May, B., Ndegwa, N., Schmidt, E., Shamovsky, V., Yung, C., Birney, E., Hermjakob, H., D'Eustachio, P., and Stein, L. (2011). Reactome: a database of reactions, pathways and biological processes. *Nucleic Acids Res.* **39**, D691–D700.

Davis, J., Goadrich, M., 2006. The relationship between precision-recall and ROC curves. *Proceedings of the 23rd International Conference on Machine Learning – ICML'06*, 233–240.

De Smet, R., and Marchal, K. (2010). Advantages and limitations of current network inference methods. Nature reviews. *Microbiology* **8**, 717–729.

di Bernardo, D., Bansal, M., and Gatta, G. D. (2006). Inference of gene regulatory networks and compound mode of action from time course gene expression profiles. *Bioinformatics* **22**, 815–822.

di Bernardo, D., Thompson, M. J., Gardner, T. S., Chobot, S. E., Eastwood, E. L., Wojtovich, A. P., Elliott, S. J., Schaus, S. E., and Collins, J. J. (2005). Chemogenomic profiling on a genome-wide scale using reverse-engineered gene networks. *Nat. Biotechnol.* **23**, 377–383.

DiMaggio, P. A., Jr., McAllister, S. R., Floudas, C. A., Feng, X. J., Rabinowitz, J. D., and Rabitz, H. A. (2008). Biclustering via optimal re-ordering of data matrices in systems biology: rigorous methods and comparative studies. *BMC Bioinformat.* **9**, 458.

Edgar, R., Domrachev, M., and Lash, A. E. (2002). Gene expression omnibus: NCBI gene expression and hybridization array data repository. *Nucleic Acids Res.* **30**, 207–210.

Efron, B., Hastie, T., Johnstone, I., and Tibshirani, R. (2004). Least angle regression. *Ann. Statist. Data* 407–451.

Elemento, O., and Tavazoie, S. (2005). Fast and systematic genome-wide discovery of conserved regulatory elements using a non-alignment based approach. *Genome Biol.* **6**, R18.

Faith, J. J., Driscoll, M. E., Fusaro, V. A., Cosgrove, E. J., Hayete, B., Juhn, F. S., Schneider, S. J., and Gardner, T. S. (2008). Many Microbe Microarrays Database: uniformly normalized Affymetrix compendia with structured experimental metadata. *Nucleic Acids Res.* **36**, D866–D870.

Faith, J. J., Hayete, B., Thaden, J. T., Mogno, I., Wierzbowski, J., Cottarel, G., Kasif, S., Collins, J. J., and Gardner, T. S. (2007). Large-scale mapping and validation of *Escherichia coli* transcriptional regulation from a compendium of expression profiles. *PLoS Biol.* **5**, 54–66.

Friedman, N., Linial, M., and Nachman, I. (2000). Using Bayesian networks to analyze expression data. *J. Comput. Biol.* **7**, 601–620.

Friedman, N., and Nachman, I. (1999). *Learning Bayesian Network Structure from Massive Datasets: The "Sparse Candidate" Algorithm.* UAI, San Fransisco, CA, pp. 206-215.

Gan, X., Liew, A. W., and Yan, H. (2008). Discovering biclusters in gene expression data based on high-dimensional linear geometries. *BMC Bioinformat.* **9**, 209.

Geier, F., Timmer, J., and Fleck, C. (2007). Reconstructing gene-regulatory networks from time series, knock-out data, and prior knowledge. *BMC Syst. Biol.* 11.

Gevaert, O., Van Vooren, S., and De Moor, B. (2007). A framework for elucidating regulatory networks based on prior information and expression data. *Ann. N Y Acad. Sci.* **1115**, 240–248.

Goll, J., Rajagopala, S. V., Shiau, S. C., Wu, H., Lamb, B. T., and Uetz, P. (2008). MPIDB: the microbial protein interaction database. *Bioinformatics* **24**, 1743–1744.

Greenfield, A., Madar, A., Ostrer, H., and Bonneau, R. (2010). DREAM4: combining genetic and dynamic information to identify biological networks and dynamical models. *PLoS ONE* **5**, e13397.

Guo, A., Villen, J., Kornhauser, J., Lee, K. A., Stokes, M. P., Rikova, K., Possemato, A., Nardone, J., Innocenti, G., Wetzel, R., Wang, Y., MacNeill, J., Mitchell, J., Gygi, S. P., Rush, J., Polakiewicz, R. D., and Comb, M. J. (2008). Signaling networks assembled by oncogenic EGFR and c-Met. *Proc. Natl. Acad. Sci. U S A* **105**, 692–697.

Hanahan, D., and Weinberg, R. A. (2000). The hallmarks of cancer. *Cell* **100**, 57–70.

Hanahan, D., and Weinberg, R. A. (2011). Hallmarks of cancer: the next generation. *Cell* **144**, 646–674.

Hartigan, J. A. (1972). Direct clustering of a data matrix. *J. Amer. Statist. Assoc.* **67**, 123–129.

Husmeier, D., and Werhli, A. V. (2007). Bayesian integration of biological prior knowledge into the reconstruction of gene regulatory networks with Bayesian networks. Computational systems bioinformatics/Life Sciences Society. *Comput. Syst. Bioinformat. Conf.* **6**, 85–95.

Huttenhower, C., Mutungu, K. T., Indik, N., Yang, W., Schroeder, M., Forman, J. J., Troyanskaya, O. G., and Coller, H. A. (2009). Detailing regulatory networks through large scale data integration. *Bioinformatics* **25**, 3267–3274.

Huynh-Thu, V. A., Irrthum, A., Wehenkel, L., and Geurts, P. (2010). Inferring regulatory networks from expression data using tree-based methods. *PLoS ONE* **5**, e12776.

Ideker, T., Thorsson, V., Ranish, J. a., Christmas, R., Buhler, J., Eng, J. K., Bumgarner, R., Goodlett, D. R., Aebersold, R., and Hood, L. (2001). Integrated genomic and proteomic analyses of a systematically perturbed metabolic network. *Science (New York, NY)* **292**, 929–934.

Ihmels, J., Bergmann, S., Berman, J., and Barkai, N. (2005). Comparative gene expression analysis by differential clustering approach: application to the Candida albicans transcription program. *PLoS Genet.* **1**, e39.

Jeong, H., Tombor, B., Albert, R., Oltvai, Z. N., and Barabási, a. L. (2000). The large-scale organization of metabolic networks. *Nature* **407**, 651–654.

Kacmarczyk, T., Waltman, P., Bate, A., Eichenberger, P., and Bonneau, R. (2011). Comparative Microbial Modules Resource: Generation and Visualization of Multi-species Biclusters. *PLoS computational biology* **7**, e1002228.

Kanehisa, M., and Goto, S. (2000). KEGG: kyoto encyclopedia of genes and genomes. *Nucleic Acids Res.* **28**, 27–30.

Kluger, Y., Basri, R., Chang, J. T., and Gerstein, M. (2003). Spectral biclustering of microarray data: coclustering genes and conditions. *Genome Res.* **13**, 703–716.

Knox, C., Law, V., Jewison, T., Liu, P., Ly, S., Frolkis, A., Pon, A., Banco, K., Mak, C., Neveu, V., Djoumbou, Y., Eisner, R., Guo, A. C., and Wishart, D. S. (2011). DrugBank 3.0: a comprehensive resource for 'omics' research on drugs. *Nucleic Acids Res.* **39**, D1035–D1040.

Lazzeroni, L., and Owen, A. (1999). Plaid models for gene expression data. *Statistica Sinica* **12**, 61–86.

Lee, S. -I., Dudley, A. M., Drubin, D., Silver, P. A., Krogan, N. J., Pe'er, D., and Koller, D. (2009). Learning a prior on regulatory potential from eQTL data. *PLoS Genet.* **5**, e1000358.

Liang, K. -C., and Wang, X. (2008). Gene regulatory network reconstruction using conditional mutual information. *EURASIP J. Bioinformat. Syst. Biol.* **2008**, 253894.

Lu, Y., Huggins, P., and Bar-Joseph, Z. (2009). Cross species analysis of microarray expression data. *Bioinformatics* **25**, 1476–1483.

Lynn, D. J., Winsor, G. L., Chan, C., Richard, N., Laird, M. R., Barsky, A., Gardy, J. L., Roche, F. M., Chan, T. H. W., Shah, N., Lo, R., Naseer, M., Que, J., Yau, M., Acab, M., Tulpan, D., Whiteside, M. D., Chikatamarla, A., Mah, B., Munzner, T., Hokamp, K., Hancock, R. E. W., and Brinkman, F. S. L. (2008). InnateDB: facilitating systems-level analyses of the mammalian innate immune response. *Mol. Syst. Biol.* **4**, 218.

Madar, A., Greenfield, A., Ostrer, H., Vanden-Eijnden, E., and Bonneau, R. (2009). The inferelator 2.0: A scalable framework for reconstruction of dynamic regulatory network models. Conference proceedings:.. Annual International Conference of the IEEE Engineering in Medicine and Biology Society. *IEEE Eng. Med. Biol. Soc. Conf.* **1**, 5448–5451.

Madar, A., Greenfield, A., Vanden-Eijnden, E., and Bonneau, R. (2010). DREAM3: network inference using dynamic context likelihood of relatedness and the inferelator. *PLoS ONE* **5**, e9803.

Marbach, D., Costello, J., Küffner, R., Vega, N., Prill, R., Camacho, D., Allison, K., Consortium, T.D., Kellis, M., Collins, J., Stolovitzky, G., submitted. Wisdom of crowds for gene network inference.

Marbach, D., Mattiussi, C., and Floreano, D. (2009a). Combining multiple results of a reverse-engineering algorithm: application to the DREAM five-gene network challenge. *Ann. N Y Acad. Sci.* **1158**, 102–113.

Marbach, D., Mattiussi, C., and Floreano, D. (2009b). Replaying the evolutionary tape: biomimetic reverse engineering of gene networks. *Ann. N Y Acad. Sci.* **1158**, 234–245.

Marbach, D., Prill, R. J., Schaffter, T., Mattiussi, C., Floreano, D., and Stolovitzky, G. (2010). Revealing strengths and weaknesses of methods for gene network inference. *Proc. Natl. Acad. Sci. U S A* **107**, 6286–6291.

Marbach, D., Schaffter, T., Mattiussi, C., and Floreano, D. (2009c). Generating realistic in silico gene networks for performance assessment of reverse engineering methods. *J. Comput. Biol.* **16**, 229–239.

Margolin, A. A., Nemenman, I., Basso, K., Wiggins, C., Stolovitzky, G., Favera, R. D., and Califano, A. (2006). ARACNE: an algorithm for the reconcstruction of gene regulatory networks in a mammalian cellular context. *BMC Bioinformat.* **15**, 1–15.

Mazur, J., Ritter, D., Reinelt, G., and Kaderali, L. (2009). Reconstructing nonlinear dynamic models of gene regulation using stochastic sampling. *BMC Bioinformat.* **10**, 448.

Mellor, J. C., Yanai, I., Clodfelter, K. H., Mintseris, J., and DeLisi, C. (2002). Predictome: a database of putative functional links between proteins. *Nucleic Acids Res.* **30**, 306–309.

Michaut, M., Kerrien, S., Montecchi-Palazzi, L., Chauvat, F., Cassier-Chauvat, C., Aude, J. -C., Legrain, P., and Hermjakob, H. (2008). InteroPORC: automated inference of highly conserved protein interaction networks. *Bioinformatics* **24**, 1625–1631.

Mirkin, B. G. (1996). *Mathematical classification and clustering.* Kluwer Academic Publishers, Dordrecht, the Netherlands.

Morgan, J. N., and Sonquist, J. A. (1963). Problems in the analysis of survey data, and a proposal. *J. Amer. Stat. Assoc.* 415–434.

Mukherjee, S., and Speed, T. P. (2008). Network inference using informative priors. *Proc. Natl. Acad. Sci. U S A* **105**, 14313–14318.

Ong, S. E., Blagoev, B., Kratchmarova, I., Kristensen, D. B., Steen, H., Pandey, A., and Mann, M. (2002). Stable isotope labeling by amino acids in cell culture, SILAC, as a simple and accurate approach to expression proteomics. *Mol. Cell Proteomics* **1**, 376–386.

Painter, M. W., Davis, S., Hardy, R. R., Mathis, D., and Benoist, C. (2011). Transcriptomes of the B and T lineages compared by multiplatform microarray profiling. *J. Immunol.* **186**, 3047–3057.

Pinna, A., Soranzo, N., and de la Fuente, A. (2010). From knockouts to networks: establishing direct cause-effect relationships through graph analysis. *PloS ONE* **5**, e12912.

Poultney, C. S., Gutiérrez, R. a., Katari, M. S., Gifford, M. L., Paley, W. B., Coruzzi, G. M., and Shasha, D. E. (2007). Sungear: interactive visualization and functional analysis of genomic datasets. *Bioinformatics (Oxford, England)* **23**, 259–261.

Prieto, C., and Rivas, J. D. L. (2006). APID: agile protein interaction dataanalyzer. *Nucleic Acids Res.* **34**, W298–W302.

Prill, R. J., Marbach, D., Saez-Rodriguez, J., Sorger, P. K., Alexopoulos, L. G., Xue, X., Clarke, N. D., Altan-Bonnet, G., and Stolovitzky, G. (2010). Towards a rigorous assessment of systems biology models: the DREAM3 challenges. *PloS ONE* **5**, e9202.

Ravasz, E., Somera, A. L., Mongru, D. A., Oltvai, Z. N., and Barabasi, A. L. (2002). Hierarchical organization of modularity in metabolic networks. *Science* **297**, 1551–1555.

Razick, S., Magklaras, G., and Donaldson, I. M. (2008). iRefIndex: a consolidated protein interaction database with provenance. *BMC Bioinformat.* **9**, 405.

Reiss, D. J., Baliga, N. S., and Bonneau, R. (2006). Integrated biclustering of heterogeneous genome-wide datasets for the inference of global regulatory networks. *BMC Bioinformat.* **7**, 280.

Remm, M., Storm, C. E., and Sonnhammer, E. L. (2001). Automatic clustering of orthologs and in-paralogs from pairwise species comparisons. *J Mol Biol* **314**, 1041–1052.

Robinson, J. T., Thorvaldsdottir, H., Winckler, W., Guttman, M., Lander, E. S., Getz, G., and Mesirov, J. P. (2011). Integrative genomics viewer. *Nat. Biotechnol.* **29**, 24–26.

Sachs, K., Gifford, D., Jaakkola, T., Sorger, P., Lauffenburger, D.A., 2002. Bayesian network approach to cell signaling pathway modeling. Science's STKE: signal transduction knowledge environment 2002, pe38.

Sachs, K., Perez, O., Pe'er, D., Lauffenburger, D. A., and Nolan, G. P. (2005). Causal protein-signaling networks derived from multiparameter single-cell data. *Science (New York, NY)* **308**, 523–529.

Saeed, A. I., Sharov, V., White, J., Li, J., Liang, W., Bhagabati, N., Braisted, J., Klapa, M., Currier, T., Thiagarajan, M., Sturn, A., Snuffin, M., Rezantsev, A., Popov, D., Ryltsov, A., Kostukovich, E., Borisovsky, I., Liu, Z., Vinsavich, A., Trush, V., and Quackenbush, J. (2003). TM4: a free, open-source system for microarray data management and analysis. *BioTechniques* **34**, 374–378.

Salwinski, L., Miller, C. S., Smith, A. J., Pettit, F. K., Bowie, J. U., and Eisenberg, D. (2004). The database of interacting proteins: 2004 update. *Nucleic Acids Res.* **32**, D449–D451.

Schmitt, W. A., Jr., Raab, R. M., and Stephanopoulos, G. (2004). Elucidation of gene interaction networks through time-lagged correlation analysis of transcriptional data. *Genome Res.* **14**, 1654–1663.

Segal, E., Shapira, M., Regev, A., Pe'er, D., Botstein, D., Koller, D., and Friedman, N. (2003). Module networks: identifying regulatory modules and their condition-specific regulators from gene expression data. *Nat. Genet.* **34**, 166–176.

Shannon, C. (1948). A mathematical theory of communication. *Bell Syst.Tech. J.* **27**, 379–423.

Shannon, P., Markiel, A., Ozier, O., Baliga, N. S., Wang, J. T., Ramage, D., Amin, N., Schwikowski, B., and Ideker, T. (2003). Cytoscape: a software environment for integrated models of biomolecular interaction networks. *Genome Res.* **13**, 2498–2504.

Shannon, P. T., Reiss, D. J., Bonneau, R., and Baliga, N. S. (2006). The Gaggle: an open-source software system for integrating bioinformatics software and data sources. *BMC Bioinformat.* **7**, 176.

Singh, A. H., Wolf, D. M., Wang, P., and Arkin, A. P. (2008). Modularity of stress response evolution. *Proc. Natl. Acad. Sci. U S A* **105**, 7500–7505.

Smith, E. A., and Corn, R. M. (2003). Surface plasmon resonance imaging as a tool to monitor biomolecular interactions in an array based format. *Appl. Spectrosc.* **57**, 320A–332A.

Snel, B., Lehmann, G., Bork, P., and Huynen, M. A. (2000). STRING: a web-server to retrieve and display the repeatedly occurring neighbourhood of a gene. *Nucleic Acids Res.* **28**, 3442–3444.

Soule, H. D., Maloney, T. M., Wolman, S. R., Peterson, W. D., Jr., Brenz, R., McGrath, C. M., Russo, J., Pauley, R. J., Jones, R. F., and Brooks, S. C. (1990). Isolation and characterization of a spontaneously immortalized human breast epithelial cell line, MCF-10. *Cancer Res.* **50**, 6075–6086.

Stark, C., Breitkreutz, B. -J., Chatr-Aryamontri, A., Boucher, L., Oughtred, R., Livstone, M. S., Nixon, J., Auken, K. V., Wang, X., Shi, X., Reguly, T., Rust, J. M., Winter, A., Dolinski, K., and Tyers, M. (2011). The BioGRID Interaction Database: 2011 update. *Nucleic Acids Res.* **39**, D698–D700.

Stolovitzky, G., Monroe, D., and Califano, A. (2007). Dialogue on reverse-engineering assessment and methods: the DREAM of high-throughput pathway inference. *Ann. N Y Acad. Sci.* **1115**, 1–22.

Supper, J., Strauch, M., Wanke, D., Harter, K., and Zell, A. (2007). EDISA: extracting biclusters from multiple time-series of gene expression profiles. *BMC Bioinformat.* **8**, 334.

Suzuki, H., Forrest, A. R. R., van Nimwegen, E., Daub, C. O., Balwierz, P. J., Irvine, K. M., Lassmann, T., Ravasi, T., Hasegawa, Y., de Hoon, M. J. L., Katayama, S., Schroder, K., Carninci, P., Tomaru, Y., Kanamori-Katayama, M., Kubosaki, A., Akalin, A., Ando, Y., Arner, E., Asada, M., Asahara, H., Bailey, T., Bajic, V. B., Bauer, D., Beckhouse, A. G., Bertin, N., Björkegren, J., Brombacher, F., Bulger, E., Chalk, A. M., Chiba, J., Cloonan, N., Dawe, A., Dostie, J., Engström, P. G., Essack, M., Faulkner, G. J., Fink, J. L., Fredman, D., Fujimori, K., Furuno, M., Gojobori, T., Gough, J., Grimmond, S. M., Gustafsson, M., Hashimoto, M., Hashimoto, T., Hatakeyama, M., Heinzel, S., Hide, W., Hofmann, O., Hörnquist, M., Huminiecki, L., Ikeo, K., Imamoto, N., Inoue, S., Inoue, Y., Ishihara, R., Iwayanagi, T., Jacobsen, A., Kaur, M., Kawaji, H., Kerr, M. C., Kimura, R., Kimura, S., Kimura, Y., Kitano, H., Koga, H., Kojima, T., Kondo, S., Konno, T., Krogh, A., Kruger, A., Kumar, A., Lenhard, B., Lennartsson, A., Lindow, M., Lizio, M., Macpherson, C., Maeda, N., Maher, C. A., Maqungo, M., Mar, J., Matigian, N. A., Matsuda, H., Mattick, J. S., Meier, S., Miyamoto, S., Miyamoto-Sato, E., Nakabayashi, K., Nakachi, Y., Nakano, M., Nygaard, S., Okayama, T., Okazaki, Y., Okuda-Yabukami, H., Orlando, V., Otomo, J., Pachkov, M., Petrovsky, N., Plessy, C., Quackenbush, J., Radovanovic, A., Rehli, M., Saito, R., Sandelin, A., Schmeier, S., Schönbach, C., Schwartz, A. S., Semple, C. A., Sera, M., Severin, J., Shirahige, K., Simons, C., St Laurent, G., Suzuki, M., Suzuki, T., Sweet, M. J., Taft, R. J., Takeda, S., Takenaka, Y., Tan, K., Taylor, M. S., Teasdale, R. D., Tegnér, J., Teichmann, S., Valen, E., Wahlestedt, C., Waki, K., Waterhouse, A., Wells, C. A., Winther, O., Wu, L., Yamaguchi, K., Yanagawa, H., Yasuda, J., Zavolan, M., Hume, D. A., Arakawa, T., Fukuda, S., Imamura, K., Kai, C., Kaiho, A., Kawashima, T., Kawazu, C., Kitazume, Y., Kojima, M., Miura, H., Murakami, K., Murata, M., Ninomiya, N., Nishiyori, H., Noma, S., Ogawa, C., Sano, T., Simon, C., Tagami, M., Takahashi, Y., Kawai, J., and Hayashizaki, Y. (2009). The transcriptional network that controls growth arrest and differentiation in a human myeloid leukemia cell line. *Nat. Genet.* **41**, 553–562.

Szklarczyk, D., Franceschini, A., Kuhn, M., Simonovic, M., Roth, A., Minguez, P., Doerks, T., Stark, M., Muller, J., Bork, P., Jensen, L. J., and von Mering, C. (2011). The STRING database in 2011: functional interaction networks of proteins, globally integrated and scored. *Nucleic Acids Res.* **39**, D561–D568.

Tanay, A., Sharan, R., Kupiec, M., and Shamir, R. (2004). Revealing modularity and organization in the yeast molecular network by integrated analysis of highly heterogeneous genomewide data. *Proc. Natl. Acad. Sci. U S A* **101**, 2981–2986.

Tatusov, R. L., Galperin, M. Y., Natale, D. A., and Koonin, E. V. (2000). The COG database: a tool for genome-scale analysis of protein functions and evolution. *Nucleic Acids Res.* **28**, 33–36.

Tirosh, I., and Barkai, N. (2007). Comparative analysis indicates regulatory neofunctionalization of yeast duplicates. *Genome Biol.* **8**, R50.

Tusher, V. G., Tibshirani, R., and Chu, G. (2001). Significance analysis of microarrays applied to the ionizing radiation response. *Proc. Natl. Acad. Sci. U S A* **98**, 5116–5121.

van Helden, J. (2003). Regulatory sequence analysis tools. *Nucleic Acids Res.* **31**, 3593–3596.

Waltman, P., Kacmarczyk, T., Bate, A. R., Kearns, D. B., Reiss, D. J., Eichenberger, P., and Bonneau, R. (2010). Multi-species integrative biclustering. *Genome Biol.* **11**, R96.

Yannakakis, M. (1981). Node-deletion problems on bipartite graphs. *SIAM J. Comput.* **10**, 310–327.

Yip, K. Y., Alexander, R. P., Yan, K. -K., and Gerstein, M. (2010). Improved reconstruction of in silico gene regulatory networks by integrating knockout and perturbation data. *PLoS ONE* **5**, e8121.

Yong-a-poi, J., Someren, E. V., Bellomo, D., and Reinders, M. (2008). Adaptive least absolute regression network analysis improves genetic network reconstruction by employing prior knowledge. *Commun. Theory* 1–14.

Zhou, J., Deng, Y., Luo, F., He, Z., Tu, Q., and Zhi, X. (2010). Functional molecular ecological networks. *mBio* **1**, e00169-00110.

Zou, H. (2006). The adaptive lasso and its oracle properties. *J. Amer. Statis. Assoc.* **101**, 1418–1429.

CHAPTER 3

Swimming Upstream: Identifying Proteomic Signals that Drive Transcriptional Changes using the Interactome and Multiple "-Omics" Datasets

Shao-shan Carol Huang[*,‡] **and Ernest Fraenkel**[*,†]

[*]Department of Biological Engineering, Massachusetts Institute of Technology, Cambridge, Massachusetts, USA

[†]Computer Science and Artificial Intelligence Laboratory, Massachusetts Institute of Technology, Cambridge, Massachusetts, USA

[‡]Current address: Plant Molecular and Cellular Biology Laboratory, Salk Institute for Biological Studies, La Jolla, California, USA

0091-679X/10 $35.00
DOI 10.1016/B978-0-12-388403-9.00003-5

Abstract

Signaling and transcription are tightly integrated processes that underlie many cellular responses to the environment. A network of signaling events, often mediated by post-translational modification on proteins, can lead to long-term changes in cellular behavior by altering the activity of specific transcriptional regulators and consequently the expression level of their downstream targets. As many high-throughput, "-omics" methods are now available that can simultaneously measure changes in hundreds of proteins and thousands of transcripts, it should be possible to systematically reconstruct cellular responses to perturbations in order to discover previously unrecognized signaling pathways.

This chapter describes a computational method for discovering such pathways that aims to compensate for the varying levels of noise present in these diverse data sources. Based on the concept of constraint optimization on networks, the method seeks to achieve two conflicting aims: (1) to link together many of the signaling proteins and differentially expressed transcripts identified in the experiments "*constraints*" using previously reported protein–protein and protein–DNA interactions, while (2) keeping the resulting network small and ensuring it is composed of the highest confidence interactions "*optimization*". A further distinctive feature of this approach is the use of transcriptional data as evidence of upstream signaling events that drive changes in gene expression, rather than as proxies for downstream changes in the levels of the encoded proteins.

We recently demonstrated that by applying this method to phosphoproteomic and transcriptional data from the pheromone response in yeast, we were able to recover functionally coherent pathways and to reveal many components of the cellular response that are not readily apparent in the original data. Here, we provide a more detailed description of the method, explore the robustness of the solution to the noise level of input data and discuss the effect of parameter values.

I. Introduction

One of the central challenges for systems biology is the reconstruction of cellular processes from high-throughput experimental data. Much of the early work in this area was driven by the development of microarray technologies that allowed relatively comprehensive measurement of changes in mRNA expression. Using these data as proxies for changes at the protein level has generated many insights into the regulatory networks of the cell (Spellman *et al.*, 1998; Segal *et al.*, 2005; Ozsolak and Milos, 2011). However, the actual correlation between the transcriptome and the proteome is unclear (Schwanhäusser *et al.*, 2011; Maier *et al.*, 2009; de Sousa Abreu *et al.*, 2009), and more direct proteomic data are likely to provide a more reliable and thorough view of cellular processes.

Recently, technological advances have made it possible to directly measure proteomic changes at the global level. Mass spectrometry (MS) techniques can quantify the relative levels of hundreds of peptides across multiple biological conditions (Choudhary and Mann, 2010; White, 2008) and focused data collection on phosphoproteins was able to reveal the regulatory dynamics of cellular signaling networks at the level of the proteome (Grimsrud *et al.*, 2010; Macek *et al.*, 2009; Yi Zhang *et al.*, 2007).

With new data come new challenges. Even in the best-characterized responses there is poor overlap between hits identified by phosphoproteomics technologies and known pathway components. For example, in a study of phosphorylation changes that occur in response to mating pheromone in yeast (Gruhler *et al.*, 2005), 112 proteins contain differentially phosphorylated sites; of these, only 11 are known components of the expected mitogen-activated protein kinase (MAPK) cascade that responds to pheromone, and 76 were not present in any of the yeast pathways annotated in the KEGG PATHWAY database (Kanehisa *et al.*, 2010). Finding new ways to interpret these data could reveal previously unrecognized cellular pathways.

A second important challenge is to integrate transcriptional and proteomic data in order to observe the interplay between different layers of cellular signaling. For example, it may be possible to detect proteomic changes in signal transduction cascades that drive expression and also to reveal the resulting feedback of transcription on the proteome. But integrating these data will require novel computational approaches. Because regulation is mediated by diverse mechanisms, even the most comprehensive proteomics technologies cannot capture all these events. For example, MS-based methods focusing on protein phosphorylation will fail to detect changes in other post-translational modifications such as acetylation, ubiquitination, and sumoylation. Computational techniques are needed to discover proteins that participate in the signaling networks but are undetected in the experiments and also to provide insight into their functional roles. One successful approach has been to map these proteins onto known metabolic and regulatory pathways such as those curated in the KEGG PATHWAY (Kanehisa *et al.*, 2010) and Reactome (Matthews *et al.*, 2009) databases. This approach can reveal functional coherence and relevant biological processes from the data. However, as mentioned above, a large fraction of the phosphoproteomic data does not map to known pathway models, so we must turn to other approaches.

The interactome provides an alternative to using well-studied pathways. Advances in high-throughput experimental mapping of protein–protein interactions as well as efforts to extract known interactions from the literature have produced a number of large databases of protein interactions (selected examples are listed in Table I). Despite being incomplete, especially for higher organisms, the quantity of interaction data in these databases is still very large. Thus, it may be possible to discover unknown pathways among these interactions. While utilizing these large interactome datasets improves our ability to find connections among a set of proteins of interest, it also presents several challenges. First, the size of the potential network explodes exponentially and quickly becomes non-interpretable, as pointed out by previous data integration efforts (Hwang *et al.*, 2005). Second, interaction records in databases come from hundreds of laboratories and many experimental techniques of

Table I
A selection of publicly available protein–protein interaction databases.

Type of interactions	Data sources	Database and references
Direct/physical	Curation of primary literature	Biological General Repository for Interaction Datasets (BioGRID) (Stark *et al.*, 2011) Human Protein Reference Database (HPRD) (Keshava Prasad *et al.*, 2009) Molecular Interaction database (MINT) (Chatr-aryamontri *et al.*, 2007) IntAct molecular interaction database (Kerrien *et al.*, 2007) Mammalian Protein-Protein Interaction Database (MIPS) (Pagel *et al.*, 2005) Database of Interacting Proteins (DIP) (Salwinski *et al.*, 2004) Biomolecular Interaction Network Database (BIND) (Bader *et al.*, 2001)
	Collection of multiple primary databases	Interaction Reference Index (iRefIndex) (Razick *et al.*, 2008) Agile Protein Interaction DataAnalyzer (APID) (Prieto *et al.*, 2006) Michigan Molecular Interactions database (MiMI) (Tarcea *et al.*, 2009) Unified Human Interactome database (UniHI) (Chaurasia *et al.*, 2007)
Direct/physical + indirect/ functional	Collection of multiple primary databases and computational predictions	STRING (von Mering *et al.*, 2005)

Note: For further details see recent summary and reviews in Turinsky *et al.* (2011), De Las Rivas *et al.* (2010), Klingström and Plewczynski (2010). Many databases in this table have adopted the Proteomic Standards Initiative Molecular Interaction (PSI-MI) data formats and implemented the PSI Common Query Interface (PSICQUIC) (Aranda *et al.*, 2011) that allows easy, programmatic access and integration of these data.

varying degrees of reliability (von Mering *et al.*, 2002), so overall the data quality is heterogeneous and should not be treated indiscriminately. Lastly, pooling these interactions together risks losing the specific context under which they were detected. It is with these issues in mind that we propose a constraint optimization approach for finding regulatory networks that are interpretable, reliable, and biologically relevant.

Our method starts with a collection of protein–protein and protein–DNA interactions, which represent known or experimentally determined signaling and regulatory connections. It considers the observed phosphorylation events and differential gene expression as connectivity constraints that the reconstructed network must satisfy. Additionally, we take into account the different confidence levels among the interaction data sources by preferentially selecting the more reliable interactions. We show that these objectives can be formulated as a constraint network optimization problem, in particular, as a prize-collecting Steiner tree (PCST) problem on the interactome graph. Since the interactions are not limited to known pathways and the phosphorylation events and differential expressed genes are not limited to known players in these pathways, there is great potential for novel discoveries. On the other hand, all the interactions were experimentally determined and therefore have mechanistic basis that might become relevant in the current context. These two features of the method strike a balance between finding novel connections and revealing the relevance of known connections. We hypothesize that since each of our input data sources provides a different view of the molecular regulatory network, by putting them together we can generate high-confidence hypotheses that have biological relevance and can be tested experimentally. This framework serves to organize these heterogeneous datasets and enhance our understanding of the cell at the systems level.

II. Computational Methods

Network optimization is an area of computer science that has recently become very useful for analyzing biological problems, and a variety of algorithms are available to solve specific optimizations. The problem we have posed consists of finding a set of edges of minimum weight in order to connect a defined set of nodes (known as termini) in a weighted network. This problem is called the Steiner tree problem. An important generalization that allows some terminal nodes to be excluded is known as the PCST problem. For our purpose, we will use a network in which edge weights reflect our confidence in the interactions and where terminal nodes represent hits from the experiments, that is, phosphorylated proteins and differentially expressed transcripts. In this setting, the solution to the PCST optimization is a set of most confident interactions that link together the hits while possibly leaving some unconnected (Fig. 1(A)).

Although the concept of Steiner tree has been previously applied to biological networks (Dittrich *et al.*, 2008; Scott *et al.*, 2005), we note that our approach is distinctive in multiple aspects. First, instead of mRNA transcript abundance we use protein-level measurements on nodes in the interactome, which provides a much more accurate representation of the underlying biological processes. Second, we explicitly model the confidence of individual edges in the interactome to account for the uncertainties in the interaction data. Third, we do not require all nodes in the solution to be detected in the experiments, allowing our approach to compensate for multiple sources of noise. This last feature is absent in an application of a Steiner tree like algorithm to build a high-confidence network with genetic screening hits as terminal nodes (Yosef *et al.*, 2009). A minimum-cost flow optimization approach connects genetic hits to differentially expressed genes (Lan *et al.*, 2011; Yeger-Lotem *et al.*, 2009) but the result is less compact than the PCST (Huang and Fraenkel, 2009). We now describe the process of constructing the optimization problem, solving it, and analyzing the results. We also offer some advice on practical matters such as tuning the parameter values and visualizing the network.

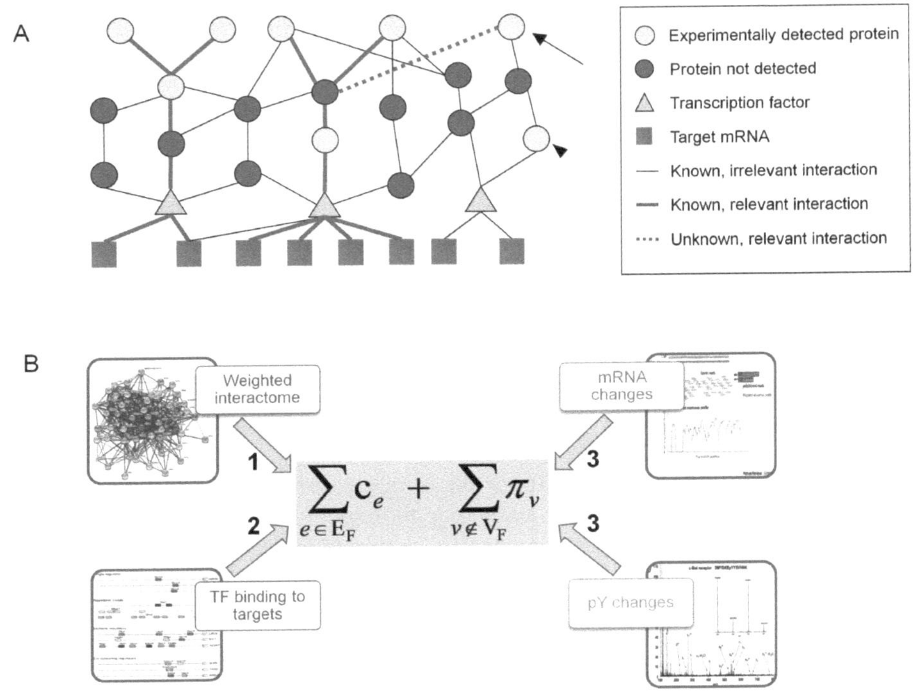

Fig. 1 (A) Finding relevant interactions as a constraint optimization problem. We seek a set of high-confidence edges present in the interactome that directly or indirectly link the proteins and genes identified in the experimental assays. Because some of the input data may be false positives (arrowhead) or may not be explained by currently known interactome (arrow), our approach does not require that all the input data be connected, but rather uses these data as constraints. Note that the protein product and mRNA transcript of the same gene are represented as separate nodes. Image reproduced with permission from Huang and Fraenkel (2009). (B) Work-flow diagram for defining the optimization objective function from input datasets. Interaction weights go into the edge cost summation term (Step 1) and the changes in tyrosine phosphorylation from MS data go into the node penalty summation term (Step 3). The transcription factors to mRNA target relationships are added to the edges to form the total interactome (Step 2), and the mRNA nodes are assigned penalty values (Step 3). (For color version of this figure, the reader is referred to the web version of this book.)

A. Setting Up the Prize-Collecting Steiner Tree

We treat the interactome as an undirected graph $G = (V, E)$ where nodes are proteins or genes and edges represent the known interactions. Each node $v \in V$ is associated with a penalty $\pi_v \geq 0$. Protein nodes to which experimental data are mapped receive positive penalty values and therefore are termini for the PCST. All other nodes receive zero penalties. As the magnitude of the penalty value increases, the more confident we are that the protein/gene was experimentally detected as relevant in the signaling response. The algorithm is forced to pay a

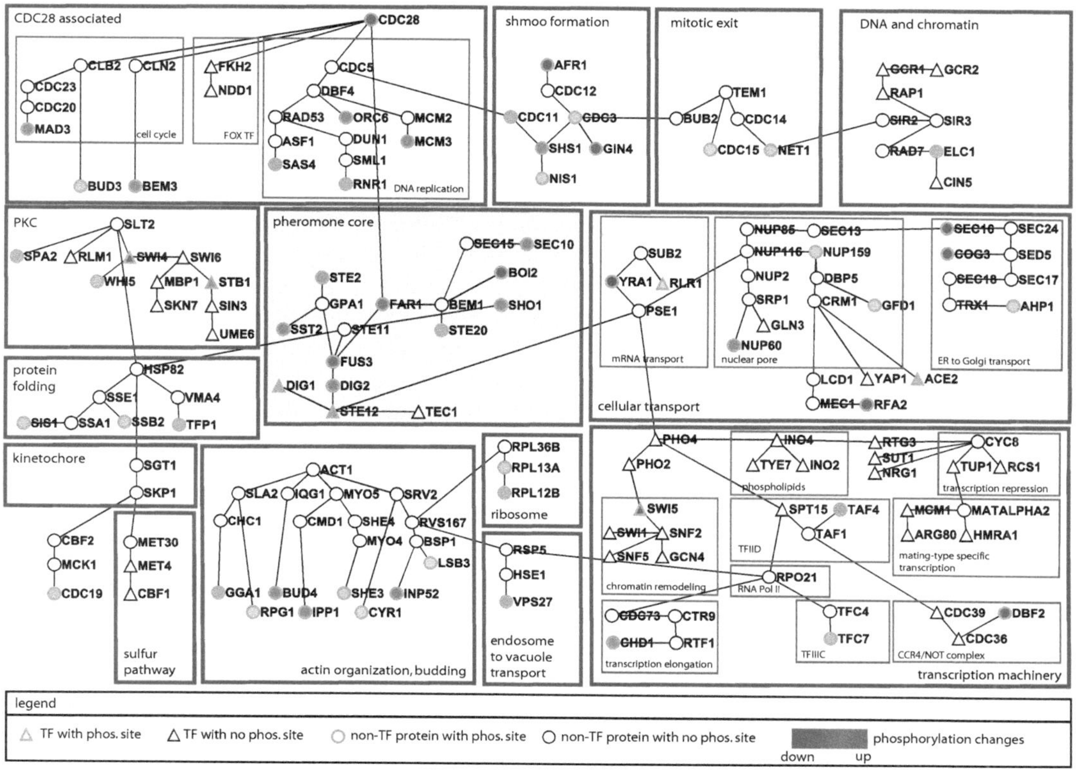

Fig. 2 The protein components of the pheromone response network constructed by the PCST approach. Note that the canonical pheromone response pathway (enclosed by dashed lines) is but a small component of the broad cellular changes revealed by applying the algorithm to the mass spectrometry and expression data. For clarity, the differentially transcribed genes included in the network are not presented. Functional groups based on GO annotation are outlined with red boxes. *PKC*, protein kinase C; *TF with phos. site*, transcription factor with at least one differentially phosphorylated sites; *TF with no phos. site*, transcription factor with no differentially phosphorylated sites; *non-TF protein with phos. site*, a protein that is not a transcription factor and with at least one differentially phosphorylated sites; *non-TF with no phos. site*, a protein that is not a transcription factor and with no differentially phosphorylated sites. Image reproduced with permission from Huang and Fraenkel (2009). (See color plate)

penalty each time it leaves a terminal out of its final network. This constraint causes the network to include as many high-confidence nodes as possible. However, this constraint alone would lead to very large networks that might contain many unreliable edges. So we also assign to each edge $e \in E$ a cost $c_e \geq 0$ that is inversely related to our confidence in each interaction.

We aim to find a subtree $F = (V_F, E_F)$ of G that minimizes the objective function $\sum_{e \in E_F} c_e + \sum_{v \notin V_F} \pi_v$. Because we incur penalties for excluding nodes while paying costs for including edges, the algorithm will be forced to favor connecting high-confidence data with high-confidence interactions. We further introduce a scaling parameter β to balance the penalties and the edge costs:

$$\sum_{e \in E_F} c_e + \sum_{v \notin V_F} \beta \pi_v.$$

We may solve this optimization problem exactly by using the branch-and-cut approach (Ljubić *et al.*, 2005) implemented in the `dhea-code` software program that calls the ILOG CPLEX mathematical programming solver. As an alternative to solving it as an integer linear program, an approach from statistical physics (Bayati *et al.*, 2008) has resulted in new heuristic algorithms based on message-passing techniques (Bailly-Bechet *et al.*, 2011). We now describe how the experimental data are transformed into input for the algorithm. An overview of the work-flow is in Fig. 1(B).

B. A Probabilistic Interactome

This is Step 1 in Fig. 1(B). The set of edges E of the input graph G consists of direct (physical) protein–protein interactions found in databases of molecular interactions such as those listed in Table I. To assign confidence values for these interactions, a few methods have been previously published (Razick *et al.*, 2008; Orchard *et al.*, 2007; Jansen *et al.*, 2003). Here we use a naïve Bayes probabilistic model (Jansen *et al.*, 2003). Interaction between two proteins is modeled as random variable $i \in \{0, 1\}$ with $i = 1$ when two proteins interact and $i = 0$ otherwise, and each kind of experimental evidence is modeled as a random variable $f_j \in \{0, 1\}$ where $f_j = 1$ indicates f_j is observed and $f_j = 0$ otherwise. From published gold standard sets of positive (Yu *et al.*, 2008) and negative interactions (Jansen *et al.*, 2003), we can compute the conditional probability table for each kind of evidence, $P(f_j|i)$. Then, for each interaction e supported by a set of experimental evidence $F_e = \{f_{e,j}|j = 1, \ldots, n\}$, assuming independence between the evidence we have

$$P(F_e|i) = \prod_j P(f_{e,j}|i),$$

and a straightforward application of Bayes rule gives the probability that this interaction is real:

$$P(i = 1|F_e) = \frac{P(F_e|i = 1)P(i = 1)}{\sum_{i' \in \{0,1\}} P(F_e|i')}.$$

The cost c_e on edge e that is input into the PCST objective function is $c_e = -\log P(i = 1|F_e), \forall e \in E$.

C. Determining Transcription Factor Targets

Transcription factor to mRNA target relationships are added to the protein–protein interactome to form the total interactome (Step 2 in Fig. 1(B)). A variety of experimental, computational techniques and combinations of both are possible. For yeast, there are published genome-wide binding sites for almost all the transcriptional regulators under multiple conditions measured by chromatin immunoprecipitation (ChIP) experiments (Harbison *et al.*, 2004; MacIsaac *et al.*, 2006). The human and mouse ENCODE projects (Birney *et al.*, 2007) represent systematic efforts to generate ChIP profiles for multiple transcription factors in a variety of human cell lines and mouse tissues. Computationally, transcription factors often have sequence specificities that allow binding sites to be predicted to some extent (Box 1). Commonly used quantitative representations of such binding patterns, also known as sequence motifs, include position weight matrices (PWM)/position-specific scoring matrices (PSSM) (D'haeseleer, 2006; Stormo, 2000) with an information theoretic perspective, and position-specific affinity matrices (PSAM) with a statistical mechanics perspective (Foat *et al.*, 2006, 2005; Manke *et al.*, 2008; Roider *et al.*, 2007). Motifs from the TRANSFAC (Wingender, 2008; Matys *et al.*, 2006) and JASPAR (Sandelin *et al.*, 2004; Bryne *et al.*, 2008) databases, which collect published transcription factor binding motifs from the literature, can be used for predicting regulatory elements. Once a genomic region is determined to be bound by a transcription factor based on experimental and/or computational evidence, nearby genes can be associated with this factor as its potential downstream targets, and we add to the interactome edges going from the transcription factor (a protein node) to these target mRNA nodes.

D. Node Penalties

This is Step 3 in Fig. 1(B). We define two kinds of penalties for proteins in the interactome: one at the protein level derived from the phosphoproteomics MS data, and the other at the mRNA level derived from mRNA expression data.

Although published phosphoproteomic MS datasets often provide the identities of the proteins that contain the peptide sequences inferred from the MS spectra, it is still advisable to map the peptides to a database of protein sequences from which the interactome dataset is derived in order to avoid issues such as inconsistencies in mapping gene identifiers and in treating protein isoforms. This can be achieved by finding protein sequences in a database that contains matches to the peptide sequences, for example, by the sequence alignment search tool BLAST (Altschul *et al.*, 1990) with parameter settings optimized for matching short peptide sequences. In an analysis comparing two conditions, proteins that contain perfect alignment to a peptide sequence receive a positive penalty value that is proportional to the absolute value of log-fold change in phosphorylation between the conditions of interest. If one peptide sequence is aligned to multiple proteins in the interaction graph, all these proteins receive the same penalty value. If multiple phosphorylated

Box 1

Aligned binding sites

```
      CGTGCATTCCctgcag
    cgCGGCATTTCCacgc
 gcttaCGGGGTTTCCa
tacatgaGGGGTTTTTC
 ccaatGGGAATTTCCc
 agcgtGCGGTATTCC
 gttgaTGGTCTTTCCa
gtatgccCGGGAATTCC
   aatCTAAAAAACCcaa
caattgtGGGGGTTTCC
    tgGGGTTTTTCCccca
agggaagGGGAACTTTCttt
       GGGAAGTACAaggc
      tGGGGCTTTCCgtggc
atccgccTGGAGTTTCC
 gtttaTGGGCTTTCCg
tggcgtgTGGGCATTCC
```

Count base frequency $f_{b,i}$ →

Position frequency matrix (PFM)

	1	2	3	4	5	6	7	8	9	10
A	0	0	1	5	6	5	1	2	0	1
C	5	1	0	1	5	1	0	0	15	16
G	8	15	15	9	3	1	0	0	0	0
T	4	1	1	2	3	10	16	15	2	0

Pseudo-count correction

$$p(b,i)=\frac{f_{b,i}+s(b)}{N+\sum_{b'\in\{A,C,G,T\}} s(b')}$$

	1	2	3	4	5	6	7	8	9	10
A	0.05	0.05	0.10	0.29	0.33	0.29	0.10	0.14	0.05	0.10
C	0.29	0.10	0.05	0.10	0.29	0.10	0.05	0.05	0.76	0.81
G	0.43	0.76	0.76	0.48	0.19	0.10	0.05	0.05	0.05	0.05
T	0.24	0.10	0.10	0.14	0.19	0.52	0.81	0.76	0.14	0.05

$$W_{b,i}=\log_2\frac{p(b,i)}{p(b)}$$

$$W_{b,i}=\frac{p(b,i)}{\max_{b'\in\{A,C,G,T\}} p(b',i)}$$

Position specific scoring matrix (PSSM)

	1	2	3	4	5	6	7	8	9	10
A	-2.39	-2.39	-1.39	0.19	0.42	0.19	-1.39	-0.81	-2.39	-1.39
C	0.19	-1.39	-2.39	-1.39	0.19	-1.39	-2.39	-2.39	1.61	1.70
G	0.78	1.60	1.60	0.93	-0.39	-1.39	-2.39	-2.39	-2.39	-2.39
T	-0.07	-1.39	-1.39	0.81	-0.39	1.07	1.70	1.61	-0.81	-2.39

Approximate position specific affinity matrix (PSAM)

	1	2	3	4	5	6	7	8	9	10
A	0.11	0.06	0.13	0.60	1.00	0.55	0.12	0.19	0.06	0.11
C	0.67	0.13	0.06	0.20	0.86	0.18	0.06	0.06	1.00	1.00
G	1.00	1.00	1.00	1.00	0.57	0.18	0.06	0.06	0.06	0.06
T	0.56	0.13	0.13	0.30	0.57	1.00	1.00	1.00	0.19	0.06

Candidate sequence

a g t t g c a a a t c g t g g a a t t t c c t c t g a c

$$S=\sum_{i=1}^{w} W_{l_i,i}$$

T	G	G	A	A	T	T	T	C	C
-0.07	1.60	1.60	0.19	0.42	1.07	1.70	1.61	1.61	1.70

$S=11.43$

$$S=\prod_{i=1}^{w} W_{l_i,i}$$

T	G	G	A	A	T	T	T	C	C
0.56	1.00	1.00	0.60	1.00	1.00	1.00	1.00	1.00	1.00

$S=0.33$

$f_{b,i}$ = counts of base b at position i; N = number of sites; $p(b,i)$ = corrected probability of base b at position i; $s(b)$ = pseudo-count function for base b; $p(b)$ = background probability of base b; $W_{b,i}$ = PSSM or PSAM score for base b at position i; l_i = the nucleotide at position i in candidate sequence; S = PSSM or PSAM score of the current window; w = width of the PSSM or PSAM

Computational representation and discovery of transcription-factor-binding sites, with an example of the human REL protein-binding profile (JASPAR MA0101.1, curated from Kunsch *et al.* (1992)) and NFκB binding site in the human IL8 promoter (TRANSFAC binding site HS$IL8_21). *in vitro* techniques such as SELEX (systematic evolution of ligands by exponential enrichment) (Stoltenburg *et al.*, 2007) can generate a set of sequences that bind to a specific transcription factor with high affinity. From an alignment of these sequences, a PFM is created to represent the base preference of this factor at each position of the binding site. After pseudo-count correction, the PSSM approach takes the base preference at each position, adjusts for background (usually genome-wide) frequency of that base, and computes a numerical value for the bases at each position that can be used to score a DNA sequence (D'haeseleer, 2006; Stormo, 2000). Alternatively, an approximate PSAM for scoring can be created from a pseudo-count corrected PFM by calculating the preference of a base relative to the most frequent base at each position (Foat *et al.*, 2006, 2005; Manke *et al.*, 2008; Roider *et al.*, 2007). See MacIsaac and Fraenkel (2006) for a more detailed treatment of the topic.

peptide sequences are perfectly aligned to one protein, the maximum fold change in phosphorylation of these peptides is used to calculate the penalty value for this protein. Other methods of assigning penalties are also possible and are discussed below.

For penalty values on mRNA nodes, some modifications to the interactome are required to make the resulting network more biologically realistic. If we simply put penalty values on the mRNA nodes, the tree structure of the solution network means that any one mRNA node is connected to at most one upstream transcription factor. Such a network cannot capture one gene being targeted by multiple transcription factors, which is a common feature of transcriptional regulation. Instead, we represent multiple transcription factors bound to the same gene with separate nodes. Let M be the set of differentially expressed transcripts, and $fc(m)$ be the fold change in mRNA abundance of each gene $m \in M$. For each m, we searched the interactome for the set of upstream transcription factors F that target m, remove m from the interactome, and add one node m_f for each transcription factor $f \in F$ and one edge between f and m_f. The fold change of m is transferred to all the m_f to compute the penalty values on m_f. Each new terminal node m_f may be interpreted as a binding site of f on m.

E. Sensitivity Analysis

Applying an optimization approach to inherently noisy biological data makes it necessary to explore the alternative or suboptimal solution space surrounding the reported optimal solution. This is to ensure that the nodes and edges selected by the algorithm, from which significant efforts will be invested to extract biological

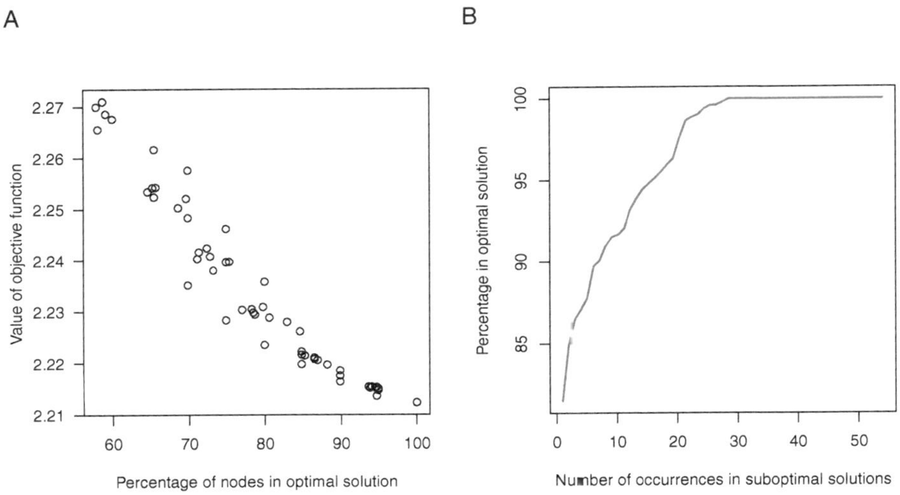

Fig. 3 Alternative or suboptimal solutions to the yeast pheromone response dataset. Because we use an optimization approach to analyze inherently noisy data, we asked whether the network we obtained was stable – are there very different networks that explain the data almost as well? For this, we compared the optimal solution network to a set of alternative solution networks obtained by finding networks that are different from the optimal one by at least a specific percentage of nodes. (A) No alternative solutions in the neighborhood of the optimal solution achieve the same objective function value. (B) Of the nodes that appear at least once in the 54 suboptimal solutions, at least 80% also appear in the optimal solution. Image reproduced with permission from Huang and Fraenkel (2009). (For color version of this figure, the reader is referred to the web version of this book.)

meaning, are relatively stable to possible sources of noise. Fig. 3 presents two ways to quantify this stability at the global level. First, starting from the optimal solution reported by the algorithm, we can re-formulate the optimization problem to find a number of suboptimal solutions – networks that are optimal under the additional constraint that they must differ from the original optimal solution by a predefined percentage of nodes. We can then compare these suboptimal solutions to the optimal one in terms of the objective function value (Fig. 3(A)) and the frequency at which the nodes in the original optimal solution are preserved in the suboptimal solutions (Fig. 3(B)) in order to decide whether the solution is robust to noise.

F. Practical Advice

Parameters: Tuning the value of parameter β essentially controls the size of the PCST solution output. With larger β values it becomes more expensive to exclude each terminal node (i.e., making the objective function larger), so the optimization algorithm will include more edges in the PCST solution. Although a larger network may include more hits from the experimental data, it is more difficult to interpret and also more likely to include false-positive hits that may connect to the real underlying network via tenuous interactions. To find a suitable value of this parameter, it is advisable to run the algorithm with a range of values and choose a solution that (1) includes any expected pathways based on prior biological knowledge, (2) is stable for the neighborhood of β values, and (3) contains as many of the hits as possible. One can also start with a small value of β to build a core network and gradually increase β to explore how more hits are connected to the core network.

It may be possible to use cross-validation to objectively choose β. In such an approach, one would randomly partition the terminal nodes into two complementary subsets, build a PCST using one subset (training set), and compute the recovery of the second subset (validation set) in that PCST. To reduce the effect of random variations, for each value of β, multiple rounds of such cross-validation can be performed and one average performance value is reported. Based on this performance measure, a β value can be selected.

While this approach has a certain appeal, we urge caution since the assumptions and requirements of cross-validation may not be satisfied by the biological datasets. First, in order for the recovery of the validation set by a PCST to be a good indicator of its performance, the training set and the validation set must be drawn from the same distribution. This criterion requires the terminal sets to be sufficiently large that each random sample contains termini from all the underlying biological processes. Since the current datasets are subject to many limitations such as the sensitivity of the MS instrument depending on protein abundance and the coverage of the interactome, we do not know *a priori* whether this assumption is appropriate. Second, it is unclear which of the conventionally used measures of predictor performance is suitable in this setting. We aim to recover intermediate nodes that are undetected in experiments, so we cannot count such nodes included in the PCST as false positives. In the absence of a false-positive definition, counting the recovery of the terminal nodes makes little sense since the optimal value of β will be the one that produces a PCST that include the most terminal nodes (weighted by penalty values).

Implementation: There are various approximation algorithms to solve the PCST problem. These have recently been reviewed (Archer *et al.*, 2011). The `dhea-code` program (Ljubić *et al.*, 2005), which can be downloaded from Dr. Ljubić's website (Ljubić, 2008), uses a branch-and-cut approach to obtain exact, optimal solutions. This program requires the ILOG CPLEX (IBM) optimization library that is available at no-charge for teaching and non-commercial research as part of the IBM Academic Initiative (IBM, 2010). In the supplement of this article, we provide a simple Python script for creating the input file for `dhea-code` from tab delimited text files of the weighted interactome and terminal nodes. The output files of `dhea-code` include the PCST solution in a DOT file (a plain text format for specifying graphs; Graphviz, 2011b). From there the solution can be rendered and viewed by the tools in Graphviz (2011a), or further manipulated and analyzed by the Python library NetworkX (Hagberg *et al.*, 2008). One standard operation is to convert the DOT file to one of the file formats supported by Cytoscape (Smoot *et al.*, 2011; Cline *et al.*, 2007) in order to utilize Cytoscape's many visualization capabilities for biological networks.

A recently published message-passing algorithm, although taking a heuristic approach, is able to find solutions with objective values comparable to `dhea-code` under much less computing time and memory (Bailly-Bechet *et al.*, 2011). It requires a depth parameter to be specified *a priori* to control the length of paths in the solution network. This appears to have the consequence of eliminating long braches in the solution. The effect of this difference on the identities and functional relevance of the recovered nodes remains to be investigated.

III. Biological Insights

The PCST solution connects together the phosphorylation events and transcriptional changes using a compact set of interactions. Since the method puts the phosphorylation events in the context of protein–protein interactions, the connections participated by these events or groups of events are suggestive of their cellular functions. The transcription factors included in the network and the connections among them point to the functional consequence of the upstream signals. These are certainly of great interest for elucidating the role of individual hits. Also interesting are the properties that emerge from the network at the systems level, and we will describe a few computational techniques for such analyses using the yeast pheromone response PCST solution as an example (Fig. 2).

A. Properties of the Full Network

The PCST solution in Fig. 2 was constructed from published phosphoproteomic (Gruhler *et al.*, 2005) and transcription profiling (Roberts *et al.*, 2000) datasets of the yeast *Saccharomyces cerevisiae* in response to the mating pheromone α-factor. This network was first reported in Huang and Fraenkel (2009). The network connects 56 of the 112 proteins with α-factor-responsive phosphorylation sites and 100 of the 201 differentially expressed genes through 94 intermediate proteins.

The solution network shows a few notable features at the global level. First, the MAPK cascade known to be induced by pheromone (labeled "pheromone core" in Fig. 2) is recovered by the algorithm. In particular, it correctly identifies the proteins GPA1, STE11, and BEM1, where no phosphorylation sites were detected, as well as their connections to other proteins in the pheromone signaling pathway. In addition, only proteins that are present in the pheromone response pathway are included. Second, beyond the MAPK cascade, the solution network partitions into highly coherent subnetworks with biological functions relevant to mating. At the transcription level, phosphorylated proteins seem highly informative in selecting interacting transcription factors. Examples include DIG1/DIG2/STE12 complex in the pheromone signaling pathway, SWI4/SWI6 and SWI6/MBP1 in the PKC pathway, and FKH2/NDD1 complex regulated by CDC28. These observations suggest the constraints imposed by the phosphorylated proteins and differentially expressed genes are sufficient to guide the selection of important players that contribute to the response.

To assess the functional significance of the intermediate nodes from the PCST solution in mating response, we examined two independent whole-genome deletion screen datasets that screen for genes whose deletion result in mating defects. One screen measures a molecular phenotype in the form of activation of FUS1-lacZ reporter (Chasse *et al.*, 2006) and the other screen measures a morphological phenotype in the form of cell cycle arrest and shmoo formation (Narayanaswamy *et al.*, 2006). For each screen, we counted the number of hits that overlap with the intermediate nodes in the PCST solution, and using all the screening genes as background we computed a hypergeometric p-value for which such overlap would appear by chance. As seen in Fig. 4, compared to networks constructed from shortest paths and

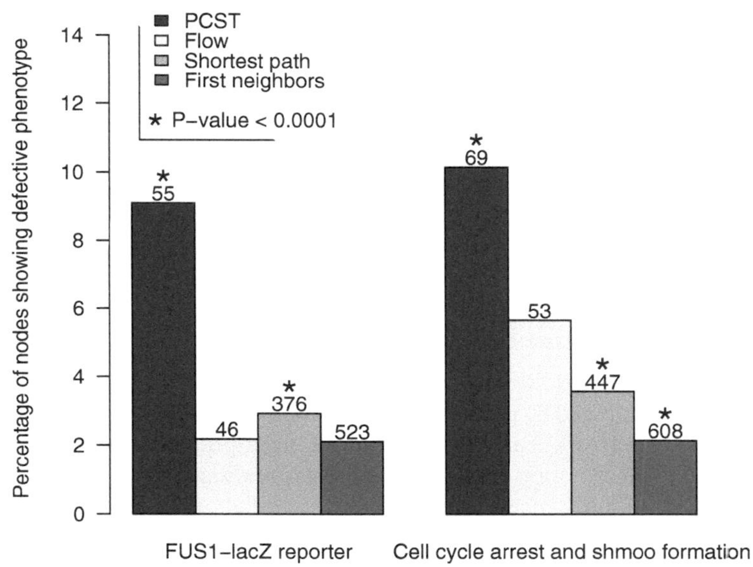

Fig. 4 The PCST pheromone response network is compact, and when compared to networks predicted by other methods, it contains higher fraction of genes that are implicated in mating response, measured by defects in activating a FUS1-lacZ reporter gene (Chasse *et al.*, 2006) and defects in cell cycle arrest and shmoo formation (Narayanaswamy *et al.*, 2006). The *Flow* network was constructed from the phosphorylated proteins and differential expressed genes by a previously published algorithm based on network flows (Yeger-Lotem *et al.*, 2009). The *Shortest path* network consists of pairwise shortest paths between the terminal nodes and the *First neighbor* network consists of nodes in the interactome that directly interact with the phosphorylated proteins. Enrichment p-values were computed by hypergeometric tests using all the genes tested in the respective genetic screen as background. The number above each bar denotes the number of nodes in the network. Image reproduced with permission from Huang and Fraenkel (2009). (For color version of this figure, the reader is referred to the web version of this book.)

first neighbors of the terminal nodes, the PCST solution is more compact while achieving higher enrichment of genes implicated in mating defects.

B. Biological Functions of Subnetwork/Modules

To objectively quantify the empirical observation that the PCST solution is partitioned into functional coherent subnetworks, we applied the Girvan–Newman algorithm (Dunn *et al.*, 2005; Girvan and Newman, 2002) to cluster the solution. This algorithm is used for detecting clusters in an interaction network that contain dense connections between nodes in the same cluster but less dense connections to nodes in other clusters. Gene *Ontology* enrichment analysis of the resulting clusters reveals that all the clusters have high degree of functional coherence (Table II). It is interesting to note that many of the clusters are not coordinately expressed at the mRNA level, as quantified by the significance of expression coherence score (Pilpel *et al.*, 2001) or by the significance of expression activity score (Ideker *et al.*, 2002). Notably, the clusters that show significant coordinated expression are involved in cell cycle processes.

Being able to recover functionally coherent clusters that are not coherent at the transcript level is a significant result. Transcriptional data, which are more readily available than proteomics data, are the focus of many computational methods for regulatory network construction. Our results suggest that methods that rely solely on expression data, including a prior Steiner tree approach (Dittrich *et al.*, 2008), will be unable to recover the full extent of a biological response.

C. Quantifying the Relevance of the Transcription Factors

In addition to the transcription factors mentioned above that are known to be induced by pheromone or function in related biological processes, the PCST solution network features many other transcriptional regulators not previously implicated in pheromone response. We use expression coherence score as a metric to quantify the significance of these transcription factors at the global level. For each transcription factor with targets in the interactome, we obtained the expression values of those targets across a set of conditions that stimulate pheromone signaling, and computed the significance p-value of the expression coherence score. Then we set a threshold on the significant p-value, and compared the percentage of transcription factors included and excluded in the PCST that pass this threshold. As shown in Fig. 5, the transcription factors included in the network are more likely to have a set of targets that are coherently expressed than the factors excluded from the network. To check if these transcription factors are condition specific, we did a similar calculation for the expression values from a set of conditions that are unrelated to pheromone: when yeast undergoes the metabolic shift from fermentation to respiration (diauxic shift). We found that coherence is specific to the conditions related to pheromone signaling but not to diauxic shift.

Table II

Biological functions and measures of coordinated mRNA expression of the clusters in the pheromone response PCST network generated from edge-betweenness clustering.

Cluster	Top three enriched GO biological process terms		Corrected *p*-value	*p*-value of EC score	*p*-value of EA score
1	GO:0046907	Intracellular transport	1.23E–09	0.711	1
	GO:0051649	Establishment of cellular localization	1.23E–09		
	GO:0051641	Cellular localization	1.71E–09		
2	GO:0006457	Protein folding	1.41E–04	0.251	0.735
	GO:0042026	Protein refolding	1.41E–04		
	GO:0000069	Kinetochore assembly	8.35E–04		
3	GO:0016193	Endocytosis	1.73E–06	0.128	1
	GO:0007114	Cell budding	1.26E–05		
	GO:0051301	Cell division	1.26E–05		
4	GO:0000074	Regulation of progression through cell cycle	2.68E–06	0.421	0.453
	GO:0051726	Regulation of cell cycle	2.68E–06		
	GO:0006270	DNA replication initiation	3.44E–06		
5	GO:0006350	Transcription	8.00E–14	0.863	1
	GO:0045449	Regulation of transcription	1.94E–12		
	GO:0019219	Regulation of nucleobase, nucleoside, nucleotide, and nucleic acid metabolism	7.15E–12		
6	GO:0007096	Regulation of exit from mitosis	3.52E–07	0.063	1
	GO:0007088	Regulation of mitosis	4.45E–07		
	GO:0000074	Regulation of progression through cell cycle	1.05E–05		
7	GO:0048856	Anatomical structure development	3.19E–14	0.35	0
	GO:0007148	Cell morphogenesis	3.19E–14		
	GO:0019236	Response to pheromone	1.26E–11		
8	GO:0006350	Transcription	1.89E–09	0.504	0.35
	GO:0006351	Transcription, DNA-dependent	7.90E–09		
	GO:0032774	RNA biosynthesis	7.90E–09		
9	GO:0000082	G1/S transition of mitotic cell cycle	2.15E–04	0.272	0.008
	GO:0051325	Interphase	1.07E–03		
	GO:0051329	Interphase of mitotic cell cycle	1.07E–03		
Full network	GO:0006350	Transcription	2.67E–23	0.729	1
	GO:0019222	Regulation of metabolism	2.73E–21		
	GO:0050791	Regulation of physiological process	1.16E–20		

Note: EC, expression coherence (Pilpel *et al.*, 2001); EA, expression activity (Ideker *et al.*, 2002). Reproduced with permission from Huang and Fraenkel (2009).

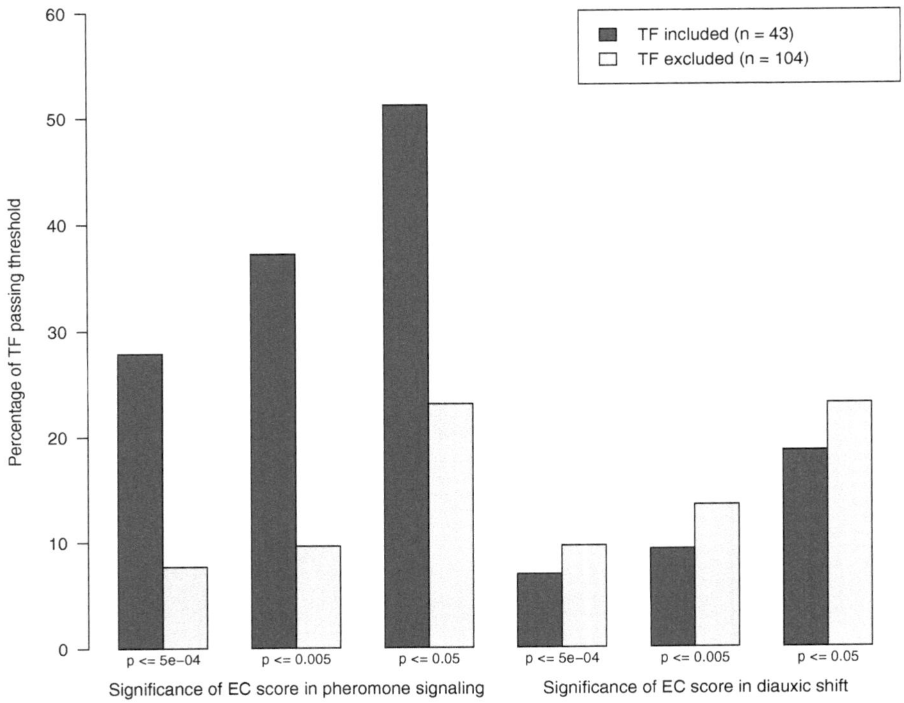

Fig. 5 Percentage of transcription factors (TF) with targets that show significant expression coherence (EC) scores computed from 50 nM α-factor time course (Roberts *et al.*, 2000) and diauxic shift conditions (DeRisi *et al.*, 1997), for transcription factors included in and excluded from the PCST solution network. The *p*-values indicate thresholds on the significance of the expression coherence score of the target genes. Image reproduced with permission from Huang and Fraenkel (2009).

IV. Open Challenges

A. Improving the Input Data

The central premise behind our constraint optimization framework is that the experimental measurements at the signaling and transcription level are sufficient for guiding selection of relevant interactions from the interactome. It is important to note, however, that many of these interactions may only occur under specific conditions that are not relevant to the problem being studied. It is not yet practical to collect condition-specific interaction data on a large scale. Nevertheless, there are a few strategies to ensure the selected interactions are indeed relevant. First, as a preprocessing step, the input interaction network can be filtered to remove nodes that are not believed to be expressed under the condition of interest, based on transcript or protein assays. With the improved sensitivity of RNA-seq to detect low-abundance transcripts compared to microarrays, this step may now be done with higher confidence. However, expression data are still noisy, and removing nodes completely risks missing important components of a network. Alternatively, we can add to the PCST formulation capacities on the nodes that represent the expression level. There are well-established procedures that transform node capacitated network flow problems to ones without the node capacities (Ahuja *et al.*, 1995).

Our current analysis defines node penalties on the phosphorylated proteins in a practical but *ad hoc* manner: the penalty values are proportional to the absolute value of log-fold changes of phosphorylation; if there are multiple phosphorylation sites on one protein, the maximum value is used. This reflects the assumption that larger changes in phosphorylation carry higher importance and thus should be given higher priority to be included. There are other, probably more principled, ways of quantifying the significance of the phosphorylation changes. We distinguish two kinds of significance: statistical significance and biological significance. The former requires the development of robust error models (Yi Zhang *et al.*, 2010) while the latter would benefit from knowledge about the context of the phosphorylation sites, such as the structural domain or binding sequence motif where the sites are located (see examples in Naegle *et al.* (2010)). But these two need not to be exclusive: once statistical significance is established, penalty values can be defined by analyzing for potential biological significance.

As phosphorylation sites are the starting point from which the PCST network solution is built, it is critical to have a good coverage of interactions involving these proteins in the interactome graph. Phosphorylation sites participate in interactions with other proteins in two ways: as substrates of kinase and phosphatases, and as binding partners of proteins that recognize the phosphorylated residues. Many of these interactions are transient and context specific and thus difficult to capture in some interaction assays. In particular, among the various high-throughput interactome mapping techniques, a modified affinity capture MS method is the most informative in identifying kinase targets, with yeast two-hybrid being second (Sharifpoor *et al.*, 2011). Many *in vivo* methods are available to link kinases to phosphorylation substrates (reviewed in Sopko and Andrews (2008)) but only for specific kinases. Taking these efforts to the global level, and using other information such as sequence motifs integrated within a computational framework such as NetworKIN (Linding *et al.*, 2007), will produce interaction datasets that greatly enhance the ability of our algorithm to connect the phosphorylated proteins.

Beyond the focused mapping of interactions involving phosphorylated proteins, the ability to discover novel signaling pathways also depends on the coverage of other parts of the interactome. Even with the combination of large experimental efforts and curated databases we are still far from a complete mapping of all possible protein–protein interactions, especially in less well-studied organisms. Therefore, many computational methods have been developed to predict possible interactions. These methods make use of features such as gene neighborhood (M. Huynen *et al.*, 2000), gene fusion (Marcotte *et al.*, 1999), sequence co-evolution (Goh *et al.*, 2000), and may incorporate several such features in a Bayesian framework (Jansen *et al.*,

2003). The probabilistic nature of edge weights in our PCST formulation provides a natural way to include these computational predictions.

B. Other Applications and Potentials

The PCST approach can be used to analyze jointly a wide variety of types of data. Cellular functions are operated by networks of molecular interactions, which include a lot more than phosphorylation-mediated signaling and transcription factor binding to target genes. But regardless of the data type, there are many situations in which we see to find a parsimonious, high-confidence interaction network satisfying a defined set of constraints. Therefore, this approach can be applied to many other levels of regulation, depending on the source of the constraints and the molecular interactions. For example, we may model the global effect of a microRNA by using the microRNA targets as constraints and including microRNA to target relationships in the interactome. Metabolomics data are another area of great interest and may become an entry point to link together protein signaling networks with metabolic networks. The detected metabolites can be used as constraints in a network of metabolic reactions catalyzed by enzymes that are also part of the protein interaction network. For all these datasets, taking a network approach such as the PCST will yield more insight than simply following up on the top hits.

One disadvantage of the PCST method is the tree structure of the resulting network: all the included terminal nodes must be connected to each other. However, it is possible that the terminal nodes belong to multiple, separate signaling pathways that are not connected to each other, either because there is no cross-talk biologically or the cross-talk interactions are not in the known interactome. Adopting a forest formulation, where multiple trees may be used to connect the terminal nodes, may remedy this drawback.

Finally, it is useful to consider this approach in the context of other types of network modeling. The strengths of our method lie in the ability to identify previously unrecognized components of a cellular response and to discover functionally coherent subsets of proteins. However, this approach is not designed to capture the dynamics of a system, including feedback regulation. A natural way to describe such feedback mathematically is by differential equations, which can be simulated numerically or analyzed. Differential-equation-based models have been applied genome-wide in a comprehensive transcriptional and translational network for *Escherichia coli* (Thiele *et al.*, 2009) and have been applied extensively to relatively small networks of mammalian proteins (Eungdamrong and Iyengar, 2004; Aldridge *et al.*, 2006; Tyson *et al.*, 2003). However, such approaches are not suitable for very large networks where there are not enough data to sufficiently constrain the necessary parameters of the models.

We believe that these two approaches may ultimately be used together to develop dynamic models of previously uncharacterized biological systems. In a first phase, proteomic, transcriptional or other "-omics" datasets would be analyzed using constraint optimization to identify a set of proteins that seem most relevant to the biological process. With the size of the problem now reduced to a more manageable level, more focused experiments together with differential equation-based modeling could reveal the dynamics of the system.

Acknowledgments

We thank Dr. Nurcan Tuncbag for comments on the manuscript, and Cindy Woolley for help with editing. This work was supported by National Institutes of Health (NIH) Grants U54-CA112967 and GM089903. S.S.H. was supported by an NCI integrative cancer biology program graduate fellowship and a National Science and Engineering Research Council of Canada Postgraduate Scholarship. E.F. is the recipient of the Eugene Bell Career Development Chair.

References

Ahuja, R.K., Magnanti, T.L., Orlin, J.B., and Weihe, K. (1995). Network flows: theory, algorithms and applications. Physica-Verlag, Wurzburg, 1972–1995.

Aldridge, B. B., Burke, J. M., Lauffenburger, D. A., and Sorger, P. K. (2006). Physicochemical modelling of cell signalling pathways. *Nat. Cell Biol.* **8**, 1195–1203.

Altschul, S. F., Gish, W., Miller, W., Myers, E. W., and Lipman, D. J. (1990). Basic local alignment search tool. *J. Mol. Biol.* **215**, 403–410.

Aranda, B., *et al.* (2011). PSICQUIC and PSISCORE: accessing and scoring molecular interactions. *Nat. Meth.* **8**, 528–529.

Archer, A., Bateni, M., Hajiaghayi, M., and Karloff, H. (2011). Improved approximation algorithms for prize-collecting Steiner tree and TSP. *SIAM J. Comput.* **40**, 309.

Bader, G. D., Donaldson, I., Wolting, C., Ouellette, B. F., Pawson, T., and Hogue, C. W. (2001). BIND – the biomolecular interaction network database. *Nucleic Acids Res.* **29**, 242–245.

Bailly-Bechet, M., Borgs, C., Braunstein, A., Chayes, J., Dagkessamanskaia, A., François, J. -M., and Zecchina, R. (2011). Finding undetected protein associations in cell signaling by belief propagation. *Proc. Natl. Acad. Sci. U S A* **108**, 882–887.

Bayati, M., Borgs, C., Braunstein, A., Chayes, J., Ramezanpour, A., and Zecchina, R. (2008). Statistical mechanics of Steiner trees. *Phys. Rev. Lett.* **101**, 37208.

Birney, E., *et al.* (2007). Identification and analysis of functional elements in 1% of the human genome by the ENCODE pilot project. *Nature* **447**, 799–816.

Bryne, J. C., Valen, E., Tang, M. -H. E., Marstrand, T., Winther, O., Piedade, I., da Krogh, A., Lenhard, B., and Sandelin, A. (2008). JASPAR, the open access database of transcription factor-binding profiles: new content and tools in the 2008 update. *Nucleic Acids Res.* **36**, D102–D110.

Chasse, S. A., Flanary, P., Parnell, S. C., Hao, N., Cha, J. Y., Siderovski, D. P., and Dohlman, H. G. (2006). Genome-scale analysis reveals Sst2 as the principal regulator of mating pheromone signaling in the yeast Saccharomyces cerevisiae. *Eukaryot. Cell* **5**, 330–346.

Chatr-aryamontri, A., Ceol, A., Palazzi, L. M., Nardelli, G., Schneider, M. V., Castagnoli, L., and Cesareni, G. (2007). MINT: the Molecular INTeraction database. *Nucleic Acids Res.* **35**, D572–D580.

Chaurasia, G., Iqbal, Y., Hänig, C., Herzel, H., Wanker, E. E., and Futschik, M. E. (2007). UniHI: an entry gate to the human protein interactome. *Nucleic Acids Res.* **35**, D590–D600.

Choudhary, C., and Mann, M. (2010). Decoding signalling networks by mass spectrometry-based proteomics. *Nat. Rev. Mol. Cell Biol.* **11**, 427–439.

Cline, M. S., *et al.* (2007). Integration of biological networks and gene expression data using Cytoscape. *Nat. Protoc.* **2**, 2366–2382.

DeRisi, J. L., Iyer, V. R., and Brown, P. O. (1997). Exploring the metabolic and genetic control of gene expression on a genomic scale. *Science* **278**, 680–686.

Dittrich, M. T., Klau, G. W., Rosenwald, A., Dandekar, T., and Müller, T. (2008). Identifying functional modules in protein-protein interaction networks: an integrated exact approach. *Bioinformatics* **24**, i223–i231.
Dunn, R., Dudbridge, F., and Sanderson, C. M. (2005). The use of edge-betweenness clustering to investigate biological function in protein interaction networks. *BMC Bioinformat.* **6**, 39.
D'haeseleer, P. (2006). What are DNA sequence motifs? *Nat. Biotechnol.* **24**, 423–425.
Eungdamrong, N. J., and Iyengar, R. (2004). Modeling cell signaling networks. *Biol. Cell* **96**, 355–362.
Foat, B. C., Houshmandi, S. S., Olivas, W. M., and Bussemaker, H. J. (2005). Profiling condition-specific, genome-wide regulation of mRNA stability in yeast. *Proc. Natl. Acad. Sci. U S A* **102**, 17675–17680.
Foat, B. C., Morozov, A. V., and Bussemaker, H. J. (2006). Statistical mechanical modeling of genome-wide transcription factor occupancy data by MatrixREDUCE. *Bioinformatics* **22**, e141–e149.
Girvan, M., and Newman, M. E. J. (2002). Community structure in social and biological networks. *Proc. Natl. Acad. Sci. U S A* **99**, 7821–7826.
Goh, C. S., Bogan, A. A., Joachimiak, M., Walther, D., and Cohen, F. E. (2000). Co-evolution of proteins with their interaction partners. *J. Mol. Biol.* **299**, 283–293.
Graphviz (2011a). Graphviz – graph visualization software. Available at: http://graphviz.org/ [Accessed August 6, 2011].
Graphviz (2011b). The DOT Language | Graphviz – graph visualization software. Available at: http://www.graphviz.org/content/dot-language [Accessed August 6, 2011].
Grimsrud, P. A., Swaney, D. L., Wenger, C. D., Beauchene, N. A., and Coon, J. J. (2010). Phosphoproteomics for the masses. *ACS Chem. Biol.* **5**, 105–119.
Gruhler, A., Olsen, J. V., Mohammed, S., Mortensen, P., Faergeman, N. J., Mann, M., and Jensen, O. N. (2005). Quantitative phosphoproteomics applied to the yeast pheromone signaling pathway. *Mol. Cell Proteomics* **4**, 310–327.
Hagberg, A. A., Schult, D. A., and Swart, P. J. (2008). Exploring network structure, dynamics, and function using networkX. *In* "Proceedings of the 7th Python in Science Conference," (G. Varoquaux, T. Vaught, J. Millman, eds.), (Pasadena)11–15.
Harbison, C. T., *et al.* (2004). Transcriptional regulatory code of a eukaryotic genome. *Nature* **431**, 99–104.
Huang, S. -S. C., and Fraenkel, E. (2009). Integrating proteomic, transcriptional, and interactome data reveals hidden components of signaling and regulatory networks. *Sci. Signal* **2**, ra40.
Huynen, M., Snel, B., Lathe, W., and Bork, P. (2000). Predicting protein function by genomic context: quantitative evaluation and qualitative inferences. *Genome Res.* **10**, 1204–1210.
Hwang, D., *et al.* (2005). A data integration methodology for systems biology: experimental verification. *Proc. Natl. Acad. Sci. U S A* **102**, 17302–17307.
IBM (2010). IBM IBM ILOG Optimization Academic Initiative – United States.
Ideker, T., Ozier, O., Schwikowski, B., and Siegel, A. F. (2002). Discovering regulatory and signalling circuits in molecular interaction networks. *Bioinformatics* **18**(Suppl 1), S233–S240.
Jansen, R., *et al.* (2003). A Bayesian networks approach for predicting protein-protein interactions from genomic data. *Science* **302**, 449–453.
Kanehisa, M., Goto, S., Furumichi, M., Tanabe, M., and Hirakawa, M. (2010). KEGG for representation and analysis of molecular networks involving diseases and drugs. *Nucleic Acids Res.* **38**, D355–D360.
Kerrien, S., *et al.* (2007). IntAct – open source resource for molecular interaction data. *Nucleic Acids Res.* **35**, D561–D570.
Keshava Prasad, T. S., *et al.* (2009). Human protein reference database – 2009 update. *Nucleic Acids Res.* **37**, D767–D772.
Klingström, T., and Plewczynski, D. (2011 Nov). Protein-protein interaction and pathway databases, a graphical review. *Brief Bioinform* **12**(6), 702–713.
Kunsch, C., Ruben, S. M., and Rosen, C. A. (1992). Selection of optimal kappa B/Rel DNA-binding motifs: interaction of both subunits of NF-kappa B with DNA is required for transcriptional activation. *Mol. Cell Biol.* **12**, 4412–4421.
Lan, A., Smoly, I. Y., Rapaport, G., Lindquist, S., Fraenkel, E., and Yeger-Lotem, E. (2011). ResponseNet: revealing signaling and regulatory networks linking genetic and transcriptomic screening data. *Nucleic Acids Res.* **39**, W424–W429.

Las Rivas, J., and De Fontanillo, C. (2010). Protein-protein interactions essentials: key concepts to building and analyzing interactome networks. *PLoS Comput. Biol.* **6**, e1000807.

Linding, R., *et al.* (2007). Systematic discovery of in vivo phosphorylation networks. *Cell* **129**, 1415–1426.

Ljubić, I. (2008). Prize-collecting Steiner tree. Available at: http://homepage.univie.ac.at/ivana.ljubic/research/pcstp/ [Accessed August 6, 2011].

Ljubić, I., Weiskircher, R., Pferschy, U., Klau, G. W., Mutzel, P., and Fischetti, M. (2005). An algorithmic framework for the exact solution of the prize-collecting Steiner tree problem. *Math. Program.* **105**, 427–449.

MacIsaac, K. D., Wang, T., Gordon, D. B., Gifford, D. K., Stormo, G. D., and Fraenkel, E. (2006). An improved map of conserved regulatory sites for *Saccharomyces cerevisiae*. *BMC Bioinformati.* **7**, 113.

MacIsaac, K. D., and Fraenkel, E. (2006). Practical strategies for discovering regulatory DNA sequence motifs. *PLoS Comput. Biol.* **2**, e36.

Macek, B., Mann, M., and Olsen, J. V. (2009). Global and site-specific quantitative phosphoproteomics: principles and applications. *Annu. Rev. Pharmacol. Toxicol.* **49**, 199–221.

Maier, T., Güell, M., and Serrano, L. (2009). Correlation of mRNA and protein in complex biological samples. *FEBS Lett.* **583**, 3966–3973.

Manke, T., Roider, H. G., and Vingron, M. (2008). Statistical modeling of transcription factor binding affinities predicts regulatory interactions. *PLoS Comput. Biol.* **4**, e1000039.

Marcotte, E. M., Pellegrini, M., Ng, H. L., Rice, D. W., Yeates, T. O., and Eisenberg, D. (1999). Detecting protein function and protein-protein interactions from genome sequences. *Science* **285**, 751–753.

Matthews, L., *et al.* (2009). Reactome knowledgebase of human biological pathways and processes. *Nucleic Acids Res.* **37**, D619–D620.

Matys, V., *et al.* (2006). TRANSFAC and its module TRANSCompel: transcriptional gene regulation in eukaryotes. *Nucleic Acids Res.* **34**, D108–D110.

Mering, C., von Jensen Lars, J., Snel, Berend., Hooper, S. D., Krupp, M., Foglierini, M., Jouffre, N., Huynen, M. A., and Bork, Peer. (2005). STRING: known and predicted protein-protein associations, integrated and transferred across organisms. *Nucleic Acids Res.* **33**, D433–D440.

Mering, C., von Krause, R., Snel, Berend., Cornell, M., Oliver, S. G., Fields, Stanley., and Bork, Peer. (2002). Comparative assessment of large-scale data sets of protein–protein interactions. *Nature* **417**, 399–403.

Naegle, K. M., Gymrek, M., Joughin, B. A., Wagner, J. P., Welsch, R. E., Yaffe, M. B., Lauffenburger, D. A., and White, F. M. (2010). PTMScout, a Web resource for analysis of high throughput post-translational proteomics studies. *Mol. Cell Proteomics* **9**, 2558–2570.

Narayanaswamy, R., Niu, W., Scouras, A. D., Hart, G. T., Davies, J., Ellington, A. D., Iyer Vishwanath, R., and Marcotte, E. M. (2006). Systematic profiling of cellular phenotypes with spotted cell microarrays reveals mating-pheromone response genes. *Genome Biol.* **7**, R6.

Orchard, S., *et al.* (2007). The minimum information required for reporting a molecular interaction experiment (MIMIx). *Nat. Biotechnol.* **25**, 894–898.

Ozsolak, F., and Milos, P. M. (2011). RNA sequencing: advances, challenges and opportunities. *Nat. Rev. Genet.* **12**, 87–98.

Pagel, P., *et al.* (2005). The MIPS mammalian protein–protein interaction database. *Bioinformatics* **21**, 832–834.

Pilpel, Y., Sudarsanam, P., and Church, G. M. (2001). Identifying regulatory networks by combinatorial analysis of promoter elements. *Nat. Genet.* **29**, 153–159.

Prieto, C., and Las Rivas, J De. (2006). APID: agile protein interaction data analyzer. *Nucleic Acids Res.* **34**, W298–W302.

Razick, S., Magklaras, G., and Donaldson, I. M. (2008). iRefIndex: a consolidated protein interaction database with provenance. *BMC Bioinformat.* **9**, 405.

Roberts, C. J., *et al.* (2000). Signaling and circuitry of multiple MAPK pathways revealed by a matrix of global gene expression profiles. *Science* **287**, 873–880.

Roider, H. G., Kanhere, A., Manke, T., and Vingron, M. (2007). Predicting transcription factor affinities to DNA from a biophysical model. *Bioinformatics* **23**, 134–141.

Salwinski, L., Miller, C. S., Smith, A. J., Pettit, F. K., Bowie, J. U., and Eisenberg, David. (2004). The database of interacting proteins: 2004 update. *Nucleic Acids Res.* **32**, D449–D450.

Sandelin, A., Alkema, W., Engström, P., Wasserman, W. W., and Lenhard, B. (2004). JASPAR: an open-access database for eukaryotic transcription factor binding profiles. *Nucleic Acids Res.* **32**, D91–D94.

Schwanhäusser, B., Busse, D., Li, N., Dittmar, G., Schuchhardt, J., Wolf, J., Chen, W., and Selbach, M. (2011). Global quantification of mammalian gene expression control. *Nature* **473**, 337–342.

Scott, M. S., Perkins, T., Bunnell, S., Pepin, F., Thomas, D. Y., and Hallett, M. (2005). Identifying regulatory subnetworks for a set of genes. *Mol. Cell Proteomics* **4**, 683–692.

Segal, E., Friedman, N., Kaminski, N., Regev, A., and Koller, D. (2005). From signatures to models: understanding cancer using microarrays. *Nat. Genet.* **37**(Suppl), S38–S45.

Sharifpoor, S., *et al.* (2011). A quantitative literature-curated gold standard for kinase-substrate pairs. *Genome Biol.* **12**, R39.

Smoot, M. E., Ono, K., Ruscheinski, J., Wang, P. -L., and Ideker, T. (2011). Cytoscape 2.8: new features for data integration and network visualization. *Bioinformatics* **27**, 431–432.

Sopko, R., and Andrews, B. J. (2008). Linking the kinome and phosphorylome – a comprehensive review of approaches to find kinase targets. *Mol. Biosyst.* **4**, 920–933.

Sousa Abreu, R., de Penalva, L. O., Marcotte, E. M., and Vogel, C. (2009). Global signatures of protein and mRNA expression levels. *Mol. Biosyst.* **5**, 1512–1526.

Spellman, P. T., Sherlock, G., Zhang, M. Q., Iyer, V. R., Anders, K., Eisen, M. B., Brown, P. O., Botstein, D., and Futcher, B. (1998). Comprehensive identification of cell cycle-regulated genes of the yeast *Saccharomyces cerevisiae* by microarray hybridization. *Mol. Cell Biol.* **9**, 3273–3297.

Stark, C., *et al.* (2011). The BioGRID interaction database: 2011 update. *Nucleic Acids Res.* **39**, D698–D700.

Stoltenburg, R., Reinemann, C., and Strehlitz, B. (2007). SELEX – a (r)evolutionary method to generate high-affinity nucleic acid ligands. *Biomol. Eng.* **24**, 381–403.

Stormo, G. D. (2000). DNA binding sites: representation and discovery. *Bioinformatics* **16**, 16–23.

Tarcea, V. G., *et al.* (2009). Michigan molecular interactions r2: from interacting proteins to pathways. *Nucleic Acids Res.* **37**, D642–D650.

Thiele, I., Jamshidi, N., Fleming, R. M. T., and Palsson, B. Ø. (2009). Genome-scale reconstruction of Escherichia coli's transcriptional and translational machinery: a knowledge base, its mathematical formulation, and its functional characterization. *PLoS Comput. Biol.* **5**, e1000312.

Turinsky, A. L., Razick, S., Turner, B., Donaldson, I. M., and Wodak, S. J. (2011). Interaction databases on the same page. *Nat. Biotechnol.* **29**, 391–393.

Tyson, J. J., Chen, K. C., and Novak, B. (2003). Sniffers, buzzers, toggles and blinkers: dynamics of regulatory and signaling pathways in the cell. *Curr. Opin. Cell Biol.* **15**, 221–231.

White, F. M. (2008). Quantitative phosphoproteomic analysis of signaling network dynamics. *Curr. Opin. Biotechnol.* **19**, 404–409.

Wingender, E. (2008). The TRANSFAC project as an example of framework technology that supports the analysis of genomic regulation. *Brief Bioinform.* **9**, 326–332.

Yeger-Lotem, E., *et al.* (2009). Bridging high-throughput genetic and transcriptional data reveals cellular responses to alpha-synuclein toxicity. *Nat. Genet.* **41**, 316–323.

Yosef, N., Ungar, L., Zalckvar, E., Kimchi, A., Kupiec, M., Ruppin, E., and Sharan, R. (2009). Toward accurate reconstruction of functional protein networks. *Mol. Syst. Biol.* **5**, 248.

Yu, H., *et al.* (2008). High-quality binary protein interaction map of the yeast interactome network. *Science* **322**, 104–110.

Zhang, Yi., Askenazi, M., Jiang, J., Luckey, C. J., Griffin, J. D., and Marto, J. A. (2010). A robust error model for iTRAQ quantification reveals divergent signaling between oncogenic FLT3 mutants in acute myeloid leukemia. *Mol. Cell Proteomics* **9**, 780–790.

Zhang, Yi., Wolf-Yadlin, A., and White, F. M. (2007). Quantitative proteomic analysis of phosphotyrosine-mediated cellular signaling networks. *Methods Mol. Biol.* **359**, 203–212.

CHAPTER 4

A Framework for Modeling the Relationship Between Cellular Steady-state and Stimulus-responsiveness

Paul M. Loriaux and Alexander Hoffmann

Signaling Systems Laboratory, San Diego Center for Systems Biology of Cellular Stress Responses, Program in Bioinformatics and Systems Biology, University of California San Diego, La Jolla, California, USA

Abstract

In cell signaling systems, the abundances of signaling molecules are generally thought to determine the response to stimulation. However, the kinetics of molecular processes, for example receptor trafficking and protein turnover, may also play an important role. Few studies have systematically examined this relationship between the resting state and stimulus-responsiveness. Fewer still have investigated the relative contribution of steady-state concentrations and reaction kinetics. Here we

0091-679X/10 $35.00
DOI 10.1016/B978-0-12-388403-9.00004-7

describe a mathematical framework for modeling the resting state of signaling systems. Among other things, this framework allows steady-state concentration measurements to be used in parameterizing kinetic models, and enables comprehensive characterization of the relationship between the resting state and the cellular response to stimulation.

I. Introduction

Cell systems respond to external stimuli through a coordinated network of biochemical reactions mediated by any number of molecular species. Although it is customary to think of these systems as being "at rest" prior to stimulation, a growing number of studies have demonstrated that the resting state of a cell prior to stimulation can be a powerful determinant of the response. For example, with regards to stimulation by the death-inducing TNF superfamily member TRAIL, studies have shown that cells may be sensitized via up-regulation of the TRAIL receptor DR5 (Dolloff *et al.*, 2011) or caspase 8 (Fulda *et al.*, 2001), down-regulation of TRADD (Kim *et al.*, 2011) or c-FILP (Li *et al.*, 2011), or alternatively desensitized by up-regulation of Bcl-XL (Hinz *et al.*, 2000) or Bcl-2 (Fulda *et al.*, 2002) (reviewed in Zhang and Fang, 2005).

By contrast, in other systems it has been shown that the kinetics of species turnover – not their outright abundance – determine the response to stimulation. For example, high turnover of the Epo receptor is required to maintain a linear, non-refractory response over a broad range of Epo concentrations (Becker *et al.*, 2010), while high turnover of the inhibitor of NFκB is required to distinguish acute inflammatory stimuli from metabolic stress conditions (O'Dea *et al.*, 2008). Studies like these highlight an important dichotomy in the resting state of a cell. In one hand are the concentrations of molecules prior to stimulation, and in the other are the rates of the biochemical processes in which they participate. How do each of these facets of the resting state affect the cellular response to stimulation?

Because systematic changes in the resting states of living cells are difficult to engineer, investigating this relationship cannot be addressed by laboratory science alone. For example, short interfering (si) RNA can be used to reduce the concentration of a specific gene product, but this reduction is effected by interfering with the translation of the product (Fire *et al.*, 1998; Izant and Weintraub, 1984). Changes in stimulus-responsiveness due to siRNA knockdown may, therefore, be caused by a reduction in the concentration of the target species, reduction in the kinetics of its turnover, or both. Furthermore, RNA dilution in rapidly dividing cells (Bartlett and Davis, 2006) or secondary induction of the mammalian interferon response (Reynolds *et al.*, 2006) may further cloud interpretation of the results.

Using a mathematical model, the behavior of a system can be studied rapidly and in isolation, providing a sort of sufficiency test for proposed mechanisms of cellular responsiveness (Faller *et al.*, 2003; Kearns and Hoffmann, 2009; and Kitano, 2002). The steady-state of a model, discussed in further detail below, is furthermore a good approximation for the resting state of the system. A complication that arises in models when trying to characterize the relationship between steady-state and stimulus-responsiveness, however, is that models of cell systems are typically nonlinear. As such, the steady-state must often be found numerically, and this compromises the modeler's ability to investigate its role in stimulus-responsiveness.

To that end, in this chapter we describe a method for deriving an analytical expression for the steady-state of a common class of models, called mass action models. From this analytical expression, we go on to give precise steps for introducing systematic changes to the steady-state concentrations of molecular species and the rates of biochemical processes in which they participate. In doing so, we demonstrate how specific hypotheses can be generated about the resting state of a system and its impact on stimulus-responsiveness. Examples include:

- Are my measurements of steady-state concentrations and kinetic *rate constants* consistent with the proposed model?
- Is a change in the steady-state concentration or activity of a particular species sufficient to explain the changes I observed in the system's response to stimulation?
- Can I expect a system at a particular resting state to exhibit a certain response to stimulation?

The remainder of this text is divided into the following sections: "Overview of Algorithm," in which we provide a verbal description of the steps required to model a system and derive a solution to its steady-state; "Biological Insights," in which we demonstrate how a model at steady-state can help generate hypotheses about the relationship between the resting state of a system and its response to stimulation; "Open Challenges," in which we describe limitations of the method and potential avenues for refinement; "Computational Methods," in which we provide step-by-step instructions for modeling a system and manipulating its steady-state; and finally "Further Reading," where we offer some references for further reading on the subject of modeling, steady-state, and parameterization, and dynamic response analysis.

II. Overview of Algorithm

For the purposes of this manuscript we assume that the system to be studied can be described by a biochemical reaction network (BRN). A BRN consists a set of a set of molecular species and a set of biochemical reactions. The set of species must contain every species consumed or produced by the reactions. Neither every species nor every reaction need be elementary – a species may refer to a complex biomolecule like a ribosome, for example, and a reaction to a multistep process like protein synthesis.

A simple BRN to illustrate the steps used in the forthcoming algorithm is the activation of the tumor suppressor p53. This network consists of two species, p53 and Mdm2, and four reactions. These reactions describe the process by which p53 and

Mdm2 self-regulate through coordinated synthesis and degradation. Specifically, p53 is constitutively synthesized but degraded in an Mdm2-dependent manner. Mdm2 is synthesized in a p53-dependent manner but constitutively degraded. An illustration of this network is given in Fig. 1, as are all steps used in the forthcoming algorithm.

A. Model the System of Interest Using Mass Action Kinetics

We further assume that the BRN used to describe our system can be modeled using mass action kinetics. Mass action assumes that the velocity of a chemical reaction – or the rate at which it converts reactants into products – is proportional to the product of each reactant raised to some power. This power is often equal to the stoichiometry of the reactant and is, therefore, simply one. Note too that when we refer to a species of a mass action model, we nearly always mean the abundance of that species. The resultant mathematical expression for the velocity of the reaction is often called a *rate equation*.

There are four reactions in the p53 model, which cumulatively describe the synthesis and degradation of p53 and Mdm2. Under mass action, the velocity of, say, p53 degradation is proportional to the product of p53 and Mdm2. Equivalently, we can say that the velocity of p53 degradation is *equal* to the product of p53, Mdm2, and a proportionality constant. This proportionality constant is commonly called a *rate constant*.

Once rate equations have been written for each chemical reaction, we apply the principle of mass balance to arrive at a set of governing equations that describes how every species behaves over time. This principle holds that the rate at which a species changes over time is equal to the sum of reaction velocities for which that species is produced, minus the sum of reaction velocities for which it is consumed. For example, p53 is produced by a zero-order synthesis reaction and consumed by a second-order degradation reaction. Consequently, we can write that the first derivative of p53 with respect to time is equal to the velocity of synthesis minus the velocity of degradation. In this way, application of mass action kinetics to any BRN yields a system of ordinary differential equations that describes the instantaneous rate of change for every species as a function of the reaction velocities.

Mass action may not always be appropriate to model the behavior of a BRN. A key assumption of the rate equation is that of spatial homogeneity. That is, there are no gradients in the concentration of any species and the local concentration at any point in space is equal to the global concentration (Grima and Schnell, 2008). This condition is violated when there are differences in the diffusivity of the species, due to either complex formation, tethering to subcellular structures, or compartmentalization (Kholodenko, 2009). Such systems are more appropriately modeled using reaction diffusion equations, reviewed in Slepchenko *et al.*, 2002. A second assumption of the rate equation is that the concentration of each participating species is sufficiently "large" (Sreenath *et al.*, 2008). If this is not the case, then random

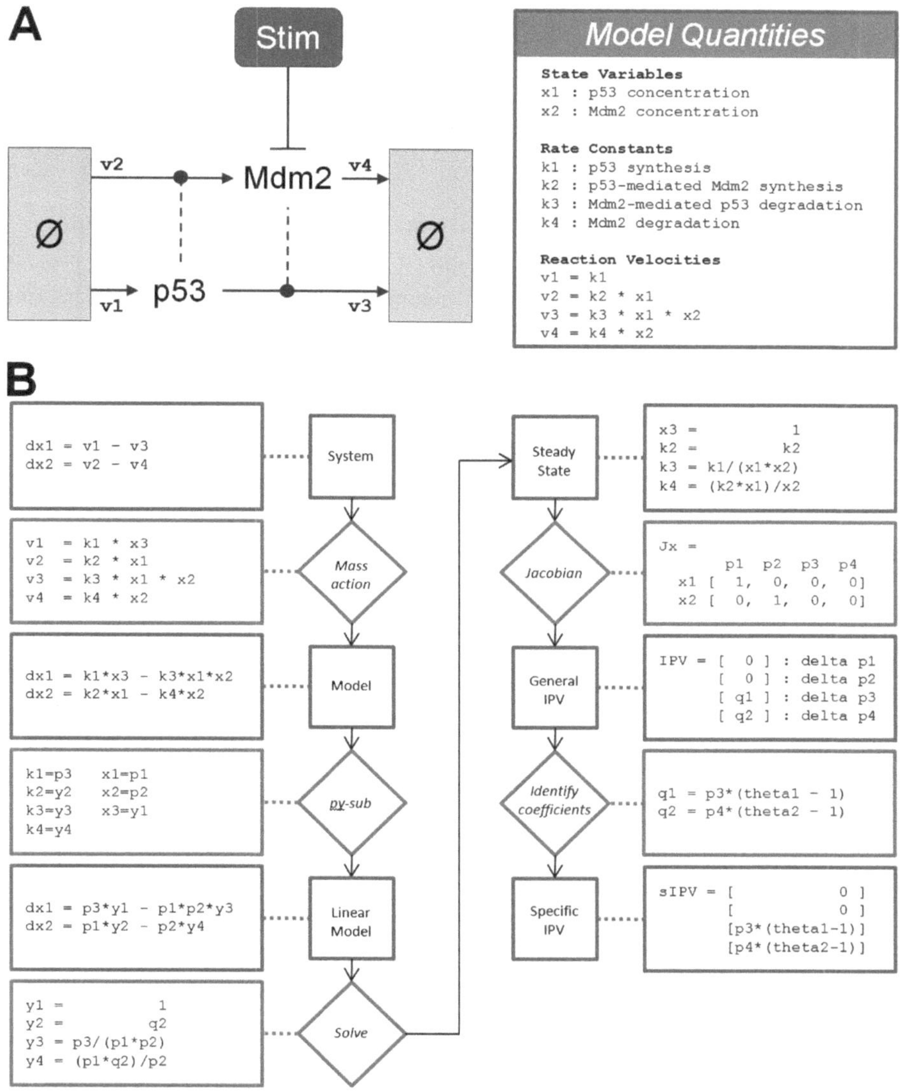

Fig. 1 Schematic of the simple model of p53 activation and regulation by Mdm2 used throughout this document. Below that, a flowchart of a mathematical framework for modeling the resting state. Diamonds represent major steps in the framework while boxes represent the outcome of those steps. The outside track illustrates the results of this framework when applied to the p53 model. (For color version of this figure, the reader is referred to the web version of this book.)

fluctuations can no longer be ignored and the reaction velocity must be modeled by a propensity function, called the chemical master equation (Gillespie, 1992). Extending the method presented here to systems where the assumptions of mass action fail is the subject of future research, and is discussed in Open Challenges.

B. Derive an Expression for the Steady-state of the Mass Action Model

For any mass action model there exists at least one set of reaction velocities where every species is being produced as quickly as it is being consumed. When this is the case, the model is said to be at *steady-state*. In this chapter, we equate steady-state with the resting state, but remark that a more sophisticated relationship between the two could be the subject of future work.

Every mass action model will have one or more trivial steady-states. These are steady-states in which all reaction velocities are zero. Closed systems, or systems that don't consider synthesis and degradation, always have a trivial steady-state in which every species' abundance is zero. Open systems also require that one or more synthesis rate be zero. An example of a trivial steady-state in the p53 model is one where there is neither p53 nor synthesis thereof. Since trivial steady-states are of little physiological interest, how might we identify nontrivial steady-states. More pointedly, in order to examine the relationship between steady-state and the dynamic response, how might all nontrivial steady-states be identified?

Mathematically, finding steady-states is equivalent to solving the system of equations that results when we set the rate of change of every species equal to zero. If the system happens to be linear in the variables of interest, then a solution can often be found quite easily. The key then is simply to find a subset of species and *rate constants* that may be treated as variables such that the resulting system of equations is linear. Ideally, the complement of that subset will be species and *rate constants* for which accurate measurements are available, since these are elements for which numerical values will need to be given prior to simulation. A detailed description of this process, which we call *py-substitution*, is given below.

1. Develop and Apply a py-substitution Strategy

From the set of all *rate constants* and species abundances, identify a substitution strategy by which elements with known values are replaced by a p and elements with unknown values are replaced by a y. We refer to these quantities as parameters and variables, respectively. Every substitution strategy must also satisfy the following conditions: (1) the resultant system of equations is linear in y, and (2) there are at least as many variables as there are linearly independent equations. The latter of these ensures that the py-substituted system of equations is not overdetermined.

Zero-order reaction velocities and velocities with exponents not equal to unity introduce a further complication: the former cannot be substituted by parameters nor the latter by variables. To do so violates the linearity constraint. If this constraint is undesirable, a work-around is to introduce a *pseudospecies*. For example, the velocity of p53 synthesis in our p53 model is independent of any species abundance. That is, the rate of synthesis is constant and equal to a single *rate constant*. If a reliable measurement exists for that *rate constant*, we may wish to substitute it with a parameter rather than a variable. But because doing so would violate the linearity requirement, we let the velocity be equal to the product of a first order *rate constant* and a pseudospecies. The latter of these we substitute normally with a variable and, once the system of steady-state equations has been solved, go back and make sure its value is unity.

A similar tactic can be used for reaction velocities that are superlinear in one of their reactants. If no reliable estimate exists for the abundance of the reactant, we may wish to substitute it with a variable rather than a parameter. Since doing so results in a superlinearity in y, we replace the reactant with a pseudospecies whose exponent is unity. The pseudospecies can then be substituted normally with a variable. After solving the system of steady-state equations, we go back and ensure that the steady-state expressions for the pseudospecies and the superlinear reactant are equal.

2. Solve the Linear System

After developing a py-substitution strategy, the system of steady-state equations is rendered linear in the variables. This allows us to rewrite the system using matrix notation. Specifically, we can write that the product formed by a matrix of parameters $\boldsymbol{C}$ with a column vector of variables $\boldsymbol{y}$ equals a column vector of zeros. We call this matrix of parameters the *coefficient matrix*:

$$\boldsymbol{Cy} = \boldsymbol{0} \tag{1}$$

The solution to this equation is precisely the null space of the coefficient matrix. Most modern mathematics software can derive a symbolic basis for the null space, so long as the matrix is not too large. If it is large (say, over 100 rows and columns, approximately equivalent to a system containing 100 species and reactions), then so too is the number of row operations needed to derive a basis. Since the elements in the matrix are symbolic, they can seldom can be reduced after each row operation. As a result, certain elements will grow geometrically in complexity and consume all the available RAM on the host device, causing a *de facto* arrest of the computation. Not all software packages handle this explosion equally well. In our experience, Maple outperforms both Mathematica and Matlab.

What is the benefit of a symbolic solution to the steady-state equation over a numerical one? With the latter, every independent parameter is a numeric value, which by the coefficient matrix is mapped efficiently to a value for each variable such that the system is at steady-state. The downside of this approach is that the contribution of each parameter to the variables is lost during the calculation. If the values of the independent parameters change, as is often required during the analysis of a mass action model, the values for the dependent variables must be calculated anew. With a symbolic solution, the contribution from each parameter to the steady-state expression of each variable is preserved. This has several advantages. (1) The relationship between a variable and an independent parameter can sometimes be identified directly from its steady-state expression. For example, the expression may reveal that a certain concentration scales linearly or nonlinearly with another species' concentration, or that the concentration does not depend at all on certain reaction rates. (2) More generally, the sensitivity of each dependent variable to each independent parameter can be calculated, so that, for example, changes in parameter values can be identified that only affect a certain subset of variables. This is precisely the approach we use below to selectively alter the steady-state turnover of p53 and Mdm2.

3. Derive a General Expression for the Vector of Variables

A basis for the null space of the coefficient matrix spans the solution to the steady-state equation. If we let the vector of variables be any linear combination of null space basis vectors, then the system will be at steady-state no matter what values we assign to the parameters. By any linear combination, we mean that the coefficient of each basis vector can be any real-valued number. If the basis vectors are arranged as columns in a matrix, this is equivalent to postmultiplying that matrix by a column vector of real-valued coefficients.

4. Resolve Any Constraints Imposed by Pseudospecies

Once a general expression is derived for the vector of variables, we must resolve any additional constraints imposed by the pseudospecies. Typically these will have the form $y_a = y_b^2$ in the case of a superlinearity, or $y_a = 1$ in the case of a sublinearity. The solution to these equations is not always straightforward, especially the former. Whichever mathematics software was employed to derive the null space for the coefficient matrix, however, can be used again here to solve the pseudospecies constraints.

Another complication that may arise during this step is that a superlinear constraint will yield two or more possible solutions. In theory, this presents a very interesting scenario where two or more values for the same species result in an otherwise identical steady-state. In other words, this may represent a bi- or multi-stability. In practice, our experience has been that when two solutions are possible, one of them is always negative and, therefore, physiologically infeasible. Furthermore, bistabilities reported in the literature typically manifest themselves in all of the species, not just one. Therefore, a practical resolution to this complication has been to keep both symbolic solutions but discard the infeasible one after numerical values are given to the parameters.

5. Reverse the py-substitution

Once the steady-state equation is solved and an expression derived for the vector of variables, one may wish to revert the substitution so that the relationships between variables and parameters are expressed in terms of species and *rate constants*. For simple systems, this can yield insight into how steady-state is achieved. For larger systems, these relationships can become intractable. Furthermore, for subsequent steps in this algorithm, the parametric description of steady-state can be the more useful of the two. For these reasons, reverting the py-substitution is an optional step.

If reversion is desired, note that a technical complication was introduced by the linear combination of null space basis vectors. Specifically, the forward py-substitution results in a linear system that is solvable but underdetermined. If this was not the case then the coefficient matrix would not have a null space. By taking a linear combination of the basis vectors, we effectively identify dimensions of the null space that are independent and thus need to be given a numerical value. In other words, the original py-substitution contained too many variables. A number of these variables equal to the dimension of the null space must become parameters. Fortunately, by scaling the basis vectors such that they are normalized with respect to the desired variable, we have a fair amount of freedom in specifying which variables are to become parameters.

Once this is done, we are left with an equation where the left hand side is the original vector of variables, and the right hand side is the product of the matrix of null space basis vectors and the vector of coefficients. Letting these elements be represented by $\boldsymbol{y}$, $\boldsymbol{N}$, and $\boldsymbol{q}$, respectively, the equation looks like the following.

$$\boldsymbol{y} = \boldsymbol{N}\boldsymbol{q} \tag{2}$$

It is precisely this equation that preserves the steady-state in our mass action model. The left hand side is within the domain of the inverse of the original py-substitution and can be reverted quite easily. The right hand side is a function of parameters and basis vector coefficients. The latter of these is not within the domain of the inverse py-substitution, so we must first convert these to variables. Fortunately, this conversion can be easily identified from the equation itself. By the derivation of the null space basis via row reduction of the coefficient matrix, there will exist at least one row in $\boldsymbol{N}$ for which only one column contains a nonzero entry. If the vector is scaled to this entry, then the row defines a one-to-one mapping between basis vector coefficients and variables. This mapping restores the right hand side to the domain of the inverse py-substitution, thus making the full reversion possible.

C. Identify the Isostatic Subspace of the Mass Action Model

Once we have derived an expression for the steady-state of our mass action model, we may wish to characterize the relationship between the dynamics of the system and its *rate constants* and steady-state abundances. The former of these is not straightforward, however, because changes in kinetic *rate constants* often result in changes to the abundances. To isolate the effects of changes in *rate constants* on system dynamics, we must derive an expression for the *isostatic subspace* of the model, that being the set of all parameter perturbations that do not in turn alter the steady-state species abundances.

1. Calculate the Jacobian Matrix of Partial Sensitivities in Abundances With Respect to Parameters

The first step in deriving the isostatic subspace is to define precisely what we mean by a perturbation in parameters that in turn does not alter the steady-state abundances. From our derivation of the steady-state above, we have that every species abundance is equal to some function of parameters and null space basis vector coefficients. It is convenient at this point to simply consider the latter of these as parameters as well. Some species equate one-to-one with a single parameter; other species are equal to complex expressions in the parameters. Either way, we are interested in a change in parameters $\Delta\boldsymbol{p}$ that, when added to the original set of parameters $\boldsymbol{p}$, results in a change in species $\Delta\boldsymbol{x}$ equal to zero. This is expressed succinctly by the following equation.

$$\Delta\boldsymbol{p} \in \{\Delta\boldsymbol{p} \neq 0 : \ \Delta\boldsymbol{x} = \boldsymbol{x}(\boldsymbol{p}) - \boldsymbol{x}(\boldsymbol{p} + \Delta\boldsymbol{p}) = 0\} \tag{3}$$

A valid change in parameters is thus any that satisfies

$$\boldsymbol{x}(\boldsymbol{p}) = \boldsymbol{x}(\boldsymbol{p} + \Delta\boldsymbol{p}) \tag{4}$$

The right hand side of this equation can be approximated by a truncated Taylor series, as follows,

$$\Delta\boldsymbol{x}(\boldsymbol{p} + \Delta\boldsymbol{p}) \approx \boldsymbol{x}(\boldsymbol{p}) + \boldsymbol{J}_x\Delta\boldsymbol{p} \tag{5}$$

where $\boldsymbol{J}_x$ is the Jacobian matrix whose elements are the partial derivatives of each species with respect to each parameter. The first step in deriving the isostatic subspace is, therefore, to calculate this matrix, which can be done efficiently using our choice of mathematics software.

2. Derive a Basis for the Null Space of the Jacobian Matrix

We are now confronted with the same situation as we were when deriving an expression for the steady-state. Since we want our new vector of species abundances to equal the old one, we require that

$$\boldsymbol{J}_x\Delta\boldsymbol{p} = \boldsymbol{0} \tag{6}$$

In other words, the change in parameters must reside within the null space of the Jacobian matrix. Equivalently, we call this particular null space the isostatic subspace, since it contains every perturbation in the parameters that does not affect the steady-state species abundances. A basis for this subspace can be derived as before.

3. Derive a General Isostatic Perturbation Vector

Every dimension in the isostatic subspace is a degree of freedom through which we can introduce an *isostatic perturbation*. A general expression for an isostatic perturbation then is simply the product of a matrix whose columns are the basis vectors of the isostatic subspace and a vector of basis vector coefficients. Notice how closely this step mirrors the derivation for an expression for the vector of variables, above. Once this multiplication is done, we are left with a general isostatic perturbation vector.

4. Derive a Specific Isostatic Perturbation Vector

Each degree of freedom in the general isostatic perturbation vector may introduce a perturbation that, in isolation, is of no physiological interest. For example, in the p53 model, we may be interested in introducing a perturbation that alters the rate of synthesis and degradation of p53. There is no guarantee that this perturbation exists as a single vector in our basis for the isostatic subspace. Therefore, the final step is to identify a specific combination of basis vectors to achieve the desired perturbation. In Section V, below, all of these steps are illustrated in detail as they are applied to our simple model of p53 activation.

III. Biological Insights

In this section, we illustrate some of the insights and hypotheses that can be generated from the steady-state and isostatic subspace of a mass action model. First, we show that statics and kinetics must cooperate to achieve steady-state. If an expression for the steady-state is known, then static parameters can be used to calculate the values for some, but not all, kinetic parameters. The fact that not all kinetic parameters can be calculated is related to the fact that the dynamic response to perturbation cannot be uniquely determined from static information alone. Using our simple model for the activation of the tumor suppressor p53, we show that the kinetics of homeostatic protein turnover determine the dynamic response of p53 to DNA damage.

A. Explicit Derivation of Kinetic *Rate Constants* From Static Measurements

A key motivation for the development of py-substitution was to calculate kinetic *rate constants* directly from static measurements (Fig. 2). For example, in the p53

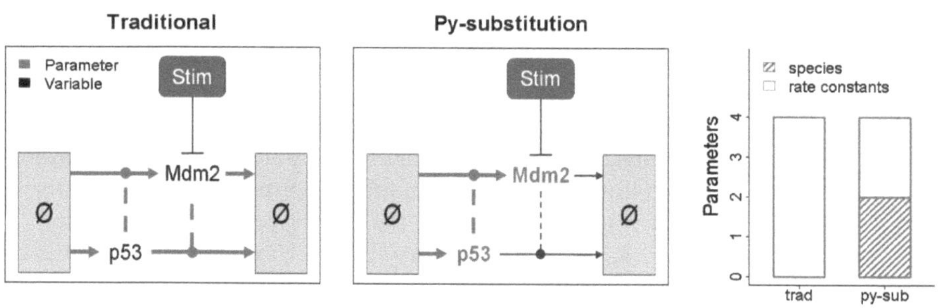

Fig. 2 A comparison of py-substitution versus a traditional parameterization strategy. A traditional strategy requires numeric values for all four *rate constants*. Using py-substitution, the steady-state abundances of p53 and Mdm2 can be given explicitly. In conjunction with the rates of synthesis of p53 and Mdm2, the degradation *rate constants* can be calculated such that steady-state is preserved. (For color version of this figure, the reader is referred to the web version of this book.)

model, values can be given for the steady-state abundances of p53 and Mdm2. Just the degradation rates of p53 and Mdm2 are then required to fully parameterize the model. The rates of synthesis can be calculated explicitly using these four parameters and the steady-state expression derived by py-substitution.

By comparison, a traditional parameterization strategy would require that all four kinetic *rate constants* be specified. The steady-state abundances of p53 and Mdm2 could then be derived numerically by integrating the model to steady-state, but this process is comparatively less efficient. Furthermore, the steady-state behavior of p53 and Mdm2 over a range of synthesis and degradation rates is unknowable except through simulation. If estimates for the steady-state abundances of p53 and Mdm2 exist, then a parameter fitting procedure must be used to infer the optimal values for the kinetic *rate constants*. This is an example of a "backward problem," in that the "forward problem" – calculating the steady-state abundances of p53 and Mdm2 given a set of four kinetic *rate constants* – must be iteratively solved until an optimal set of *rate constants* is identified. If, however, an expression for the steady-state is known, this backward problem is turned into a forward problem: given the steady-state abundances for p53 and Mdm2 and their rates of degradation, a simple calculation gives the rates of synthesis required to support that steady-state.

The significance of this difference is that making kinetic measurements can be a considerable technical challenge. Typically kinetic parameters must be determined with purified proteins using *in vitro* assays (Nutiu *et al.*, 2011; Tanious *et al.*, 2008) or must be derived from biochemical assays requiring millions of cells (Schwanhäusser *et al.*, 2011). By contrast, static measurements are often more sensitive and can be performed using fixed cells (Itzkovitz and van Oudenaarden, 2011; Jain *et al.*, 2011). As a result, measuring static variables is easier and more accurate than measuring kinetic ones, and py-substitution allows kinetic *rate constants* to be derived explicitly from simpler, static measurements.

B. Static Control of the Dynamic Response

Another benefit of py-substitution is that we can systematically evaluate the relationship between dynamic responsiveness and steady-state abundances. This is made possible by the fact that py-substitution allows steady-state abundances to be treated as independent parameters. For example, the dynamic response of p53 to DNA damage is affected by the steady-state abundance of Mdm2. Because we have modeled this abundance as an input parameter, it is straightforward to vary it over a range of values and simulate the p53 response at each value.

In Fig. 3, we let Mdm2 vary from 0.1 to 10 times its nominal wildtype value. As Mdm2 increases, p53 exhibits a faster and stronger dynamic response. As it decreases, p53 becomes slower and weaker. This is because the rate of p53 degradation scales with the steady-state abundance of Mdm2. As the latter increases, so does the former. Since we have not varied the steady-state abundance of p53 but rather kept it fixed, the rate of p53 synthesis must also scale with the abundance of Mdm2. In other words, a higher steady-state abundance of Mdm2 results in a higher

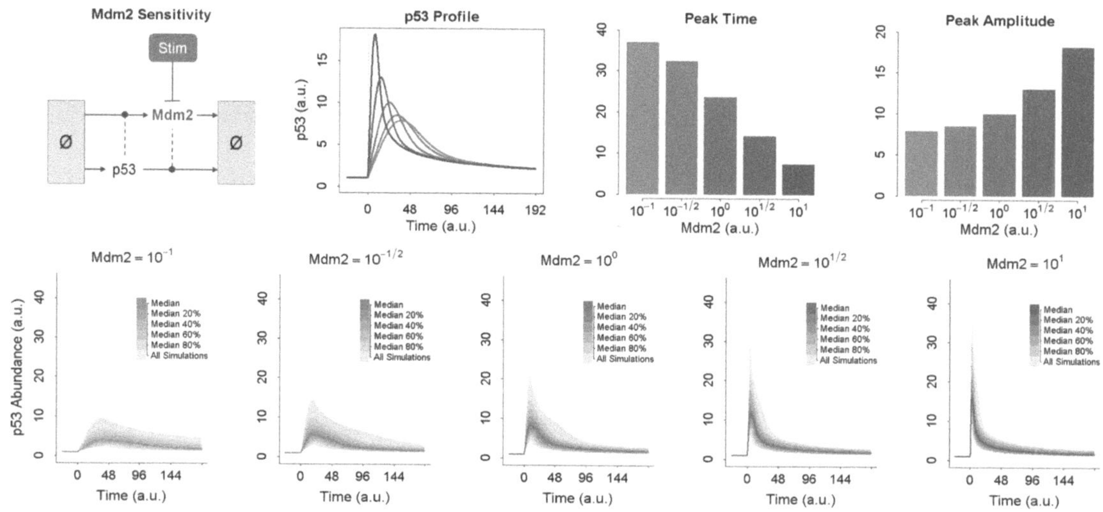

Fig. 3 The effect of the steady-state abundance of Mdm2 on the dynamic response of p53 to stimulation. At top, the steady-state abundance of Mdm2 is varied from 0.1 (light gray) to 10 (dark gray). The result of this variation on the time and amplitude of the p53 response are shown as bar graphs on the right. At bottom, Mdm2 is again varied from 0.1 to 10. Each of the five panels represents a distinct but constant abundance of Mdm2. The abundance of p53 is always 1. The rates of p53 and Mdm2 synthesis and degradation are then allowed to take a random value from a uniform distribution over 0.1 to 10 times their nominal wildtype values. The p53 response to perturbation is simulated for 1000 samples in each panel and the median dynamics plotted. (For color version of this figure, the reader is referred to the web version of this book.)

steady-state turnover of p53. The velocity of this turnover partially dictates the dynamics of the p53 response. Homeostatic turnover will be examined in more detail in the next subsection.

Interestingly, even though steady-state abundances affect the dynamic response, they do not uniquely determine it. Put another way, the dynamic response to perturbation is underdetermined with respect to the steady-state abundances. This is illustrated in Fig. 3, bottom. Here, each panel depicts the median behavior of 1000 simulations of the p53 model. For a given panel, every simulation has the same steady-state abundance of Mdm2 and p53. The rates of homeostatic turnover of Mdm2 and p53, however, are allowed to take a uniform random value between 0.1 and 10 times their nominal wildtype value. We say that these simulations are isostatic but *anisokinetic* – their steady-state abundances are identical but their kinetic *rate constants* are not. This variability in the kinetics causes variability in the dynamics, but is entirely opaque with respect to the steady-state abundances.

C. Kinetic Control of the Dynamic Response

As Fig. 3 shows, isostatic systems can exhibit significant variability in their response to perturbation. This is a consequence of the fact that the steady-state of a mass action model is degenerate with respect to its kinetics; an infinite number of kinetic *rate constants* can support the same set of steady-state abundances. We call a change in kinetic *rate constants* that does not affect any steady-state abundances an isostatic perturbation to the parameters, or an isostatic perturbation for short.

Special cases of isostatic perturbations are those that simultaneously alter the homeostatic rates of synthesis and degradation – or *flux* – of a particular species. Above we saw that changing the steady-state abundance of Mdm2 altered the flux of p53 and thereby its dynamic responsiveness. However, we can alter the flux of p53 without altering the steady-state abundance of Mdm2 as well. This is shown in Fig. 4. Similar to the subsection above, increasing the flux of p53 results in a faster, stronger response. Decreasing the p53 flux results in a slower, weaker response, and to a greater degree than observed when changing Mdm2 alone.

In addition to p53, we can alter the homeostatic flux of the negative regulator, Mdm2. This is shown in Fig. 4, bottom. In contrast to p53, increasing the flux of Mdm2 results in a faster but weaker p53 response. This result highlights the fact that while the homeostatic flux of a species within a biochemical reaction or gene regulatory network can affect the dynamic response to perturbation, the precise nature of the effect depends on the function of that species within the network.

D. Precise Control of the Dynamic Response by Homeostatic Flux

The distinct effects of homeostatic p53 versus Mdm2 flux on the dynamic response of p53 raise the possibility that these fluxes can be used to precisely control the shape of the p53 trajectory. Using the time and amplitude of the peak of the p53 trajectory as descriptors of the shape, we can look for isostatic perturbations that

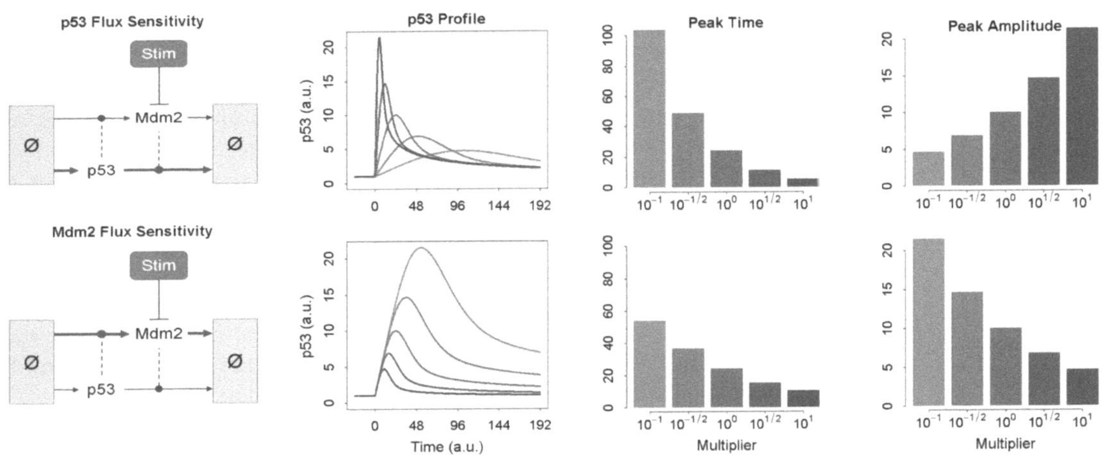

Fig. 4 The effect of p53 and Mdm2 flux on the dynamic response of p53 to stimulation. At top, the flux of p53 is varied from 0.1 (light gray) to 10 (dark gray) times its nominal wildtype value. At bottom, the flux of Mdm2 is varied from 0.1 (light gray) to 10 (dark gray) times its wildtype vale. The effects of each on the time and amplitude of the p53 response are shown as bar graphs on the right. (For color version of this figure, the reader is referred to the web version of this book.)

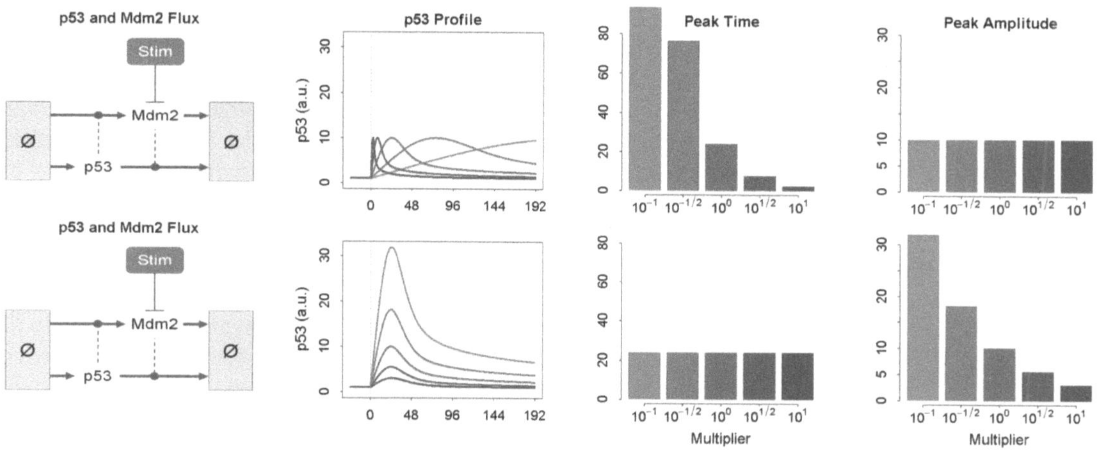

Fig. 5 Precise tuning of the dynamic response of p53 to stimulation by homeostatic flux. At top, the flux of p53 is varied from 0.1 (light gray) to 10 (dark gray) times its nominal wildtype value, while the flux of Mdm2 is varied from 10 (light gray) to 0.1 (times its wildtype value). The result of this modulation is that amplitude of the p53 response is held constant. At bottom, the flux of Mdm2 is varied from 0.1 (light gray) to 10 (dark gray) times its wildtype value. A modification to the flux of p53 is then derived numerically such that the time of the p53 response is held constant. (For color version of this figure, the reader is referred to the web version of this book.)

affect the flux of both p53 and Mdm2 such that the peak time is altered independently of the amplitude, or the amplitude independently of the time.

In Fig. 5, we see that this is indeed possible. In fact, in Fig. 4 we can see that the p53 and Mdm2 fluxes have an equal but opposite effect on the peak amplitude. This suggests that an isostatic perturbation that pairs an increase in one flux with an equal but opposite decrease in the other will preserve the amplitude of p53. This is shown to be the case in Fig. 5 top. Since this same phenomenon is not manifested in the p53 peak time, it is less straightforward to derive the desired isostatic perturbation. However, given a particular change in Mdm2 flux, we can indeed find a change in p53 flux such that the p53 peak time is preserved (Fig. 5, bottom). Together, these results demonstrate that the dynamic response of p53 can be precisely controlled by homeostatic flux, independently of the steady-state abundances of either p53 or Mdm2.

IV. Open Challenges

Because the assumptions of spatial homogeneity and high concentrations remain prevalent in the systems biology and modeling literature, we believe there is ample opportunity to use py-substitution to generate novel hypotheses about the impact of steady-state on stimulus responsiveness. Nevertheless, even within a mass action framework there are limitations to the method as described here. Chief among these is that of model size. Deriving symbolic bases for the solution space to the steady-state equation and isostatic subspace of a large model can yield elements with hundreds and sometimes thousands of terms. An attractive solution to this problem would be *a priori* identification of network modules (Bowsher, 2011; Hartwell *et al.*, 1999). In the ideal case, this would result in block diagonal coefficient and Jacobian matrices. Since each block can be treated independently, the algebraic complexity of the resultant basis vectors would be much more manageable. Identifying modules would also offer the benefit of allowing some species to be in disequilibrium, as the case might be when a signaling network experiencing ambient, tonic signaling is coupled to a periodic oscillator such as the cell cycle.

For systems in which the assumptions for mass action are not supported, some work will have to be done to extend the py-substitution framework. For spatially heterogeneous systems, the mass balance equations include a diffusion term in addition to the standard mass action rate equations. It remains to be shown whether such a system of equations can be linearized in the same manner as described here. If indeed it can, this could lead to new insights regarding the interplay between reaction kinetics and diffusivity in establishing spatial gradients and responding to spatially heterogeneous signals. When the assumption of high concentrations is violated and a system loses its deterministic behavior, the inference of kinetic parameters from steady-state concentrations or dynamic response measurements becomes a probabilistic one. Additional work will be done to extend py-substitution to these stochastic systems.

More generally, the class of models that can be addressed using py-substitution remains to be determined. Are their structural motifs within a BRN that are particularly challenging to linearize? Can more exotic reaction rate equations be entertained, notably Michaelis–Menten kinetics and Hill functions? Precisely defining the domain of py-substitution will not only guide its further development, but perhaps also dissuade the use of exotic reaction kinetics to achieve a certain dynamical behavior, at the expense of a knowable steady-state.

V. Computational Methods

In this section, we give step-by-step instructions for identifying the steady-state of the p53 model. Although the size of this model makes it unnecessary to employ the rigorous treatment described here, that the results can be reproduced by hand makes the steps tractable and easy to follow. For information on how the method scales to larger models, see Loriaux *et al.*, 2012. Once we identify a solution to the steady-state of the p53 model, we show how to derive a basis for its isostatic subspace. Finally, from the isostatic subspace we show how to derive specific isostatic perturbation vectors for modifying the homeostatic flux of p53 and Mdm2.

All of the steps below are performed using Matlab. As noted earlier, Matlab is not the best choice of software for symbolic calculations, but because it enjoys the most familiarity, we use it here for clarity. In the passages that follow, commands are identified by a double arrow prompt while output from the Matlab terminal is identified by a boldface font. Finally, it should be noted that the following code is in no way optimal; a more efficient implementation would make use of matrices, but again this efficiency comes at the expense of clarity.

A. Identifying an Expression for the Steady-state of a Mass Action Model

The p53 model consists of two species and four reactions. These must all be declared as symbolic variables using the `syms` keyword. Following convention, we use `x` to denote species, `v` for reaction velocities, and `k` for reaction *rate constants*. The `real` keyword identifies these variables as being real-valued. The semi-colon suppresses Matlab output.

```
syms  x1  x2          real;

syms  k1  k2  k3  k4  real;

syms  v1  v2  v3  v4  real;
```

By mass balance, we let the rate of change of each species be equal to the sum of reaction velocities in which that species is produced minus the sum of reaction velocities in which it is consumed. This yields the following.

```
dx1 = v1 - v3;

dx2 = v2 - v4;
```

Assuming mass action, we let the velocity of each reaction be equal to the product of its reactants and the corresponding *rate constant*.

```
v1  = k1;

v2  = k2 * x1;

v3  = k3 * x1 * x2;

v4  = k4 * x2;
```

Substituting in the reaction velocities yields a system of mass balance equations expressed in terms of species and *rate constants*.

```
>> dx1 = subs(dx1)

   dx1 =

k1 - k3*x1*x2

>> dx2 = subs(dx2)

   dx2 =

k2*x1 - k4*x2
```

We must now linearize the system by imposing a py-substitution strategy. Even for a model of this size, several strategies exist. Here we'll implement a strategy that assumes we have accurate measurements for the abundances of p53 and Mdm2 and the rate of p53 synthesis. The degradation *rate constants* and rate of Mdm2 synthesis will be left variable. Note that substituting for the rate of p53 synthesis requires the use of a pseudospecies, x3, which we introduce now.

```
>> syms  x3  real;

>> dx1 = subs(dx1, k1, k1*x3)

   dx1 =

k1*x3 - k3*x1*x2
```

As before, we must declare all symbolic parameters and variables prior to substitution. Once a strategy is defined, we can use the same `subs` command to generate the py-substituted mass balance equations.

```
>> syms  p1  p2  p3  p4  real;

>> syms  y1  y2  y3  y4  real;

>>

>>  kx = [k1,k2,k3,k4,x1,x2.x3];

>>  py = [p3,y2,y3,y4,p1,p2,y1];

>>

>> dx1 = subs( dx1, kx, py )

   dx1 =

p3*y1 - p1*p2*y3

>> dx2 = subs( dx2, kx, py )

   dx2 =


 p1*y2 - p2*y4
```

As expected, py-substitution results in a linear system of mass balance equations. As such, we can express it as the product between a coefficient matrix of parameters and a vector of variables. To derive the coefficient matrix, we use the `jacobian` command.

```
>> dx = [dx1,dx2];

>> y  = [y1,y2,y3,y4];

>> C  = jacobian( dx, y )

   C =

[ p3,  0, -p1*p2,   0]

[  0, p1,      0, -p2]
```

Now we'd like to find all vectors that, when left-multiplied by the coefficient matrix, equal zero. In other words, the vector must be in the null space of the coefficient matrix. To ensure this is true, we need to find a basis for the null space. This can be done using the `null` command. We'll store the results of this operation in a second matrix, N.

```
>> N = null(C)

   N =

[ (p1*p2)/p3,     0]

[          0, p2/p1]

[          1,     0]

[          0,     1]
```

We now let the vector of variables equal any linear combination of column vectors in N. Because N has two columns, we'll need two additional parameters, q1 and q2. These will be the coefficients of the null space basis vectors.

```
>> syms  q1  q2  real;

>> y = N*[q1;q2]

   y =

(p1*p2*q1)/p3

   (p2*q2)/p1

           q1

           q2
```

Next, we must resolve the pseudospecies such that its value is one. Since y1 is the variable that corresponds to the pseudospecies x3, this means we must find a value for q1 such that y1=1. Also note that the coefficient q2 maps to y4. This indicates that the *rate constant* k4 must in fact be a parameter. Here we will assume that this is not desirable, and that we would prefer to let the rate of Mdm2 synthesis be a parameter instead. To do this, we scale the second null space vector N(:,2) by a factor a2 such that N(2,:)*[q1;a2* q2] =q2.

```
>> syms a2 real;
>> eq1      = strcat( char(N(1,:)*[q1;a2*q2]), '=1' );
>> eq2      = strcat( char(N(2,:)*[q1:a2*q2]), '=q2' );
>> [a2,q1] = solve( eq1, eq2, 'q1,a2' )
        a2 =
        p1/p2
        q1 =
        p3/(p1*p2)
```

The final expression for the vector of variables is as follows.

```
>> y = simplify(subs(N*[q1;a2*q2]))
   y =
            1
           q2
   p3/(p1*p2)
   (p1*q2)/p2
```

To be prudent, we verify that this vector is in the null space of the coefficient matrix.

```
>> simplify(C*y)
ans =
  0
  0
  0
  0
```

Finally, we may wish to revert the substitution so that our steady-state expression is in terms of species and *rate constants*. Notice that the linear combination `[ q1;q2]` effectively identifies variables that, because the coefficient matrix was underdetermined, turn out to be parameters. These variables map one-to-one with null space basis vector coefficients. Thus in our steady-state expression for `y`, to the left hand side we simply reverse the substitution from `py` back to `kx`. To the right hand side we first perform the one-to-one substitution from `q` to `y`, then the reverse substitution from `py` to `kx`.

```
>> lhs = subs([y1;y2;y3;y4],py,kx);

>> rhs = subs(subs(y,q2,y2),py,kx);

>> [ lhs rhs ]

   ans =

[ x3,           1]

[ k2,          k2]

[ k3, k1/(x1*x2)]

[ k4, (k2*x1)/x2]
```

The result of the inverse substitution is a relationship between dependent and independent species and *rate constants* that, if satisfied, guarantees steady-state. Note that this relationship is particular to our choice of py-substitution strategies and null space basis vector coefficients. As illustrated above, by scaling the appropriate basis vector, we were able to choose which variables remain dependent. Finally, it is worth verifying that our solution for y does indeed guarantee steady-state.

```
>> subs( subs(dx',py,kx), lhs, rhs )

ans =

 0

 0
```

B. Identifying the Isostatic Subspace of a Mass Action Model

As illustrated in Figs. 2–4, there are many advantages to having an analytical expression for the steady-state of a mass action model: (a) static measurements of species abundances can be used to calculate kinetic *rate constants*, (b) the total number of parameters required is often reduced (Loriaux *et al.*, 2012), and perhaps most importantly, (c) we can characterize the relationship between dynamic responsiveness and the abundances of species at steady-state. However, as seen in Fig. 3, steady-state abundances do not uniquely determine the dynamic response; the kinetics of the system are also important.

To study the effects of kinetics on dynamic responsiveness in isolation, we would like to identify any and all changes that can be made to the kinetic *rate constants* that do not alter the steady-state species abundances. The set of all such changes is called the isostatic subspace. To identify this subspace, it is first easier to return to the parametric description of the steady-state. At this point, we'll also replace the null space basis vector coefficient `q2` with the parameter, `p4`.

```
>> kxp   = [x1,x2,k1,k2];
>> p     = [p1,p2,p3,p4];
>>
>> kxss = subs( subs(kx',lhs,rhs), kxp, p );
>> kss   = kxss(1:4)
    kss =
          p3
          p4
  p3/(p1*p2)
  (p1*p4)/p2
>> xss   = kxss(5:6)
 xss =
  p1
  p2
```

As expected, every element in our model is a function of the four parameters used in the py-substitution strategy. Now recall that a Taylor expansion can be used to approximate how these elements change in response to changes in parameters. The first order term of this expansion requires a matrix of partial derivatives of each element with respect to each parameter. This matrix is also called the Jacobian matrix, and can be calculated in Matlab using the `jacobian` command.

```
>> Jx = jacobian( xss, p )
  Jx =
[ 1, 0, 0, 0]
[ 0, 1, 0, 0]
```

In the Jacobian Jx, the rows correspond to steady-state species abundances and the columns to parameters. Element Jx(1,1)=1 indicates that a change in parameter p1 results in an equal change in species x1. This of course is not surprising since our py-substitution strategy had that x1=p1. It is also not surprising that Jx(1,2)=Jx(1,3)=Jx(1,4)=0; the species x1 doesn't depend on any other parameter. The second row of Jx has a similar structure; the species x2 depends only on the parameter p2. This Jacobian matrix is extraordinarily simple because both steady species abundances were modeled as independent parameters. In general, there will be species whose steady-state abundances are variable expressions of the parameters, and this significantly complicates the Jacobian.

It now remains to identify the set of all vectors that, when left multiplied by Jx, result in zero. By our Taylor expansion, any such vector identifies a change in parameters that results in no changes to the species abundances. By the same argument as above, any such vector must be in the null space of the Jacobian, and as before, a basis for this null space is found using the null command.

```
>> Jx = jacobian( xss, p )

   Jx =

[ 1, 0, 0, 0]

[ 0, 1, 0, 0]
```

```
>> Nx = null( Jx )

   Nx =

[ 0, 0]

[ 0, 0]

[ 1, 0]

[ 0, 1]
```

In this matrix, each row corresponds to a parameter and every column to a degree of freedom in the null space. That the first two rows are comprised entirely of zeros indicates that we can alter neither p1 nor p2 without altering at least one steady-state abundance. Again, this is not surprising since x1=p1 and x2=p2. The bottom two rows indicate that we can alter either p3 or p4 independently of one another. This too is not surprising; neither p3 nor p4 appear in the steady-state expressions for x1 and x2. As with the Jacobian matrix, the null space basis will typically have a more complicated expression.

To derive a general expression for the isostatic subspace of our p53 model, we take a linear combination of the null space basis vectors. The null space is two-dimensional so two coefficients are required, q1 and q2. And because we already used these variables in the previous subsection, we'll clear them prior to using them again.

```
>> clear  q1  q2;

>> syms   q1  q2  real;


>> isox = Nx*[q1;q2]

isox =

0

0

q1

q2
```

The vector `isox` states simply that we may make any change to the parameters `p3` and `p4` without altering the steady-state abundances `x1` and `x2`. To verify that this is true, we map the parameter perturbation `isox` into a species abundance perturbation `delx` using the Jacobian, `Jx`. As expected, the parameter perturbation resides in the null space of the Jacobian, indicating that the perturbation causes no change in species abundances.

```
>> delx = Jx*isox

delx =

0

0
```

How does the general isostatic perturbation `isox` affect the *rate constants* of our model? And how do we identify a specific perturbation such that only certain *rate constants* are altered? To calculate the effect of the general perturbation `isox` on the set of *rate constants* we use the same procedure as above, but using the Jacobian matrix of *rate constants* with respect to the parameters instead of species abundances.

```
>> Jk = jacobian( kss, p )

Jk =

[              0,              0,          1,     0]

[              0,              0,          0,     1]

[ -p3/(p1^2*p2), -p3/(p1*p2^2), 1/(p1*p2),     0]

[         p4/p2, -(p1*p4)/p2^2,          0, p1/p2]
```

As we saw with `Jx`, the first two rows indicate that changes to `p3` and `p4` result in equivalent changes to `k1` and `k2`, respectively. This simply reflects the fact that `k1=p3` and `k2=p4`, and that our py-substitution strategy was designed to make the rates of synthesis of p53 and Mdm2 independent parameters. The degradation *rate constants* `k3` and `k4` are variable and constrained by steady-state, and are thus each sensitive to changes in three out of four parameters. The product of `Jx` and `isox` maps this perturbation into a change in *rate constants*.

```
>> delk = Jk*isox

delk =

          q1

          q2

q1/(p1*p2)

(p1*q2)/p2
```

As we observed in the Jacobian, a change `q1` in parameter `p3` results in an equivalent change in the *rate constant* `k1`. A change `q2` in parameter `p4` results in an equivalent change in `k2`. The resultant changes in the degradation *rate constants*, however, are scaled by the species abundances `p1` and `p2`. We can calculate the vector of *rate constants* that results from the perturbation `isox` by executing the first sum in the Taylor expansion.

```
>> kprime = kss + delk

kprime =

                         p3 + q1

                         p4 + q2

p3/(p1*p2) + q1/(p1*p2)

(p1*p4)/p2 + (p1*q2)/p2
```

Finally, what if we are interested in not just any isostatic perturbation but a specific one? In Fig. 5, we saw that the homeostatic flux of p53 and Mdm2 can precisely control the dynamic response of p53 to DNA damage. Altering the homeostatic flux is just a special case of the general perturbation `isox`. We need only find values for `q1` and `q2` such that the *rate constants* `k1` and `k3` and `k2` and `k4` take on values theta1 and theta2 times their nominal wildtype values, respectively. To do this, we first declare the symbolic variables theta1 and theta2. We then express our desired outcome as a system of equations. Specifically, letting k1prime and k2prime be the altered values of k1 and k2, the ratio of the k1 prime to k1 should be theta1, and the ratio of k2 prime to k2 should be theta2. Once expressed as such, we can solve for the requisite values of q1 and q2.

```
>> syms  theta1  theta2  real;

>> eq1 = strcat( char(kprime(1)/kss(1)), '=theta1' );

>> eq2 = strcat( char(kprime(2)/kss(2)), '=theta2' );

>> [q1sub,q2sub] = solve( eq1, eq2, 'q1,q2' )

 q1sub =

p3*(theta1 - 1)

 q2sub =

 p4*(theta2 - 1)
```

Substituting these values into the general isostatic perturbation isox results in the desired, specific isostatic perturbation that scales the homeostatic flux of p53 and Mdm2.

```
>> ipv = subs( isox, [q1,q2], [q1sub,q2sub] )

  ipv =

                 0

                 0

 p3*(theta1 - 1)

 p4*(theta2 - 1)
```

Finally, it remains to verify that this perturbation results in the desired change. Again this is done by executing the first sum in the Taylor expansion.

```
>> kprime = simplify( kss + Jk*ipv )

  kprime =

          p3*theta1

          p4*theta2

 (p3*theta1)/(p1*p2)

  (p1*p4*theta2)/p2

>> simplify( kprime./kss )

     ans =

 theta1

 theta2

 theta1

 theta2
```

In summary, from the parametric expression for the steady-state of our model, we have identified a specific isostatic perturbation that alters the homeostatic flux of either or both p53 and Mdm2 to the user-specified parameters `theta1` and `theta2`, respectively.

VI. Further Reading

Another method for deriving expressions for the steady-states of mass action models was introduced by King and Altman in 1956 (King and Altman, 1956). This graphical method was greatly improved upon in Volkenstein and Goldstein (1966) and again in Thomson and Gunawardena (2009), and enjoys a robust and sophisticated implementation in Matlab (Qi *et al.*, 2009).

The application of linear algebra to dynamical networks has a similarly rich history, especially as it pertains to flux balance analysis (Gianchandani *et al.*, 2010) and systems biology (Palsson, 2006). For a deeper understanding of the relevant concepts in linear algebra, see Poole (2010) and Cooperstein (2010).

Evaluating the effects of perturbations on network dynamics and steady-state has long been the subject of metabolic control analysis, or MCA (Heinrich and Rapoport, 1974; Fell, 2005). Succinctly, MCA can be used to quantify the steady-state change in a reaction velocity or species concentration due to a change in an independent parameter. Recently this framework was extended to dynamical states as well (Ingalls and Sauro, 2003). For an excellent review of quantitative modeling of network dynamics, see Sauro (2009).

References

Bartlett, D. W., and Davis, M. E. (2006). Insights into the kinetics of siRNA-mediated gene silencing from live-cell and live-animal bioluminescent imaging. *Nucleic Acids Res.* **34**, 322–333.

Becker, V., Schilling, M., Bachmann, J., Baumann,Raue, A., Maiwald, T., Timmer, J., and Klingmüller, U. (2010). Covering a broad dynamic range: information processing at the erythropoietin receptor. *Science* **328**, 1404–1408.

Bowsher, C. G. (2011). Automated analysis of information processing, kinetic independence and modular architecture in biochemical networks using MIDIA. *Bioinformatics* **27**, 584–586.

Cooperstein, B. (2010). Advanced Linear AlgebraVol. 1, CRC Press, p. 364.

Dolloff, N. G., Mayes, P. A., Hart, L. S., Dicker, D. T., Humphreys, R., and El-Deiry, W. S. (2011). Off-target lapatinib activity sensitizes colon cancer cells through TRAIL death receptor up-regulation. *Sci. Transl. Med.* **3**, 86ra50.

Faller, D., Klingmuller, U., and Timmer, J. (2003). Simulation methods for optimal experimental design in systems biology. *Simulation* **79**, 717–725.

Fell, D. (2005). Metabolic control analysis. *In* "Systems Biology," (L. Alberghina, and H. V. Westerhoff, eds.), pp. 397–424. Springer, Berlin/Heidelberg.

Fire, A., Xu, S., Montgomery, M. K., Kostas, S. A., Driver, S. E., and Mello, C. C. (1998). Potent and specific genetic interference by double-stranded RNA in Caenorhabditis elegans. *Nature* **391**, 806–811.

Fulda, S., Küfer, M. U., Meyer, E., van Valen, F., Dockhorn-Dworniczak, B., and Debatin, K. M. (2001). Sensitization for death receptor- or drug-induced apoptosis by re-expression of caspase-8 through demethylation or gene transfer. *Oncogene* **20**, 5865–5877.

Fulda, S., Meyer, E., and Debatin, K. (2002). Inhibition of TRAIL-induced apoptosis by Bcl-2 over-expression. *Oncogene* **21**, 2283–2294.

Gianchandani, E. P., Chavali, A. K., and Papin, J. A. (2010). The application of flux balance analysis in systems biology. *Wiley Interdiscip Rev Syst Biol Med* **2**, 372–382.

Gillespie, D. (1992). A rigorous derivation of the chemical master equation. *Physica A* **188**, 404–425.

Grima, R., and Schnell, S. (2008). Modeling reaction kinetics inside cells. *Essays Biochem.* **45**, 41–56.

Hartwell, L. H., Hopfield, J. J., Leibler, S., and Murray, A. W. (1999). From molecular to modular cell biology. *Nature* **402**, C47–C52.

Heinrich, R., and Rapoport, T. A. (1974). A linear steady-state treatment of enzymatic chains. Critique of the crossover theorem and a general procedure to identify interaction sites with an effector. *Eur. J. Biochem.* **42**, 97–105.

Hinz, S., Trauzold, A., Boenicke, L., Sandberg, C., Beckmann, S., Bayer, E., Walczak, H., Kalthoff, H., and Ungefroren, H. (2000). Bcl-XL protects pancreatic adenocarcinoma cells against CD95- and TRAIL-receptor-mediated apoptosis. *Oncogene* **19**, 5477–5486.

Ingalls, B. P., and Sauro, H. M. (2003). Sensitivity analysis of stoichiometric networks: an extension of metabolic control analysis to non-steady state trajectories. *J. Theor. Biol.* **222**, 23–36.

Itzkovitz, S., and van Oudenaarden, A. (2011). Validating transcripts with probes and imaging technology. *Nat. Methods* **8**, S12–S19.

Izant, J. G., and Weintraub, H. (1984). Inhibition of thymidine kinase gene expression by anti-sense RNA: a molecular approach to genetic analysis. *Cell* **36**, 1007–1015.

Jain, A., Liu, R., Ramani, B., Arauz, E., Ishitsuka, Y., Ragunathan, K., Park, J., Chen, J., Xiang, Y. K., and Ha, T. (2011). Probing cellular protein complexes using single-molecule pull-down. *Nature* **473**, 484–488.

Kearns, J. D., and Hoffmann, A. (2009). Integrating computational and biochemical studies to explore mechanisms in NF-κB signaling. *J. Biol. Chem.* **284**, 5439–5443.

Kholodenko, B. N. (2009). Spatially distributed cell signalling. *FEBS Lett.* **583**, 4006–4012.

Kim, J. -Y., Lee, J. -Y., Kim, D. -G., Koo, G. -B., Yu, J. -W., and Kim, Y. -S. (2011). TRADD is critical for resistance to TRAIL-induced cell death through NF-κB activation. *FEBS Lett.* **585**, 2144–2150.

King, E. L., and Altman, C. (1956). A schematic method of deriving the rate laws for enzyme-catalyzed reactions. *J. Phys. Chem.* **60**, 1375–1378.

Kitano, H. (2002). Systems biology: a brief overview. *Science* **295**, 1662–1664.

Li, H., Cao, Y., Petzold, L. R., and Gillespie, D. T. (2011). Algorithms and software for stochastic simulation of biochemical reacting systems. *Biotechnol. Prog.* **24**, 56–61.

Loriaux, P. M., Tesler, G., and Hoffmann, H. (2012). An algebraic method for deriving analytical expressions for steady states of mass action models. Submitted.

Nutiu, R., Friedman, R. C., Luo, S., Khrebtukova, I., Silva, D., Li, R., Zhang, L., Schroth, G. P., and Burge, C. B. (2011). Direct measurement of DNA affinity landscapes on a high-throughput sequencing instrument. *Nat. Biotechnol.* **29**, 659–664.

O'Dea, E. L., Kearns, J. D., and Hoffmann, A. (2008). UV as an amplifier rather than inducer of NF-kappaB activity. *Mol. Cell.* **30**, 632–641.

Palsson, B. Ø. (2006). *Systems Biology: Properties of Reconstructed Networks.* Cambridge University Press, 334 p.

Poole, D. (2010). Linear Algebra: A Modern Introduction Vol. 3, Brooks Cole, 768p.

Qi, F., Dash, R. K., Han, Y., and Beard, D. A. (2009). Generating rate equations for complex enzyme systems by a computer-assisted systematic method. *BMC Bioinformatics* **10**, 238.

Reynolds, A., Anderson, E. M., Vermeulen, A., Fedorov, Y., Robinson, K., Leake, D., Karpilow, J., Marshall, W. S., and Khvorova, A. (2006). Induction of the interferon response by siRNA is cell type- and duplex length-dependent. *RNA* **12**, 988–993.

Sauro, H. M. (2009). Network Dynamics. *In* "Computational Systems Biology," (R. Ireton, K. Montgomery, R. Bumgarner, R. Samudrala, J. McDermott, eds.), pp. 269–309. Humana Press, Totowa, NJ.

Schwanhäusser, B., Busse, D., Li, N., Dittmar, G., Schuchhardt, J., Wolf, J., Chen, W., and Selbach, M. (2011). Global quantification of mammalian gene expression control. *Nature* **473**, 337–342.

Slepchenko, B. M., Schaff, J. C., Carson, J. H., and Loew, L. M. (2002). Computational cell biology: spatiotemporal simulation of cellular events. *Annu. Rev. Biophys. Biomol. Struct.* **31**, 423–441.

Sreenath, S. N., Cho, K. -H., and Wellstead, P. (2008). Modelling the dynamics of signalling pathways. *Essays Biochem.* **45**, 1–28.

Tanious, F. A., Nguyen, B., and Wilson, W. D. (2008). Biosensor-surface plasmon resonance methods for quantitative analysis of biomolecular interactions. *Methods Cell Biol.* **84**, 53–77.

Thomson, M., and Gunawardena, J. (2009). The rational parameterization theorem for multisite post-translational modification systems. *J. Theor. Biol.* **261**, 626–636.

Volkenstein, M., and Goldstein, B. (1966). A new method for solving the problems of the stationary kinetics of enzymological reactions. *Biochim Biophys Acta* **115**, 471–477.

Zhang, L., and Fang, B. (2005). Mechanisms of resistance to TRAIL-induced apoptosis in cancer. *Cancer Gene Ther.* **12**, 228–237.

CHAPTER 5

Stochastic Modeling of Cellular Networks

Jacob Stewart-Ornstein and Hana El-Samad

Department of Biochemistry and Biophysics, California Institute for Quantitative Biosciences, University of California, San Francisco, CA

Abstract

Noise and stochasticity are fundamental to biology because they derive from the nature of biochemical reactions. Thermal motions of molecules translate into randomness in the sequence and timing of reactions, which leads to cell–cell variability ("noise") in mRNA and protein levels even in clonal populations of genetically

0091-679X/10 $35.00
DOI 10.1016/B978-0-12-394620-1.00005-9

identical cells. This is a quantitative phenotype that has important functional repercussions, including persistence in bacterial subpopulations challenged with antibiotics, and variability in the response of cancer cells to drugs. In this chapter, we present the modeling of such stochastic cellular behaviors using the formalism of jump Markov processes, whose probability distributions evolve according to the chemical master equation (CME). We also discuss the techniques used to solve the CME. These include kinetic Monte Carlo simulations techniques such as the stochastic simulation algorithm (SSA) and method closure techniques such as the linear noise approximation (LNA).

I. Introduction

Cells are microscopic reactors where multitudes of chemical reactions occur. Biochemical reactions are probabilistic collisions between randomly moving molecules, with each event resulting in the increment or decrement of molecular species by integer amounts (Hasty and Collins, 2002; McAdams and Arkin, 1999; Rao *et al.*, 2002; Raser and O'Shea, 2005). As many crucial biological species including RNA and DNA are present in small quantities (ones or tens) per cell, these stochastic events can have measurable effects. The amplified effect of fluctuations in a molecular reactant, or the compounded of fluctuations across many molecular reactants, referred to as "molecular noise," often can accumulate as an observable phenotype, endowing the cell with individuality and generating nongenetic cell-to-cell variability in a population.

Observations of such nongenetic variation date back to the 1940s when it was determined that bacterial cultures were not completely killed by antibiotic treatment—a small fraction of cells "persist" (Bigger, 1944). The insensitivity to antibiotics exhibited by these persister cells was nonheritable (Moyed and Broderick, 1986), and persister cells spontaneously switched back to the nonpersistent state, regaining sensitivity to antibiotics. The advent of optical measurement methods, which monitor fluorescent reporter expression in single cells using flow cytometry or fluorescence microscopy, further illustrated that isogenic populations of cells can show great variability or "noise" in their gene expression (Cai *et al.*, 2006; Thattai and van Oudenaarden, 2001). By measuring the fluorescence intensity of single cells, probability distributions representing variability in a process across a population of cells can be constructed (Fig. 1). A broad distribution indicates a large dispersion of expression levels across the population. Recently, genome scale assays of variability in gene expression revealed that specific types of genes—those involved in energy metabolism and stress response—showed heightened variability (Bar-Even *et al.*, 2006; Newman *et al.*, 2006b). These data were used to lend support to the hypothesis that variability in protein content among cells might be a regulated trait that confers a selective advantage through a "bet-hedging" strategy with respect to future environmental shifts (Avery *et al.*, 2007; Blake *et al.*, 2006). Such stochastic fate specification has also been postulated in other contexts. For example, each cell in the mouse olfactory bulb must select only one olfactory receptor to express, and is thought to implement this decision by stocastically selecting to express a gene which then mediates global repression of the other ~1300

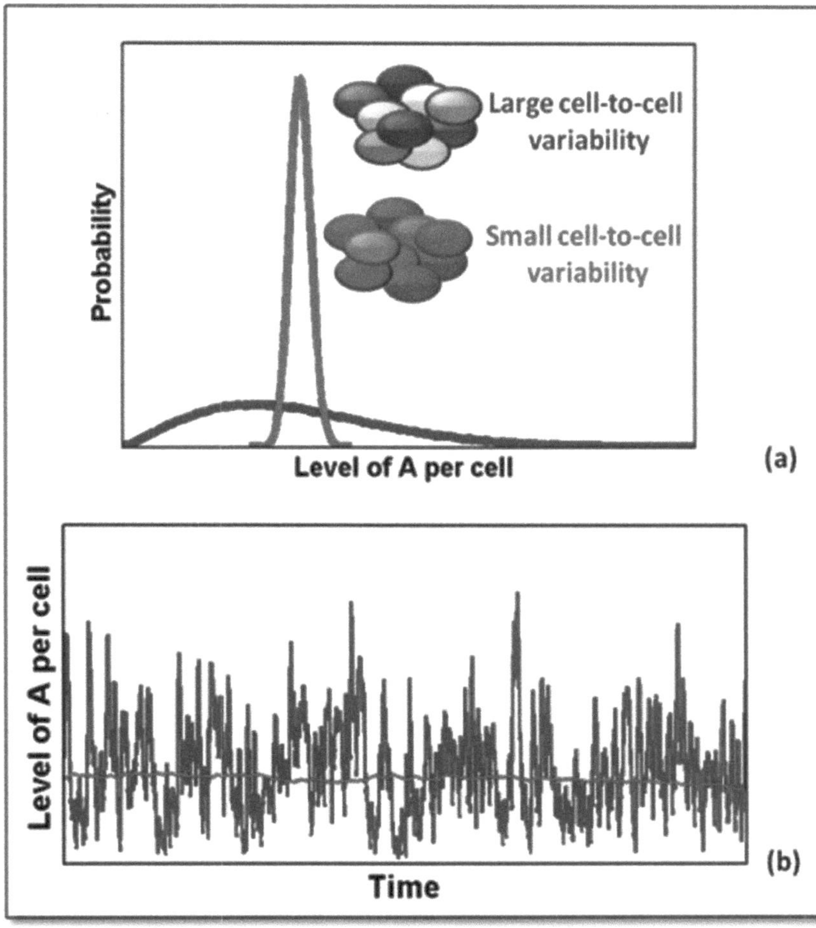

Fig. 1 Biochemical noise. (a) Distribution of a cellular component A for large cell–cell variability (blue, noisy) and small cell-to-cell variability (red). (b) Fluctuations of A as a function of time in one cell. (For interpretation of the references to color in this figure legend, the reader is referred to the web version of this book.)

receptors (Serizawa *et al.*, 2003). A similar model exists in two precursor cells in *Caenorhabditis elegans*, called Z1.ppp and Z4.aaa. In 50% of embryos, Z1.ppp differentiates into the AC cell, whereas Z4.aaa adopts the VU cell fate. In the other 50% of embryos, the opposite occurs. Through a random process, one cell adopts the VU cell fate, and then inhibits that choice in the other through a Notch signaling mechanism (Karp and Greenwald, 2003).

Variability, however, is not always beneficial. In the cell cycle for example, numerous feedback loops exist to ensure a tightly regulated and orderly transition through DNA replication and cell division (Tsai *et al.*, 2008). Similarly, in the *Drosophila* embryos, variability in the pattern of the bicoid protein results in undesirable developmental alterations, and studies suggest that the system is poised at the fundamental limit of the precision it can achieve (Gregor *et al.*, 2007a, b). In all these cases, understanding the roots and consequences of variability in the cell through careful measurements and quantitative modeling was of paramount importance for understanding the functioning of the underlying biological networks.

II. The Need for a Stochastic Modeling Framework

Most often than not, mathematical models represent the dynamic operation of cellular networks as deterministic processes with continuous variables. This continuous and deterministic approach may be warranted when large numbers of molecules justify a continuous valued concentration description. This is, for example, the case in metabolic networks where the concentration of reactants is in the millimolar range. There, chemical reactions can be modeled as reaction diffusion processes, and their dynamics described by partial differential equations (PDEs). When the reacting chemical solutions are well-mixed, these PDEs can then be well approximated with ordinary differential equations (ODEs).

There are many situations where this continuous deterministic modeling fails and stochastic models are necessary to capture biologically relevant properties of the systems under study. One such scenario is one where continuous models fail to describe quantitatively the behavior of a system because key regulatory molecules are found in very small integer populations. For example, the Lac operon in *Escherichia coli* is regulated by lactose binding to the repressor *Lac*I, which needs to be inactivated to allow for transcription of the operon. In this system, the key regulatory event in sensing lactose is the stochastic expression of a very small number of copies of the lactose permease lacY. As a result, the switching rate of *E. coli* to a lactose metabolizing state is governed by small number fluctuations of lacY (Choi *et al.*, 2008), necessitating a discrete stochastic model of the chemical species involved in this regulation.

A second situation where stochastic models are needed arises when fluctuations induce dynamical behaviors, which cannot be captured even qualitatively using deterministic models. For example, stochastic fluctuations in excitable systems cause them to undergo large excursions away from their equilibrium point. Such excitable behavior occurs in the prokaryote *Bacillus subtilis* when it transitions between low- and high-"competence" states that have differential abilities to absorb DNA from their environments. Periods of high competence occur stochastically when the master regulators comK and comS exceed a certain concentration. After an individual cell has passed the threshold, strong positive feedback loops drive the cell toward competence, followed by a slower negative feedback loop, which switches the system off after a defined period (Suel *et al.*, 2006, 2009). These dynamics occur nonsynchonously in a small population of cells, and therefore cannot be recapitulated using a deterministic mode, which, in this instance, can only settle into its only stable equilibrium. In contrast, accounting quantitatively for stochastic variation in protein concentrations is needed to reproduce this behavior.

III. Overview of Computational Approach

A quantitative modeling framework that takes into account the inherent stochasticity of biochemical interactions occurring inside a cell should handle discrete systems, should be adaptable to many different problems, and should be computationally tractable. A widely used such approach that we review in this chapter is one developed to address the chemical kinetics of well-mixed homogeneous systems. In this approach the cell is treated as a well-mixed bag of chemical species (Gillespie, 1977; Mcquarri, 1967). A model then probabilistically describes the chemical interactions of a subset of these species as a Markov (memoryless) jump process. After such a model is initiated from a defined state (in terms of the number of molecules of different species), reactions are allowed to occur between the chemical species. These reactions are represented by state transitions in a Markov chain, and transitions occur in discrete steps after a random time period, with the change and the time both depending only on the previous state. In this way, the transitions model the change in the number of each type of biological molecule in accordance with the stoichiometry of the chemical reaction (Fig. 2).

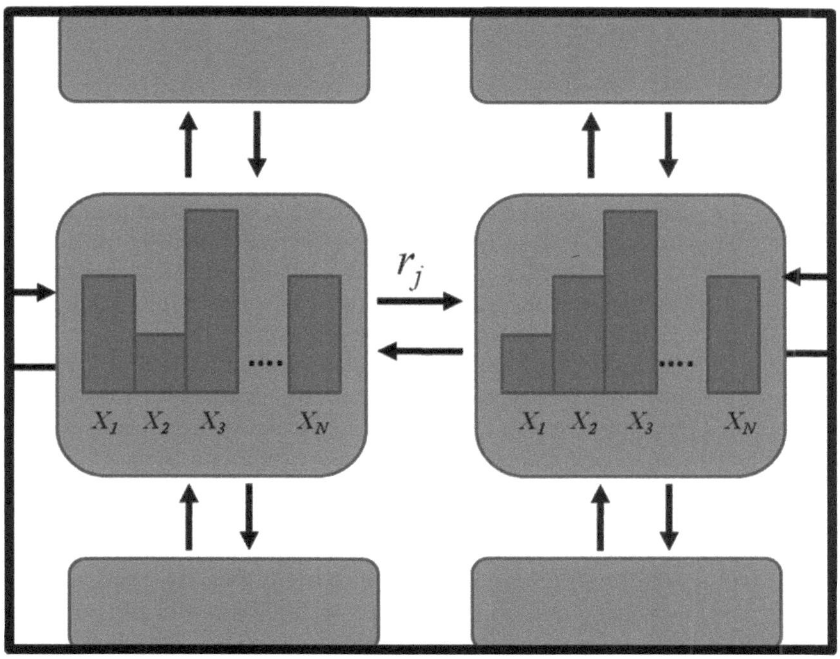

Fig. 2 Markov chain model for chemical kinetics. The states of the Markov chain are defined by the numbers of biological molecules of each chemical species, labeled $X_1, X_2, \ldots, X_N$. Transitions between these states model the individual chemical reactions which may occur in the system. The transition corresponding to the chemical reaction of type k is labeled by R_k. (For color version of this figure, the reader is referred to the web version of this book.)

The chemical master equation (CME) is a differential equation that governs the time evolution of the probability for observing the Markov chain in a given state at a given time. The CME is generally derived using the Markov property, by writing the Chapman–Kolmogorov equation, an identity that must be obeyed by the transition probability of any Markov process (Gillespie, 1992; Mcquarri, 1967). Although the CME is straightforward to write, it cannot be analytically solved for any but the simplest problems. Therefore, numerical simulations on a computer are the key tool used for understanding the behavior of a system described by a CME. Monte Carlo simulation techniques are routinely used. Specifically, in this context, an algorithm known as the stochastic simulation algorithm (SSA, but more commonly known as the Gillespie algorithm) is used to generate exact realizations (or "runs") of the Markov jump process (Gillespie, 1977). The algorithm generates time course trajectories of the system states over a given time window, starting from a given initial system state. Each such run is "exact" in the sense that it is an independent realization from the true underlying process. However, each realization is also stochastic and is therefore different for each simulation run. A construction of the probability distributions of the underlying stochastic processes can then be done by executing and compiling a sufficient number of such runs.

IV. Biological Insights from Computational Approaches

Cell-to-cell variability (molecular noise) is ubiquitous in the cellular world where typical transcription factors can exist in as a few as 10 copies per cell and bind to promoters of individual genes, which produce bursts of a few mRNAs. Although the functional repercussions of this variability were observed in bacterial persistence as early as 1944 (Bigger, 1944), it is only recently that this aspect of cellular physiology has captured the imagination of both theorists and biologists. As a result, the last few decades have witnessed many discoveries about how cells and organisms attenuate or exploit their molecular fluctuations, and what implications these bear on cellular phenotypes. Computational methods based on the formalism presented in this chapter continue to play a central role in these investigations.

Cellular decision making has been one area where stochastic models have made a crucial contribution. One of the earlier landmark works to apply the Gillespie algorithm (Gillespie, 1977) for modeling a natural gene network included a comprehensive model of the Lambda switch (McAdams *et al.*, 1998). This seminal work described how the Lamba phage balanced lytic and lysogenic outcomes of bacterial infection and illustrated how stochastic molecular events, originating from the random movement of molecules, can trigger decisions on a much larger scale leading to divergent cellular fates. A flurry of subsequent work used the same approach to investigate stochastic cellular switching and decision making in a number of biological contexts. For example, theoretical work illustrated that a population of cells capable of random phenotypic switching can have an advantage in a fluctuating environment (Kussell *et al.*, 2005; Thattai and van Oudenaarden, 2001; Wolf *et al.*, 2005). Some of these predictions have subsequently been confirmed, showing that noise can aid survival in severe stress (Blake *et al.*, 2006) and can optimize the efficiency of resource uptake during starvation (Suel *et al.*, 2009) and survival in fluctuating environments (van Oudenaarden *et al.*, 2008).

In addition to their role in unraveling the functional repercussions of molecular noise, computational methods that capture biological fluctuations have been instrumental in pinpointing their origins and the cellular mechanisms that modulate them. Stochasticity in gene expression received special attention. There, the synergy between quantitative measurements at the single cell or single molecule level and appropriate quantitative models deepened our understanding of the processes involved in transcription and translation and yielded some unexpected observations (Cai *et al.*, 2006; Chubb *et al.*, 2006; Cluzel *et al.*, 2005; Golding *et al.*, 2005; Raj *et al.*, 2006; Yu *et al.*, 2006). For example, it was demonstrated that transcription of genes in *E. coli* is not as simple as RNA polymerases transcribing with a constant flux. Instead, the process is highly variable and proceeds in bursts rather than continuously. The origin of this behavior is still unknown, although possible candidates include global fluctuations of chromosome supercoiling states and RNA polymerase availability. A discrete stochastic framework accounting for all possible promoter states was also necessary to interpret experimental measurements of stochastic expression from eukaryotic promoters (Murphy *et al.*, 2007). Quantitative computational approaches of the type we discuss in this chapter and high-resolution measurement technologies are poised to further reveal the workings of these fundamental cellular processes.

Synthetic biology is a nascent branch of biological investigation where accurate predictive modeling is of crucial importance. The aim of synthetic biology is to bring together ideas from biology and engineering to design and build biological networks that can achieve novel functions inside cells. It is now appreciated that the robust operation of synthetic cellular networks requires an understanding of molecular fluctuations, and that this understanding stems from rigorous probing of their stochastic dynamics. Analysis of stochastic models of the type we will tackle in this chapter has, for example, enabled the design and construction of synthetic oscillators that are robust to expected cellular variability (Tigges *et al.*, 2009).

V. Computational Methods

A. A Simple Example

In a simple model of transcription, a gene is transcribed to generate a mRNA at a constant rate k, and each mRNA molecule is independently degraded at a rate γ. The mRNA copy number is then a random variable $M(t)$, which can assume positive integer values m. These interactions can be written using chemical reaction notation as:

$$\varnothing \xrightarrow{k} M$$

$$M \xrightarrow{\gamma m} \varnothing$$

From a deterministic perspective, the mean mRNA copy number per cell across a population can be described with the differential equation:

$$\frac{dM}{dt} = k - \gamma M$$

At steady state, $dM/dt = 0$ and hence the mean mRNA copy number is then given by:

$$M^{ss} = k/\gamma$$

This result gives the mean mRNA per cell as a ratio of synthesis and decay rates. Note that this mean value does not necessarily represent the number of mRNA in any given cell. It is just the average expected value of mRNA at steady state across the population.

In a stochastic context, we are concerned with finding the distribution of mRNA numbers across a population of cells. That is, we want to document the number of mRNA molecules in individual cells, and use this information to determine how many cells in a population are expected to contain a given number of mRNA molecules. To do this, we begin by writing an equation governing the time evolution of $p(m, t)$, the probability that $M(t) = m$. We can start with $p(m, t + dt)$, the probability that the system achieves m mRNA molecules at time $t + dt$. This probability is intuitively computed by enumerating the number of scenarios through which this outcome could be achieved. For example, the system could achieve m molecules at time $t + dt$ if it had $m - 1$ molecules at time t, and then one molecule is transcribed during time interval dt. This probability is simply given by $P(m - 1, t)kdt$. Similarly, the probability that the system has $m + 1$ molecules and loses one by degradation in time dt is given by $P(m + 1, t)\ (m + 1)\gamma dt$, whereas the probability of the system to have exactly m mRNA molecules at time t and not lose or gain any additional molecules in the time interval dt is given by $P(m, t)(1-kdt)(1-m\gamma dt)$. As a result, $P(m, t + dt)$ can be written as:

$$P(m,t+dt) = P(m-1,t)kdt + P(m+1,t)(m+1)\gamma dt + P(m,t)(1-kdt)(1-m\gamma dt) \quad (1)$$

Multiplying out and rearranging terms in Eq. (1), we get:

$$P(m,t+dt) - P(m,t) = P(m-1,t)kdt + P(m+1,t)(m+1)\gamma dt - P(m,t)(K+m\gamma)dt + \varphi(dt^2) \quad (2)$$

Dividing Eq. (2) by dt and taking the limit as $dt \rightarrow 0$, we get:

$$\frac{d}{dt}P(m,t) = kP(m-1,t) + (m+1)\gamma P(m+1,t)dt - (K+m\gamma)P(m,t)dt \quad (3)$$

Eq. (3) is known as the CME. Although the derivation of the CME was illustrated for this specific example, similar derivations can be done for any biomolecular network described by a system of chemical reactions. Below, we provide a general formulation of the CME.

B. The General Formulation for Building Discrete Stochastic Models for Biomolecular Networks Using the Chemical Master Equation

In this section we describe the discrete state, continuous time Markov process model for well-stirred chemical reaction systems. First, we consider a system of chemical reactions with N molecular species (S_1, S_2, .., S_N) occurring in a volume V. We make two key assumptions. The first is that the system is well-mixed, that is the probability of finding any molecule in the volume V is given by dV/V. In many biological systems this is a reasonable assumption. For example, the length of a bacterial cell is around 1 μm and the diffusion coefficient of a protein *in vivo* has been measured to be on the order of 10 μm/s. Therefore, complete mixing of the bacterial cytosolic protein pool can possibly occur on the milisecond to second time scale (Konopka *et al.*, 2006). However, the diffusion constant of many proteins moving in 2D on membranes may be much less than the area over which reactions occur, causing local depletion or enrichment of chemical species that renders the well-mixed assumption invalid (Vrljic *et al.*, 2002). The second assumption we make is that the system is at thermal equilibrium. As a result, the velocity v of a molecule moving due to thermal energy is given by the Boltzman distribution:

$$f = \sqrt{\frac{m}{2\pi k_B T}} e^{-(m/2k_B T)v^2}$$

where T is the constant system temperature. We use the state $X(t) \in Z_+^N$ to denote the vector whose elements $X_i(t)$ are the number of molecules of the ith species at time t. If there are M elementary chemical reactions that can occur among these N species, then we associate with each reaction r_j ($j = 1, \ldots M$) a nonnegative *propensity function* a_j defined such that $a_j(X(t))\tau + o(\tau^2)$ is the probability that reaction r_j will happen in the next small time interval $(t, t + \tau)$ as $\tau \to 0$. The polynomial form of the propensities $a_j(x)$ may be derived from fundamental physical principles under certain assumptions (Gillespie, 1977). If r_j is the unimolecular reaction $S_1 \to product$, then a quantum mechanical argument dictates the existence of some constant c_j such that $c_j dt$ gives the probability that any particular S_1 molecule will transform into product in the next infinitesimal time dt. If there are currently n_1 such S_1 molecules in the system, then the probability that one of them will undergo the reaction in the next dt is $n_1 c_j dt$. Therefore, the propensity function of this unimolecular reaction is $a_j = n_1 c_j$. By contrast, if r_j is a bimolecular reaction of the form $S_1 + S_2 \to product$, then kinetic arguments can be used to assert the presence of a constant c_j such that $c_j dt$ is the probability that a randomly chosen pair of molecules S_1 and S_2 will react in the next infinitesimal time interval dt. Therefore, if n_1 molecules of S_1 and n_2 molecules of S_2 exist in volume V, then a reaction r_j will occur in the next dt with a probability $a_j dt = n_1 n_2 c_j dt$ (a_j is again called the propensity function of this reaction). Propensity functions for different types of reactions are summarized in Table I.

Table I
Stochastic Reaction Propensities

Reaction	Propensity $a_j(x)$
$\emptyset \xrightarrow{c_j}$ product	c_j
$S_i \xrightarrow{c_j}$ product	$c_j n_i$
$S_i + S_j \xrightarrow{c_j}$ product	$c_j n_i n_j$
$S_i + S_i \xrightarrow{c_j}$ product	$c_j \frac{n_i(n_i-1)}{2}$

The occurrence of a reaction r_j leads to a stoichiometric change of ϑ_j for the state X of the reactants involved. ϑ_j is therefore a stoichiometric vector that reflects the integer change in reactant species due to a reaction r_j.

It is useful to define these quantities:

Probability that reaction r_j fires one in $[t, t + dt] = a_j(x)dt + O(dt^2)$

Probability that no reaction in the system fires in $[t, t + dt] = 1 - \sum_{j=1}^{M} a_j(x)dt + O(dt^2)$

Probability that more than one reaction fires in $[t, t + dt] = O(dt^2)$

As in the simple example above, the CME for this system can be written by inspection using these quantities. Specifically, the probability of achieving state $X = x$ at time $t + dt$, $p(x, t + dt)$, is the sum of the following terms:

$$p(x, t + dt) = p(x, t)\left[1 - \sum_{j=1}^{M} a_j(x)dt + O(dt^2)\right] + \sum_{j=1}^{M}[p(x - \vartheta_j, t)a_j(x - \vartheta_j)dt + O(dt^2)] + O(dt^2) \tag{4}$$

The first term in Eq. (4) is simply the probability that the system was already in state x in terms of the number of its molecules for different species, and remained in that state with no reactions occurring during dt. The second term is the probability that the system was a ϑ_j step away from state x, and then was brought into that state by the occurrence of a reaction. Obviously, one has to account for all the reactions that can drive the system into that state, hence the summation.

Rearranging Eq. (4) we obtain:

$$p(x, t + dt) - p(x, t) = -p(x, t)\sum_{j=1}^{M} a_j(x)dt + \sum_{j=1}^{M}[p(x - \vartheta_j, t)a_j(x - \vartheta_j)dt] + O(dt^2) \tag{5}$$

Dividing Eq. (5) by dt and taking the limit as $dt \rightarrow 0$ gives the differential form

$$\frac{dP(x, t)}{dt} = \sum_{j=1}^{M} a_j(x - \vartheta_j)P(x - \vartheta_j) - a_j(x)P(x, t) \tag{6}$$

Eq. (6) is the CME for a general set of chemically reacting species in a constant, well-stirred volume.

C. Stationary Solutions of the CME

The stationary (steady state) distribution of the CME is solved for by setting $dP(x, t)/dt = 0$. For the simple model of transcription described by the CME in Eq. (3), this translates to: $kp(m-1) + (m+1)\gamma p(m+1) = (K + m\gamma)p(m)$

Solution of this balance equation can be done by induction. We observe that:

$$kp(0) = \gamma p(1)$$

$$kp(1) = 2\gamma p(2)$$

$$kp(m-1) = m\gamma p(m)$$

As a result, $p(m)$ can be expressed as a function of $p(0)$ as:

$$p(m) = \left(\frac{k}{\gamma}\right)^m \frac{1}{m!} p(0) \tag{7}$$

We can solve for $p(0)$ from Eq. (7) using the fact that $\sum_m p(m) = 1$. Therefore, $1 = \sum_m (k/\gamma)^m \frac{1}{m!} p(0) = e^{k/\gamma} p(0)$. As a result, $p(0) = e^{-(k/\gamma)}$ and $p(n) = e^{-a(a^m/m!)}$ with $= k/\gamma$. This corresponds to a Poisson distribution with equal mean and variance $\mu = \sigma^2 = a$.

This model has recently been validated using RNA fluorescence *in situ* hybridization (FISH) for ~100 well-expressed bacterial genes. These measurements conformed reasonably well to the predicted Poisson distribution, showing a relationship $\mu = 1.6\sigma^2$ (in contrast, protein expression in *Saccharomyces cerevisiae* scales as $\mu = 1200\sigma^2$ (Bar-Even *et al.*, 2006)). However, the subtle quantitative deviation from the Poisson relationship also suggested that other processes beyond simple production/degradation model might be at play to account for all the variability occurring in bacterial gene expression (Taniguchi *et al.*, 2010).

In general for a typical biological problem with several species and parameters neither the time evolution nor the stationary distribution described by the CME are analytically solvable. Therefore, one has to resort to numerical techniques to determine these quantities through sample path computations.

1. The Stochastic Simulation Algorithm: Generating Sample Paths

The approach here is to run a simulation describing the fluctuating behavior of a set of interacting chemical reactions in a single cell over time, and then to repeat this procedure multiple times to build an ensemble of behaviors across a population of cells.

To each of the chemical reactions $r_j (j = 1, \ldots, M)$ occurring among species (S_1, S_2, ..., S_N) in a well-stirred volume, we attribute a random variable τ_j defined as the time to the firing of the next reaction r_j. Based on this formulation, τ_j is exponentially distributed with parameter $a_j(x)$ (a_j is the propensity function of this reaction). It can be shown that the time to the next reaction, defined as the random variable $\tau = \min\{\tau_j\}$, is exponentially distributed with parameter $\sum_{j=1}^{M} a_j(x)$. The random variable representing the index of the next reaction to occur $\mu = \text{argmin}\{\tau_j\}$ can also be shown to be uniformly distributed with $(\mu = j) = a_j(x) / \sum_{j=1}^{M} a_j(x)$. Using these quantities, one can then simulate the system with the four simple steps:

Initialize time t_0 and state x_0

Draw a sample $\hat{\tau}$ from $P(\tau)$, the distribution of τ

Draw a sample $\hat{\mu}$ from $P(\mu)$, the distribution of μ

Update time $t\, t + \hat{\tau}$ and state $x\, x + \hat{\mu}$ and repeat if final time is not reached.

This method is known as the SSA, and belongs to a wider class of numerical techniques known as Kinetic Monte Carlo algorithms. Every run of the algorithm above will generate a sample path of the stochastic process described by the CME (see for example Fig. 1(b)). To generate the probability distributions, one can run a large number of such sample paths.

D. Moment Computations

The CME is an equation for the probability distribution and can therefore be used in a straightforward manner to derive an expression for the evolution of the mean and higher order moments of these distributions. Simply put, for the first-order moment, $E(X_i)$, we can multiply the CME by x_i and then sum over all values of x. That is, $E[X_i] = \Sigma x_i p(x, t)$, and $(d\Sigma x_i p(x, t)/dt) = (dE[X_i]/dt)$. Similarly, for the second moment $E[X_i X_j]$, we can multiply the CME by $x_i x_j$ and sum over values of x. If we define $A(X) = [a_1(X),\ a_1(X),\ \ldots.\ a_M(X)]^T$ as the vector of propensity functions, and $S = [\vartheta_1 \vartheta_2 \ldots. \vartheta_M]$ as the stoichiometry matrix, then we can derive (using some straightforward algebraic manipulations that we will omit here) the following equations for the mean and second-order moments:

$$\frac{dE[X]}{dt} = SE[A(X)] \tag{8}$$

$$\frac{dE[XX^T]}{dt} = SE[A(X)X^T] + E[XA^T(X)]S^T + S\,diag(E[A(X)])S^T \tag{9}$$

1. Moment Equations for a System With Affine Propensities

An especially tractable form of the moment equations derived above arises when the propensity functions are affine, that is $A(X) = WX + w_0$, where W is an $N \times N$ matrix and w_0 is an $N \times 1$ vector. In this case, $E[A(X)] = W\,E[X] + w_o$ and

$E[A(X)X^T] = W\,E[XX^T] + w_o E[X^T]$. Replacing these expressions in Eqs. (8) and (9) above gives the moments equations:

$$\frac{dE[X]}{dt} = SWE[X] + Sw_o \tag{10}$$

$$\frac{dE[XX^T]}{dt} = SWE[XX^T] + E[XX^T]W^T S^T + S\,diag(WE[X] + w_o)S^T$$
$$+Sw_o E[X^T] + E[X]{w_o}^T S^T \tag{11}$$

Eq. (11) is for the uncentered second moment. The covariance matrix (containing the centered second-order moments) is defined as $C = E[(X - E[X])(X - E[X])^T]$. Therefore, an expression for its time evolution can be derived by manipulation of Eqs. (11) and (12) to give:

$$\frac{dC}{dt} = SWC + CW^T S^T + S\,diag(WE[X] + w_o)S^T$$

The steady state means and covariances can be obtained by solving the linear algebraic equations corresponding to setting $(dE[X]/dt) = 0$ and $(dC/dt) = 0$. Let $\overline{X} = lim_{t\to\infty}\,E[X(t)]$ and $\overline{C} = lim_{t\to\infty}\,C(t)$. Then,

$$SW\overline{X} = -Sw_o \tag{12}$$

$$SW\overline{C} + \overline{C}W^T S^T + S\,diag(W\overline{X} + w_o)S^T = 0 \tag{13}$$

Now, if we define $M = SW$, $B = S\sqrt{diag(W\overline{X} + w_o)}$, and $D = BB^T$, then the steady state covariance given by Eq. (13) becomes

$$M\overline{C} + \overline{C}M^T + D = 0$$

This is the well-known Lyapunov equation, which characterizes the steady state covariance of the output of the linear dynamical system

$$\frac{dY}{dt} = MY + Bw$$

where w is the unit intensity white Gaussian noise.

E. An Example Where Calculations of Means and Covariances Generated Rich Biological Insight

Consider as extension of our initial model of transcription to include translation of a protein product from an mRNA (Figure 3). mRNA and protein can also decay with first-order kinetics. The simplest representation of this module contains four biochemical reactions:

$$R_1 : \emptyset \xrightarrow{k_r} mRNA$$

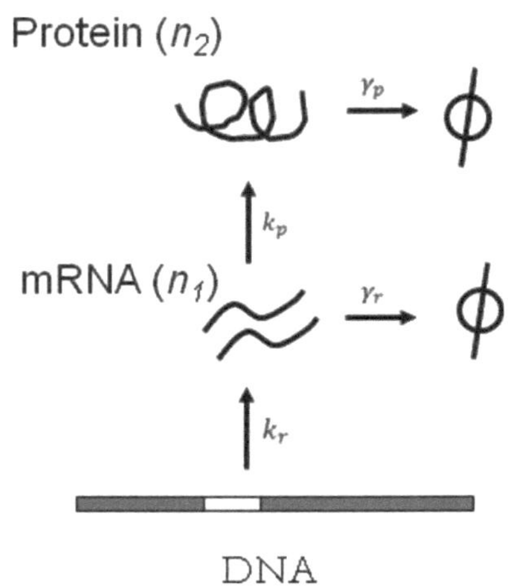

Fig. 3 Simple transcription and translation module. (For color version of this figure, the reader is referred to the web version of this book.)

$$R_2 : mRNA \xrightarrow{\gamma_r} \emptyset$$

$$R_3 : mRNA \xrightarrow{k_p} protein + mRNA$$

$$R_4 : protein \xrightarrow{\gamma_p} \emptyset$$

If we denote the number of molecules of mRNA by $X_1(t)$ and that of the protein by $X_2(t)$, then $X(t) = [X_1(t)X_2(t)]^T$. Also, the stoichiometry matrix is given by:

$$S = \begin{bmatrix} 1 & -1 & 0 & 0 \\ 0 & 0 & 1 & -1 \end{bmatrix}$$

Although the propensity vector is given by:

$$A(X) = \begin{bmatrix} k_r \\ \gamma_r X_1 \\ k_p X_1 \\ \gamma_p X_2 \end{bmatrix} = \begin{bmatrix} 0 & 0 \\ \gamma_r & 0 \\ k_p & 0 \\ 0 & \gamma_p \end{bmatrix} \begin{bmatrix} X_1 \\ X_2 \end{bmatrix} + \begin{bmatrix} k_r \\ 0 \\ 0 \\ 0 \end{bmatrix} = WX + w_0$$

Therefore, $M = SW = \begin{bmatrix} -\gamma_r & 0 \\ k_p & -\gamma_p \end{bmatrix}$ and $Sw_0 = \begin{bmatrix} k_r \\ 0 \end{bmatrix}$. As a result, the steady state as given by Eq. (12) is: $\overline{X} = -M^{-1}Sw_0 = \begin{bmatrix} \frac{k_r}{\gamma_r} \\ \frac{k_p k_r}{\gamma_p \gamma_r} \end{bmatrix}$.

The steady-covariance matrix can also be computed using Eq. (13). Specifically,

$$BB^T = S\,diag(W\overline{X} + w_o)S^T = \begin{bmatrix} 2k_r & 0 \\ 0 & \dfrac{2k_p k_r}{\gamma_r} \end{bmatrix}$$

As a result, the steady state covariance matrix $\overline{C}$ is given by:

$$\overline{C} = \begin{bmatrix} \dfrac{k_r}{\gamma_r} & \dfrac{k_p k_r}{\gamma_r(\gamma_r + \gamma_p)} \\ \dfrac{k_p k_r}{\gamma_r(\gamma_r + \gamma_p)} & \dfrac{k_p k_r}{\gamma_p \gamma_r}(1 + \dfrac{k_p}{\gamma_r + \gamma_p}) \end{bmatrix} \quad (14)$$

Notice that for the mRNA in Eq. (14), we have exactly recapitulated the result derived based on the exact solution of the CME above, namely that its stationary distribution has an equal mean and variance given by k_r/γ_r. The mean of the protein is given by: $\overline{X}_2 = (k_p k_r/\gamma_p \gamma_r)$, while its variance is $\overline{C}_{22} = (k_p k_r/\gamma_p \gamma_r)(1 + (k_p/(\gamma_r + \gamma_p)))$. Therefore, the coefficient of variation for the protein (a unitless quantity to be intuitively thought of as a normalized standard deviation) is given by:

$$CV = \frac{\sqrt{\overline{C}_{22}}}{\overline{X}_2} = \frac{1}{\sqrt{\overline{X}_2}}\left(1 + \frac{k_p}{\gamma_r + \gamma_p}\right)^{1/2} \quad (15)$$

This equation confirms our intuition that as the number of molecules increases, the CV ("noise") of the system would decrease. Most importantly, it assigns a very specific pattern for this decrease in that it should follow an inverse square-root function of the mean with a scaling constant dependant on the translation rate of the mRNA and decay rates of the protein and mRNA. Experimental investigations of noise in gene expression of a large set of genes in the yeast *S. cerevisae* and bacterium *E. coli* subsequently confirmed this prediction (Newman *et al.*, 2006a). However, does a large $\overline{X}_2$ necessarily imply a small CV?

Notice that:

$$\begin{aligned} CV^2 &= \frac{1}{\overline{X}_2}\left(1 + \frac{k_p}{\gamma_r + \gamma_p}\right) = \frac{1}{\frac{k_p k_r}{\gamma_p \gamma_r}}\left(1 + \frac{k_p}{\gamma_r + \gamma_p}\right) \geq \frac{1}{\frac{k_p k_r}{\gamma_p \gamma_r}}\left(\frac{k_p}{\gamma_r + \gamma_p}\right) \\ &= \frac{\gamma_p \gamma_r}{k_r} \cdot \frac{1}{\gamma_r + \gamma_p} \end{aligned} \quad (16)$$

Therefore, for some values of γ_r, γ_p, and γ_r, CV^2 in Eq. (16) can be arbitrarily large. Simultaneously, through choice of k_p, $\overline{X}_2 = (k_p k_r/\gamma_p \gamma_r)$ can be set independently of CV^2 to be arbitrarily large. Therefore, large mean does NOT necessarily imply small fluctuations. This model of gene expression predicts that decreased translation rates should decrease noise in gene expression, a result that was confirmed experimentally (Ozbudak *et al.*, 2002). More generally this framework suggests cellular contexts where noise might be expected to be particularly problematic. For example, this model predicts that when proteins are rapidly degraded and expressed at low copy number, such as the cyclins in the cell cycle, high variability would ensue. Given this insight, many recent investigations of the cell cycle focused precisely on what control strategies implemented through interlinked positive and negative feedback loops can compensate for this effect to provide robust noise free oscillations (Tsai *et al.*, 2008).

Exceptions aside, Eq. (15) and some of its variations have guided many investigations that delineated fundamental properties of noise in gene expression. Researchers have used this equation to infer promoter, mRNA and protein dynamics based on snapshots of protein distributions (see Paulsson (2005) for a review). Furthermore, these analyses proved particularly useful in describing the effect of chromatin features on gene expression. In one recent such study, a viral vector was used to integrate a green fluorescent protein (GFP) reporter construct randomly in a mammalian cell line and the CV of each integrant was measured. Fitting the data to a two-state gene expression model similar to Eq. (16), with the addition that a promoter can transition between OFF and ON states, suggested that the chromatin state of the integration site affects the stability and productivity of the ON state, but not the frequency of activation (Skupsky et al., 2010). It is worth noting here that these static snapshots of noise in gene expression are not always sufficient to resolve all the parameters involved in the process. For example, in the study mentioned above, these distributions were sufficient to determine the promoter activation frequency but not its active duration. Dynamic measurements might be necessary to resolve such parameters.

F. Linearization of Macroscopic Dynamics and the Linear Noise Approximation: Computing Approximate Moments for Nonlinear Propensity Functions

Although computation of first and second moments at steady state could be done using an algebraic equation when the propensity functions that appear in the CME are affine, no such calculation is possible when these propensity functions are nonlinear as is the case for many biological reactions. The reason is rather simple; close inspection of Eqs. (10) and (11) reveals that in this case, every moment depends on higher order moments, resulting in an infinite hierarchy of ODEs to solve. The Linear Noise Approximation (LNA) is a procedure to truncate this hierarchy. Before we present the LNA, we review selected parts of the standard treatment of linearized dynamics around a steady state (Strogatz, 1994). First, we remind the reader that the system of reaction rate equations describing the macroscopic behavior of the concentration of N biochemical species interacting through a set of M biochemical reactions is given by the coupled ODEs (Cornish-Bowden, 1979):

$$\frac{dx}{dt} = SA(x) \tag{17}$$

where $x(t) = [x_1(t)x_2(t) \ldots x_N(t)]^T$ is the vector of macroscopic concentrations, and S is the $N \times M$ stoichiometry matrix. If a steady state $\bar{x}$ exists for the macroscopic dynamics, it follows from solving the algebraic system of equations:

$$0 = SA(\bar{x})$$

Linearization of Eq. (17) around the steady state vector $\bar{x} = [\bar{x}_1\bar{x}_2 \ldots \bar{x}_n]^T$ leads to a matrix equation for the deviations $\delta x = [\delta x_1 \delta x_2 \ldots \delta x_N]^T$ from $\bar{x}$ given by:

$$\frac{d}{dt}\delta x = M\delta x$$

M is the Jacobian matrix, with the elements:

$$M_{ij} = \frac{\partial[S_i A(x)]}{\partial x_k}\Big|_{x=\bar{x}}$$

Therefore, in compact notation $M = S\frac{\partial A(x)}{\partial x}|_{x=\bar{x}}$.

Going back to the stochastic representation, we assume that the distribution of the chemical species is tightly distributed around its mean. We also assume that $x(t) = (X(t)/V)$ (where $X(t)$ is the mean of the distribution) is identical to the solution $\varphi(t)$ of the reaction rate equations (Eq. (17)) that describe the macroscopic concentrations of molecular species in the system. Notice that $\varphi(t)$ is a vector of concentrations, while $X(t)$ is a vector containing the number of molecules, hence the need for a volume scaling factor V.

More formally, let $X(t) = V\varphi(t) + \varepsilon(t)$, where $\varepsilon(t)$ is the zero mean random variable denoting the deviation from the deterministic term $V\varphi(t)$ (Tomioka *et al.*, 2004). Expanding in Taylor series around $\varphi(t)$ in Eq. (10), we get

$$\frac{dE[X]}{dt} = \frac{dV\varphi}{dt} + \frac{dE[\varepsilon]}{dt} = VSA(\varphi) + S\frac{\partial A(Vx)}{\partial Vx}\Big|_{x=\varphi} E(\varepsilon) + O(\varepsilon^2) \quad (18)$$

The assumptions on the distributions imply that $O(\varepsilon^2)$ can be neglected in Eq. (18) above. Therefore, recovering the equation: $\frac{d\varphi}{dt} = SA(\varphi)$. Furthermore, we obtain:

$$\frac{dE[\varepsilon]}{dt} = S\frac{\partial A(Vx)}{\partial Vx}\Big|_{x=\varphi} E[\varepsilon]$$

Rewriting Eq. (11) similarly in terms of Taylor series expansion and truncating the $O(\varepsilon^2)$ terms generates the following equation for the time evolution of the noise covariance matrix $C_\varepsilon = E[\varepsilon\varepsilon^T] - E[\varepsilon]E[\varepsilon^T]$:

$$\frac{dC_\varepsilon}{dt} = M(\varphi)C_\varepsilon + C_\varepsilon M^T(\varphi) + D(V\varphi) + \frac{\partial D(Vx)}{\partial Vx}\Big|_{x=\varphi} E[\varepsilon] \quad (19)$$

In Eq. (19), we defined $M = S(\partial A(Vx)/\partial Vx)|_{x=\varphi}$ as the Jacobian matrix and $D = S\,diag[A(V\varphi)]S^T$. Now, we have the closed simultaneous questions for the time evolution of mean and covariance of the random fluctuations around the macroscopic solution. We assume that the macroscopic solution is stable around $\varphi(t)$. That is, the eigenvalues of the Jacobian matrix M are negative for all t. This assumption is necessary to justify the linearization.

We also assume that the macroscopic rate equations converge to a stable steady state $\overline{\varphi}$. Under these assumptions, there exists a distribution around $\overline{\varphi}$ with mean $E[\varepsilon] = 0$ and covariance matrix C_ε that satisfies the following equation:

$$M(\overline{\varphi})\overline{C} + \overline{C}M^T(\overline{\varphi}) + D(V\overline{\varphi}) = 0 \tag{20}$$

Notice again that Eq. (20) is a Lyapunov equation, with $M(\overline{\varphi})$ being the Jacobian matrix obtained by linearizing the system around its macroscopic steady state.

In summary, one can obtain the covariance matrix of this distribution around a macroscopic steady state by taking the following simple procedure:

Find the stoichiometry matrix S and the propensity vector $A(X)$

Find a stable equilibrium of the reaction rate equations of the system

Calculate two matrices $M(\overline{\varphi})$ and $D(V\overline{\varphi})$

Solve Lyapunov equation (Eq. (20))

Above, we have presented a multivariable and compact derivation of the LNA. Multiple forms of this derivation exist under alternative names such as the system size expansion (Elf and Ehrenberg, 2003; Kampen, 1992).

Due to its minimal computation costs, the LNA makes rapid analytical investigation of noise features for different models and parameter sets possible. For example, LNA analysis of all possible three node networks over a wide range of parameter sets has recently been used to show that both positive and negative feedback motifs can buffer noise from an upstream node, but that only positive feedback loops can do so while maintaining network responsiveness. This insight was confirmed by a detailed analysis of nitrogen metabolism in yeast, which suggested that coupled positive and negative feedback in this system may indeed act to buffer noise (Hornung and Barkai, 2008).

G. Other Closure Techniques for the Moment Equations

As discussed above, the solution to the CME can be expanded in a Taylor series about the macroscopic deterministic trajectory. The first-order terms correspond to the macroscopic rate equations, and the second-order terms approximate the system noise. Variations on this procedure exist. For example, mass fluctuations kinetics (MFK) calculations take a similar approach to the LNA except that the computation of the mean is coupled with that of the variances (Gomez-Uribe and Verghese, 2007). Therefore, the MFK approach allows one to capture situations where the mean of the stochastic distributions may deviate from the solution of the macroscopic rate equations. This is particularly important for systems that exhibit emergent stochastic phenomena such as, for example, excitability (Suel *et al.*, 2006, 2007) and stochastic resonance or focusing (Paulsson *et al.*, 2000).

Other moment closure techniques proceed by assuming specific probability distributions for the underlying stochastic processes, and then using this assumption to express higher order moments as a function of the lower order ones to effectively truncate the dynamics. This has been done for well-known classes of distributions, such as normal (Whittle, 1957), lognormal (Keeling, 2000), Poisson binomial

(Nasell, 2003). Moment closure techniques that do not make explicit assumptions about the shape of the distribution also exist. One such moment closure approximation known as separable derivative matching (Singh and Hespanha, 2007) approximates the $(N+1)$th moment as a polynomial function of the first N moments. This approach matches time derivatives between the approximate closed system and the exact nonclosed system at the initial time t_0 and the given initial conditions. This allows the exponents (which remain constant over the simulation) in the polynomial function to be uniquely determined, and the solution turns out to be consistent with the underlying distribution probability distribution being lognormal. It is worth noting here that the derivation of the moment equations implicitly assumes the presence of a single macroscopic steady state. Hence, the distributions are unimodal and the process is well characterized by the first few moments. However, problems that exhibit multimodal distributions will require many higher order moments, and the applicability of these methods may quickly degrade. Usually, the choice between accurate numerical approaches and approximation analytical approaches (such as the LNA and moment closure techniques) is done on a case-by-case basis to balance computational cost versus accuracy.

VI. Open Challenges

Stochastic modeling of biological dynamics, especially at the cellular level, is increasingly making its way to the mainstream of quantitative biology investigation. The CME and its accompanying SSA have proven to be invaluable computational tools for such studies. There are, however, many challenges that need to be addressed in order to make stochastic modeling a widely applicable tool for realistic biological problems. Below, we discuss some of these challenges and recent developments in the literatures to address them.

A. Efficient Stochastic Simulation and Analysis for Systems Evolving at Disparate Temporal and Spatial Scales

For many cellular networks of biological importance, the chemical reactions occur at significantly different rates. As a motivating example, consider gene regulation in the bacterium *Escherichia coli.* There, a typical time scale for mRNA transcription is on the order of minutes, whereas the time scale for protein degradation/dilution is on the order of an hour (Alon, 2007). This suggests that the protein concentrations do not depend strongly on the instantaneous number of mRNAs but rather on their average over time. Even more drastically, posttranslational modifications of the protein (e.g., phosphorylation) often occur on the time scale of seconds. These disparate time scales in the chemical reactions pose great challenges for efficient numerical simulation of these processes. These challenges arise from having to resolve the stochastic dynamics on the fastest characteristic time scales of the system. Take for example a model in which a kinase activates a transcription factor by phosphorylating it, while a phosphotase removes the phosphate. We are interested in understanding the fluctuations in the expression of the gene that is regulated by the transcription factor. It is often the case that the competing phosphorylation and dephosphorylation reactions occur rapidly (fast reactions), whereas gene expression is relatively slow. In this situation a stochastic simulation of the system will spend most of its computational time fruitlessly adding and removing phosphates from the transcription factor and relatively little time on reactions that result in gene expression, our actual interest.

Multiple approaches have emerged to address this problem. On the analytical side, the strategy is often to derive reduced models by explicitly representing the chemical species having dynamics with relatively slow characteristic time scales while eliminating representations of the chemical species having dynamics with relatively fast characteristic time scales (Atzberger *et al.*, 2011; Cao *et al.*, 2005; Haseltine and Rawlings, 2002; Rao and Arkin, 2003). Roughly speaking, these methods parallel quasi-steady state approximations for deterministic chemical kinetics where a subset of species is assumed to be asymptotically at steady state on the time scale of interest. One commonly used example is the Hill function (a[TF/(TF + Kd)]), which describes the expression of a gene for a given concentration of a transcription factor (TF), affinity of the transcription factor for the promoter (K_d), and maximal activation (a). This expression is derived using the assumption that transcription factor binding and unbinding events are rapid relative to the rate of gene expression, and so one can approximate them as an average occupancy rather than explicitly model every individual event (Nemenman *et al.*, 2009).

On the numerical side, several approximate methods have been developed to speed up simulations while sacrificing some of the exactness of the SSA. The basic idea behind these approximate methods is that instead of simulating a single reaction per step, a number of reactions can occur in each simulation step. These approximate methods are known as leap methods including the τ-leap method (Gillespie, 2001; Gillespie and Petzold, 2003), the binomial τ-leap method (Chatterjee *et al.*, 2005; Rathinam and El Samad, 2007), and the K-leap method (Cai and Xu, 2007).

Despite such productive work on the subject, the efficient analysis and simulation of stochastic cellular dynamics for realistic problems is still very difficult. For example, there is little theory that can provide reassurance about the accuracy of the approximate SSAs in challenging scenarios. Furthermore, quasi-steady state approximations of stochastic fast scales are done based on intuition and assumptions derived from deterministic chemical kinetics. For these methods to be broadly applicable, they need to be placed on more solid theoretical footing in terms of the assumptions that can and cannot be made in a stochastic context and rigorous proofs need to be generated for their accuracy in different realistic contexts.

The holistic understanding of biological systems often involves the probing of cellular biochemical networks in the context of the cell, of cells in the context of a tissue, and of a tissue in the context of the organism. How to account for and move between these spatial scales remains an open problem for stochastic modeling. This "multiscale" problem is of poignant relevance to pharmacological studies, which need to integrate effects of small molecules therapies at the single cell level with global metabolic processes within the body such as prodrug activation, degradation of the active molecules, and off-target toxicities (Eissing *et al.*, 2011).

B. Efficient Spatiotemporal Simulations

Previous sections cover the stochastic algorithms for modeling biological pathways with no spatial information. However, biological networks in practice consist of components that interact in a three-dimensional space and are not necessarily distributed homogeneously as they diffuse between different cellular compartments. For example, even within *E. coli* (the prototypical cell-as-a-bag modeling system) membrane invaginations can dramatically alter the diffusive properties of molecules (Weisshaar *et al.*, 2006). In eukaryotic neuronal cells, axons can be meters long raising immense barriers to diffusive mixing. Thus, the basic assumption of spatial homogeneity and large concentration diffusion may be challenged in some biological systems. In this context, stochastic spatiotemporal representations are required.

Roughly speaking, discrete spatial stochastic simulations can be separated into lattice and off-lattice particle based methods. In off-lattice methods, the Brownian movements of the individual molecules are accounted for and all particles in the system have explicit spatial coordinates (Bartol, 2002). At each time step, molecules with nonzero diffusion coefficients are able to move, in a random walk fashion, to new positions. In this case, the motion and direction of the molecules are determined by using random numbers during the simulation. Similarly, collisions with potential binding sites and surfaces are detected and handled by using only random numbers with a computed binding probability. Particle methods can provide very detailed simulations of highly complex systems at the cost of exceedingly large amounts of computational effort.

For lattice methods, the two- or three-dimensional volume used to represent a cellular compartment (organelles or membranes) is covered by a computational mesh (Morton-Firth and Bray, 1998; Schnell *et al.*, 2004). The lattice is then "populated" with particles of the different molecular species that comprise the system. Particles with nonzero diffusion coefficient are able to diffuse by jumping to an empty neighboring domain. If the domain is assumed to accommodate only one molecule, chemical reactions can take place with a certain probability among molecules in adjacent domains. Another scenario is one in which subvolumes can host many molecules, with well-mixedness assumed in each subvolume. In both cases, diffusion steps are treated as treated first-order reactions, with a reaction rate constant proportional to the diffusion coefficient (Ander *et al.*, 2004; Baras and Mansour, 1996; Elf *et al.*, 2010; Stundzia and Lumsden, 1996). As a result, diffusion can be treated as an additional chemical reaction, and one is back to the SSA formalism.

Many caveats of these methods exist. For example, the artificial nature of the lattice may introduce lattice anisotropy (Ridgway *et al.*, 2009). Furthermore, in many physiologically relevant situations, molecular crowding can prevent reacting molecules from reaching regions of the domain due to the high concentration of macromolecules impeding their passage (Ridgway *et al.*, 2009). A particularly striking example of this is diffusive motion in the context of the eukaryotic nucleus where densely packed nucleoli and hetrochromatin structures greatly reduce diffusive rates, suggesting one mechanism whereby heterchormatin prevents active transcription (Bancaud et al., 2009). Therefore, despite their conceptual appeal, these spatiotemporal algorithms need to be updated to capture the full scope of biological reality. Furthermore, even in their current approximate forms, these algorithms require substantial and sometimes prohibitive computational power and have only been successfully applied to small systems with finite number of molecular species. As a result, many computational innovations are still needed to enable the quantitative probing of the spatial stochastic dynamics of biological systems.

C. Parametrization and Sensitivity Analysis of Stochastic Models

Stochastic models of biological systems typically depend on a set of kinetic parameters whose values are often unknown or fluctuate due to an uncertain environment. These parameters determine the dynamic behavior of the model, and changes in them may alter the system's output in nonintuitive ways. Typically, many of the parameters in a biological system have not been measured or are unmeasurable. For example, a typical assay for measuring the affinity of a transcription factor for its promoter by gel shift will describe this interaction in terms of a disassociation constant (K_d), which gives the ratio of binding and unbinding rates. A stochastic model, however, requires explicit ON and OFF rates that are rarely available. In this case, one strategy would be to estimate the ON and OFF rates under the assumption that binding of two molecules is "diffusion limited." However, a more commonly encountered situation is one in which no direct measurement exists from which to base a choice of parameters. In this case, it becomes imperative to establish that specific choices for the value of these parameters do not substantially change the model behavior of interest.

Assessing the change in a system output pursuant to perturbations in its kinetic parameters is carried out using sensitivity analysis. Traditionally, the concept of sensitivity analysis has been applied largely to continuous deterministic systems, for example, systems described by differential (or differential-algebraic) equations. Much of these analyses have focused on the effects of infinitesimal perturbations of certain parameters. In deterministic chemical kinetics, the infinitesimal sensitivities are represented using the first-order sensitivity coefficients, given by (Varma *et al.*, 2005):

$$S_{ij}(t) = \frac{\partial x_i(t)}{\partial \theta_j} \tag{21}$$

where x_i denotes that i^{th} output of the system at time t (e.g., the concentration of chemical species as given by Eq. (17)) and θ_j is the jth parameter. This equation assumes implicitly that the output $\boldsymbol{x}_i$ is continuous with respect to the parameter θ_j. Using the definition in Eq. (21), dynamic evolution equations can be derived for $S_{ij}(t)$ and solved along with the original system equation. In the context of biological systems modeling, sensitivity analysis has been indispensable to deduce important system properties, such as robustness in an uncertain environment (Stelling *et al.*, 2004). In large networks, sensitivity analysis can pinpoint critical or rate limiting pathways and aid in reduced order modeling. Despite their usefulness, these sensitivities report on changes of model behavior changes as parameters change *locally*, but do not address the outcome of large changes to parameters or simultaneous perturbations to multiple parameters. Assessing the effect of large perturbations is typically carried out numerically by recomputing the reaction rate equations for the perturbed parameter values and comparing these to the nominal parameter values.

The most common approach for sensitivity analysis in stochastic systems resembles the simulation-based strategy. Monte Carlo (SSA) simulations are run for various values of the parameter whose sensitivity is of interest, and the variation in the outcome of these simulations for a variable of interest, such as mean, quantified. The sensitivity at time T to a finite perturbation h of a parameter θ about its nominal value $\theta = \theta_0$ can be computed via a finite difference of the expected value, such as

$$S = \frac{E[X(T, \theta_0 + h)] - E[X(T, \theta_0)]}{h}$$

Basically, one uses SSA to compute these expected values by generating many samples of $X(T, \theta_0 + h)$ and $X(T, \theta_0)$, usually using two independent streams of random numbers to generate samples of $X(T, \theta_0 + h)$ and $X(T, \theta_0)$. This is called the independent random number (IRN) approach and has been recently used in combination with the Fisher information matrix to generate several different sensitivity measures (Gunawan *et al.*, 2005). Evidently, Monte Carlo simulations need to be carried out for the nominal and perturbed parameter value making this approach often computationally expensive. Furthermore, the use of IRNs usually results in a statistical estimator with large variance, thereby increasing the computational effort as large samples may be required. Recent work has shown that using the same stream of common random numbers (CRNs) to generate samples of $X(T, \theta_0 + h)$ and $X(T, \theta_0)$ can typically result in an estimator with low variance and thus requires far fewer samples (Rathinam *et al.*, 2010). Approaches based on the Girasnov measure have also been proposed to smooth the sensitivity estimates and reduce their bias (Plyasunov and Arkin, 2006). Finally, more tractable but approximate approaches to computing sensitivities of stochastic models have also been formulated based on the LNA (Hornung and Barkai, 2008).

The application of sensitivity analysis, nonetheless, is still prohibitive for most realistic models of stochastic cellular networks. This problem is further compounded by the aforementioned challenge posed by large numbers of unknown model parameters, which need to be identified from data. Many parameter identifiability analyses use the concept of sensitivity to determine *a priori* whether certain parameters can be estimated from experimental data and to search for these parameters using iterative algorithms. Efficient computation of parameter sensitivities is therefore a topic of great interest and bearing on the applicability of stochastic methods, and one where many challenges still lie ahead.

VII. Conclusions

Stochastic modeling methods are generating many important insights into the operation and organizational principles of cellular networks. Challenges remain before the full power of these methods can be unleashed in the study of many complex biological dynamics. This is an area of great promise, and one where progress will greatly deepen our understanding of the stochastic underpinnings of life.

References

Alon, U. (2007). *An Introduction to Systems Biology: Design Principles of Biological Circuits.* Chapman & Hall/CRC, Boca Raton, FL.

Ander, M., Beltrao, P., and Di Ventura, B., *et al.* (2004). SmartCell, a framework to simulate cellular processes that combines stochastic approximation with diffusion and localisation: analysis of simple networks. *Syst. Biol.* **1**, 129–138.

Atzberger, P. J., Pahlajani, C. D., and Khammash, M. (2011). Stochastic reduction method for biological chemical kinetics using time-scale separation. *J. Theor. Biol.* **272**, 96–112.

Avery, S. V., Smith, M. C. A., and Sumner, E. R. (2007). Glutathione and Gts1p drive beneficial variability in the cadmium resistances of individual yeast cells. *Mol. Microbiol.* **66**, 699–712.

Bancaud, A., Huet, S., Daigle, N., Mozziconacci, J., Beaudouin, J., and Ellenberg, J. (2009). Molecular crowding affects diffusion and binding of nuclear proteins in heterochromatin and reveals the fractal organization of chromatin. *EMBO J.* **28**(24), 3785–3798.

Bar-Even, A., Paulsson, J., Maheshri, N., Carmi, M., O'Shea, E., Pilpel, Y., and Barkai, N. (2006). Noise in protein expression scales with natural protein abundance. *Nat. Genet.* **38**, 636–643.

Baras, F., and Mansour, M. M. (1996). Reaction-diffusion master equation: a comparison with microscopic simulations. *Phys. Rev. E* **54**, 6139–6148.

Bartol, TM., Stiles, JR. (2002). MCell: A Monte Carlo Simulation of Cellular Physiology.

Bigger, W. B. (1944). Treatment of staphylococcal infections with penicillin by intermittent sterilization. *Lancet* **2**, 497–500.

Blake, W. J., Balazsi, G., Kohanski, M. A., Isaacs, F. J., Murphy, K. F., Kuang, Y., Cantor, C. R., Walt, D. R., and Collins, J. J. (2006). Phenotypic consequences of promoter-mediated transcriptional noise. *Mol. Cell* **24**, 853–865.

Cai, L., Friedman, N., and Xie, X. S. (2006). Stochastic protein expression in individual cells at the single molecule level. *Nature* **440**, 358–362.

Cai, X. D., and Xu, Z. Y. (2007). K-leap method for accelerating stochastic simulation of coupled chemical reactions. *J. Chem. Phys.* **126**.

Cao, Y., Gillespie, D. T., and Petzold, L. R. (2005). The slow-scale stochastic simulation algorithm. *J. Chem. Phys.* **122**, 14116.

Chatterjee, A., Vlachos, D. G., and Katsoulakis, M. A. (2005). Binomial distribution based tau-leap accelerated stochastic simulation. *J. Chem. Phys.* **122**.

Choi, P. J., Cai, L., Frieda, K., and Xie, S. (2008). A stochastic single-molecule event triggers phenotype switching of a bacterial cell. *Science* **322**, 442–446.

Chubb, J. R., Trcek, T., Shenoy, S. M., and Singer, R. H. (2006). Transcriptional pulsing of a developmental gene. *Curr. Biol.* **16**, 1018–1025.

Cluzel, P., Le, T. T., Harlepp, S., Guet, C. C., Dittmar, K., Emonet, T., and Pan, T. (2005). Real-time RNA profiling within a single bacterium. *Proc. Natl. Acad. Sci. U S A* **102**, 9160–9164.

Cornish-Bowden, A. (1979). *Fundamentals of Enzyme Kinetics.* Butterworths, London, Boston.

Eissing, T., Kuepfer, L., Becker, C., Block, M., Coboeken, K., Gaub, T., Goerlitz, L., Jaeger, J., Loosen, R., and Ludewig, B., *et al.* (2011). A computational systems biology software platform for multiscale modeling and simulation: integrating whole-body physiology, disease biology, and molecular reaction networks. *Front. Physiol.* **2**, 4.

Elf, J., and Ehrenberg, M. (2003). Fast evaluation of fluctuations in biochemical networks with the linear noise approximation. *Genome Res.* **13**, 2475–2484.

Elf, J., Fange, D., Berg, O. G., and Sjoberg, P. (2010). Stochastic reaction-diffusion kinetics in the microscopic limit. *Proc. Natl. Acad. Sci. U S A* **107**, 19820–19825.

Gillespie, D. T. (1977). Exact stochastic simulation of coupled chemical reactions. *J. Phys. Chem.* **81**, 2340–2361.

Gillespie, D. T. (1992). A rigorous derivation of the chemical master equation. *Physica A* **188**, 404–425.

Gillespie, D. T. (2001). Approximate accelerated stochastic simulation of chemically reacting systems. *J. Chem. Phys.* **115**, 1716–1733.

Gillespie, D. T., and Petzold, L. R. (2003). Improved leap-size selection for accelerated stochastic simulation. *J. Chem. Phys.* **119**, 8229–8234.

Golding, I., Paulsson, J., Zawilski, S. M., and Cox, E. C. (2005). Real-time kinetics of gene activity in individual bacteria. *Cell* **123**, 1025–1036.

Gomez-Uribe, C. A., and Verghese, G. C. (2007). Mass fluctuation kinetics: capturing stochastic effects in systems of chemical reactions through coupled mean-variance computations. *J. Chem. Phys.* **126**, .

Gregor, T., Tank, D. W., Wieschaus, E. F., and Bialek, W. (2007a). Probing the limits to positional information. *Cell* **130**, 153–164.

Gregor, T., Wieschaus, E. F., McGregor, A. P., Bialek, W., and Tank, D. W. (2007b). Stability and nuclear dynamics of the bicoid morphogen gradient. *Cell* **130**, 141–152.

Gunawan, R., Cao, Y., Petzold, L., and Doyle, F. J. (2005). Sensitivity analysis of discrete stochastic systems. *Biophys. J.* **88**, 2530–2540.

Haseltine, E. L., and Rawlings, J. B. (2002). Approximate simulation of coupled fast and slow reactions for stochastic chemical kinetics. *J. Chem. Phys.* **117**, 6959–6969.

Hasty, J., and Collins, J. J. (2002). Translating the noise. *Nat. Genet.* **31**, 13–14.

Hornung, G., and Barkai, N. (2008). Noise propagation and signaling sensitivity in biological networks: a role for positive feedback. *PLOS Comput. Biol.* **4**.

Kampen, N. (1992). *Stochastic Processes in Chemistry and Physics.* Elsevier, .

Karp, X., and Greenwald, I. (2003). Post-transcriptional regulation of the E/Daughterless ortholog HLH-2, negative feedback, and birth order bias during the AC/VU decision in *C. elegans*. *Genes Dev.* **17**, 3100–3111.

Keeling, M. J. (2000). Multiplicative moments and measures of persistence in ecology. *J. Theor. Biol.* **205**, 269–281.

Konopka, M. C., Shkel, I. A., Cayley, S., Record, M. T., and Weisshaar, J. C. (Sep 2006). Crowding and confinement effects on protein diffusion in vivo. *J Bacteriol.* **188**(17), 6115–6123.

Kussell, E., Kishony, R., Balaban, N. Q., and Leibler, S. (2005). Bacterial persistence: a model of survival in changing environments. *Genetics* **169**, 1807–1814.

McAdams, H. H., and Arkin, A. (1999). It's a noisy business! Genetic regulation at the nanomolar scale. *Trends Genet.* **15**, 65–69.

McAdams, H. H., Arkin, A., and Ross, J. (1998). Stochastic kinetic analysis of developmental pathway bifurcation in phage lambda-infected *Escherichia coli* cells. *Genetics* **149**, 1633–1648.

Mcquarri, D. A. (1967). Stochastic approach to chemical kinetics. *J. Appl. Probability* **4**, 413–477.

Morton-Firth, C. J., and Bray, D. (1998). Predicting temporal fluctuations in an intracellular signalling pathway. *J. Theor. Biol.* **192**, 117–128.

Moyed, H. S., and Broderick, S. H. (1986). Molecular-cloning and expression of *HIPA*, a gene of *Escherichia coli* K-12 that affects frequency of persistence after inhibition of murein synthesis. *J. Bacteriol.* **166**, 399–403.

Murphy, K. F., Balazsi, G., and Collins, J. J. (2007). Combinatorial promoter design for engineering noisy gene expression. *Proc. Natl. Acad. Sci. U S A* **104**, 12726–12731.

Nasell, I. (2003). An extension of the moment closure method. *Theor. Popul. Biol.* **64**, 233–239.

Nemenman, I., Sinitsyn, N. A., and Hengartner, N. (2009). Adiabatic coarse-graining and simulations of stochastic biochemical networks. *Proc. Natl. Acad. Sci. U S A* **106**, 10546–10551.

Newman, J. R., Ghaemmaghami, S., Ihmels, J., Breslow, D. K., Noble, M., DeRisi, J. L., and Weissman, J. S. (2006a). Single-cell proteomic analysis of *S. cerevisiae* reveals the architecture of biological noise. *Nature* **441**, 840–846.

Newman, J. R. S., Ghaemmaghami, S., Ihmels, J., Breslow, D. K., Noble, M., DeRisi, J. L., and Weissman, J. S. (2006b). Single-cell proteomic analysis of *S. cerevisiae* reveals the architecture of biological noise. *Nature* **441**, 840–846.

Ozbudak, E. M., Thattai, M., Kurtser, I., Grossman, A. D., and van Oudenaarden, A. (2002). Regulation of noise in the expression of a single gene. *Nat. Genet.* **31**, 69–73.

Paulsson, J. (2005). Models of stochastic gene expression. *Phys. Life Rev.* **2**, 157–175.

Paulsson, J., Berg, O. G., and Ehrenberg, M. (2000). Stochastic focusing: Fluctuation-enhanced sensitivity of intracellular regulation. *Proc. Natl. Acad. Sci. U S A* **97**, 7148–7153.

Plyasunov, S., and Arkin, A. P. (2006). Averaging methods for stochastic dynamics of complex reaction networks: description of multiscale couplings. *Multiscale Modeling Simulation* **5**, 497–513.

Raj, A., Peskin, C. S., Tranchina, D., Vargas, D. Y., and Tyagi, S. (2006). Stochastic mRNA synthesis in mammalian cells. *PLOS Biol.* **4**, 1707–1719.

Rao, C. V., and Arkin, A. P. (2003). Stochastic chemical kinetics and the quasi-steady-state assumption: application to the Gillespie algorithm. *J. Chem. Phys.* **118**, 4999–5010.

Rao, C. V., Wolf, D. M., and Arkin, A. P. (2002). Control, exploitation and tolerance of intracellular noise. *Nature* **420**, 231–237.

Raser, J. M., and O'Shea, E. K. (2005). Noise in gene expression: origins, consequences, and control. *Science* **309**, 2010–2013.

Rathinam, M., and El Samad, H. (2007). Reversible-equivalent-monomolecular tau: a leaping method for "small number and stiff" stochastic chemical systems. *J. Comput. Phys.* **224**, 897–923.

Rathinam, M., Sheppard, P. W., and Khammash, M. (2010). Efficient computation of parameter sensitivities of discrete stochastic chemical reaction networks. *J. Chem. Phys.* **132**.

Ridgway, D., Broderick, G., Lopez-Campistrous, A., Ru'aini, M., Winter, P., Hamilton, M., Boulanger, P., Kovalenko, A., and Ellison, M. J. (2009). Coarse-grained molecular simulation of diffusion and reaction kinetics in a crowded virtual cytoplasm. *Biophys. J.* **96**, 2548.

Schnell, S., Turner, T. E., and Burrage, K. (2004). Stochastic approaches for modelling in vivo reactions. *Comput. Biol. Chem.* **28**, 165–178.

Serizawa, S., Miyamichi, K., Nakatani, H., Suzuki, M., Saito, M., Yoshihara, Y., and Sakano, H. (2003). Negative feedback regulation ensures the one receptor-one olfactory neuron rule in mouse. *Science* **302**, 2088–2094.

Singh, A., and Hespanha, J. P. (2007). A derivative matching approach to moment closure for the stochastic logistic model. *Bull. Math. Biol.* **69**, 1909–1925.

Skupsky, R., Burnett, J. C., Foley, J. E., Schaffer, D. V., and Arkin, A. P. (2010). HIV promoter integration site primarily modulates transcriptional burst size rather than frequency. *PLOS Comput. Biol.* **6**.

Stelling, J., Doyle, F. J., and Gilles, E. D. (2004). Robustness properties of circadian clock architectures. *Proc. Natl. Acad. Sci. U S A* **101**, 13210–13215.

Strogatz, S. H. (1994). *Nonlinear Dynamics and Chaos: With Applications to Physics, Biology, Chemistry, and Engineering.* Addison-Wesley, Reading, MA.

Stundzia, A. B., and Lumsden, C. J. (1996). Stochastic simulation of coupled reaction-diffusion processes. *J. Comput. Phys.* **127**, 196–207.

Suel, G. M., Cagatay, T., Turcotte, M., Elowitz, M. B., and Garcia-Ojalvo, J. (2009). Architecture-dependent noise discriminates functionally analogous differentiation circuits. *Cell* **139**, 512–522.

Suel, G. M., Garcia-Ojalvo, J., Liberman, L. M., and Elowitz, M. B. (2006). An excitable gene regulatory circuit induces transient cellular differentiation. *Nature* **440**, 545–550.

Suel, G. M., Kulkarni, R. P., Dworkin, J., Garcia-Ojalvo, J., and Elowitz, M. B. (2007). Tunability and noise dependence in differentiation dynamics. *Science* **315**, 1716–1719.

Taniguchi, Y., Choi, P. J., Li, G. W., Chen, H. Y., Babu, M., Hearn, J., Emili, A., and Xie, X. S. (2010). Quantifying *E. coli* proteome and transcriptome with single-molecule sensitivity in single cells. *Science* **329**, 533–538.

Thattai, M. T., and van Oudenaarden, A. (2001). Intrinsic noise in gene regulatory networks. *Biophys. J.* **80**, 479a.

Tigges, M., Marquez-Lago, T. T., Stelling, J., and Fussenegger, M. (2009). A tunable synthetic mammalian oscillator. *Nature* **457**, 309–312.

Tomioka, R., Kimura, H., Kobayashi, T. J., and Aihara, K. (2004). Multivariate analysis of noise in genetic regulatory networks. *J. Theor. Biol.* **229**, 501–521.

Tsai, T. Y. C., Choi, Y. S., Ma, W. Z., Pomerening, J. R., Tang, C., and Ferrell, J. E. (2008). Robust, tunable biological oscillations from interlinked positive and negative feedback loops. *Science* **321**, 126–129.

van Oudenaarden, A., Acar, M., and Mettetal, J. T. (2008). Stochastic switching as a survival strategy in fluctuating environments. *Nat. Genet.* **40**, 471–475.

Varma, A., Morbidelli, M., and Wu, H. (2005). Parametric Sensitivity in Chemical Systems. Cambridge University Press, Cambridge, New York.

Vrljic, M., Nishimura, S. Y., Brasselet, S., Moerner, W. E., and McConnell, H. M. (2002). Uncorrelated diffusion of MHC class II proteins in the plasma membrane. *Biophys. J.* **82**, 523a.

Weisshaar, J. C., Konopka, M. C., Shkel, I. A., Cayley, S., and Record, M. T. (2006). Crowding and confinement effects on protein diffusion in vivo. *J. Bacteriol.* **188**, 6115–6123.

Whittle, P. (1957). On the use of the normal approximation in the treatment of stochastic-processes. *J. R. Stat. Soc. Ser. B-Stat. Methodol.* **19**, 268–281.

Wolf, D. M., Vazirani, V. V., and Arkin, A. P. (2005). Diversity in times of adversity: probabilistic strategies in microbial survival games. *J. Theor. Biol.* **234**, 227–253.

Yu, J., Xiao, J., Ren, X. J., Lao, K. Q., and Xie, X. S. (2006). Probing gene expression in live cells, one protein molecule at a time. *Science* **311**, 1600–1603.

Further reading

Avery, S. V. (2006). Microbial cell individuality and the underlying sources of heterogeneity. *Nat. Rev. Microbiol.* **4**, 577–587.

Gillespie, D. T. (1977a). Exact stochastic simulation of coupled chemical reactions. *J. Phys. Chem.* **81**, 2340–2361.

Kampen, N. (1992a). *Stochastic Processes in Chemistry and Physics.* Elsevier.

Maheshri, N., and O'Shea, E. K. (2007). Living with noisy genes: how cells function reliably with inherent variability in gene expression. *Annu. Rev. Biophys. Biomol. Struct.* **36**, 413–434.

CHAPTER 6

Quantifying Traction Stresses in Adherent Cells

Casey M. Kraning-Rush, Shawn P. Carey, Joseph P. Califano and Cynthia A. Reinhart-King

Department of Biomedical Engineering, Cornell University, Ithaca, New York, USA

METHODS IN CELL BIOLOGY, VOL 110

0091-679X/10 $35.00
DOI 10.1016/B978-0-12-388403-9.00006-0

Abstract

Contractile force generation plays a critical role in cell adhesion, migration, and extracellular matrix reorganization in both 2D and 3D environments. Characterization of cellular forces has led to a greater understanding of cell migration, cellular mechanosensing, tissue formation, and disease progression. Methods to characterize cellular traction stresses now date back over 30 years, and they have matured from qualitative comparisons of cell-mediated substrate movements to high-resolution, highly quantitative measures of cellular force. Here, we will provide an overview of common methods used to measure forces in both 2D and 3D microenvironments. Specific focus will be placed on traction force microscopy, which measures the force exerted by cells on 2D planar substrates, and the use of confocal reflectance microscopy, which can be used to quantify collagen fibril compaction as a metric for 3D traction forces. In addition to providing experimental methods to analyze cellular forces, we discuss the application of these techniques to a large range of biomedical problems and some of the significant challenges that still remain in this field.

I. Introduction

Cellular traction forces have been shown to drive cell adhesion (Reinhart-King *et al.*, 2003), spreading (Reinhart-King *et al.*, 2005), migration (Dembo and Wang, 1999; Pelham and Wang, 1997), and extracellular matrix (ECM) deposition and remodeling (Lemmon *et al.*, 2009). To migrate, a cell must undergo changes in cellular force production to modify both its shape and its internal tension to interact with the surrounding ECM, which provides both a substrate for the cell to adhere to as it moves forward, but also a barrier through which the cell must advance (Ehrbar *et al.*, 2011). In most adherent cells, forward movement is initiated by actin polymerization, causing a pseudopod to extend from the leading edge of the cell. Cell extensions interact with the surrounding ECM and initiate binding through transmembrane integrin receptors, forming focal complexes and focal adhesions (Hynes, 2002). Contractile force caused by actomyosin contraction generates both intracellular tension and extracellular tension transmitted to the substrate, ultimately causing the cell's posterior focal adhesions to release and allowing the cell to move forward (Lauffenburger and Horwitz, 1996). During migration, changes in the cytoskeleton alter cell–matrix dynamics and cellular force generation. These processes can also be altered in disease states. For example, during malignant transformation, cellular forces have been shown to increase (Paszek *et al.*, 2005; Rosel *et al.*, 2008). Because migration is fundamental to many essential biological processes including development, immune response, inflammation and wound healing, and cells must exert force to migrate, many groups have described methods to characterize force generation of adherent cells (Dembo and Wang, 1999; du Roure *et al.*, 2005; Galbraith and Sheetz, 1997; Harris *et al.*, 1980; Tan *et al.*, 2003).

The earliest technique used to describe the traction forces exerted by cells was developed by Harris and colleagues in the 1980s. In this landmark article, cells were seeded on top of an elastomeric silicone rubber substrate (Harris *et al.*, 1980). As the cells adhered and migrated, they generated wrinkles within the substrate during contraction. Although both the cells and the wrinkles they produced were easily visualized, it was difficult to extract out quantitative information regarding the cellular forces as the wrinkles were typically nonlinear and irregularly shaped. Therefore, although informative as a probe of cellular forces, this technique yielded only semiquantitative data in the form of number and length of wrinkles (Harris *et al.*, 1981). In later work using wrinkling substrates, cellular force data were extracted by using flexible microneedles to exert a known force onto the substrate, reversing the wrinkle caused by the cell (Burton and Taylor, 1997).

Building upon the work pioneered by Harris *et al.* in the early 1990s, the idea of observing and measuring bead displacement in an elastic substratum, rather than wrinkles, was introduced. A thin layer of latex beads was airbrushed over a non-wrinkling elastic film created by cross-linking silicone oil to the sides of a rigid vessel. This created a tightly stretched film upon which cellular forces transmitted to the surface could be more directly detected through bead displacements (Lee *et al.*, 1994; Oliver *et al.*, 1995). This technique had the advantage of being more sensitive than the silicone wrinkling technique, in that relatively small forces ($\sim$20 nN) could be detected based on bead movements. Additionally, the silicone oil could be cross-linked to varying degrees to produce a range of substrate moduli. However, this approach did suffer from some limitations. For instance, more compliant silicone substrates were unable to completely recover from cellular deformation (Lee *et al.*, 1994). Moreover, these substrates were nonporous and poorly adhesive, and their mechanical properties could not be sufficiently tuned to match the strength of the majority of mammalian cell types (Dembo and Wang, 1999).

These limitations were overcome in the late 1990s with the advent of polyacrylamide (PA) hydrogels as a substrate on which to plate cells (Brandley *et al.*, 1987; Wang and Pelham, 1998). The mechanical and chemical properties of PA gels are ideal for the study of cellular forces. First, similar to the earlier generation of silicone films, PA gels are optically transparent, allowing cells cultured on them to be easily imaged, and they can also have fluorescent markers embedded into them, allowing the user to measure deformations caused by cell migration using standard fluorescent microscopy. More importantly, PA gels are elastic and will deform in direct proportion to a broad range of applied force. Once this force is removed, PA gels immediately and reproducibly recover to an unstressed conformation. Moreover, the stiffness and ECM protein ligand density presented on the surface of PA gels can be independently tuned, allowing for precise control over experimental conditions. PA gels are also nontoxic, and create a more physiological environment than glass or silicone rubber for short-term culture of a wide variety of adherent mammalian cell types (Wang and Pelham, 1998).

The PA system was rapidly adapted for use in quantifying the traction forces of adherent cells by Dembo and co-workers, giving rise to the technique of traction force microscopy (TFM), which we currently use in our own lab and is the primary focus of this chapter (Dembo and Wang, 1999). TFM can be used to calculate the traction stresses of individual cells based on the displacement of fluorescent bead markers embedded within PA gels to produce sensitive, quantitative data characterizing intracellular force generation (Reinhart-King *et al.*, 2003). The goal of this chapter is to describe the basic theory underlying TFM and to describe in detail this technique for quantifying cellular forces during 2D migration. Additionally, we will examine the current state of the field, and discuss the current transitions from 2D to 3D TFM methods. We will describe a technique used in our own lab to visualize force generation of cells embedded within 3D collagen matrices: confocal reflectance microscopy. In summary, in this chapter we will describe (1) the steps to fabricate substrates for TFM, (2) a protocol for acquiring and analyzing TFM data, (3) a novel method for qualitatively assessing force generation in 3D collagen gels, and (4) applications of the techniques described herein.

II. Overview of Method

A. Polyacrylamide Gel Substrates

As previously described, TFM in its current form utilizes a well-characterized PA gel substrate system (Wang and Pelham, 1998). The most unique and useful property of PA gels is the ability to independently adjust their mechanical and chemical properties. The Young's modulus (*E*) of PA gels is tunable simply by altering the ratio of acrylamide to bis-acrylamide (for ratios used in our lab, Table I). Additionally, the density of protein ligand available to the cell on the surface of

Table I
Components required to synthesize polyacrylamide gels of stiffness 0.2–300 kPa.

E (kPa)	Percentage of Acrylamide	Acrylamide (mL)	Percentage of Bis-acrylamide	Bis-acryl-amide (mL)	250 mM HEPES (mL)	MilliQ water (mL)	TEMED (μL)
0.2	3.0	1.50	0.040	0.40	2.60	13.99	10
0.5	3.0	1.50	0.050	0.50	2.60	13.89	10
1[a]	3.0	1.50	0.100	1.00	2.60	13.39	10
2.5[a]	5.0	2.50	0.100	1.00	2.60	12.39	10
5[a]	7.5	3.75	0.175	1.75	2.60	10.39	10
10[a]	7.5	3.75	0.350	3.50	2.60	8.64	10
15	12.0	6.00	0.130	1.30	2.60	8.59	10
20	12.0	6.00	0.190	1.90	2.60	7.99	10
30	12.0	6.00	0.280	2.80	2.60	7.09	10
300	15.0	7.50	1.200	12.00	0.65[b]	0.00	10

[a] Indicates PA gel stiffness used successfully with TFM.
[b] Substitute 1 M HEPES buffer.

the gel is controlled by reacting varying concentrations of ligand with a bifunctional linker, which is added to the polymer mix. The ligand can be any protein or amino acid sequence desired, as the cross-linker within the gel is designed to form a stable amide linkage to any molecule with a primary amine group (Pless *et al.*, 1983). Given the mechanical and chemical flexibility of PA gels, they are useful to recreate various physiological conditions as well as disease states *in vitro*, which will be discussed later in the chapter (Califano and Reinhart-King, 2010; Rosel *et al.*, 2008; Yeung *et al.*, 2005).

B. Traction Force Microscopy

1. Rationale

TFM is a technique that allows for the precise quantification of traction stresses generated by cells adherent to an underlying two-dimensional substrate, most often a PA gel (Dembo *et al.*, 1996). As cells adhere to and migrate over sufficiently compliant substrates, traction forces generated by the cell create deformations. These deformations are detected by the inclusion of fiduciary markers (usually submicron diameter fluorescent beads) within the PA gel that relax back to their original position when the cell is either released chemically from the substrate (e.g., with trypsin) (Fig. 1A and B) or when it migrates away from the field of view during a time-course study.

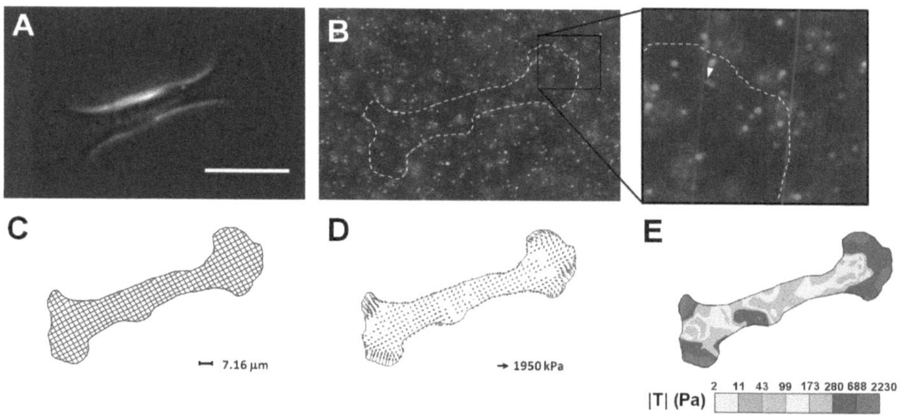

Fig. 1 Traction force microscopy is used to quantify traction forces in 2D. MDA-MB-231 cells were seeded onto a polyacrylamide substrate (A). To quantify traction forces, fluorescent images are acquired of the bead field beneath the cell during force generation (B, red) and after the cell has been released with trypsin (B, green). An overlay of these two images indicates the regions of greatest bead displacement (inset, white arrow). To calculate the most likely traction field causing the observed bead displacement, the cell is first discretized into a mesh (C). Individual tractions are then calculated for each node of the mesh (D). From these tractions, a color contour plot can be generated indicating regions of highest and lowest traction stresses (E). Scale bar = 50 μm. (See color plate.)

Experimentally, calculation of the substrate strain field requires images of the bead field in both its stressed state (with the cell present) and relaxed state (without the cell). The image of the beads in their relaxed state is typically captured after the cell is removed and the beads in the field of view have returned to their "unstressed" position due to the elastic nature of the PA gel substrate. The displacements caused by the cells are computed by comparing the stressed and relaxed images of the bead field.

The calculation of traction stresses is regarded as an inverse problem, that is, measurements of substrate deformation are used to statistically compute the most likely traction stress field that can give rise to the observed deformations. The traction field is thus derived from numerical integrals to determine the maximum likelihood tractions based on the displacement field. These tractions are tied to chi-square and Bayesian statistics that iterate until convergence. TFM thus involves both methods that determine the substrate displacements caused by adherent cells and algorithms that convert these displacements into a traction stress field (Fig. 1C–E). Although the precise details of TFM theory are discussed at length elsewhere (Dembo *et al.*, 1996; Dembo and Wang, 1999), we put forth here a brief summary describing key equations leading to the determination of the traction stress field.

2. Theory

The theory of TFM is founded on the isotropic and linearly elastic material properties of the PA gel. First, a basic stress–strain relationship describing homogeneous deformation of the PA gel is established in Eq. (1),

$$\sigma_{ik} = \frac{E}{1+\upsilon}\left(\varepsilon_{ik} + \frac{\upsilon}{1-2\upsilon}\varepsilon_{ll}\delta_{ik}\right) \tag{1}$$

where σ_{ik} are the components of the stress tensor, ε_{ik} are the components of the strain tensor, E is the Young's Modulus, υ is the Poisson's ratio, and δ_{ik} is the Kronecker delta (Dembo *et al.*, 1996; Landau *et al.*, 1986). Next, the assumption is made that our system is a longitudinal plate that is sufficiently thin, such that the deformation is regarded as uniform over its thickness and the strain tensor is dependent only on x and y (with the x–y plane being that of the plate or PA gel) (Landau *et al.*, 1986). The boundary conditions on both surfaces of the plate are then $\sigma_{ik}n_k = 0$, where n_k is the normal vector. Because the normal vector is parallel to the z axis in this case, $\sigma_{iz} = 0$ (i.e. $\sigma_{xz} = \sigma_{yz} = \sigma_{zz} = 0$). It is also important to note that because $\sigma_{iz} = 0$ at the surface, the quantities σ_{xz}, σ_{yz}, and σ_{zz} must be small throughout the thickness of the plate, and we will approximate them as zero everywhere within the plate (Landau *et al.*, 1986). These boundary conditions can then be substituted into Eq. (1) to get the nonzero components of stress, Eqs. (2)–(4),

$$\sigma_{xx} = \frac{E}{1-\upsilon^2}(\varepsilon_{xx} + \upsilon\varepsilon_{yy}) \tag{2}$$

$$\sigma_{yy} = \frac{E}{1-\upsilon^2}(\varepsilon_{yy} + \upsilon\varepsilon_{xx}) \tag{3}$$

$$\sigma_{xy} = \frac{E}{1+\upsilon}\varepsilon_{xy} \tag{4}$$

If the plate is considered as a 2D elastic plane of zero thickness, then a displacement vector ***d*** can be considered as a two-dimensional vector with components d_x and d_y. If T_x and T_y are the components of the external body force per unit area of the plate, then the general equations of equilibrium are Eqs. (5) and (6),

$$h\left(\frac{\partial\sigma_{xx}}{\partial x} + \frac{\partial\sigma_{xy}}{\partial y}\right) + T_x = 0 \tag{5}$$

$$h\left(\frac{\partial\sigma_{yx}}{\partial x} + \frac{\partial\sigma_{yy}}{\partial y}\right) + T_y = 0 \tag{6}$$

where h is the thickness of the PA gel. Next, if the stress components from Eqs. (2) to (4) are substituted in, the results are the equations of equilibrium in the form of Eqs. (7) and (8) (Dembo *et al.*, 1996),

$$Eh\left\{\frac{1}{1-\upsilon^2}\frac{\partial^2 d_x}{\partial x^2} + \frac{1}{2(1+\upsilon)}\frac{\partial^2 d_x}{\partial y^2} + \frac{1}{2(1-\upsilon)}\frac{\partial^2 d_y}{\partial x\partial y}\right\} + T_x = 0 \tag{7}$$

$$Eh\left\{\frac{1}{1-\upsilon^2}\frac{\partial^2 d_y}{\partial y^2} + \frac{1}{2(1+\upsilon)}\frac{\partial^2 d_y}{\partial x^2} + \frac{1}{2(1-\upsilon)}\frac{\partial^2 d_x}{\partial x\partial y}\right\} + T_y = 0 \tag{8}$$

Because the response of the PA gel substrate to deformation is linear, the displacement of the pth bead marker can be related to the traction field *via* an integral transform, Eq. (9),

$$d_{p\alpha} = \int\int g_{\beta\alpha}(\mathbf{m}_p - \mathbf{r})T_\beta(\mathbf{r})dr_1 dr_2 \tag{9}$$

where $g_{\beta\alpha}(\mathbf{m}_p - \mathbf{r})$ represent the coefficients of a Green's tensor that give the substrate displacement in the α direction at location $\mathbf{m}$ induced by a force in the β direction acting at location $\mathbf{r}$ (Dembo and Wang, 1999). As the thickness of our PA gel substrate (~70 μm) can be considered infinite compared to the greatest bead displacement (~1 μm), the coefficients of $g_{\beta\alpha}$ can be approximated using the Boussinesq theory for an elastic solid in the half-space beneath the cell, Eqs. (10)–(18),

$$g_{11} = \frac{1+\nu}{2\pi E}\left\{\frac{(2(1-\nu)r - x_3)}{r(r-x_3)} + \frac{(2r(\nu r - x_3) + x_3^2)x_1^2}{r^3(r-x_3)^2}\right\} \tag{10}$$

$$g_{21} = \frac{1+\nu}{2\pi E}\left\{\frac{(2r(\nu r - x_3) + x_3^2)x_1x_2}{r^3(r-x_3)^2}\right\} \tag{11}$$

$$g_{31} = \frac{1+\nu}{2\pi E}\left\{\frac{x_1x_3}{r^3} + \frac{(1-2\nu)x_1}{r(r-x_3)}\right\} \tag{12}$$

$$g_{12} = \frac{1+\nu}{2\pi E}\left\{\frac{(2r(\nu r - x_3) + x_3^2)x_1x_2}{r^3(r-x_3)^2}\right\} \tag{13}$$

$$g_{22} = \frac{1+\nu}{2\pi E}\left\{\frac{(2(1-\nu)r - x_3)}{r(r-x_3)} + \frac{(2r(\nu r - x_3) + x_3^2)x_2^2}{r^3(r-x_3)^2}\right\} \tag{14}$$

$$g_{32} = \frac{1+\nu}{2\pi E}\left\{\frac{x_2x_3}{r^3} + \frac{(1-2\nu)x_2}{r(r-x_3)}\right\} \tag{15}$$

$$g_{13} = \frac{1+\nu}{2\pi E}\left\{\frac{x_1x_3}{r^3} - \frac{(1-2\nu)x_1}{r(r-x_3)}\right\} \tag{16}$$

$$g_{23} = \frac{1+\nu}{2\pi E}\left\{\frac{x_2x_3}{r^3} - \frac{(1-2\nu)x_2}{r(r-x_3)}\right\} \tag{17}$$

$$g_{33} = \frac{1+\nu}{2\pi E}\left\{\frac{2(1-\nu)}{r} + \frac{x_3^2}{r^3}\right\} \tag{18}$$

The functions for the coefficients of the Green's tensor incorporate both ν and E of the PA substrate (Dembo and Wang, 1999), which can be determined experimentally (Boudou *et al.*, 2006; Li *et al.*, 1993). One technique for measuring the parameter E will be described in a later section (Lo *et al.*, 2000). Additionally, it is important to note that Boussinesq theory predicts negligible coupling of in-plane displacements to out-of-plane tractions at or near the surface of an incompressible substrate (i.e., $g_{13} = g_{23} = 0$). In this manner, bead displacement in the direction normal to the PA substrate is ignored (Dembo and Wang, 1999).

To produce a traction image from the displacement data, the projected area of the cell must be imposed onto the traction field (Fig. 1A). The cell is outlined with a series of points that generate a list of pixel coordinates defining the cell boundary. These points define a bounded region that corresponds to the projected cell area. In the TFM theory described by Dembo and Wang, the assumption is made that all tractions occur within this domain (Fig. 1B). The outline of the cell is divided into a quadrilateral mesh utilizing a paving algorithm (Dembo and Wang, 1999) (Fig. 1C). The x and y components of the traction located at each node of the mesh can then be determined (Fig. 1D), and a representation of the continuum of forces that occurs over the entire mesh interior can be constructed (Fig. 1E).

Within the bounded domain the in-plane traction components are first approximated using standard bilinear shape functions, $H_k(\boldsymbol{r})$, as in Eq. (19),

$$T_\beta(\boldsymbol{r}) \approx T_{k\beta} H_k(\boldsymbol{r}) \tag{19}$$

where $T_{k\beta}$ are now the components of the nodal traction vectors (Dembo *et al.*, 1996; Dembo and Wang, 1999). Next, for this mesh, any choice of $T_{k\beta}$ corresponds to an allowable traction image. By substituting Eq. (19) into Eq. (9), this traction image can make a definite prediction about the calculated marker displacements, Eq. (20),

$$d_{p\alpha} = d_\alpha(\boldsymbol{m}_p) = T_{k\beta} \iint g_{\alpha\beta}(\boldsymbol{m}_p - \boldsymbol{r}) H_k(\boldsymbol{r}) dr_1 dr_2 = A_{k\beta p\alpha} T_{k\beta} \tag{20}$$

where the index p runs over all of the bead markers (i.e., $p = 1, 2, \ldots, N_p$) (Dembo and Wang, 1999). It is important to note that here, $A_{k\beta p\alpha}$ depends only on the imposed mesh, the location of the bead markers, and the material properties of the PA gel.

Next, the ability of this traction image to explain the observed bead marker displacements can be quantified by using the chi-square statistic, Eq. (21),

$$\chi^2 \equiv (\hat{d}_{p\alpha} - d_{p\alpha})^2 \sigma_{p\alpha}^{-2} = (\hat{d}_{p\alpha} - A_{k\beta p\alpha} T_{k\beta})^2 \sigma_{p\alpha}^{-2} \tag{21}$$

where $\hat{d}_{p\alpha}$ is the experimental displacement of the pth marker particle along the α-coordinate axis, $\sigma_{p\alpha}$ is the error of $\hat{d}_{p\alpha}$, and summation over all repeated indices is implied (Dembo and Wang, 1999). Additionally, Dembo *et al.* quantify the intrinsic "complexity" of a traction image using the scalar invariant shown in Eq. (22),

$$c^2 \equiv \int_\Omega (\partial_\alpha T_\beta + \partial_\beta T_\alpha)(\partial_\alpha T_\beta + \partial_\beta T_\alpha) dr_1 dr_2 \tag{22}$$

Again, by substituting Eq. (19) into Eq. (22), the complexity of the traction image can be written as a quadratic form in the nodal degrees of freedom, Eq. (23),

$$c^2 = C_{i\alpha j\beta} T_{i\alpha} T_{j\beta} \tag{23}$$

where $C_{i\alpha j\beta}$ are constants dependent on the geometry of the mesh (Dembo and Wang, 1999). Finally, by combining Eq. (21) and Eq. (23), the Bayesian likelihood of the $T_{k\beta}$ is found to be represented by Eq. (24),

$$L_b(T_{k\beta} | \hat{d}_{p\alpha}) = \exp[-(\chi^2 + \lambda c^2)] \tag{24}$$

where λ is a positive real number determined by obtaining the simplest traction image consistent with the given data set (Dembo and Wang, 1999). λ is progressively increased, and new values of $\chi^2$and c^2 are found by minimizing the linear combination $\chi^2 + \lambda c^2$. These new values are substituted into Eq. (21) and Eq. (23), becoming $\overline{\chi}^2$ and $\overline{c}^2$. As λ is increased, $\overline{\chi}^2$ increases and $\overline{c}^2$ decreases. Finally, once $\overline{\chi}^2 \approx N_p + \sqrt{N_p}$, that threshold image is the simplest distribution of traction forces consistent with the experimental data (Dembo and Wang, 1999).

Once solved, the magnitude and direction of traction forces, and other parameters including the bead displacement vector field and strain energy density field may be examined. For many cells types, the magnitude of traction forces is on the order of ~0.05–2 μN (Califano and Reinhart-King, 2010; Gaudet *et al.*, 2003; Paszek *et al.*, 2005; Reinhart-King *et al.*, 2003). It is important to recognize that others have used a similar experimental system (PA gels embedded with fluorescent beads), but have solved the inverse problem using Fourier's method to solve the general equations of equilibrium relating displacements to tractions (Butler *et al.*, 2002).

3. Calculating Substrate Displacements: Correlation-based Optical Flow

All current methods to calculate cell-generated traction stresses first require calculation of the underlying substrate deformations. In TFM, substrate deformations are calculated based on the movements of beads embedded within the PA gel substrate. While it is possible, in theory, to map individual bead movements by hand, this would be cumbersome. To automate this process, Marganski *et al.* developed an algorithm based on *correlation-based optical flow*, which has been refined since its original description (Marganski *et al.*, 2003b). This algorithm takes two images (the stressed and relaxed bead field images described above) as the input. Bead tracking is done by systematically scanning all pixels in the relaxed image to find the pixel coordinates of the fluorescent beads (identified as strict pixel intensity maxima after image intensity normalization). For each bead that is tracked, a box of pixels centered on the local maximum intensity pixel is defined and the relative pixel intensities in that box serve as a "fingerprint" for the tracked bead. This search box is used to determine the coordinates of the corresponding "fingerprint" in the stressed image. This process is iterated for every pixel in the image and is able to determine the bead displacements with submicron resolution.

An important parameter considered when comparing the pixel coordinates between two images from the same region of interest is the *registration error* introduced into the images by the misalignment of the relaxed and stressed images. This error is usually inadvertently introduced as rectilinear motion by mechanical vibrations of the microscope stage or imperfect stage return during multipoint acquisition of multiple fields of view. Rectilinear motion introduces a constant vector on the displacements between images that would normally be absent in a perfect experiment. A histogram of the raw displacements of tracked beads provides an elegant way to identify and remove the registration error (the most frequent constant vector tracked by the optical flow algorithm). The current TFM algorithm in use accounts for the presence of a translational, but not a rotational, drift. Rotational drift is rarely introduced as long as the cell sample sits firmly in the microscope stage, and the PA gel is not disturbed during trypsinization. Problems encountered with correlation-based optical flow include occasional bead mistracking between images, which will result in erroneous displacement vectors. In the majority of cases these displacement vectors are easily identified by a single large vector pointing in an unexpected direction, and can be discounted. Additionally, some cell types may also phagocytose beads from the substrate, which causes similar tracking errors. Decreasing incubation time prior to TFM imaging can reduce the occurrence of phagocytosis.

It should be noted that in addition to the correlation-based optical flow algorithm described here, a number of other algorithms have also been developed that can be applied to bead tracking for TFM, including algorithms based on digital image correlation (Qin *et al.*, 2007; Sutton *et al.*, 1983) and, more recently, a combination of particle image velocimetry (PIV) and particle tracking velocimetry (PTV) (Sabass *et al.*, 2008; Tseng *et al.*, 2011). Digital image correlation is a widely used method for the detection of optical displacements, and has undergone numerous refinements in the last 20 years (Huang *et al.*, 2009; Schreier *et al.*, 2000). In digital image correlation, markers are tracked by searching the matching pixel-matrix of intensities in a pair of fluorescent images in order to numerically correlate a selected subset of markers. Digital image correlation has recently been adapted for quantifying 3D tractions exerted by cells on a 2D substrate (Franck *et al.*, 2011). PIV is a technique that has been widely utilized to track bulk particle movement through fluid flow, which does not generally require individual particle tracking. Recently, Sabass *et al.* have paired PIV with PTV in a technique termed correlation-based PTV. PIV is first used to determine the deformation of a PA gel on a coarse scale before PTV is used to segment individual bead displacement. In contrast to the correlation-based optical flow algorithm described above, several variations of these algorithms are available through open source or free software, or else through commercial sources.

4. Key Assumptions

The calculation of traction forces is based on the Boussinesq equations, which describe the relationship between the deformation of a material due to forces applied to its free surface. In this regard, the elastic PA gel substrate is assumed to be uniform, isotropic, and linearly elastic, and it is assumed that the inclusion of marker beads in the substrate does not perturb this elastic behavior. The external loads acting on the substrate surface are assumed to be solely tangential with negligible displacements in the z-direction.

Inherent to TFM is the ability of cells to deform the substrate. This assumption ultimately limits the range of substrate stiffness that is testable using TFM or any traction method that requires the substrate to deform a detectable amount. Although there is considerable variation between the strength of different cell types, TFM may have an upper limit of $E \sim 10$–30 kPa. At the other end of the spectrum, compliant substrates (<1 kPa) may not adequately support the weight of adherent cells and will allow significant bead displacement in the z direction, disrupting the in-plane bead marker displacements quantified in the TFM calculations. It is important to note that recent work (Franck *et al.*, 2011) has demonstrated that cells are able to displace beads in the z-direction even on stiff substrates. These displacements are not accounted for using the protocols and algorithms described in this chapter, introducing a small degree of error into the final calculation.

C. Alternative Methods for Measuring 2D Cell Tractions

In addition to TFM, several other techniques have been developed to quantify traction forces during cell migration. In the late 1990's, Galbraith and Sheetz developed a micromachined device consisting of an array of lithographically patterned silicon cantilever pads coated with ECM protein (Galbraith and Sheetz, 1997). This method allows for the quantification of isolated subcellular tractions, and as a result is quite sensitive, measuring stresses on the order of single nN/μm^2. However, an individual cell can only depress a limited number of cantilevers at a time, limiting the spatial resolution of forces. Additionally, because the cantilevers can move in only one direction, they can only be used to quantify traction forces in that direction. For cells that do not cross the cantilever beam at a 90° angle, forces are calculated based on the assumption that traction stresses are directed along only the long axis of the cell, which is not necessarily always valid (Califano and Reinhart-King, 2010; Dembo and Wang, 1999). Moreover, production of the device requires an elaborate fabrication procedure that requires specialized technology that may not be readily accessible for many labs.

The most commonly used alternative to TFM is the use of microfabricated post-array detection systems (mPADs) (du Roure *et al.*, 2005; Tan *et al.*, 2003). In this method, cylindrical microposts are fabricated out of polydimethylsiloxane (PDMS), and ECM protein is adsorbed to the top of the posts to enable cell adhesion. Traction forces are based on the extent of deflection of the posts from their original position. Post deflection can then be linearly correlated to the local traction forces exerted by the cell using classical beam bending theory. In this system, the height of the micropost can be varied to adjust the rigidity of the posts, and thus to adjust the stiffness of the substrate sensed by the cell. Each post acts as an individual vertical cantilever, sensing force at a discrete location beneath the cell. Moreover, unlike the silicon cantilevers described above, mPADs are able to detect forces generated in all directions of the x–y plane. A more detailed description of the fabrication process and supporting theory and computation can be found elsewhere (Fu *et al.*, 2010; Sniadecki and Chen, 2007; Yang *et al.*, 2011).

Although elastomeric microposts offer several advantages over the original silicon cantilever system, there are several disadvantages that are important to note, especially when comparing this technique to PA gel-based TFM. First, there is considerable controversy over the appropriateness of culturing cells on a topographical landscape which is very distinct from the native environment of mammalian cells. mPADs restrict adhesions to distinct circular patches, imposing arbitrary constrictions on the size, shape, and location of focal adhesions, thus controlling where and how the cell transmits force (Yang *et al.*, 2007). Although this system has been used to elegantly determine the amount of force that individual focal adhesions can exert (Fu *et al.*, 2010), it remains unclear how these calculations relate to the forces actually transmitted in the native physiological environment. Additionally, while the elastomeric posts may be more easily fabricated than the silicon cantilevers, and protocols have been published describing this process in great detail (Yang *et al.*, 2011), an advanced microfabrication facility is still required to reproducibly fabricate the posts. PA gels, on the other hand, are easily produced using standard laboratory chemicals and equipment.

Another significant limitation of mPAD technology is the lower limit of E that can be produced. Posts have been successfully fabricated with a lower limit of $E \sim 1.5$ kPa (Fu *et al.*, 2010), which is considerably higher than that of PA gels ($E \sim 0.1$ kPa), although it should be fairly noted that performing TFM on PA gels of $E < 1$ kPa has its own limitations. PA gels can also be used to examine the effect of mechanical communication of multiple cells through the underlying substrate (Califano and Reinhart-King, 2010; Reinhart-King *et al.*, 2008), a technique that could not be done using microposts, which effectively isolate cells from one another. On the other hand, the ability to mechanically isolate cells can be advantageous. For example, microposts were recently used to determine specific point forces at cell–cell junctions (Liu *et al.*, 2010). In summary, elastomeric microposts and PA gels each have their own distinct advantages and disadvantages, which must be considered when determining the appropriate system to use for quantifying traction forces in a given experiment.

D. Quantifying Cell Force in 3D

Like cells on 2D substrates, cells within 3D microenvironments encounter biochemical, biomechanical, and physical cues that affect basic cellular processes such as adhesion, spreading, and migration. As in 2D, these 3D cell behaviors are closely tied to cellular biomechanics and the generation of cell forces. However, because of the spatial complexity and dimensionality of the three-dimensional microenvironment, both the control and manifestation of cell forces are likely more complex in 3D (Dikovsky *et al.*, 2008; Fraley *et al.*, 2010; Gunzer *et al.*, 2000; Mierke *et al.*, 2008). For example, while the interactions among traction forces and regulators of cell migration such as cell adhesion, the cytoskeleton, and ECM deposition are increasingly well characterized in 2D, there are many additional factors, including ECM steric hindrance and proteolytic path-making, that are unique to cells within 3D microenvironments (Wolf *et al.*, 2003). Such factors critically impact cell migration and are also intimately tied to traction forces (Zaman *et al.*, 2006).

1. Overview of 3D Methods

A variety of techniques have been developed to assess single-cell tractions in models that recapitulate the 3D *in vivo* microenvironment, and the majority of these methods rely upon microscopic visualization and tracking of either embedded beads (Fraley *et al.*, 2010; Poincloux *et al.*, 2011; Shih and Yamada, 2010; Tamariz and Grinnell, 2002) or the structural components of the microenvironment (Friedl *et al.*, 1997; Hartmann *et al.*, 2006; Kim *et al.*, 2006). Briefly, cells are embedded within collagen, fibrin, Matrigel, or synthetic hydrogel matrices or allowed to invade into 3D matrices. To probe 3D cell tractions, the displacements of randomly dispersed, micron-scale beads or ECM components are tracked over time or compared between the stressed and relaxed states (as in 2D TFM) with widefield, confocal, or multiphoton microscopy. These displacement fields can be used themselves as quantitative metrics of 3D cell traction forces or can be used to compute strain energy and traction stress fields.

2. Bead Tracking

A common technique to assess 3D cell traction forces is to track the displacement of fluorescent beads embedded within a 3D hydrogel scaffold. The resulting strain maps can be used to describe 3D cell contractility and traction-mediated matrix reorganization (Fraley *et al.*, 2010; Poincloux *et al.*, 2011; Shih and Yamada, 2010; Tamariz and Grinnell, 2002). To compute the traction stresses that give rise to the observed 3D displacements, current techniques require that the mechanics of the hydrogel matrix be well characterized and isotropic. Using a PEG hydrogel and confocal microscopy, Legant *et al.* showed that cells exerted traction stresses ranging from 0.1 to 5 kPa and that the strongest forces were generated primarily at the tips of long, thin pseudopodia (Legant *et al.*, 2010). Additionally, Maskarinec *et al.* used confocal microscopy to quantify traction stresses in the z direction in fibroblasts plated on 2D PA gel substrates, indicating that 3D forces may also play a significant role in 2D cell migration (Franck *et al.*, 2011; Maskarinec *et al.*, 2009). These findings, which are based upon many of the same principles and assumptions as 2D TFM, represent the most quantitative description of fully 3D traction forces to date. However, although the mechanically defined synthetic hydrogels required for computation of numerical traction stresses can be engineered to be degraded, modified, and remodeled by cells, they often lack the fibrillar structure and full bioactivity of native ECM. Notably, these are the very factors that impart complex mechanical properties to 3D matrices and are normally involved in the critical mechanical and biochemical feedback networks that determine many cell behaviors both *in vivo* and in natural fibrillar extracellular matrices *in vitro* (Wolf and Friedl, 2009). For these reasons, there is presently increasing interest in quantifying the functional outcomes of 3D cell traction by monitoring the dynamic microstructure in physiologically relevant microenvironments rather than translating measured strains into numerical traction stresses.

3. Matrix Tracking

The use of natural biopolymer matrices for 3D *in vitro* cell culture presents unique opportunities and challenges to mechanobiologists: on the one hand, these matrices are analogous to the microenvironment in which cells reside *in vivo*, and on the other hand, they are heterogeneous and mechanically complex. Importantly, use of such *in vitro* tissue models allows multiscale interactions between cells and the fibrillar matrix and cell behaviors such as native ECM deposition, ECM remodeling, path-making, and path-finding, all of which have been shown to occur *in vivo* (Wolf and Friedl, 2009; Wolf *et al.*, 2003). Furthermore, microscopy techniques such as differential interference contrast (DIC) and confocal reflectance microscopy can be used to visualize the dynamic fibrillar structural elements of these matrices due to differences in refractive index between the fibrils and the surrounding media. As it is a functional outcome of 3D traction forces and provides direct visualization of how mechanical loads are bidirectionally transferred between a cell and its microenvironment, our lab and others have used ECM reorganization as a metric of cell force generation in 3D environments (Kim *et al.*, 2006; Kraning-Rush *et al.*, 2011; Pang *et al.*, 2009).

Qualitative work with confocal reflectance and DIC microscopy has enabled the visualization of ECM fibers during cell migration and demonstrated how cell–matrix adhesions dynamically associate with the ECM, enabling remodeling (Friedl *et al.*, 1997; Gunzer *et al.*, 2000; Hartmann *et al.*, 2006; Petroll *et al.*, 2004). Recently, more rigorous techniques have been developed to assess local ECM remodeling at the single-cell level. Quantitative analysis of ECM fiber alignment around cell pseudopodia using Fourier transforms provides insight into the spatiotemporal development of 3D traction forces and matrix reorganization (Kim *et al.*, 2006; Pang *et al.*, 2009). Similar orientation-based strategies are used to assess fiber alignment in gels subjected to exogenous forces, which can help elucidate the interdependence of the ECM, external factors such as interstitial flow and macroscale strain, and cell behaviors like migration and remodeling (Ng and Swartz, 2006; Vader *et al.*, 2009). As dynamic, local matrix alignment and remodeling events ultimately lead to ECM compaction around single cells, optical measurement of collagen density has emerged as another metric of 3D cell traction (Kim *et al.*, 2006; Ng and Swartz, 2006; Pang *et al.*, 2009).

Our lab recently developed an image-processing technique based on local changes in collagen compaction that allows us to quantitatively describe the extent of ECM remodeling that a cell has induced through traction forces (Kraning-Rush *et al.*, 2011). This method is based on the principle that 3D cell tractions result in pericellular matrix compaction, which manifests in a higher density of ECM fibers and thus, increased confocal reflectance signal in proximity to the cell. Using either live or fixed and stained samples, sparsely seeded cells in fibrillar 3D collagen gels are simultaneously imaged with reflected light and either fluorescence or DIC. Using ImageJ, the cell area, which is determined from the fluorescence or DIC/phase contrast image, is subtracted from the reflectance channel and a 40–50 μm selector line is drawn from the cell's centroid into the surrounding matrix. A custom-written ImageJ script rotates the selector line around the entire cell at 1-degree increments, capturing an intensity profile at each step. Zero-intensity values are removed, which defines the cell membrane as the origin and effectively normalizes the data for cell shape. Reflectance intensities are averaged as a function of distance from the cell membrane and the resulting collagen intensity profiles are normalized by subtracting the baseline intensity measured far from the cell membrane. Collagen intensity profiles are fit to an exponential decay model and the half-length of the exponential decay, λ, is extracted to describe how far from the cell the collagen has been remodeled. This method has allowed us to assess the dynamics and evolution of matrix remodeling (Fig. 2) as well as to indirectly assess 3D traction forces generated

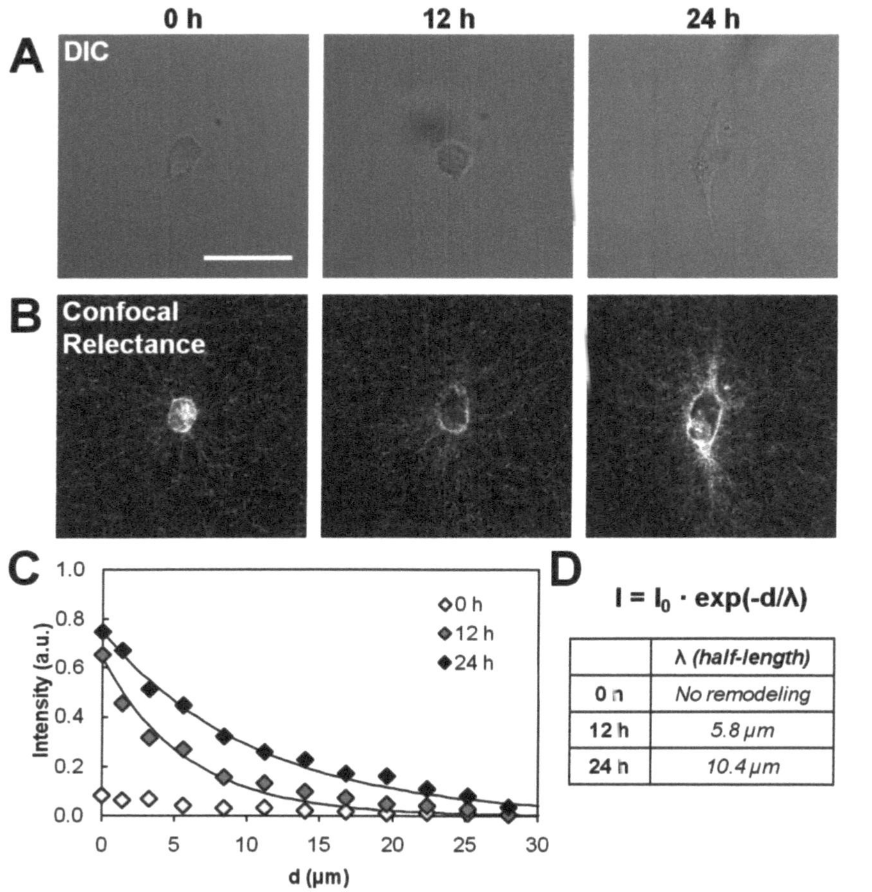

Fig. 2 MDA-MB-231 metastatic breast cancer cell seeded in 1.5 mg/mL collagen gel for 24 h. (A) DIC and (B) confocal reflectance images show coordinated changes in cell morphology and collagen matrix reorganization, respectively, over time. Confocal reflectance image intensity increases with collagen fibril compaction. (C) Quantification of collagen fibril compaction. Data points show normalized, baseline-subtracted, average reflectance intensity as a function of distance from the cell membrane; solid lines are best-fit exponential decays for 12 and 24 h. Consistent with no compaction, 0 h images show no increase in reflectance intensity over baseline. (D) Collagen intensity decreases exponentially as a function of distance from the cell membrane and can be modeled by the equation $I = I_0 \cdot \exp(-d/\lambda)$, where I is the average intensity, I_0 is the intensity at the cell membrane, d is the distance from the cell membrane in microns, and λ is the half-length of the exponential decay, which describes how far from the cell the collagen has been remodeled. Confocal reflectance images are 1 μm slices; scale bar = 50 μm.

by cells treated with various cytoskeleton-perturbing agents (Kraning-Rush *et al.*, 2011). Our results show that increased traction forces in 2D and bulk collagen gel contraction correlate with 3D ECM remodeling as quantified through the above method. The matrix compaction metric that our lab uses is outlined in more detail in the Computational Methods section.

III. Biological Insights from Traction Methods

PA gel substrates and TFM have been widely utilized to study cell forces and other behaviors in a variety of contexts, in both physiologically normal and disease states. These behaviors include morphology (Tang *et al.*, 2010; Yeung *et al.*, 2005), differentiation (Engler *et al.*, 2006), single-cell (Dembo and Wang, 1999) and collective cell migration (Trepat *et al.*, 2009), cell–cell interactions (Califano and Reinhart-King, 2010; Reinhart-King *et al.*, 2008), cell–ECM interactions (Maskarinec *et al.*, 2009), and focal adhesion assembly (Balaban *et al.*, 2001; Rape *et al.*, 2011; Stricker *et al.*, 2011). In this section, we will briefly describe an overview of some of this work. However, this is by no means considered to be all-inclusive, but rather is designed to spark further interest in these topics. For excellent reviews on these and other related topics, see the Further Reading section.

A. Using PA Gel Patterning to Study Force Generation

In addition to manipulating the stiffness of PA gels to assess the effects of matrix mechanics on cell behavior, work has also been done using these gels to generate cell adhesive "islands" where cell morphology is controlled by the geometry of a patterned substrate. Several similar techniques have been developed to pattern protein ligands onto PA gels using microcontact printing (Li *et al.*, 2008a; Rape *et al.*, 2011; Wang *et al.*, 2002). Additionally, other substrates have also been employed, including PDMS (Balaban *et al.*, 2001) and glass (Chen *et al.*, 1997), patterned with ECM proteins such as collagen and fibronectin, although these substrates are limited in their mechanical stiffness range. Using these methods, cell spreading is constricted to the patterned shape, and the shape can be manipulated to induce a desired morphology. For example, patterning elongated cell geometries has been shown to enhance the differentiation and maturation of myotubes (Li *et al.*, 2008a), and also to increase the expression of type I collagen in human tendon fibroblasts (Li *et al.*, 2008b). Moreover, by embedding beads within the PA gel, contractile forces exerted by patterned cells can be examined using TFM (Wang *et al.*, 2002). Recently, a study by Rape *et al.* has shown that the magnitude and spatial distribution of traction forces are not necessarily dependent on cell size, but on the distance from the cell centroid to the perimeter, such that when comparing cells of equal area, the more elongated cell will generate stronger traction forces (Rape *et al.*, 2011). Legant *et al.* identified a similar pattern of traction force generation in 3D, with cells generating the strongest inward traction stresses at the tips of long, matrix-probing pseudopodia (Legant *et al.*, 2010).

Patterned PA and PDMS gels have also been used to elucidate the exact nature of the relationship between focal adhesions and force generation. In a study by Balaban *et al.*, the size and elongated shape of mature focal adhesions correlated with the magnitude and direction of traction forces exerted by a cell (Balaban *et al.*, 2001). Building on this work, Rape *et al.* also manipulated the size of focal adhesions the cell could form by patterning 4 and 200 μm^2 adhesive squares within a 2500 μm^2 square region. The cells that could only form small focal adhesions exerted significantly less force than those with larger focal adhesions, regardless of the fact that the cells themselves were the same size (Rape *et al.*, 2011). Interestingly, while Balaban *et al.* report a linear relationship between force and focal adhesion size in unconstrained spread cells, Rape *et al.* note an increase in amount of force exerted per focal adhesion as the distance from cell centroid to perimeter increases. Recent work by Stricker *et al.* in unconstrained cells indicates that the correlation between focal adhesion size and traction force generation may exist only in the early stages of focal adhesion formation. Once mature, they find that this correlation is abolished, and these adhesions can now generate a broad range of forces (Stricker *et al.*, 2011).

Because cell adhesion and morphology differ between 2D and 3D environments (Cukierman *et al.*, 2001), there is also interest in exploring the relationship between cell traction and cell–matrix adhesion in three-dimensional environments. Early work by Friedl *et al.* linked 3D tumor cell migration with 3D matrix reorganization and redistribution and shedding of cell adhesions (Friedl *et al.*, 1997). More recent studies have compared the temporal and spatial dynamics of zyxin-positive cell adhesions with 3D ECM deformation (Petroll and Ma, 2003) and demonstrated that ECM density can, in part, determine the extent of matrix reorganization (Pizzo *et al.*, 2005). Further, Fraley *et al.* identified several specific matrix adhesion molecules that are involved in traction generation during 3D cell migration of HT-1080 fibrosarcoma cells (Fraley *et al.*, 2010). Interestingly, the authors found that, while classical 2D focal adhesion proteins such as talin, VASP, and FAK contribute to elastic matrix deformation by HT-1080 cells, these molecules are not significantly involved in inelastic matrix remodeling in 3D. Together, these studies suggest that, as in 2D, the size, morphology, and composition of 3D cell adhesions may be related to cell traction as assessed through local ECM strains and matrix remodeling.

B. TFM for the Study of Cell Migration in 2D

TFM has been widely utilized to study the specific mechanisms by which cells use force to migrate. Performing TFM on cells while manipulating the actomyosin cytoskeleton has revealed that actin stress fibers are critical for the transmission of forces to the substrate (Kraning-Rush *et al.*, 2011; Pelham and Wang, 1999). In a study by Kumar *et al.*, a single actin stress fiber within a living cell was ablated using multiphoton laser nanoscissors, resulting in large-scale cytoskeletal rearrangement, particularly on compliant substrates (Kumar *et al.*, 2006). This study and others lend support to the tensegrity model of cellular architecture, wherein a network of cytoskeletal elements maintains a prestress within the cell, which drives its adhesion and migration behavior (Wang *et al.*, 1993, 2001).

Studies in fibroblasts have also revealed that traction forces tend to be spatially concentrated at the periphery of the cell, with little to no force being exerted beneath the nucleus of the cell. Force generation is also generally greater at the leading edge, or anterior of the cell, with weaker, more passive forces located in the posterior of the cell (Dembo and Wang, 1999; Munevar *et al.*, 2001; Pelham and Wang, 1999). Interestingly, this trend is reversed in neutrophil migration, when forces are concentrated in the uropod of the cell during migration (Smith *et al.*, 2007). Additionally, in fibroblast (Gaudet *et al.*, 2003) and endothelial cell (Califano and Reinhart-King, 2010; Reinhart-King *et al.*, 2003) models, increasing the density of ECM protein conjugated to the surface of PA gels has been shown to increase both the spread area of cells and the magnitude of the force generated by these cells, although whether this phenomenon is driven by stronger cells spreading more or by larger cells inherently exerting greater forces remains an area of debate. Moreover, in this fibroblast model, the increase in force and area was also directly correlated to an increase in migration speed with increasing collagen density, suggesting that stronger traction forces drive increased cell motility (Gaudet *et al.*, 2003).

In addition to the widespread use in the study of mammalian cells described here, 2D TFM has also been used to study the forces generated by several other unique cell types, for example, during the unique single-cell and multicellular stages in the life of the amoeba *Dictyostelium discoideum* (Delanoe-Ayari *et al.*, 2008; Lombardi *et al.*, 2007), during migration of the malarial parasite *Plasmodium berghei* in the stage after it is injected into the host's skin during a mosquito bite (Munter *et al.*, 2009), and during the migration of fish keratocytes on compliant substrates (Lee, 2007).

C. Cellular Force Generation in 3D Migration

Evaluation of cell forces in 3D environments has revealed several insights into the molecular mechanisms of three-dimensional force generation and cell migration. Although ROCK-mediated traction can enable 3D matrix reorganization through mechanisms analogous to contractility in 2D (Kim *et al.*, 2006), the role of cell forces in driving cell migration in 2D and 3D are unique. By tracking the displacement of beads around cells in 3D, Shih *et al.* demonstrated that myosin-IIA-dependent retrograde flow at the cell cortex exerts traction forces against the anterior ECM, propelling the cell body forward during amoeboid migration of MDCK epithelial cells (Shih and Yamada, 2010). Through assessment of ECM displacement fields around amoeboid-migrating MDA-MB-231 breast cancer cells in Matrigel, Poincloux *et al.* identified regions of the cell that generate distinct cell tractions in a RhoA/ROCK/myosin II-dependent manner (Poincloux *et al.*, 2011). Similar traction-dependent patterns of matrix deformation have been defined during mesenchymal migration of HT-1080 cells (Bloom *et al.*, 2008). Finally, there has been substantial interest in exploring the requirement for proteolysis during 3D cell migration (Dikovsky *et al.*, 2008; Wolf and Friedl, 2009; Wolf *et al.*, 2003), and ROCK- and myosin-dependent matrix deformation has been identified as a primary facilitator of protease-independent 3D tumor cell migration both *in vitro* and *in vivo* (Wyckoff *et al.*, 2006).

D. Force Generation and Cancer Progression

Given the intimate role that traction forces play in cell adhesion and migration, two key behaviors that have been shown to be disrupted during certain disease states, it is logical that cell contractility may be affected by disease, or perhaps even a factor driving the condition. Indeed, TFM has been used as a tool for examining the effect of several diseases on cellular force generation, including hypertensive heart disease and arthritis (Bakker *et al.*, 2009; Marganski *et al.*, 2003a). Most notably, Marganski *et al.* found that hypertensive cardiac fibroblasts were excessively contractile compared to their healthy counterparts, and that the hypertensive cells were unable to effectively regulate their contractions (Marganski *et al.*, 2003a).

Perhaps the most comprehensive disease state in which TFM research has been done is in cancer progression. In the seminal work on tensional homeostasis during tumor progression, Paszek *et al.* found that increasing the stiffness of the 3D microenvironment surrounding mammary epithelial cells drives malignant progression by clustering integrins, increasing focal adhesion formation, disrupting adherens junctions, and increasing cell proliferation (Paszek *et al.*, 2005). Likewise, Tang *et al.* found that increasing 2D stiffness promotes a metastasis-like phenotype in colon carcinoma cells (Tang *et al.*, 2010), suggesting that increased mechanical stiffness may be an important driving factor in a wide range of cancer models. Additionally, Paszek *et al.* examined the relationship between malignancy and contractile force generation. Using the human isogenic nonmalignant S-1 mammary epithelial cells and malignant T4–2 cell lines, they found that tractions forces were significantly elevated in the malignant cancer cells, and that these forces were RhoA-dependent. Likewise, Rosel *et al.* found that in a rat sarcoma model of protease-independent amoeboid migration, highly metastatic A3 cells generated traction forces that were five times greater than the spontaneously transformed, nonmetastatic K2 cells, with traction forces at the leading edge found to be even higher (Rosel *et al.*, 2008). Moreover, using a Deformation Quantification and Analysis (DQA) algorithm to quantify collagen fiber deformation, Wyckoff *et al.* found that during nonproteolytic amoeboid migration, metastatic MTLn3E murine mammary tumor cells generated increased force, and were thus able to push through collagen fibers and invade into the ECM in 3D, while their nonmetastatic parental cells were unable to invade (Wyckoff *et al.*, 2006). However, more recent research has called into question the existence of protease-independent migration in native collagen environments, and this remains an area of great controversy in three-dimensional tumor migration research (Sabeh *et al.*, 2009).

Surprisingly, in contrast to these four studies, Indra *et al.* recently found that in yet a different set of murine mammary tumor cells, traction forces actually decreased as the metastatic potential of the subpopulations increased (Indra *et al.*, 2011). Similarly, using patterned PA gels and inducing tumoral transformation, Tseng *et al.* found that increased contractility appeared to be dependent on the method of transformation, with TGFβ-treated mammary epithelial cells generating increased force, while ErbB2 receptor-activated cells and CK2b-knockdown cells exerted weaker forces (Tseng *et al.*, 2011). Given these conflicting results, the precise role of force generation in cancer progression, and particularly in relation to proteolytic activity, remains somewhat unclear. It may be that the effect of malignant transformation on force generation is specific to the type of cancer and the underlying genetic mutations. Regardless, this remains an exciting area of study, and holds great potential for future diagnostic and therapeutic applications.

IV. Open Challenges

Quantification of three-dimensional cell tractions within a physiologically relevant ECM is a complex problem, and there are still several challenges to overcome before it will be possible to translate 3D displacement fields of beads or ECM fibers into true traction stresses. Toward this, Legant *et al.* have developed a technique to numerically quantify three-dimensional tractions of cells embedded in PEG (Legant *et al.*, 2010). Additionally, techniques have recently been developed to probe the three-dimensional forces exerted by cells plated on a 2D substrate using laser scanning confocal microscopy (Franck *et al.*, 2011; Maskarinec *et al.*, 2009). Although these methods have the inherent limitations as previously discussed, they should serve as a foundation for the development of increasingly quantitative models for 3D cell tractions that incorporate the viscoelastic fibrillar architecture and bioactivity of natural ECM.

If we are to use native ECM as a probe for 3D cell force, we will need to better understand its dynamic biomechanics. Already, several computational biophysical approaches have been developed to dynamically assess local matrix deformations (Mierke *et al.*, 2008; Roeder *et al.*, 2004; Vanni *et al.*, 2003; Wyckoff *et al.*, 2006). For example, automated tracking of individual fibers can be achieved by using a DQA algorithm, which transforms fiber deformations into a displacement field (Vanni *et al.*, 2003). This technique has been used to track both the rate and spatial dependence of cell-mediated matrix remodeling (Wyckoff *et al.*, 2006). Roeder *et al.* have incorporated three-dimensional biomechanics into their 3D Incremental Digital Volume Correlation algorithm, enabling them to relate macroscale stresses to the resulting microscale changes in ECM architecture (Roeder *et al.*, 2004), track collagen fibers during 3D matrix reorganization, and quantify local cell-induced volumetric strains (Pizzo *et al.*, 2005). Although these strategies provide a more detailed assessment of how the ECM microstructure is changing under cellular traction forces, there remains a need for a more robust definition of the mechanical properties of the 3D ECM. Importantly, unlike most 2D substrates used to assess cell tractions, the mechanics of 3D matrices are subject to cell-induced changes as cell forces are transmitted to the 3D microenvironment. Ultimately, because of the deformation, degradation, and secretion processes that constitute a cell's interaction with its physiological 3D microenvironment, computation of a true traction stress may not be a suitable stand-alone metric by which to assess cell contractility. However, a better understanding of the microscopic mechanics of the ECM will contribute to improved biomechanical models of 3D force generation.

Finally, an important open challenge to quantifying cell forces in 3D is the variety of cellular outcomes that 3D tractions can elicit. As discussed in this chapter, 3D cell tractions have been shown to enable elastic matrix deformation, permanent matrix remodeling, and both amoeboid and mesenchymal migration. While the molecular players involved in these behaviors are beginning to be revealed (Fraley *et al.*, 2010; Wyckoff *et al.*, 2006), there is still a need for more extensive evaluation of the molecular mechanisms of 3D traction force generation and transmission. Such work may provide an explanation for the diverse phenotypic manifestations of three-dimensional contractility and would ultimately improve models for 3D cell traction forces.

V. Methods

A. 2D Polyacrylamide Gel Preparation and Functionalization

Using the following steps, PA gels can be created that have a Young's Modulus (E) between 0.2 and 300 kPa. A number of groups have published methods for making PA gels as model substrates to investigate the effects of matrix stiffness on cell behavior, a method originally described by Wang and Pelham (Beningo and Wang, 2002; Klein *et al.*, 2007; Wang and Pelham, 1998; Yeung *et al.*, 2005). Here, we include our own method that was adapted and modified from Pelham and Wang, and has successfully been used to make PA gels for use with TFM. One of the more significant differences between our methods and the methods of most other groups is the protocol for conjugation of proteins to the PA gels. Most groups have used photo cross-linkers, whereas we use a bifunctional linker that is polymerized into the gel. We find that this method produces a more uniform coating of protein on the gel surface.

We have successfully performed TFM within a stiffness range of 1–10 kPa. PA gels with a stiffness lower than 1 kPa tend to have significant exogenous bead movement to be easily tracked using our system. On the other hand, the cell types that we have used have not been able to exert enough force to deform a PA gel with E greater than 10 kPa. However, these PA gels can be very useful for measuring the effects of stiffness on other cell behaviors such as morphology (Yeung *et al.*, 2005), migration (Dembo and Wang, 1999), and proliferation (Klein *et al.*, 2009). Once prepared, these gels can be stored in phosphate-buffered saline (PBS, Invitrogen, Carlsbad, CA) at 4°C for up to 2 weeks. Storing these gels in a dehydrated state is not recommended.

1. Coverslip Activation

This step can be performed in advance of the polymerization process. Dry activated coverslips can be stored for over a month, preferably under desiccation. For TFM studies, we recommend an activated glass coverslip size of 43 × 50 mm (VWR, West Chester, PA). Other PA gel studies that do not require a traction chamber can use square 22 × 22 mm glass coverslips, which will fit easily into a 6-well dish for cell seeding, or other sizes as desired. The following steps assume the use of 43 × 50 mm slips.

a. Line up coverslips on top of inverted Petri dishes (2 slips/dish). This aids in the ease of handling of the coverslips.
b. In a chemical fume hood, holding a corner of the coverslip with forceps, briefly pass the coverslip through the flame of a Bunsen burner. Using a clean cotton swab, immediately apply 0.1N NaOH (Sigma-Aldrich, St. Louis, MO) to the flamed side. Be careful not to overheat the glass, or it will break. If the glass is not heated enough, the NaOH will not spread well. If this occurs, repeat flaming step and reapply NaOH.
c. Allow coverslips to dry completely inside the fume hood, about 10–20 min.
d. Reapply 0.1N NaOH with clean cotton swab until the whole coverslip appears coated. Allow the coverslips to dry.
e. In fume hood, add ~60 μL of 3-aminopropyl-trimethyoxysilane (APTMS, Sigma-Aldrich) to each coverslip and spread quickly by rolling the thin end of a glass Pasteur pipette over the coverslip surface.

   i. Work in groups of two coverslips at a time, use 120 μL of APTMS and deposit half on each coverslip. Use one Pasteur pipette for two coverslips, and spread drop until it looks evenly coated and glossy. Note that once the drop is deposited on the coverslip, it should be spread quickly, as the APTMS will dry rapidly. APTMS is corrosive, and care should be taken to avoid skin contact. We recommend discarding gloves after this step.

f. Allow coverslips to dry for 5 min inside the fume hood. Do not allow the coverslips to dry for more than 10 min.
g. Place each coverslip in a separate Petri dish filled with 18.2 MΩ cm purified deionized (MilliQ) water. Wait until APTMS layer starts to crack and lift off from the surface of the coverslip. Shake dishes to dislodge APTMS from each slip and discard water.
h. Rinse three more times with MilliQ water, incubating for 5 min between each rinse.

   ii. If coverslips are not thoroughly rinsed, the gluteraldehyde in the next step will react with any remaining APTMS and form an orange–red precipitate on the coverslip. These coverslips must be discarded.

i. In fume hood, prepare a 0.5% gluteraldehyde solution (70% aqueous gluteraldehyde stock solution, Sigma-Aldrich) in 1× PBS (pH 7.1, without Ca^{2+} or Mg^{2+}).

Each 43 × 50 mm coverslip requires 1 mL. Vortex the solution to ensure thorough mixing.

j. Tape down a piece of Parafilm to the benchtop long enough for all coverslips to be laid down side by side. Pipette a 1-mL drop of diluted gluteraldehyde solution onto the Parafilm for each coverslip. Remove coverslips from Petri dishes and invert onto the gluteraldehyde drop. Incubate for 30 min.
k. Remove coverslips from parafilm and return them to the Petri dishes. Dispose of gluteraldehyde waste in specified container. Wash each coverslip three times with MilliQ water, incubating for 5 min between each rinse.
l. Remove coverslips from dishes and place on a clean paper towel. Allow coverslips to dry inside fume hood, ~30–45 min. This step can be performed overnight, and coverslips can be stored after this point as described above.

2. Polyacrylamide Gel Polymerization

There are many different formulations of acrylamide/bis-acrylamide that can be used to make PA gels with a similar stiffness. The formulations described in this chapter have been adapted from (Yeung *et al.*, 2005) and their Young's moduli have been measured using the protocol described later in this chapter. The volumes describe herein will create a gel that has a height of ~70 μm. It is important to note that gels will shrink after polymerization, and because of this, the height of the gel cannot be directly calculated from the volume of polymerization solution used. Because the extent of polymer swelling varies with the polymer formulation (Charest *et al.*, 2011) and cannot be easily predicted based on modulus alone, it is important to measure the height of the resulting gel that is used for TFM. The gel must be sufficiently thick such that the gel can freely deform due to cellular forces without the influence of the underlying glass (Sen *et al.*, 2009).

a. Using a clean cotton swab, coat one 18 mm diameter circular glass coverslip for each 43 × 50 mm activated coverslip with Rain-X (ITW Global Brands, Houston, TX). Allow circular coverslips to dry for at least 5 min. Buff off excess Rain-X with a Kimwipe, making sure to buff the edges well. Remove dust and debris using canned air to obtain a clean surface. It is particularly important to minimize particles that may appear on the glass and be transferred to the gel as they can interfere with even polymerization and imaging.
b. Mix 30 μL of 0.5 μm diameter fluorescent polystyrene beads (Invitrogen) and 90 μL of MilliQ water per gel formulation to be made in a 1.5 mL microcentrifuge tube and sonicate for 10 min to create a homogenous mixture.
c. In a 50 mL tube, for each desired stiffness, combine in order acrylamide (40% w/v aqueous stock solution, Bio-Rad, Hercules, CA), *N,N′*-methylene-bis-acrylamide (2% w/v aqueous stock solution, Bio-Rad), 250 mM 4-(2-hydroxyethyl)-1-piperazineethanesulfonic acid (HEPES, pH 6.0, Sigma-Aldrich), MilliQ water, and *N,N,N,N*-tetramethylethylenediamine (TEMED, Bio-Rad), according to Table I. Mix thoroughly before and after TEMED addition.

d. Adjust the pH of the solution to 6.0 by adding ~40–50 μL of 2N HCl (Sigma-Aldrich).
e. Remove 845 μL of acrylamide mixture and place in a 5 mL plastic culture tube. Add 80 μL of the sonicated fluorescent bead mixture and mix thoroughly. If not performing TFM, add 80 μL of MilliQ water to the mixture instead. Place the tube in a vacuum flask and cover flask with aluminum foil to prevent bleaching of the beads. Degas the solution for at least 30 min. Insufficient degassing will affect the extent of polymerization. One tube contains enough mixture to polymerize ~35 PA gels.
f. Add 70 μL of 200 proof ethyl alcohol (Sigma-Aldrich) to 5.6 mg of N-6 ((acryloyl)amido)hexanoic acid ((N-6), synthesized in our lab according to the method of (Pless *et al.*, 1983)) for each gel formulation. Pipette until N-6 is well distributed throughout the ethyl alcohol and add it to the degassed acrylamide mixture.
g. To initiate polymerization, add 5 μL of freshly prepared 10% ammonium persulfate (APS, Bio-Rad) in MilliQ water to acrylamide mixture and mix gently by pipetting up and down with a 1-mL pipettor, being careful not to introduce bubbles.
h. Add 25 μL of gel solution to activated coverslips from Section 1. Gently apply the Rain-X-coated circular coverslip by carefully touching the round coverslip to the edge of the drop and lowering it slowly using forceps, being careful to avoid bubbles. For TFM gels, invert the coverslip sandwich onto a 35 mm dish to allow the beads to form a uniform layer at the top of the gel.
i. Allow polymerization to occur for 25–45 min. More compliant gels (<2.5 kPa) will require more time to polymerize (~45 min), while less compliant gels (>5 kPa) will require less time (~25 min). The edges of the gel should begin to recede beneath the top coverslip.
j. Peel off the top coverslip from each gel using a clean razor blade.

3. Functionalization with Protein Ligand

Cells cannot adhere directly to an unmodified PA gel. However, PA gels can be readily functionalized with a variety of different protein ligands, depending on the cell type used and desired experimental conditions. In our lab, we have successfully functionalized PA gels with collagen, laminin, fibronectin, and RGD peptide, at concentrations ranging from 0.1 to 1000 μg/mL, although other proteins could be easily substituted. We most commonly use a concentration of 100 μg/mL as the ligand density.

a. Just before the PA gels have finished polymerizing, dilute the desired concentration of protein ligand in 50 mM HEPES (pH 8.0, Sigma-Aldrich) on ice. You will need 200 μL of protein solution for each coverslip.
b. Tape down a piece of Parafilm onto a plastic tray long enough to hold each coverslip. Pipette a 200 μL drop of protein solution onto the Parafilm for each coverslip.

c. Once the top coverslips have been removed, immediately invert coverslips over the protein solution, taking care to ensure that the surface of the PA gel is covered entirely, with no air bubbles.
d. Incubate at 4°C for 2 h.
e. Remove gels from parafilm, and place each gel into a labeled Petri dish for storage.
f. In a tube, mix a 1:1000 volume of ethanolamine (Sigma-Aldrich) with 50 mM HEPES (pH 8.0, Sigma-Aldrich). You will need 500 μL of ethanolamine solution for each gel.
g. Deposit 500 μL of the ethanolamine solution directly onto each gel, making sure the volume covers the entire gel surface. Incubate at room temperature for 30 min.
h. Rinse gels with MilliQ water. Place gels in PBS and store at 4°C. For best results, use gels within 2 weeks of polymerization. To prevent bacterial growth, gels can also be stored in 1× penicillin/streptomycin.

4. Validating the Young's Modulus of PA Gels

To measure the Young's Modulus of PA gels, two primary methods have been reported: the use of Atomic Force Microscopy (AFM) and the "steel ball" method. Although the use of AFM poses some advantage over the steel ball method, it is much more technically challenging and requires a skilled user. Here, we focus on the steel ball method, as it is a tractable method that requires no specialized tools. It is important to note that batch to batch variations in acrylamide and bis-acrylamide stock solutions will result in some variation in the PA gels. Additionally, there is some error inherent in this measurement in determining the precise focal plane of the beads before and after deformation. Moreover, E should be measured only in PA gels that have been equilibrated with media and incubated at 37°C in order to most closely mimic the cell culture conditions in which they are typically used.

a. On the stage of an epifluorescent microscope, place a steel ball (radius (r) = 0.32 mm, Abbott Ball Co., West Hartford, CT) on a gel with embedded fluorescent beads. Focus the microscope at 20× magnification. This is most easily done by placing the ball on the gel first, and then moving the stage until the ball is in the field of view. The ball is visible because it blocks the light path and can be seen as a shadow once it is in the field of view. Focus on the top layer of beads directly beneath the center of the steel ball and note z position (Z_1).
b. Remove the steel ball using a magnet. Focus the microscope on the top layer of beads once they have returned to their original, unstressed, position. Note the z position (Z_0).
c. Using Hertz theory (Lo *et al.*, 2000), calculate E using Eq. (25),

$$E = \frac{3(1-\nu^2)f}{4\sqrt{r}\sqrt{\delta^3}} \tag{25}$$

where δ is the indentation depth of the steel ball ($\delta = |Z_1 - Z_0|$) exerting a buoyancy-corrected force f on the surface of a gel with Poisson's ratio $\nu = 0.3$–0.5 (Dembo and Wang, 1999; Li *et al.*, 1993). f can be calculated by subtracting the buoyant force, F_b, of the ball from the weight of the ball. The buoyant force is calculated as $F_b = \rho V g$, where ρ is the density of the ball, V is the volume of the spherical cap submerged into the substrate surface with depth δ (from above), and g is the acceleration due to gravity.

B. Traction Force Microscopy Data Collection

The seeding and analysis described in this section are for isolated, single-cell studies, although they can be adapted for quantifying forces of multiple cells in contact (Califano and Reinhart-King, 2010). It is important for the cells to be fairly isolated so that bead movement from one cell does not interfere with the bead movement from a neighboring cell. Additionally, cells can transmit forces through PA substrates and affect the behavior of nearby cells (Reinhart-King *et al.*, 2008), a phenomenon that is dependent on the stiffness and ligand density of the substrate. To avoid this effect, when quantifying forces on stiffer substrates ($\geq$5 kPa), cells should be at least 50 μm apart, while on softer substrates (<2.5 kPa), cells should be greater than 200 μm apart (generally limit one cell per field of view under standard 20× magnification). Additionally, care should be taken to avoid bead movement from cells outside the viewing region, as this can also negatively affect the accurate quantification of traction forces of your target cell. A schematic of the chamber setup is shown in Fig. 3A.

This procedure uses a custom-made traction chamber that fits into a custom stage manufactured by Zeiss (see Fig. 3B). The chambers used in our lab are 7 cm long, 8 cm wide, and 1 cm deep, and contain a central hole, 3 cm in diameter. Additionally, our insets are designed with a ridge around the opening that fits the bottom of a 35-mm petri dish, for use as a lid. Each chamber is designed to accommodate one circular PA gel with an 18-mm diameter polymerized on a 43 × 50 mm glass coverslip, as described above.

1. Chamber Setup and Cell Seeding

The following steps should be performed within a sterile biosafety cabinet.

a. UV sterilize one traction chamber and one 35 mm petri dish for each PA gel, as well as several KimWipes (Kimberly-Clark, Neenah, WI), paper towels, and a syringe filled with vacuum grease (Dow Corning, Midland, MI) for 20 min.
b. Remove the coverslip containing the PA gel from PBS using a razor blade and place on paper towel. Using a KimWipe, gently dry off the excess PBS surrounding the gel surface. Be careful not to touch the gel surface with the KimWipe.
c. Invert the traction insert so that the ridged side is facing down. Apply a thin line of vacuum grease around the circular opening.

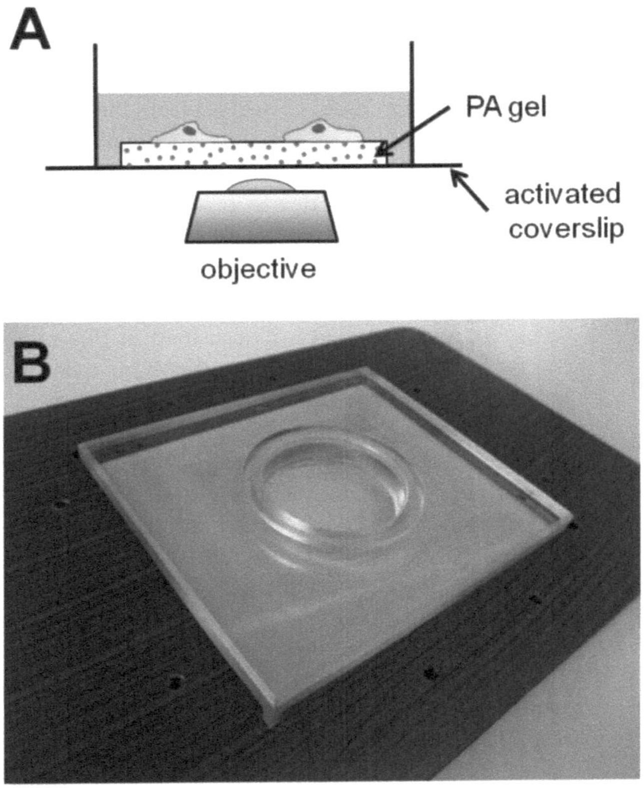

Fig. 3 Traction chamber setup. (A) A polyacrylamide gel on an activated coverslip is attached with vacuum grease to a traction chamber and seeded with a low density of the desired cell type. Measurements are acquired using an inverted microscope. (B) A custom-made traction chamber and stage insert used in our lab. (For color version of this figure, the reader is referred to the web version of this book.)

d. Invert the coverslip with the PA gel over the traction insert such that the PA gel is centered over the opening.
e. Gently press down on the coverslip until the vacuum grease has formed a tight seal around the entire opening. If the chamber leaks, it is likely that the coverslip was not properly sealed to the traction insert, and more vacuum grease is required.
f. Flip over the traction insert and add 2 mL sterile PBS to the gel to keep it hydrated until cell seeding. Repeat these steps until all PA gels are attached to their respective traction inserts. At this point, gels can be stored overnight prior to cell seeding if desired.
g. Passage the desired cell population and determine the cell count. Seed 2000–4000 cells onto each PA gel, depending on the size of the cells and the duration of incubation prior to imaging. Typical incubation times in our lab

range from 6 to 18 h, although this time can be extended if desired. However, note that if the cells begin to proliferate, finding isolated cells to analyze becomes more difficult.

2. Acquiring Traction Images

In our lab, cells are imaged inside a temperature, humidity, and CO_2-controlled automated stage of a Zeiss Axio Observer Z1m inverted phase contrast microscope with a Texas Red fluorescent filter, using a Hamamatsu ORCA-ER camera. Cells are incubated on the microscope stage at 37°C, 40% humidity, and 5% CO_2 until trypsinization. Images are acquired using AxioVision software (v. 4.6.3, Carl Zeiss), and many of the steps described below may be specific to this software, but should be readily adaptable to other microscope systems. Note that these steps can be performed without stage top incubation, as long as the length of time the cells spend outside of the incubator prior to trypsinization is minimized, preferably no longer than 20–30 min. An automated stage is required for imaging more than one cell per PA gel, and can greatly increase the efficiency of data collection. If a time course study is desired, a phase and stressed image can be acquired at each time point for each cell, and one final image of the relaxed bead field can be acquired at the end of the study.

a. Place traction insert containing cell-seeded PA gel onto the stage of the microscope.
b. Using a 10× objective, identify isolated cells and mark their positions.
c. Using a 20× objective, acquire a phase contrast image of the cell, focusing primarily on the cell boundaries. Immediately acquire a fluorescent image of the uppermost layer of beads directly beneath the cell. It is important that these two images are taken as close in time as possible, as cells can change position (and thus change the underlying bead placement) fairly rapidly, and both images need to be consistent for quantification. Repeat for each cell. Note that for compliant gels (<2.5 kPa), it may be appropriate to use a 10× objective to acquire these images, in order to obtain a significant population of fluorescent beads with negligible movement.
d. Aspirate media from the well, being careful not to touch the PA gel surface with the pipette, as this will cause distortion of the bead layer and become unusable.
e. Rinse well three times with 3 mL PBS.
f. Apply 1 mL of trypsin–EDTA. Let sit for 5–10 min depending on cell type.
g. Aspirate trypsin and wash with PBS. Check all fields of view to confirm that the cells of interest have been removed. If not, repeat PBS rinse. Once cells are removed, keep the gel hydrated with PBS.
h. Return to the location of the first cell. Open the corresponding stressed image. Align the beads in the x and y directions such that they line up at a distance far from the cell. It is generally easier to do this in a corner. Some error is acceptable in this adjustment, as the tracking software described above will ignore the most common bead displacement (Marganski *et al.*, 2003b).

i. Adjust the *z* direction such that the same layer of beads is in focus. Acquire an image of the unstressed bead field.
j. Repeat for each marked location.

C. Traction Force Microscopy Data Analysis

There are a number of methods now described to convert the images that are gathered experimentally into stress and strain fields (Angelini *et al.*, 2010; Del Alamo *et al.*, 2007; Yang *et al.*, 2006). The original method was invented and described by Dr. Micah Dembo (Boston University) and is the method we use in our lab. The analysis is done through the LIBTRC software package that includes algorithms for bead tracking, calculation of the traction stresses, and a graphical interface to organize and display the output. Complete detailed instructions for running the software can be found in the LIBTRC Users Guide.

Prior to analysis, the stressed and relaxed images of the bead field should be compared to ensure that they overlap uniformly at distances far removed from the target cell. If a bead is lost or if there is bead movement caused by a cell outside of the field of view, the images can be cropped to remove the offending area as long as there is still a large portion of the image that remains unstressed (minimal bead displacement). If this phenomenon occurs, it is important that all three images (the phase image of the cell, and the two images of the bead field) are cropped identically. At no point should beads actively displaced by the target cell be removed from the image set.

The LIBTRC software outputs a wealth of information regarding the traction forces exerted by the cell. One common way to represent this force data is by plotting total force, |F|, which is an integral of the magnitude of the traction field over the cell area, Eq. (26),

$$|F| = \int\int \sqrt{T_x^2(x,y) + T_y^2(x,y)}\,dxdy \tag{26}$$

where

$$T(x,y) = [T_x(x,y), T_y(x,y)] \tag{27}$$

is the continuous field of traction vectors defined at any *(x,y)* position within the projected cell area (Reinhart-King *et al.*, 2005). The polarization of the force distribution within the cell is also computed in LIBTRC as the integral of the absolute value of the traction magnitudes dotted with a unit vector directed along the long axis of the cell or the perpendicular short axis of the cell (Kraning-Rush *et al.*, 2011). Forces can also be analyzed by plotting the average traction stress, or total force divided by the cell area (Califano and Reinhart-King, 2010). Additionally, strain energy (erg), or the total energy transferred from the cell to the elastic displacement of the substrate, can also be used as a metric of force:

$$erg = \frac{1}{2}\boldsymbol{T} \cdot \boldsymbol{D} \tag{28}$$

where **D** is the displacement field at the substrate surface caused by the traction field **T**.

D. Imaging Collagen Remodeling Using Confocal Reflectance Microscopy

Quantitative matrix remodeling as a functional measure of cell traction forces can provide critical information about the spatial and temporal nature of cell contractility and mechanical cell–matrix interactions in 3D. Importantly, these metrics can be coupled with fluorescent tagging of cell components such as the cytoskeleton, focal adhesions, and regulatory molecules to identify cell structures and phenotypes that are uniquely involved in the generation of 3D traction forces (Wolf *et al.*, 2003; Zaman *et al.*, 2006), ultimately allowing us to better probe the dynamic interactions between cell contractility and the ECM. Here we present one metric used in our lab to quantitatively describe collagen matrix remodeling using confocal reflectance microscopy (Kraning-Rush *et al.*, 2011).

1. Preparation of collagen gels

Our method of quantifying 3D cell traction forces utilizes confocal reflectance microscopy to probe the structure and organization of collagen fibers surrounding cells embedded within a collagen matrix. In our lab, we maintain a 10-mg/mL stock of acid-solubilized collagen type I isolated from rat-tail tendon (Bornstein, 1958), and dilute this stock to a final gel collagen concentration of 1.5 mg/mL. Collagen density can be varied to alter matrix sterics and ligand density if desired. To limit cell–cell interactions, cells are seeded sparsely at 50,000–100,000 cells/mL. After cells are embedded within collagen gels, they can be cultured under a variety of conditions and imaged live or after fixation and staining.

a. Calculate the following:
   i. Volume of collagen stock to be used = (Final gel volume · Final collagen concentration)/Collagen stock concentration
   ii. Volume of 1N NaOH to neutralize the solution = Volume of collagen stock · 0.023
   iii. Volume of cell suspension and media to be added = Final gel volume – Volume of collagen stock – Volume of 1N NaOH
   The following steps should be performed within a sterile biosafety cabinet.
b. Place on ice a sterile 15 mL conical tube to hold the final mixture. All reagents should be kept on ice until use, as collagen gel polymerization is pH- and temperature-dependent.
c. Trypsinize and resuspend cells in cold media such that the volume of cell suspension is approximately 25% that of the volume of culture media.

d. Place desired volume of collagen stock in cold 15 mL conical tube.
e. Add acellular culture media; if using media containing phenol red, solution will turn yellow due to the acetic acid–collagen stock solution. Keeping the tube on ice, thoroughly and quickly mix the solution without introducing air bubbles until it turns uniformly yellow.
f. Add cold 1N NaOH and mix as in step (e) until solution is uniformly pink. Use of different types of media may require adjustment of NaOH volume; neutralization to pH 7.2 should be verified initially.
g. Add cell suspension to collagen solution and mix.
h. Aliquot the collagen–cell solution into desired volume in a glass-bottom Petri dish or multiwell plate (MatTek, Ashland, MA) and allow gel to polymerize at 37 °C for 30–60 min. We find that collagen solution batches of 1–5 mL work best as this range of volumes permits accuracy and precision of volume measurements and thorough mixing.
i. Gently add prewarmed media to gel and culture at 37 °C.
Optional:
j. To study cell contractility, allow cells to adhere and spread for 4–12 h before experimental treatments. This time period is used to allow establishment of cell–matrix adhesions while minimizing cell tractions and collagen reorganization prior to treatment. This incubation period can be optimized depending on the cell type and the cellular mechanisms being studied.

2. Confocal Reflectance Microscopy

To probe the organization of the fibrillar collagen microenvironment, the cell-seeded collagen gels prepared above can be imaged with confocal reflectance microscopy. While microscopy systems vary widely and a variety of imaging parameters can be used successfully, consistency is critical to enable quantitative comparison of collagen remodeling. We will discuss the equipment and parameters used in our lab.

Since macroscale stresses induce changes in the microscopic structure of fibrillar hydrogels, care should be taken to not handle or disturb the collagen gel. Therefore, it is best to image the collagen gel in the container in which it was originally polymerized. We use a Zeiss 710 laser scanning confocal on an Axio Observer.Z1 inverted stand. This microscope has an interchangeable main beam splitter, which is critical for sequential fluorescence and reflectance imaging. For confocal reflectance acquisition, samples are illuminated through an 80/20 dichroic mirror with low power laser light, which is reflected off of collagen fibrils and detected by a photomultiplier tube (PMT). A 40× water-immersion lens (C-Apochromat 40×/1.2 W Corr, Zeiss) provides sufficient magnification for visualization of collagen fibrils as well as a correction collar to facilitate use of glass-bottom dishes. We use 488 nm light to minimize phototoxicity, and we adjust the laser power and PMT gain to utilize the entire dynamic range of the detector. Using either live samples in a microscope incubator or fixed samples, we capture 1 μm slices at the axial center of cells. Notably, we choose isolated cells that are 150–300 μm above the bottom surface of the gel to avoid gel inconsistencies at the surface and mechanical edge effects. For live-cell experiments, cells can be visualized with DIC through a transmitted light-PMT (Fig. 2A) or with confocal fluorescence by labeling with a vital dye such as CellTracker (Invitrogen).

3. Image Analysis and Quantification of Collagen Compaction

Our metric of cell tractions is based upon the assumption that increased local collagen fiber density increases the confocal reflectance signal. Thus, as cells generate 3D traction forces, they compact the pericellular ECM and there is an increase in reflectance intensity. The output of this method is the average collagen reflectance intensity as a function of distance from the cell. The following procedure is used to quantify collagen fiber compaction around isolated cells.

a. Using ImageJ, subtract the cell area, determined from the fluorescence or DIC/phase contrast image (Fig. 2A), from the reflectance image (Fig. 2B).
b. On the reflectance image, draw a 40–50 μm selector line from the cell's centroid into the surrounding matrix. We use a custom-written ImageJ script to rotate the selector line around the cell at 1-degree increments and capture an intensity profile at each step.
c. Remove zero-intensity values from the intensity profile to define the cell membrane as the origin and normalize for differences in cell size and shape.
d. Average all of the "zeroed" reflectance intensity profiles to create a single intensity profile for the cell.
e. Normalize the intensity profile to the peak intensity and subtract the baseline reflectance value (average intensity of matrix 45–50 μm from the cell centroid, where the intensity profile reaches an asymptote). If more extensive matrix remodeling occurs, it may be necessary to extend the initial selector line into the ECM such that the baseline intensity can be assessed. Representative reflectance intensity profiles are shown as symbols in Fig. 2C.
f. Fit the intensity profile to an exponential decay model, Eq. (29),

$$I = I_0 \cdot e^{(-d/\lambda)} \tag{29}$$

allowing I_0 and λ to vary to minimize the sum of squared error. In this equation, I is the intensity of collagen reflectance, d is the distance from the cell membrane, I_0 is the normalized, baseline-subtracted intensity of collagen reflectance at the cell membrane ($d = 0$), and λ is the half-length of the exponential decay, which describes how far from the cell the collagen has been remodeled. Representative fits are shown as solid lines in Fig. 2C.
g. Extract the half-length of the exponential decay, λ (Fig. 2D). A longer decay, fit by a relatively larger λ, is indicative of more substantial collagen compaction and remodeling.
h. To compare 3D traction force and matrix reorganization among cells, half-lengths from several cells per treatment group can be compared directly (Kraning-Rush *et al.*, 2011).

VI. Summary

The study of traction forces has yielded valuable insights into key cellular behaviors including cell–cell communication, cell–ECM interactions, adhesion, and migration in both healthy and disease states. In this chapter we have described the various methods by which traction forces have been quantified in the past and at present. Moreover, we have provided a detailed description and protocol for synthesizing PA gels and performing TFM experiments. Additionally, we have discussed current techniques for qualitatively and quantitatively describing traction forces in 3D environments, and shared a technique used by our lab to extract quantitative data from confocal reflectance microscopy of collagen matrices.

Acknowledgments

This work was supported in part by the Cornell Center on the Microenvironment & Metastasis through Award Number U54CA143876 from the National Cancer Institute and funding through the NIH award number R21RR025801. This work was also supported by National Science Foundation Graduate Research Fellowships to C.M.K. and S.P.C.

References

Angelini, T. E., Hannezo, E., Trepat, X., Fredberg, J. J., and Weitz, D. A. (2010). Cell migration driven by cooperative substrate deformation patterns. *Phys. Rev. Lett.* **104**, 168104.

Bakker, A. D., Silva, V. C., Krishnan, R., Bacabac, R. G., Blaauboer, M. E., Lin, Y. C., Marcantonio, R. A., Cirelli, J. A., and Klein-Nulend, J. (2009). Tumor necrosis factor alpha and interleukin-1beta modulate calcium and nitric oxide signaling in mechanically stimulated osteocytes. *Arthritis Rheum.* **60**, 3336–3345.

Balaban, N. Q., Schwarz, U. S., Riveline, D., Goichberg, P., Tzur, G., Sabanay, I., Mahalu, D., Safran, S., Bershadsky, A., Addadi, L., and Geiger, B. (2001). Force and focal adhesion assembly: a close relationship studied using elastic micropatterned substrates. *Nat. Cell. Biol.* **3**, 466–472.

Beningo, K. A., and Wang, Y. L. (2002). Flexible substrata for the detection of cellular traction forces. *Trends Cell Biol.* **12**, 79–84.

Bloom, R. J., George, J. P., Celedon, A., Sun, S. X., and Wirtz, D. (2008). Mapping local matrix remodeling induced by a migrating tumor cell using three-dimensional multiple-particle tracking. *Biophys. J.* **95**, 4077–4088.

Bornstein, M. B. (1958). Reconstituted rattail collagen used as substrate for tissue cultures on coverslips in Maximow slides and roller tubes. *Lab. Invest.* **7**, 134–137.

Boudou, T., Ohayon, J., Picart, C., and Tracqui, P. (2006). An extended relationship for the characterization of Young's modulus and Poisson's ratio of tunable polyacrylamide gels. *Biorheology* **43**, 721–728.

Brandley, B. K., Weisz, O. A., and Schnaar, R. L. (1987). Cell attachment and long-term growth on derivatizable polyacrylamide surfaces. *J. Biol. Chem.* **262**, 6431–6437.

Burton, K., and Taylor, D. L. (1997). Traction forces of cytokinesis measured with optically modified elastic substrata. *Nature* **385**, 450–454.

Butler, J. P., Tolic-Norrelykke, I. M., Fabry, B., and Fredberg, J. J. (2002). Traction fields, moments, and strain energy that cells exert on their surroundings. *Am. J. Physiol. Cell Physiol.* **282**, C595–C605.

Califano, J. P., and Reinhart-King, C. A. (2010). Substrate stiffness and cell area predict cellular traction stresses in single cells and cells in contact. *Cell. Mol. Bioeng.* **3**, 68–75.

Charest, J. M., Califano, J. P., Carey, S. P., and Reinhart-King, C. A. (2011). Fabrication of substrates with defined mechanical properties and topographical features for the study of cell migration. *Macromol. Biosci.* **12**, 12–20.

Chen, C. S., Mrksich, M., Huang, S., Whitesides, G. M., and Ingber, D. E. (1997). Geometric control of cell life and death. *Science* **276**, 1425–1428.

Cukierman, E., Pankov, R., Stevens, D. R., and Yamada, K. M. (2001). Taking cell-matrix adhesions to the third dimension. *Science* **294**, 1708–1712.

Del Alamo, J. C., Meili, R., Alonso-Latorre, B., Rodriguez-Rodriguez, J., Aliseda, A., Firtel, R. A., and Lasheras, J. C. (2007). Spatio-temporal analysis of eukaryotic cell motility by improved force cytometry. *Proc. Natl. Acad. Sci. U S A* **104**, 13343–13348.

Delanoe-Ayari, H., Iwaya, S., Maeda, Y. T., Inose, J., Riviere, C., Sano, M., and Rieu, J. P. (2008). Changes in the magnitude and distribution of forces at different Dictyostelium developmental stages. *Cell. Motil. Cytoskeleton.* **65**, 314–331.

Dembo, M., Oliver, T., Ishihara, A., and Jacobson, K. (1996). Imaging the traction stresses exerted by locomoting cells with the elastic substratum method. *Biophys. J.* **70**, 2008–2022.

Dembo, M., and Wang, Y. L. (1999). Stresses at the cell-to-substrate interface during locomotion of fibroblasts. *Biophys. J.* **76**, 2307–2316.

Dikovsky, D., Bianco-Peled, H., and Seliktar, D. (2008). Defining the role of matrix compliance and proteolysis in three-dimensional cell spreading and remodeling. *Biophys. J.* **94**, 2914–2925.

du Roure, O., Saez, A., Buguin, A., Austin, R. H., Chavrier, P., Silberzan, P., and Ladoux, B. (2005). Force mapping in epithelial cell migration. *Proc. Natl. Acad. Sci. U S A* **102**, 2390–2395.

Ehrbar, M., Sala, A., Lienemann, P., Ranga, A., Mosiewicz, K., Bittermann, A., Rizzi, S. C., Weber, F. E., and Lutolf, M. P. (2011). Elucidating the role of matrix stiffness in 3D cell migration and remodeling. *Biophys. J.* **100**, 284–293.

Engler, A. J., Sen, S., Sweeney, H. L., and Discher, D. E. (2006). Matrix elasticity directs stem cell lineage specification. *Cell* **126**, 677–689.

Fraley, S. I., Feng, Y., Krishnamurthy, R., Kim, D. H., Celedon, A., Longmore, G. D., and Wirtz, D. (2010). A distinctive role for focal adhesion proteins in three-dimensional cell motility. *Nat. Cell. Biol.* **12**, 598–604.

Franck, C., Maskarinec, S. A., Tirrell, D. A., and Ravichandran, G. (2011). Three-dimensional traction force microscopy: a new tool for quantifying cell-matrix interactions. *PLoS One* **6**, e17833.

Friedl, P., Maaser, K., Klein, C. E., Niggemann, B., Krohne, G., and Zanker, K. S. (1997). Migration of highly aggressive MV3 melanoma cells in 3-dimensional collagen lattices results in local matrix reorganization and shedding of alpha2 and beta1 integrins and CD44. *Cancer Res.* **57**, 2061–2070.

Fu, J., Wang, Y. K., Yang, M. T., Desai, R. A., Yu, X., Liu, Z., and Chen, C. S. (2010). Mechanical regulation of cell function with geometrically modulated elastomeric substrates. *Nat. Methods* **7**, 733–736.

Galbraith, C. G., and Sheetz, M. P. (1997). A micromachined device provides a new bend on fibroblast traction forces. *Proc. Natl. Acad. Sci. U S A* **94**, 9114–9118.

Gaudet, C., Marganski, W. A., Kim, S., Brown, C. T., Gunderia, V., Dembo, M., and Wong, J. Y. (2003). Influence of type I collagen surface density on fibroblast spreading, motility, and contractility. *Biophys. J.* **85**, 3329–3335.

Gunzer, M., Friedl, P., Niggemann, B., Brocker, E. B., Kampgen, E., and Zanker, K. S. (2000). Migration of dendritic cells within 3-D collagen lattices is dependent on tissue origin, state of maturation, and matrix structure and is maintained by proinflammatory cytokines. *J. Leukoc. Biol.* **67**, 622–629.

Harris, A. K., Stopak, D., and Wild, P. (1981). Fibroblast traction as a mechanism for collagen morphogenesis. *Nature* **290**, 249–251.

Harris, A. K., Wild, P., and Stopak, D. (1980). Silicone rubber substrata: a new wrinkle in the study of cell locomotion. *Science* **208**, 177–179.

Hartmann, A., Boukamp, P., and Friedl, P. (2006). Confocal reflection imaging of 3D fibrin polymers. *Blood Cells Mol. Dis.* **36**, 191–193.

Huang, J., Qin, L., Peng, X., Zhu, T., Xiong, C., Zhang, Y., and Fang, J. (2009). Cellular traction force recovery: an optimal filtering approach in two-dimensional Fourier space. *J. Theor. Biol.* **259**, 811–819.

Hynes, R. O. (2002). Integrins: bidirectional, allosteric signaling machines. *Cell* **110**, 673–687.

Indra, I., Undyala, V., Kandow, C., Thirumurthi, U., Dembo, M., and Beningo, K. A. (2011). An in vitro correlation of mechanical forces and metastatic capacity. *Phys. Biol.* **8**, 015015.

Kim, A., Lakshman, N., and Petroll, W. M. (2006). Quantitative assessment of local collagen matrix remodeling in 3-D culture: the role of Rho kinase. *Exp. Cell Res.* **312**, 3683–3692.

Klein, E. A., Yin, L., Kothapalli, D., Castagnino, P., Byfield, F. J., Xu, T., Levental, I., Hawthorne, E., Janmey, P. A., and Assoian, R. K. (2009). Cell-cycle control by physiological matrix elasticity and in vivo tissue stiffening. *Curr. Biol.* **19**, 1511–1518.

Klein, E. A., Yung, Y., Castagnino, P., Kothapalli, D., and Assoian, R. K. (2007). Cell adhesion, cellular tension, and cell cycle control. *Methods Enzymol.* **426**, 155–175.

Kraning-Rush, C. M., Carey, S. P., Califano, J. P., Smith, B. N., and Reinhart-King, C. A. (2011). The role of the cytoskeleton in cellular force generation in 2D and 3D environments. *Phys. Biol.* **8**, 015009.

Kumar, S., Maxwell, I. Z., Heisterkamp, A., Polte, T. R., Lele, T. P., Salanga, M., Mazur, E., and Ingber, D. E. (2006). Viscoelastic retraction of single living stress fibers and its impact on cell shape, cytoskeletal organization, and extracellular matrix mechanics. *Biophys. J.* **90**, 3762–3773.

Landau, L. D., Lifshitz, E. M., Kosevich, A. M., PitaevskiÄ-, L. P., 1986. Theory of Elasticity. Butterworth-Heinemann.

Lauffenburger, D. A., and Horwitz, A. F. (1996). Cell migration: a physically integrated molecular process. *Cell* **84**, 359–369.

Lee, J. (2007). The use of gelatin substrates for traction force microscopy in rapidly moving cells. *Methods Cell Biol.* **83**, 297–312.

Lee, J., Leonard, M., Oliver, T., Ishihara, A., and Jacobson, K. (1994). Traction forces generated by locomoting keratocytes. *J. Cell Biol.* **127**, 1957–1964.

Legant, W. R., Miller, J. S., Blakely, B. L., Cohen, D. M., Genin, G. M., and Chen, C. S. (2010). Measurement of mechanical tractions exerted by cells in three-dimensional matrices. *Nat. Methods* **7**, 969–971.

Lemmon, C. A., Chen, C. S., and Romer, L. H. (2009). Cell traction forces direct fibronectin matrix assembly. *Biophys. J.* **96**, 729–738.

Li, B., Lin, M., Tang, Y., Wang, B., and Wang, J. H. (2008a). A novel functional assessment of the differentiation of micropatterned muscle cells. *J. Biomech.* **41**, 3349–3353.

Li, F., Li, B., Wang, Q. M., and Wang, J. H. (2008b). Cell shape regulates collagen type I expression in human tendon fibroblasts. *Cell. Motil. Cytoskeleton.* **65**, 332–341.

Li, Y., Hu, Z. B., and Li, C. F. (1993). New method for measuring Poisson ratio in polymer gels. *J. App. Polym. Sci.* **50**, 1107–1111.

Liu, Z., Tan, J. L., Cohen, D. M., Yang, M. T., Sniadecki, N. J., Ruiz, S. A., Nelson, C. M., and Chen, C. S. (2010). Mechanical tugging force regulates the size of cell-cell junctions. *Proc. Natl. Acad. Sci. U S A* **107**, 9944–9949.

Lo, C. M., Wang, H. B., Dembo, M., and Wang, Y. L. (2000). Cell movement is guided by the rigidity of the substrate. *Biophys. J.* **79**, 144–152.

Lombardi, M. L., Knecht, D. A., Dembo, M., and Lee, J. (2007). Traction force microscopy in Dictyostelium reveals distinct roles for myosin II motor and actin-crosslinking activity in polarized cell movement. *J. Cell Sci.* **120**, 1624–1634.

Marganski, W. A., De Biase, V. M., Burgess, M. L., and Dembo, M. (2003a). Demonstration of altered fibroblast contractile activity in hypertensive heart disease. *Cardiovasc. Res.* **60**, 547–556.

Marganski, W. A., Dembo, M., and Wang, Y. L. (2003b). Measurements of cell-generated deformations on flexible substrata using correlation-based optical flow. *Methods Enzymol.* **361**, 197–211.

Maskarinec, S. A., Franck, C., Tirrell, D. A., and Ravichandran, G. (2009). Quantifying cellular traction forces in three dimensions. *Proc. Natl. Acad. Sci. U S A* **106**, 22108–22113.

Mierke, C. T., Rosel, D., Fabry, B., and Brabek, J. (2008). Contractile forces in tumor cell migration. *Eur. J. Cell Biol.* **87**, 669–676.

Munevar, S., Wang, Y. L., and Dembo, M. (2001). Distinct roles of frontal and rear cell-substrate adhesions in fibroblast migration. *Mol. Biol. Cell.* **12**, 3947–3954.

Munter, S., Sabass, B., Selhuber-Unkel, C., Kudryashev, M., Hegge, S., Engel, U., Spatz, J. P., Matuschewski, K., Schwarz, U. S., and Frischknecht, F. (2009). Plasmodium sporozoite motility is modulated by the turnover of discrete adhesion sites. *Cell Host Microb.* **6**, 551–562.

Ng, C. P., and Swartz, M. A. (2006). Mechanisms of interstitial flow-induced remodeling of fibroblast-collagen cultures. *Ann. Biomed. Eng.* **34**, 446–454.

Oliver, T., Dembo, M., and Jacobson, K. (1995). Traction forces in locomoting cells. *Cell Motil. Cytoskeleton.* **31**, 225–240.

Pang, Y., Ucuzian, A. A., Matsumura, A., Brey, E. M., Gassman, A. A., Husak, V. A., and Greisler, H. P. (2009). The temporal and spatial dynamics of microscale collagen scaffold remodeling by smooth muscle cells. *Biomaterials* **30**, 2023–2031.

Paszek, M. J., Zahir, N., Johnson, K. R., Lakins, J. N., Rozenberg, G. I., Gefen, A., Reinhart-King, C. A., Margulies, S. S., Dembo, M., Boettiger, D., Hammer, D. A., and Weaver, V. M. (2005). Tensional homeostasis and the malignant phenotype. *Cancer Cell* **8**, 241–254.

Pelham Jr., R. J., and Wang, Y. (1997). Cell locomotion and focal adhesions are regulated by substrate flexibility. *Proc. Natl. Acad. Sci. U S A* **94**, 13661–13665.

Pelham Jr., R. J., and Wang, Y. (1999). High resolution detection of mechanical forces exerted by locomoting fibroblasts on the substrate. *Mol. Biol. Cell* **10**, 935–945.

Petroll, W. M., Cavanagh, H. D., and Jester, J. V. (2004). Dynamic three-dimensional visualization of collagen matrix remodeling and cytoskeletal organization in living corneal fibroblasts. *Scanning* **26**, 1–10.

Petroll, W. M., and Ma, L. (2003). Direct, dynamic assessment of cell-matrix interactions inside fibrillar collagen lattices. *Cell Motil. Cytoskeleton* **55**, 254–264.

Pizzo, A. M., Kokini, K., Vaughn, L. C., Waisner, B. Z., and Voytik-Harbin, S. L. (2005). Extracellular matrix (ECM) microstructural composition regulates local cell-ECM biomechanics and fundamental fibroblast behavior: a multidimensional perspective. *J. Appl. Physiol.* **98**, 1909–1921.

Pless, D. D., Lee, Y. C., Roseman, S., and Schnaar, R. L. (1983). Specific cell adhesion to immobilized glycoproteins demonstrated using new reagents for protein and glycoprotein immobilization. *J. Biol. Chem.* **258**, 2340–2349.

Poincloux, R., Collin, O., Lizarraga, F., Romao, M., Debray, M., Piel, M., and Chavrier, P. (2011). Contractility of the cell rear drives invasion of breast tumor cells in 3D Matrigel. *Proc. Natl. Acad. Sci. U S A* **108**, 1943–1948.

Qin, L., Huang, J., Xiong, C., Zhang, Y., and Fang, J. (2007). Dynamical stress characterization and energy evaluation of single cardiac myocyte actuating on flexible substrate. *Biochem. Biophys. Res. Commun.* **360**, 352–356.

Rape, A. D., Guo, W. H., and Wang, Y. L. (2011). The regulation of traction force in relation to cell shape and focal adhesions. *Biomaterials* **32**, 2043–2051.

Reinhart-King, C. A., Dembo, M., and Hammer, D. A. (2003). Endothelial cell traction forces on RGD-derivatized polyacrylamide substrata. *Langmuir* **19**, 1573–1579.

Reinhart-King, C. A., Dembo, M., and Hammer, D. A. (2005). The dynamics and mechanics of endothelial cell spreading. *Biophys. J.* **89**, 676–689.

Reinhart-King, C. A., Dembo, M., and Hammer, D. A. (2008). Cell-cell mechanical communication through compliant substrates. *Biophys. J.* **95**, 6044–6051.

Roeder, B. A., Kokini, K., Robinson, J. P., and Voytik-Harbin, S. L. (2004). Local, three-dimensional strain measurements within largely deformed extracellular matrix constructs. *J. Biomech. Eng.* **126**, 699–708.

Rosel, D., Brabek, J., Tolde, O., Mierke, C. T., Zitterbart, D. P., Raupach, C., Bicanova, K., Kollmannsberger, P., Pankova, D., Vesely, P., Folk, P., and Fabry, B. (2008). Up-regulation of Rho/ROCK signaling in sarcoma cells drives invasion and increased generation of protrusive forces. *Mol. Cancer Res.* **6**, 1410–1420.

Sabass, B., Gardel, M. L., Waterman, C. M., and Schwarz, U. S. (2008). High resolution traction force microscopy based on experimental and computational advances. *Biophys. J.* **94**, 207–220.

Sabeh, F., Shimizu-Hirota, R., and Weiss, S. J. (2009). Protease-dependent versus -independent cancer cell invasion programs: three-dimensional amoeboid movement revisited. *J. Cell Biol.* **185**, 11–19.

Schreier, H. W., Braasch, J. R., and Sutton, M. A. (2000). Systematic errors in digital image correlation caused by intensity interpolation. *Optical Eng.* **39**, 2915–2921

Sen, S., Engler, A. J., and Discher, D. E. (2009). Matrix strains induced by cells: computing how far cells can feel. *Cell Mol. Bioeng.* **2**, 39–48.

Shih, W., and Yamada, S. (2010). Myosin IIA dependent retrograde flow drives 3D cell migration. *Biophys. J* **98**, L29–L31.

Smith, L. A., Aranda-Espinoza, H., Haun, J. B., Dembo, M., and Hammer, D. A. (2007). Neutrophil traction stresses are concentrated in the uropod during migration. *Biophys. J.* **92**, L58–L60.

Sniadecki, N. J., and Chen, C. S. (2007). Microfabricated silicone elastomeric post arrays for measuring traction forces of adherent cells. *Methods Cell Biol.* **83**, 313–328.

Stricker, J., Aratyn-Schaus, Y., Oakes, P. W., and Gardel, M. L. (2011). Spatiotemporal constraints on the force-dependent growth of focal adhesions. *Biophys. J.* **100**, 2883–2893.

Sutton, M. A., Wolters, W. J., Peters, W. H., Ranson, W. F., and McNeill, S. R. (1983). Determination of displacements using an improved digital correlation method. *Image Vision Comput.* **1**, 133–139.

Tamariz, E., and Grinnell, F. (2002). Modulation of fibroblast morphology and adhesion during collagen matrix remodeling. *Mol. Biol. Cell* **13**, 3915–3929.

Tan, J. L., Tien, J., Pirone, D. M., Gray, D. S., Bhadriraju, K., and Chen, C. S. (2003). Cells lying on a bed of microneedles: an approach to isolate mechanical force. *Proc. Natl. Acad. Sci. U S A* **100**, 1484–1489.

Tang, X., Kuhlenschmidt, T. B., Zhou, J., Bell, P., Wang, F., Kuhlenschmidt, M. S., and Saif, T. A. (2010). Mechanical force affects expression of an in vitro metastasis-like phenotype in HCT-8 cells. *Biophys. J.* **99**, 2460–2469.

Trepat, X., Wasserman, M. R., Angelini, T. E., Millet, E., Weitz, D. A., Butler, J. P., and Fredberg, J. J. (2009). Physical forces during collective cell migration. *Nat. Phys.* **5**, 426–430.

Tseng, Q., Wang, I., Duchemin-Pelletier, E., Azioune, A., Carpi, N., Gao, J., Filhol, O., Piel, M., Thery, M., and Balland, M. (2011). A new micropatterning method of soft substrates reveals that different tumorigenic signals can promote or reduce cell contraction levels. *Lab. Chip* **11**, 2231–2240.

Vader, D., Kabla, A., Weitz, D., and Mahadevan, L. (2009). Strain-induced alignment in collagen gels. *PLoS One* **4**, e5902.

Vanni, S., Lagerholm, B. C., Otey, C., Taylor, D. L., and Lanni, F. (2003). Internet-based image analysis quantifies contractile behavior of individual fibroblasts inside model tissue. *Biophys. J.* **84**, 2715–2727.

Wang, N., Butler, J. P., and Ingber, D. E. (1993). Mechanotransduction across the cell surface and through the cytoskeleton. *Science* **260**, 1124–1127.

Wang, N., Naruse, K., Stamenovic, D., Fredberg, J. J., Mijailovich, S. M., Tolic-Norrelykke, I. M., Polte, T., Mannix, R., and Ingber, D. E. (2001). Mechanical behavior in living cells consistent with the tensegrity model. *Proc. Natl. Acad. Sci. U S A* **98**, 7765–7770.

Wang, N., Ostuni, E., Whitesides, G. M., and Ingber, D. E. (2002). Micropatterning tractional forces in living cells. *Cell Motil. Cytoskeleton* **52**, 97–106.

Wang, Y. L., and Pelham Jr., R. J. (1998). Preparation of a flexible, porous polyacrylamide substrate for mechanical studies of cultured cells. *Methods Enzymol.* **298**, 489–496.

Wolf, K., and Friedl, P. (2009). Mapping proteolytic cancer cell-extracellular matrix interfaces. *Clin. Exp. Metastasis* **26**, 289–298.

Wolf, K., Mazo, I., Leung, H., Engelke, K., von Andrian, U. H., Deryugina, E. I., Strongin, A. Y., Brocker, E. B., and Friedl, P. (2003). Compensation mechanism in tumor cell migration: mesenchymal-amoeboid transition after blocking of pericellular proteolysis. *J. Cell Biol.* **160**, 267–277.

Wyckoff, J. B., Pinner, S. E., Gschmeissner, S., Condeelis, J. S., and Sahai, E. (2006). ROCK- and myosin-dependent matrix deformation enables protease-independent tumor-cell invasion in vivo. *Curr. Biol.* **16**, 1515–1523.

Yang, M. T., Fu, J., Wang, Y. K., Desai, R. A., and Chen, C. S. (2011). Assaying stem cell mechanobiology on microfabricated elastomeric substrates with geometrically modulated rigidity. *Nat. Protoc.* **6**, 187–213.

Yang, Z., Lin, J. S., Chen, J., and Wang, J. H. (2006). Determining substrate displacement and cell traction fields–a new approach. *J. Theor. Biol.* **242**, 607–616.

Yang, M. T., Sniadecki, N. J., and Chen, C. S. (2007). Geometric considerations of micro- to nanoscale elastomeric post arrays to study cellular traction forces. *Adv. Mater.* **19**, 3119–3123.

Yeung, T., Georges, P. C., Flanagan, L. A., Marg, B., Ortiz, M., Funaki, M., Zahir, N., Ming, W., Weaver, V., and Janmey, P. A. (2005). Effects of substrate stiffness on cell morphology, cytoskeletal structure, and adhesion. *Cell Motil. Cytoskeleton.* **60**, 24–34.

Zaman, M. H., Trapani, L. M., Sieminski, A. L., Mackellar, D., Gong, H., Kamm, R. D., Wells, A., Lauffenburger, D. A., and Matsudaira, P. (2006). Migration of tumor cells in 3D matrices is governed by matrix stiffness along with cell-matrix adhesion and proteolysis. *Proc. Natl. Acad. Sci. U S A* **103**, 10889–10894.

Further Reading

For a more in depth description of some of the methods and applications presented in this chapter, please see these references as a starting point for further reading.

Traction Force Microscopy Theory:

Butler, J. P., Tolic-Norrelykke, I. M., Fabry, B., and Fredberg, J. J. (2002a). Traction fields, moments, and strain energy that cells exert on their surroundings. *Am. J. Physiol. Cell Physiol.* **282**, C595–605.

Dembo, M., Oliver, T., Ishihara, A., and Jacobson, K. (1996). Imaging the traction stresses exerted by locomoting cells with the elastic substratum method. *Biophys. J.* **70**, 2008–2022.

Dembo, M., and Want, Y. L. (1999). Stresses at the cell-to-substrate interface during locomotion of fibroblasts. *Biophys. J.* **76**, 2307–2316.

Landau, L. D., Lifshitz, E. M., Kosevich, A. M., and Pitaevskii, L. P. (1986). *Theory of Elasticity.* Butterworth-Heinemann, Oxford, UK.

Marganski, W. A., Dembo, M., and Wang, Y. L. (2003). Measurements of cell-generated deformations on flexible substrata using correlation-based optical flow. *Methods Enzymol.* **361**, 197–211.

Schwarz, U. S., Balaban, N. Q., Riveline, D., Bershadsky, A., Geiger, B., and Safran, S. A. (2002). Calculation of forces at focal adhesions from elastic substrate data: the effect of localized force and the need for regularization. *Biophys. J.* **83**, 1380–1394.

Timoshenko, S. (1934). Theory of Elasticity. McGraw-Hill Inc, New York.

Biological Insights:

Discher, D., Janmey, P., and Wang, Y. L. (2005). Tissue cells feel and respond to the stiffness of their substrate. *Science* **310**, 1139–1143.

Friedl, P., and Wolf, K. (2010). Plasticity of cell migration: a multiscale tuning model. *J. Cell Biol.* **188**, 11–19.

Huynh, J., Califano, J. P., and Reinhart-King, C. A. (2011). Cell-generated forces in tissue assembly and function. *In* "Mechanobiology of Cell–Cell & Cell–Matrix Interactions," (A. J. Wagoneer-Johnson, and B. Harley, eds.), Springer, .

Kumar, S., and Weaver, V. (2009). Mechanics, malignancy, and metastasis: the force journey of a tumor cell. *Cancer Metastasis Rev.* **28**, 113–127.

Levental, I., Georges, P., and Janmey, P. (2007). Soft biological materials and their impact on cell function. *Soft Matter* **3**, 299–306.

Mierke, C. T., Rosel, D., Fabry, B., and Brabek, J. (2008). Contractile forces in tumor cell migration. *Eur. J. Cell Biol.* **87**, 669–676.

Paszek, M. J., Zahir, N., Johnson, K. R., Lakins, J. N., Rozenberg, G. I., Gefen, A., Reinhart-King, C. A., Margulies, S. S., Dembo, M., Boettiger, D., Hammer, D. A., and Weaver, V. M. (2005). Tensional homeostasis and the malignant phenotype. *Cancer Cell* **8**, 241–254.

Pelham Jr., R. J., and Wang, Y. (1999). High resolution detection of mechanical forces exerted by locomoting fibroblasts on the substrate. *Mol. Biol. Cell* **10**, 935–945.

Sen, S., Engler, A. J., and Discher, D. E. (2009). Matrix strains induced by cells: computing how far cells can feel. *Cell. Mol. Bioeng* **2**, 39–48.

Wyckoff, J. B., Pinner, S. E., Gschmeissner, S., Condeelis, J. S., and Sahai, E. (2006). ROCK- and myosin-dependent matrix deformation enables protease-independent tumor-cell invasion in vivo. *Curr. Biol.* **16**, 1515–1523.

Quantifying Traction Forces in 3D:

Legant, W. R., Miller, J. S., Blakely, B. L., Cohen, D. M., Genin, G. M., and Chen, C. S. (2010). Measurement of mechanical tractions exerted by cells in three-dimensional matrices. *Nat. Methods* **7**, 969–971.

Maskarinec, S. A., Franck, C., Tirrell, D. A., and Ravichandran, G. (2009). Quantifying cellular traction forces in three dimensions. *Proc. Natl. Acad. Sci. U S A.* **106**, 22108–22113.

Petroll, W. M., and Ma, L. (2003). Direct, dynamic assessment of cell-matrix interactions inside fibrillar collagen lattices. *Cell Motil. Cytoskeleton.* **55**, 254–264.

Roeder, B. A., Kokini, K., Robinson, J. P., and Voytik-Harbin, S. L. (2004). Local, three-dimensional strain measurements within largely deformed extracellular matrix constructs. *J. Biomech. Eng.* **126**, 699–708.

Vanni, S., Lagerholm, B. C., Otey, C., Taylor, D. L., and Lanni, F. (2003). Internet-based image analysis quantifies contractile behavior of individual fibroblasts inside model tissue. *Biophys. J.* **84**, 2715–2727.

CHAPTER 7

CellOrganizer: Image-Derived Models of Subcellular Organization and Protein Distribution

Robert F. Murphy[*,†]

[*]Lane Center for Computational Biology, Carnegie Mellon University, Pittsburgh, Pennsylvania, USA

[†]Freiburg Institute for Advanced Studies, University of Freiburg, Freiburg, Germany

Abstract

This chapter describes approaches for learning models of subcellular organization from images. The primary utility of these models is expected to be from incorporation into complex simulations of cell behaviors. Most current cell simulations do not consider spatial organization of proteins at all, or treat each organelle type as a single, idealized compartment. The ability to build generative models for all proteins in a proteome and use them for spatially accurate simulations is expected to improve

0091-679X/10 $35.00
DOI 10.1016/B978-0-12-388403-9.00007-2

the accuracy of models of cell behaviors. A second use, of potentially equal importance, is expected to be in testing and comparing software for analyzing cell images. The complexity and sophistication of algorithms used in cell-image-based screens and assays (variously referred to as high-content screening, high-content analysis, or high-throughput microscopy) is continuously increasing, and generative models can be used to produce images for testing these algorithms in which the expected answer is known.

I. Introduction

As traditional reductionist paradigms of biomedical research increasingly give way to systems approaches, the need to build predictive models that synthesize large amounts of information from potentially diverse sources is becoming critical. Most such current models take the form of transcriptional regulatory networks, protein–protein interaction maps, or biochemical reaction simulations. These typically do not consider spatial organization of cells or tissues. Important advances came with systems such as MCell (Stiles *et al.*, 1998), which allowed models to be constructed using mesh representations of cells built from electron microscope images, and the Virtual Cell (Loew and Schaff, 2001), which allowed appropriately processed images to provide surface area and volume for its compartmental models. Ontologies such as the genome ontology (GO) can be used to describe protein attributes, including location, primarily at a major organelle level. Such assignments can also be used to create compartmental models (e.g., http://biologicalnetworks.net/tutorials). However, compartmental models suffer from some important limitations, in that they treat all molecules within each compartment as being homogenously distributed, and they do not allow appearance, disappearance, fission or fusion of compartments.

Given the energy expended by cells to maintain their subcellular organization, and the many defects that are associated with alterations in it, models that do not accurately reflect subcellular organization are unlikely to perform satisfactorily at predicting complex cell behaviors or how they respond to changes in conditions. There is therefore a need for computational models that accurately represent the number, size, shape, and positions of subcellular structures, the spatial relationships between different structures, and how proteins (and other molecules) are distributed between them (Murphy, 2010, 2011). In addition, there is a need for a mechanism for representing how all of these vary within a population of cells of a single cell type, within a single cell type under different conditions, among different cell types, and among different organisms. Such models can not only *capture* cell behavior but can also be an important step in *understanding* that behavior, since, for example, a sufficiently detailed model helps distinguish aspects that are conserved and presumably necessary from those that are highly variable and potentially not necessary.

In considering how to build such models, we can distinguish *descriptive* models, which allow one to recognize what state a particular cell is in, from *generative* models, which can also synthesize new examples of cells in particular states. We can also distinguish *theoretical* or *conceptual* models, which posit a particular structure based on a generalized understanding, from *data-driven* models that are learned from data and capture both general behavior and variation in that behavior.

My focus in this chapter will be primarily on methods developed in my group that have been used to learn generative models of cell organization and protein distribution from two-dimensional and three-dimensional fluorescence microscope images (Zhao and Murphy, 2007; Rohde *et al.*, 2008a,b; Peng *et al.*, 2009; Shariff *et al.*, 2010a, 2011; Peng and Murphy, 2011). We have recently grouped these methods as part of the open source CellOrganizer project (http://cellorganizer.org), which includes collaborations with a number of investigators studying particular cell systems.

II. Components of a Model of Subcellular Organization and Protein Distribution

Although there are a number of ways to break down the tasks necessary for creating such models, we can distinguish at least three major components of a model of the distribution of proteins within cells of a given type under a given condition:

- A model of subcellular organization, including distributions of the number, size, shape, and position of each subcellular structure, any of which may be conditional on the model(s) for other structures;
- A model representing the probability that a cell of a given type will contain a certain number of molecules of a given protein, the expected fraction of those molecules in each subcellular structure, and a measure of the variation in that fraction from cell to cell;
- A model of how each protein is distributed within each structure, which may consist of a self-organizing model that specifies only the affinities between pairs of proteins within each structure.

Higher order models can then be built to specify how any of these models change over time and condition: for example, during the cell cycle, in the presence of perturbagens, for cells expressing mutations, or for different cell types.

I will focus below on work on the first two types of components.

III. Models of Subcellular Organization

At a conceptual level, the most complete model of subcellular organization is probably the GO cellular component ontology (Ashburner *et al.*, 2000). A significant effort has been made to capture the vast majority of terms used to describe subcellular structures. The terms in this ontology can be assigned to proteins in order to represent the results of experimental or computational analyses. The advantage of this approach is precisely its disadvantage: general terms such as "mitochondria" can be associated with a protein while leaving many questions about what mitochondria are unanswered. However, to be useful for spatially realistic modeling, ontology terms must be associated with a representation of each organelle's number, structure, and distribution within cells. Currently, such representations are abstract and implicit rather than concrete and they often leave unspecified how the organelle would look in different cell types. For example, the abstract concept of a mitochondrion is well understood by biologists but most would be hard pressed to accurately describe how mitochondria vary in number, size, shape, and distribution from cell type to cell type or organism to organism.

In building generative models, we refer to an individual image, stack, or movie to be an *instance* drawn from an underlying model, whether an actual image or a synthetic image. These instances are considered to have been generated by particular *values* for the *parameters* of the model. The model is *generative* if it captures how parameter values can be chosen for new instances.

A critical concept in creating models of subcellular organization is the conditional relationships that exist among different components. This is easily illustrated by considering the task of building generative models of nuclear and cell shape (i.e., the positions of the nuclear and plasma membranes). We could build one generative model from many examples of nuclear shapes, and build another generative model from many examples of cell shapes. If we want to synthesize a new example of a cell containing a nucleus, we can imagine drawing a random example of a nuclear shape from the first model, and drawing a random example of a cell shape from the second. However, there is nothing that would prevent the example nuclear shape from being too wide to fit inside the example cell shape, and nothing to tell us where within the cell shape to put the nuclear shape. We must therefore connect the generation processes, which we do by making the models dependent, or *conditional*, upon each other. In our work, we have chosen to make the cell shape model conditional upon the nuclear shape. As we will see below, this means that during the learning process the relationship between the shapes is captured, and during the generation process, an example nuclear shape is first generated and used to generate an appropriate cell shape.[1] An alternative is to make the models *joint*, in which we learn simultaneously a model for both shapes.

Another major consideration is whether to make the models *parametric*, in which the values of model parameters explicitly describe various aspects of the sizes and shapes of cell components, or *nonparametric*, in which sizes and shapes are implicitly described by the relationships between examples. This distinction will be made clearer in the next sections where we consider models of cell components and how they can be made conditional upon each other. In each case, we will consider

[1] Of course, we might also have chosen to make the nuclear shape conditional upon the cell shape. Which order is better will need to be determined by future work.

- the inputs necessary for training the model,
- the means of assessing how adequately the model describes the data,
- what types of outputs the model can generate.

A. Nuclear and Cell Shape Models

1. Nuclear Shape – Medial Axis Models

Nuclear shape is often represented in theoretical models as a sphere or more generally an ellipsoid. Examination of only a few images of some cell types (especially adherent cultured cells) reveals how inaccurate this model can be. A somewhat more accurate model can be learned directly from images (Zhao and Murphy, 2007) using a *medial axis* approach (Blum, 1973). As illustrated in Fig. 1, medial axis construction typically begins by first orienting all nuclear shapes (instances) so that their major axes point in the same direction). Each instance is then represented by the position of a curve bisecting the shape perpendicular to the major axis, and by the width at each position along that curve. These curves can be fit using splines, such

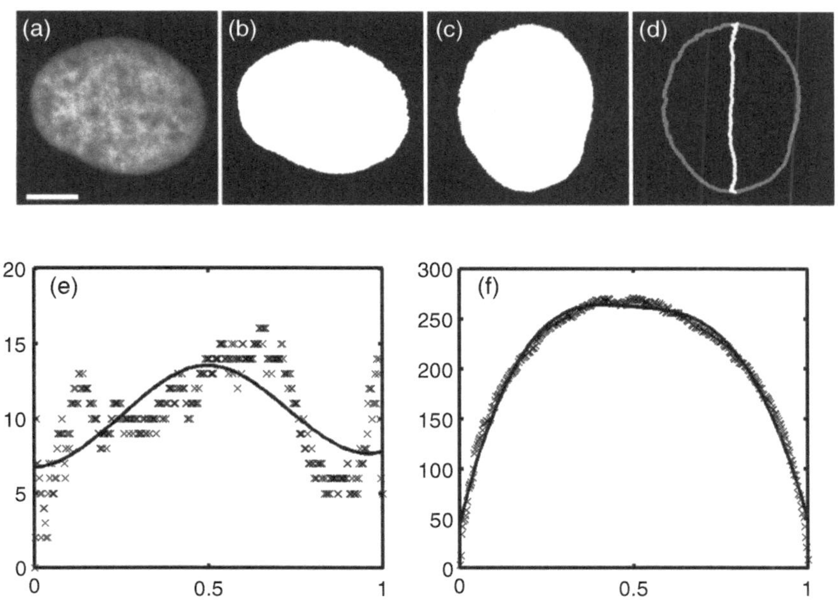

Fig. 1 Illustration of a medial axis method for modeling a 2D nuclear shape instance. The original nuclear image (a) was binarized (b) and rotated so that its major axis is vertical (c). The position of the curve that divides the shape in half horizontally at each vertical position is then found (d). The horizontal positions of the medial axis as a function of the fractional vertical distance are shown by the symbols (e), along with a B-spline fit (solid curve). The width as a function of fractional distance is shown by the symbols (f), along with the corresponding fit (solid curve). Scale bar, 5 um. From Zhao and Murphy (2007).

that a set of 11 spline coefficients describes each instance. The distribution(s) of these parameters over many instances can then be learned. In this case, two multivariate Gaussian distributions, one for the medial axis position and one for the width, were shown to provide a good representation of nuclear shape in two-dimensional images (Zhao and Murphy, 2007). Sampling from these distributions using a random number generator can be done in order to create synthetic examples from the learned model.

2. Nuclear Shape – Cylindrical Spline Surface Model

For three-dimensional images, the medial axis method can result in an oversimplified shape model. An alternative is to convert the nuclear shape to cylindrical coordinates and then fit a periodic spline surface (Peng and Murphy, 2011). This is illustrated in Fig. 2. In this case, there is one parameter for the nuclear height and 32 parameters for the coefficients of the spline surface. For a collection of three-dimensional images of HeLa cells, these parameters were also shown to be well represented by a multivariate Gaussian distribution. As before, parameter values can be randomly sampled from this distribution to generate new nuclear shape instances.

3. Nuclear Shape – Large Deformation Diffeomorphic Metric Mapping

These parametric models of nuclear shape have two significant advantages: first, they can be computed fairly quickly, and second, the parameters (and parameter distributions) can be stored compactly. However, they make assumptions about the characteristics of nuclear shape that need to be captured (e.g., that small bumps can be ignored) and do not handle well many concave or branched shapes. An important alternative therefore is to use nonparametric models such as the large deformation

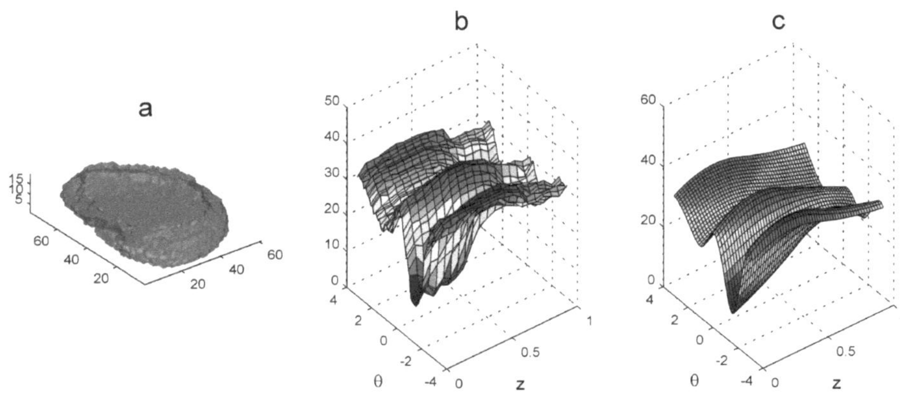

Fig. 2 Illustration of cylindrical spline surface method for modeling a three-dimensional nuclear shape instance. (a) Surface plot of a 3D HeLa cell nucleus. (b) Unfolded surface of the nuclear shape in a cylindrical coordinate system. The surface plot shows the radius r as a function of azimuth u and height z. (c) B-spline surface fitted to the unfolded nuclear surface. From Peng and Murphy (2011). (For color version of this figure, the reader is referred to the web version of this book.)

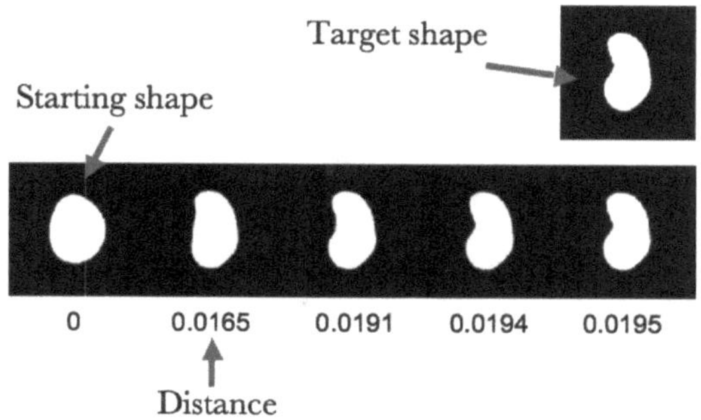

Fig. 3 Determining the distance between two shapes using large deformation metric mapping. The goal is to measure the distance between the starting shape and the target shape. This is done by gradually deforming the starting shape to become more similar to the target shape while recording how much perturbation is necessary at each step. From Rohde *et al.* (2008a).

diffeomorphic metric mapping (LDDMM) framework developed by Miller and colleagues (Beg *et al.*, 2005). In this framework, shape is represented implicitly by measuring differences between pairs of shape instances (see Fig. 3). The distance matrix is then used to create a shape space in which similar shapes are near each other. This approach has been demonstrated to provide an excellent representation of nuclear shape in HeLa cells (Rohde *et al.*, 2008a), and the method can be applied to two-, three-, or four-dimensional images. This power comes at a price: saving the shape model requires storing both the distance matrix (or the shape space) and the example images used to create it. Generating new shape instances can be achieved by interpolating between the original examples (Peng *et al.*, 2009), but this can be computationally expensive.

An important additional use of non-rigid registration methods is to identify positions *within* nuclei. In an exciting example, the positions of different chromosome regions have been mapped to a common frame of reference using a multiresolution non-rigid registration approach (Yang *et al.*, 2008). Potentially, position mapping could be combined with modeling of the nuclear shape itself as described above.

4. Cell Shape – Circular and Spherical Coordinate Ratiometric Models

Cell shape can also be represented using diffeomorphic methods, using exactly the same approach as used for nuclei. This is appropriate when modeling only the cell shape is desired, but if nuclei are to be included, as discussed above, the nuclear and cell shape models must be conditionally related. This can be achieved using diffeomorphic methods by creating indexed images in which pixels/voxels that are part of the background have one value (e.g., 0), pixels/voxels in the nucleus have a second value (e.g., 1), and pixels/voxels inside the cell but not in the nucleus have a third value. Finding the distance between such indexed images is a bit more computationally demanding.

To create more compact conditional models of cell shape, a simple approach can be used. For two-dimensional images, the coordinates of the cell and nuclear boundary are first mapped to polar coordinates, and then the ratio between the two is calculated for a fixed number of angles (e.g., every degree over 360 degrees) (Zhao and Murphy, 2007). For three-dimensional images, these ratios are calculated for each two-dimensional slice (Peng and Murphy, 2011). The model is then simplified by keeping only a certain number of principal components (for HeLa cells, 10 components were used for two-dimensional images and 25 for three-dimensional images). The distributions of these components have been shown to follow a multivariate Gaussian, providing a very compact conditional model. To generate instances from the model, a nuclear shape is first generated using one of the methods above, principal component coefficients are chosen using random numbers and converted to the cell/nuclear ratio as a function of angle, and then these ratios are multiplied by the corresponding position on the synthetic nuclear boundary to generate the synthetic cell boundary.

B. Models of Vesicular Organelles: Shape

1. Gaussian Object Models

Many vesicular organelles, such as lysosomes, show a roughly spherical shape in both electron microscope and fluorescent microscope images. Such shapes can be easily modeled if the organelles are well resolved from each other in images. However, vesicular organelles are frequently found quite close to each other, and they can appear to overlap when imaged in two dimensions. Furthermore, sampling noise may make them appear irregularly shaped. One approach to this problem is to assume that the organelles are all spherical (or ellipsoidal) and try to estimate what configuration of organelles gave rise to a particular cell image. This can be done by thresholding the image of an organelle marker to identify connected components that may consist of more than one organelle. As shown in Fig. 4, image processing

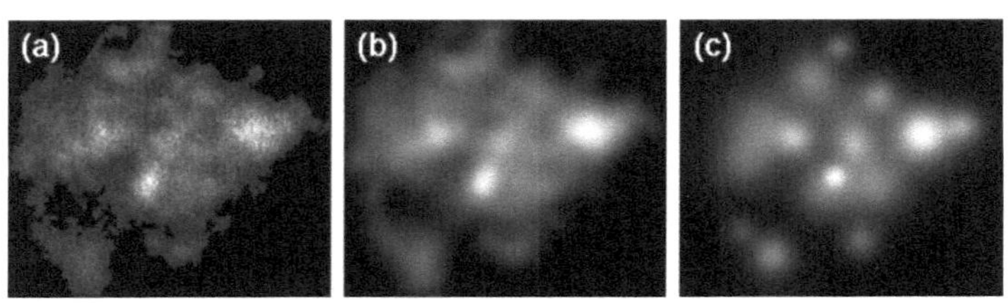

Fig. 4 Illustration of fitting objects using a 2D Gaussian mixture model. A region of a cell containing a single composite object (found by thresholding and connecting above threshold pixels) (a) is smoothed by a Gaussian low pass filter (b) to facilitate detection of local maxima (peaks) in the composite object. Fitting using a spherical covariance matrix(c) yields the estimated positions and sizes of the Gaussian objects assumed to have given rise to the original image. A similar approach is used for 3D images. After Zhao and Murphy (2007).

and parameter estimation can then be used to find the positions and sizes of the individual organelles. A statistical model of the distribution of the number of objects per cell, and the distribution of the Gaussian parameters (covariance matrix) can then be constructed. This method can be used for both two-and three-dimensional images, although distinguishing different organelles is easier in three-dimensional images.

2. Outline Models

More accurate models can be obtained using methods that seek to estimate the position of the *outline* of vesicular organelles. For example, piece-wise linear closed splines have been used to describe the shape of endosomes (Helmuth *et al.*, 2009). Such methods could be combined with eigenshape or diffeomorphic methods to create generative models.

3. Object Type Models

Even more detailed (but not necessarily more accurate!) models can be obtained by finding all objects in a large set of cell images and clustering them to identify distinct object types. This approach has been applied to a large collection of HeLa cell images, and the resulting object types were found to enable recognition of different subcellular patterns (Zhao *et al.*, 2005). As discussed below, this approach has been used to estimate the amount of a given probe in different organelles. However, it could also be used as part of a generative model by modeling the number and shape of each object type.

C. Models of Vesicular Organelles: Position

Regardless of which method is used for estimating object number and shape, a model of the *position* of each object within the cell is also needed. This clearly needs to be conditional upon the cell and nuclear shape model. One simple approach is to represent the position of each observed object in a normalized polar or spherical coordinate system (depending on whether the image is two- or three-dimensional). To do this, the distance of the center of each object from the nuclear boundary is expressed as a fraction of the sum of the distance from the nuclear boundary and the distance from the cell boundary (this normalized distance can be negative if the object is inside the nucleus). The angle (or angles) of the object's center to the center of the nucleus are also found. An empirical probability density map is then formed by tabulating these positions for many objects from many cells. To use this model to synthesize an image, the number of objects is drawn from the appropriate distribution, a size and shape are drawn for each (depending on which shape model is being used), and distances and angles are chosen randomly according to the density map for each and converted to actual coordinates for particular cell and nuclear shape instances.

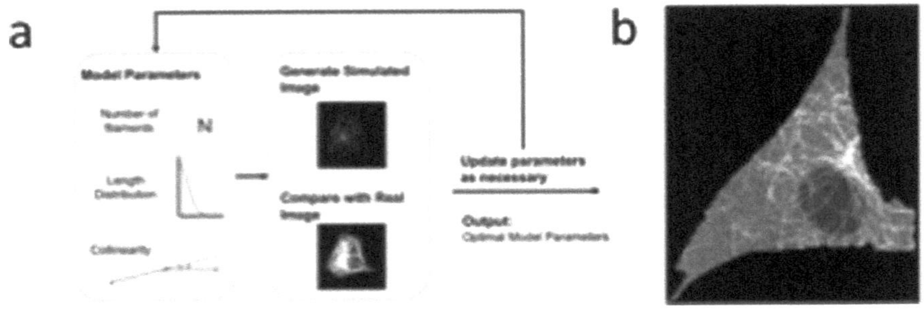

Fig. 5 (a) Overview of inverse modeling approach for estimating parameters of the microtubule generative model. From Sharif *et al.* (Shariff *et al.*, 2010a). (b) Example of two-dimensional slice from three-dimensional synthetic image generated by tubulin model.

D. Models of Cytoskeletal Structures

The methods described above for building nuclear, cell, and organelle models all make direct estimates of model parameters from real images. Although decomposing a cluster of organelles into individual objects may be difficult, it is usually possible. Some organelles or structures are much more difficult to resolve into individual elements. For example, two- or three-dimensional images of the distribution of tubulin by either wide-field or confocal microscopy typically show individual microtubules at the cell periphery but a tangle of crossing microtubules near the centrosome. Estimating the number of individual microtubules or their individual paths is nearly impossible. One solution is to use specialized microscope methods, such as speckle microscopy, to resolve individual microtubules. An alternative is to use inverse modeling methods to try to estimate the parameters of a microtubule model, as illustrated in Fig. 5a. A generative model is created and then instances of that model are created for many different sets of parameters. These instances are compared to a real image and the parameters corresponding to the best match are chosen. This approach has been used to study kinetochore-microtubule dynamics (Sprague *et al.*, 2003). We have used a similar approach to build a generative model of microtubules in interphase HeLa cells and 3T3 cells (Shariff *et al.*, 2010a, 2011). An example of a synthetic microtubule distribution is shown in Fig. 5b.

E. Putting it all Together

Once the various components of a model have been created, it is a simple matter to construct synthetic cell instances. Figs. 6 and 7 show *idealized* images (with no blurring or noise) for instances created from two- or three-dimensional models, respectively. As discussed below, these idealized images can also be used to estimate how that cell might look if imaged in a particular microscope.

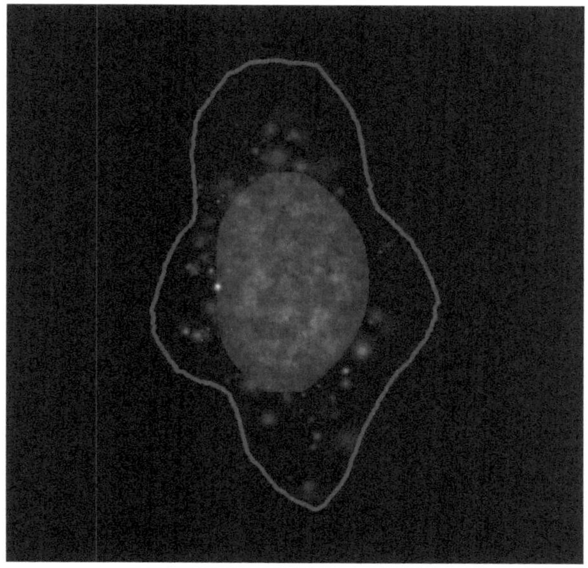

Fig. 6 Example of synthetic image generated by a two-dimensional model learned from images of the lysosomal protein LAMP2. The DNA distribution is shown in red, the cell outline in blue, and LAMP2-containing objects in green. From http://murphylab.web.cmu.edu/data/2007_Cytometry_GenModel.html. (For interpretation of the references to color in this figure legend, the reader is referred to the web version of this book.)

Fig. 7 Example synthetic image generated by a three-dimensional model learned from images of the lysosomal protein LAMP2. The nuclear surface is shown in red, the cell surface in blue, and LAMP2-containing objects in green. (See color plate.)

IV. Protein Distributions Across Subcellular Structures

The models described above capture how cellular organelles are arranged within a cell, but do not address the critical question of how the tens of thousands of proteins in each cell are distributed among these organelles. Images, especially fluorescence microscope images, can be a major source of information on the subcellular distributions of proteins, and, as mentioned above, may be used directly in cell simulations. The feasibility of using automated pattern recognition approaches to recognize the subcellular patterns of proteins that localize primarily to one organelle has been well demonstrated (for reviews see (Chen *et al.*, 2006; Conrad and Gerlich, 2010; Shariff *et al.*, 2010b)). However, many proteins are found to varying extents in more than one organelle, and therefore a means of determining that distribution is needed.

A. Boolean Vectors: GO Terms

Some information about protein subcellular location can be obtained from protein databases, which have at least some GO terms associated with most proteins. However, there are a number of limitations of these annotations, most of which derive from the absence of enough experimental data. For example, these databases do not attempt to capture changes in GO terms for different conditions or cell types or distinguish between subcellular locations of different splice isoforms. Nonetheless, when no other information is available, GO terms can be represented as a Boolean vector describing whether a particular protein is or is not found in each organelle.

B. Dirichlet Distributions: Pattern Unmixing

What is really needed for accurate modeling of a protein is a Dirichlet distribution – a probability distribution (that sums to one) for each molecule of that protein over the different organelles. We can convert the Boolean vector for a particular protein derived from GO terms into a Dirichlet distribution by dividing by the number of organelles it is thought to be found in. This assumes, in the absence of any other information, that it is equally likely to be in each of them. A much better alternative is to try to estimate the amount of a given protein in each organelle or structure. To do this, we define a set of *fundamental patterns* to be a set from which all composite patterns can be constructed. This might correspond to the set of all organelle patterns, but, depending on the extent to which they are distinct, might contain multiple subpatterns for a given organelle. For example, protein distributions in the nucleus have been divided into at least eight nuclear subdomains (Bauer *et al.*, 2011). For a collection of images of a particular protein, we seek to find the Dirichlet distribution over these fundamental patterns. In other words, we estimate how much of the protein would have to be in each pattern in order for the overall image to appear as it does. This task can be viewed as *unmixing* an image formed by mixing fundamental patterns.

We have described two approaches for estimating this: one in which we specify the fundamental patterns in advance and just try to estimate the fractions (referred to as supervised unmixing), and one in which we try to find the fundamental patterns as well as the fractions (referred to as unsupervised unmixing). Using a test set of images created by an automated high content imaging system, we have demonstrated that good estimates of the fractions can be obtained by both the supervised (Peng *et al.*, 2010) and unsupervised (Coelho *et al.*, 2010) approaches.

V. Use of Models for Testing Algorithms

A classic problem in testing algorithms for microscope images is that the correct results are frequently not known. A generative model for a desired pattern or structure can be combined with a model of image formation in a particular microscope to generate test images (phantoms) for which the correct results from image analysis are known. The process by which an image is formed in a microscope is quite well understood, so accurate models of point-spread functions and sampling noise can be constructed and applied to the idealized images generated by the methods described above. This approach has been applied previously for nuclei (Yang *et al.*, 2008; Svoboda *et al.*, 2009); the paper by Svoboda *et al.* (Svoboda *et al.*, 2009) provides a particularly good image formation model.

The phantom approach can be extended to any combination of the tools in the CellOrganizer project to generate test images with known cell boundaries, object locations, and/or subcellular patterns. The accuracy of algorithms can also be determined as a function of the parameters of the generative model, such as cell size or extent of nuclear elongation. Collections of already synthesized synthetic cell images can be found at http://CellOrganizer.org.

VI. Conclusion

In this chapter, I have described current approaches for building accurate models of cell organization directly from fluorescent microscope images. These models capture variation in cell organization at the level of the nucleus, cell membrane, and individual organelles, and can capture how particular proteins are distributed among cellular components. They represent a significant advance over the use of words (such as GO terms) as the means by which results of experiments on subcellular localization and organization are captured and communicated. Nonetheless, the field is at the beginning, and it is hoped that many investigators will develop and make available tools that improve and extend the approaches described here. Examples of future work that can be anticipated include methods for merging images at different resolutions (especially light and electron microscope images) and methods for describing the interplay between localization and structure for proteins involved in creating subcellular structures.

Acknowledgments

I express my thanks to Michael Boland, Meel Velliste, Ting Zhao, Tao Peng, Luis Coelho, Wei Wang, and especially Gustavo Rohde for their contributions to previous collaborative work described here and for many helpful discussions with them and with Jieyue Li, Taraz Buck, Ivan Cao-Berg, Baek Hwan Cho, Klaus Palme, Hagit Shatkay, Joel Stiles, Leslie Loew, Ion Moraru, Christoph Wülfing, Eric Xing, Gaudenz Danuser, Karl Rohr, and Ivo Sbalzarini. Much of the original work reviewed here was supported by NIH grants GM068845, GM075205, and GM090033, and by NSF grant EF-0331657. Part of the discussions and writing of this article were supported by a Research Award from the Alexander von Humboldt Foundation and an External Senior Research Fellowship from the Freiburg Institute of Advanced Studies.

References

Stiles, J. R., Bartol Jr, T. M., Salpeter, E. E., and Salpeter, M. M. (1998). Monte Carlo simulation of neurotransmitter release using MCell, a general simulator of cellular physiological processes. *In* "Computational Neuroscience," (J. M. Bower, ed.), pp. 279–284. Plenum, NY.

Loew, L. M., and Schaff, J. C. (2001). The virtual cell: a software environment for computational cell biology. *Trends Biotechnol.* **19**, 401–406.

Murphy, R. F. (2010). Communicating subcellular distributions. *Cytometr. Part A* **77**, 686–692.

Murphy, R. F. (2011). An active role for machine learning in drug development. *Nature Chem. Biol.* **7**, 327–330.

Zhao, T., and Murphy, R. F. (2007). Automated learning of generative models for subcellular location: building blocks for systems biology. *Cytometr. Part A.* **71A**, 978–990.

Rohde, G. K., Ribeiro, A. J., Dahl, K. N., and Murphy, R. F. (2008a). Deformation-based nuclear morphometry: capturing nuclear shape variation in HeLa cells. *Cytometr. Part A.* **73A**, 341–350.

Rohde, G. K., Wang, W., Peng, T., and Murphy, R. F. (2008b). Deformation-based nonlinear dimension reduction: applications to nuclear morphometry. *Proc. 2008 Int. Symp. Biomed. Imaging.* 500–503.

Peng, T., Wang, W., Rohde, G. K., and Murphy, R. F. (2009). Instance-based generative biological shape modeling. *Proc. 2009 Int. Symp. Biomed. Imaging.* 690–693.

Shariff, A., Murphy, R. F., and Rohde, G. K. (2010a). A generative model of microtubule distributions, and indirect estimation of its parameters from fluorescence microscopy images. *Cytometr. Part A.* **77A**, 457–466.

Shariff, A., Murphy, R. F., and Rohde, G. K. (2011). Automated estimation of microtubule model parameters from 3-D live cell microscopy images. *Proc. IEEE Int. Symp. Biomed. Imaging.* **2011**, 1330–1333.

Peng, T., and Murphy, R. F. (2011). Image-derived, three-dimensional generative models of cellular organization. *Cytometr. Part A* **79A**, 383–391.

Ashburner, M., *et al.* (2000). Gene ontology: tool for the unification of biology. The Gene Ontology Consortium. *Nat. Genet.* **25**, 25–29.

Blum, H. (1973). Biological shape and visual science. I. *J. Theor. Biol.* **38**, 205–287.

Beg, M. F., Miller, M. I., Trouve, A., and Younes, L. (2005). Computing large deformation metric mappings via geodesic flows of diffeomorphisms. *Int. J. Comput. Vis.* **61**, 139–157.

Yang, S., *et al.* (2008). Nonrigid registration of 3-D multichannel microscopy images of cell nuclei. *IEEE Trans. Image Process.* **17**, 493–499.

Helmuth, J. A., Burckhardt, C. J., Greber, U. F., and Sbalzarini, I. F. (2009). Shape reconstruction of subcellular structures from live cell fluorescence microscopy images. *J. Struct. Biol.* **167**, 1–10.

Zhao, T., Velliste, M., Boland, M. V., and Murphy, R. F. (2005). Object type recognition for automated analysis of protein subcellular location. *IEEE Trans. Image Process.* **14**, 1351–1359.

Sprague, B. L., *et al.* (2003). Mechanisms of microtubule-based kinetochore positioning in the yeast metaphase spindle. *Biophys. J.* **84**, 3529–3546.

Chen, X., Velliste, M., and Murphy, R. F. (2006). Automated interpretation of subcellular patterns in fluorescence microscope images for location proteomics. *Cytometr. Part A.* **69A**, 631–640.

Conrad, C., and Gerlich, D. W. (2010). Automated microscopy for high-content RNAi screening. *J. Cell Biol.* **188**, 453–461.

Shariff, A., Kangas, J., Coelho, L. P., Quinn, S., and Murphy, R. F. (2010b). Automated image analysis for high-content screening and analysis. *J. Biomol. Screen. Off. J. Soc. Biomol. Screen.* **15**, 726–734.

Bauer, D. C., *et al.* (2011). Sorting the nuclear proteome. *Bioinformatics* **27**, i7–i14.

Peng, T., *et al.* (2010). Determining the distribution of probes between different subcellular locations through automated unmixing of subcellular patterns. *Proc. Natl. Acad. Sci. U. S. A.* **107**, 2944–2949.

Coelho, L. P., Peng, T., and Murphy, R. F. (2010). Quantifying the distribution of probes between subcellular locations using unsupervised pattern unmixing. *Bioinformatics* **26**, i7–i12.

Svoboda, D., Kozubek, M., and Stejskal, S. (2009). Generation of digital phantoms of cell nuclei and simulation of image formation in 3D image cytometry. *Cytometr. Part A.* **75A**, 494–509.

CHAPTER 8

Spatial Modeling of Cell Signaling Networks

Ann E. Cowan, Ion I. Moraru, James C. Schaff, Boris M. Slepchenko and Leslie M. Loew

R. D. Berlin Center for Cell Analysis and Modeling, University of Connecticut Heath Center, Farmington, CT, USA

Abstract

The shape of a cell, the sizes of subcellular compartments, and the spatial distribution of molecules within the cytoplasm can all control how molecules interact to produce a cellular behavior. This chapter describes how these spatial features can be included in mechanistic mathematical models of cell signaling. The Virtual Cell computational modeling and simulation software is used to illustrate the considerations required to build a spatial model. An explanation of how to appropriately choose between physical formulations that implicitly or explicitly account for cell geometry and between deterministic versus stochastic formulations for molecular dynamics is provided, along with a discussion of their respective strengths and weaknesses. As a first step toward constructing a spatial model, the geometry needs

0091-679X/10 $35.00
DOI 10.1016/B978-0-12-388403-9.00008-4

to be specified and associated with the molecules, reactions, and membrane flux processes of the network. Initial conditions, diffusion coefficients, velocities, and boundary conditions complete the specifications required to define the mathematics of the model. The numerical methods used to solve reaction–diffusion problems both deterministically and stochastically are then described and some guidance is provided in how to set up and run simulations. A study of cAMP signaling in neurons ends the chapter, providing an example of the insights that can be gained in interpreting experimental results through the application of spatial modeling.

I. Introduction

The cell is distinctly nonhomogeneous and the spatial distribution of molecules can be of critical importance to cellular pathways. Signaling events initiated within the two-dimensional plane of the membrane move through the three-dimensional volume of the cytosol and propagate through multiple intracellular compartments. Spatial segregation of interacting molecules, whether by localization to different cellular compartments or by associations with supramolecular complexes, is a common mechanism of regulating pathway activity. Mathematical modeling and simulation in these situations requires spatial simulation methods that incorporate actual cell geometry, compute local concentrations, and account for changes that arise from transport processes (diffusion and active processes).

Spatial modeling of signaling pathways has already begun to provide unique insights into how cellular geometry intersects with the kinetic behavior of signaling components to create spatially encoded information in the cell. We are now beginning to understand at a quantitative level not only how surface to volume effects impact signaling pathways that arise on a membrane (e.g., Fink *et al.* (2000), but also how geometry effects are transmitted to downstream components. Spatial modeling studies have demonstrated that the creation of signaling molecules at the membrane and their destruction or inhibition throughout the cytosol can create gradients that vary as the local geometry of the cell changes (Kholodenko *et al.*, 2010). Local gradients likewise can have significant downstream effects. For example, regulation of calcium levels during repetitive firing of synapses is highly dependent on the specialized geometry of the neuronal spine, leading to new hypotheses for coincidence detection localized to individual synapses, a key aspect of learning and memory (Brown *et al.*, 2008; Hernjak *et al.*, 2005). Also in neuronal cells, experiments coupled to spatial simulations demonstrated that while microdomains of elevated cAMP arise from the localization of receptors and adenyl cyclase to the membrane, the kinetics of negative regulators localized to the cytosol creates spatially distinct regions of activity of downstream targets such as PKA and MAPK (Neves *et al.*, 2008).

Events at the plasma membrane that dictate polarized cellular responses, such as chemotaxis and cell migration as well as yeast budding and cell division, are among the more obvious cases where spatial modeling can lead to new insights into how signaling pathways evoke spatially discrete responses in the cell. Already modeling efforts have led to a number of new hypotheses in these fields. Some examples include the local excitation, global inhibition (LEGI) model to explain how cells respond to shallow gradients of chemoattractants (Ma *et al.*, 2004), a hypothesis that Turing-type activator–inhibitor dynamics involving the small Rho GTPase Cdc42 can explain the selection of only a single budding site in yeast (Goryachev and Pokhilko, 2008), and a proposed mechanism by which spatial gradients of two regulatory molecules evaluates cell size in yeast mitotic checkpoints (Vilela *et al.*, 2010).

In addition to testing and developing new hypotheses, spatial modeling also provides an exceptional tool for analyzing and interpreting the ever expanding arsenal of fluorescence-based microscope imaging methods. Spatial simulations help one to extract quantitative information about the dynamic behavior of molecules and the detailed kinetics of molecular interactions and enzymatic events within the exact geometry of experimental cells. This allows direct comparison of simulation results of different models and parameters with experimental image time series. Most current methods for quantitative analysis of dynamic fluorescence imaging experiments rely on analytic solutions that assume simple analytic geometries for the cell. The ability of numerical simulation approaches to account for exact morphologies of real cells dramatically broadens the range of these experimental techniques. Simulation-based approaches have been used to analyze many different types of experiments including uncaging experiments (Roy *et al.*, 2001) and fluorescence photobleaching or photoactivation experiments (Holt *et al.*, 2004; Kapustina *et al.*, 2010; Moissoglu *et al.*, 2006; Shen *et al.*, 2008). Indeed, any experimental data based on changes in fluorescence distributions over time and space can be amenable to analysis by spatial simulation methods. Particularly exciting is the promise of spatial simulation-based analysis to extract high temporal and spatial resolution information on pathway dynamics from the array of new fluorescence biosensors for kinase and phosphatase activities (Saucerman *et al.*, 2006; Zhong *et al.*, 2009).

This chapter provides a discussion of the problems that spatial modeling can effectively address in cell signaling, and different overall strategies for developing models of cellular pathways. Using our web-based Virtual Cell (VCell) modeling environment to illustrate the process (http://vcell.org), we discuss some of the important issues that need to be addressed in order to build a useful spatial model and how to negotiate the important choices and parameters involved in running numerical simulations of the models. This is followed by working through a specific example of a VCell spatial model and exploring how the model can be used to simulate specific experimental or conceptual conditions to generate predictions of the model (i.e., simulation data) that can be tested experimentally.

II. Overview of Spatial Modeling

A spatial model is a mathematical system that accounts for processes such as reactions kinetics, diffusion, advection, and membrane transport. A pair of equations serves to summarize the physical chemistry of cell signaling systems with explicit consideration of the voyage of a molecule from one region of the cell to another:

$$\frac{\partial C_i}{\partial t} = -div\,\vec{F}_i + R_i(C_j, C_k, \ldots, \Phi) \tag{1}$$

$$\vec{F}_i = -D_i \nabla C_i - C_i \vec{V}_i - z_i \mu_i C_i \nabla \Phi \tag{2}$$

The first equation describes the change in concentration, C_i, of a molecular species, i as a function of time at some point within the cell. It is a partial derivative of C_i with respect to time; C_i can also vary over spatial coordinates, x,y,z. On the right hand side of Eq. (1), R_i is the rate expression for the formation or destruction of species i; it can be a function of the concentration of any of the other molecular species in the system, as well as the electrical potential across the membrane, Φ, for a voltage-sensitive membrane bound species. The first term on the right side is the divergence of the flux of i, F_i, which is further described in the second equation. Eq. (2) is the Nernst–Planck flux equation with an added advection term, showing the factors that govern the net flow of molecules: the gradient of concentration times the diffusion coefficient, D_i; the velocity field, $\vec{V}_i$ (possibly driven by molecular motors); and an electrical term, where z_i is the charge, μ_i is the electrical mobility, and $\nabla\Phi$ is the voltage gradient (i.e., the electric field; the electrical term is unimportant within the cytosol, but can, of course, control ionic transport across membranes). The geometry specification for the model should include all the morphological features of the cell that might influence the molecular processes; it should also account for the heterogeneous distribution of the molecules within this geometry. The resulting set of partial differential equations (PDEs) for all the molecular species represent a continuous deterministic mathematical description of the system and also serve to summarize all the biophysical mechanisms hypothesized to govern the biological process under study (Slepchenko *et al.*, 2003). However, when the number of molecules involved in the process is small (<100), a deterministic mathematical description may prove to be inaccurate because it fails to account for the probabilistic nature of the reactions of single molecular species and of the Brownian dynamics of single molecules that underlie diffusion. In such a case, a stochastic mathematical formulation needs to be employed.

But it is important to ask how much detail is actually required to address a specific cell biological problem. Clearly, the more detail, the more likely that the investigator will not overlook a key contribution to the biology. However, the more detail, the greater the computer power and the longer the computation time required for the numerical methods to compute a simulation. Furthermore, the simpler the model, the easier it is to analyze and understand the simulation results; that is, when simulations from simple models fail to reproduce an experimental result, it is easier to uncover what may be missing or incorrect in the model. Arguably, this process of interacting with experiment is the most important reason for building a model and running simulations. We will therefore discuss varying choices for posing a model mathematically, in order of increasing computational intensity, describing the limitations of each. These can all be modeled and simulated with the VCell software system, as illustrated in Fig. 1, which shows simulation results for the passive flux of a molecule through the nuclear membrane from four different mathematical models.

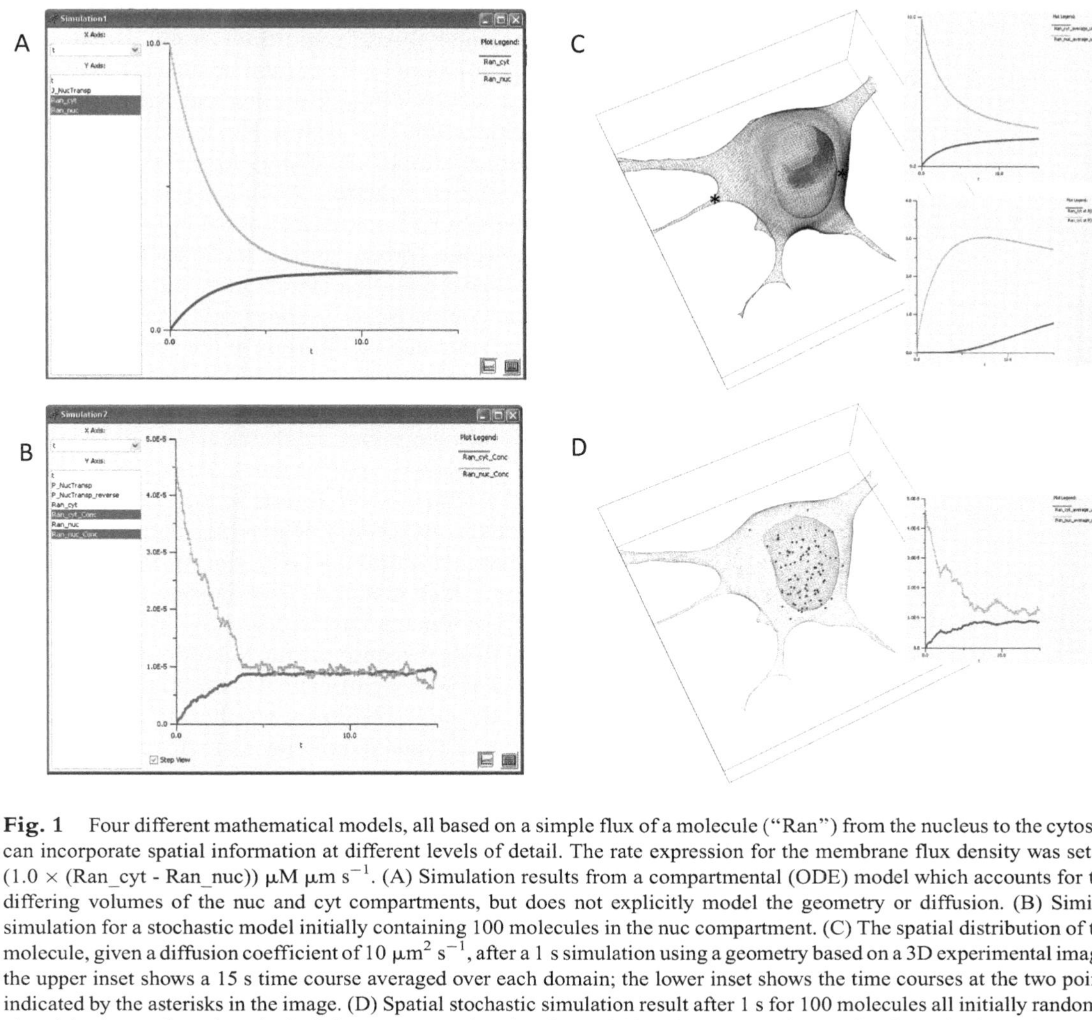

Fig. 1 Four different mathematical models, all based on a simple flux of a molecule ("Ran") from the nucleus to the cytosol, can incorporate spatial information at different levels of detail. The rate expression for the membrane flux density was set as (1.0 × (Ran_cyt - Ran_nuc)) μM μm s^{-1}. (A) Simulation results from a compartmental (ODE) model which accounts for the differing volumes of the nuc and cyt compartments, but does not explicitly model the geometry or diffusion. (B) Similar simulation for a stochastic model initially containing 100 molecules in the nuc compartment. (C) The spatial distribution of the molecule, given a diffusion coefficient of 10 $\mu m^2\ s^{-1}$, after a 1 s simulation using a geometry based on a 3D experimental image; the upper inset shows a 15 s time course averaged over each domain; the lower inset shows the time courses at the two points indicated by the asterisks in the image. (D) Spatial stochastic simulation result after 1 s for 100 molecules all initially randomly placed in the nucleus domain (red dots are cytosol molecules and blue dots are nuclear molecules); the inset shows the concentration over each compartment for the entire 15 s simulation. (See color plate.)

At the simplest level, if all diffusive and advective processes are fast compared to any of the reaction rates in the system, the flux term in equation (1) can be ignored; that is, the cell is behaving like a well-mixed reaction vessel. Instead of PDEs, this would result in a set of ordinary differential equations (ODEs) describing all the changes in species concentrations as a result of reactions or membrane transport processes. By definition, an ODE model is classified as a nonspatial model because it cannot simulate spatial gradients within volumes or surfaces. However, the geometry can still be represented in ODE models by accounting for the sizes of compartments and membranes; indeed, the surface areas of membranes and the volumes of compartments can influence the dynamics of the molecular components of the system. For example, consider a molecular flux of a molecule from a small compartment into a larger compartment, for example, from the nucleus to the cytosol. Because of the difference in volume, a flux through the nuclear membrane will produce a larger change in concentration within the nucleus than in the cytosol; these changes can be represented in the math as simple rate expressions for the species in each compartment, scaled by their relative volumes. A screenshot of the VCell simulation results for this compartmental ODE model are shown in Fig. 1A, where the nuclear concentration of our molecule (green curve), set initially to 10 μM, decays much more than the cytosolic concentration (violet curve) increases. This is because the volume of the cytosolic compartment is about five times larger than the volume of the nucleus. The two curves reach equilibrium at the same concentrations after about 10 s. These simulations were carried out with the "Combined Stiff Solver," one of eight numerical solvers for ODEs available in VCell.

The same compartmental model can also be solved stochastically, as illustrated in Fig. 1B. The results for this single trajectory for 100 molecules initially in the nucleus, are qualitatively similar to the deterministic results in Fig. 1A. To be able to make this comparison, the stochastic results are plotted in terms of concentrations rather than number of molecules, where an initial concentration in the nucleus of 45 pM corresponds to 100 molecules. Notice that at the steady state there are still fluctuations of concentration and these fluctuations are greater for the nuclear species than for the cytosolic species; this is because the nucleus contains a smaller number of molecules, so fluctuations are more significant. This simulation used the Gibson–Bruck variation (Gibson and Bruck, 2000b) of the Gillespie next reaction step algorithm (Gillespie, 1977, 2001) to calculate the trajectory; this is one of four stochastic solvers available in VCell. VCell also provides a utility for running multiple stochastic trajectories and generating a histogram for the final time point to evaluate the distribution of the species numbers. Stochastic simulations can capture behaviors due to the intrinsic fluctuations of molecular processes, but they are more computationally intensive than ODE simulations. For large numbers of molecules, such simulations become both impractical, because of the long computing times, and unnecessary, because fluctuations are relatively insignificant.

Diffusion is important in a cell biological process when it is slower than the reaction rates producing or consuming the diffusing species. This produces spatial gradients in concentration. A number of examples were given in the introduction and a specific example will be analyzed in detail in the final section of this chapter. Here, in Fig. 1C, we explore how our simple model of nucleocytoplasmic transport behaves in a real 3D cell geometry. We used a cell geometry based on a 3D confocal microscope image that had the same nucleus and cytoplasm volumes as the compartmental model used for the simulations of Fig. 1A and 1B (actually, VCell derived the size parameters in the compartmental model from the real geometry). Fig. 1C shows a volume rendering of the distribution of the concentration at the 1 s time point with the standard rainbow color scheme corresponding to the full range of concentration (red~9 μM; blue~1 μM). As can be seen in this simulation, which used a diffusion coefficient of 10 $\mu m^2 s^{-1}$ (typical for a protein in cytoplasm), the distribution of the species is far from the uniform distribution that is assumed in a compartmental model. Furthermore, the overall kinetics are strongly affected, as demonstrated by the time plots shown in the inset for the same 15 s duration used in Fig. 1A. The upper inset shows the concentration of the nuclear and cytosolic species not reaching steady state even after 15 s (compare Fig. 1A); this is because the slow diffusion prevents the molecule from instantly equilibrating within each compartment. The lower inset shows time plots at the two spatial points indicated by the asterisks in the cell image; the cytoplasmic concentration for the point (green curve) where the nucleus is close to the outer membrane actually overshoots the steady state value because of the restricted diffusion in this crowded region of the cell. The other point, at the mouth of a process at the left side of the cell, shows (violet curve) a several second lag period in the appearance of our molecule – again a behavior that cannot be captured in a compartmental model. Clearly spatial models provide details that may be missed in a compartmental model. But not always: if the diffusion coefficient in our model was an order of magnitude greater (as for a metabolite or a nucleotide), the results of the spatial model would be virtually identical to those of the compartmental ODE model. Of course the disadvantage of a spatial simulation is that it is computationally intensive; the model in Fig. 1C, with 367,000 grid points, took about 100 s to simulate on a single processor compared to essentially instantaneous for the ODE model of 1A. We used the fully implicit adaptive time step finite volume solver in VCell to run this simulation; it is fully described in Section IV on spatial simulation methods.

The most detailed (and computationally intensive) mathematical model to simulate for cell biology is a spatial stochastic model. The results of a spatial stochastic simulation for 100 molecules are shown in Fig. 1D. These simulations are performed with the Smoldyn algorithm developed by Steven Andrews (Andrews *et al.*, 2010). This model accounts for both probabilistic reactions of individual molecular species and Brownian dynamics of the motion of individual molecules in solution. The image in Fig. 1D is taken at the 1 s time point and therefore can be directly compared to 1C. The blue molecules are in the nucleus and the red molecules are in the cytosol (some of the red molecules appear to be in the nuclear region but they are actually above or below the nucleus in this 3D rendering). The inset shows the time course of change in concentration for the entire nucleus and cytosol – a noisy version of the upper inset in Fig. 1C. The computing time for the Smoldyn algorithm was about

600 s (compared to 100 s for the PDE simulation of Fig. 1C). However, there is significant overhead for special geometry handling required by Smoldyn for non-analytic geometries such as the one we used (as opposed to analytical geometries such as a sphere or cylinder); such image-based geometries could add hours to the computation time. But even without the geometry issue, spatial stochastic simulations are the most computationally expensive; furthermore, multiple simulations will often be required to develop statistics for the overall behavior of the system. Alternative algorithms and software packages for spatial stochastic simulations are described in Section IV.

III. Building a Spatial Model

In the VCell, a biological model is described in a layered branched fashion within the graphical user interface know as the "BioModel Workspace." The trunk is the "*Physiology*," describing the underlying network of quantitative reaction and transport mechanisms that are associated with volumetric and membrane cellular compartments (Fig. 2). These mechanisms makes no explicit reference to spatial coordinates, but rather describes the local time rate of change of concentration for reaction rates and the local molecular flux density for membrane transport mechanisms (*flux reactions*) in terms of the local environment. The Physiology in Fig. 2C consists of two compartments, "extra" and "cyto" separated by a membrane, "cyto_mem." The small circles represent species, four of which are highlighted; note how many species are required to fully represent all the states and interactions represented by the cartoon diagram of (B). The yellow squares each contain a rate expression for the quantitative mechanism of the reaction specified by the connecting lines. This Physiology is taken from a public model found in the VCell database as "susana:neves_cell_2008"; it contains the model and simulations published by Neves *et al.* (2008), which is discussed in Section E. This type of description is very flexible, allowing a single *Physiology* to simultaneously form the basis of multiple spatial, nonspatial, deterministic, and stochastic computational experiments (*Applications*), such as the four mathematical models already discussed in Section II (Fig. 1). An Application together with its parent Physiology is sufficient to completely define the mathematical system. The remainder of this section will discuss the geometry definition and other specifications required to define an *Application* in VCell.

The cellular distributions of organelles, fixed structures, and free and bound molecules are far from homogeneous. To begin, consider those cellular compartments that are encapsulated by membranes, and are thus capable of maintaining distinct cellular environments with specialized composition (e.g., the organelles). Some organelles are small, punctate, and numerous and could be considered as either discrete objects, spatially resolved compartments, or as a continuous average density (e.g., volume fraction of cytoplasm). The endoplasmic reticulum (ER) presents the problem of fine structure that is contiguous and distributed throughout the cytoplasm. The fine structure of the ER is difficult to spatially resolve, and therefore it is

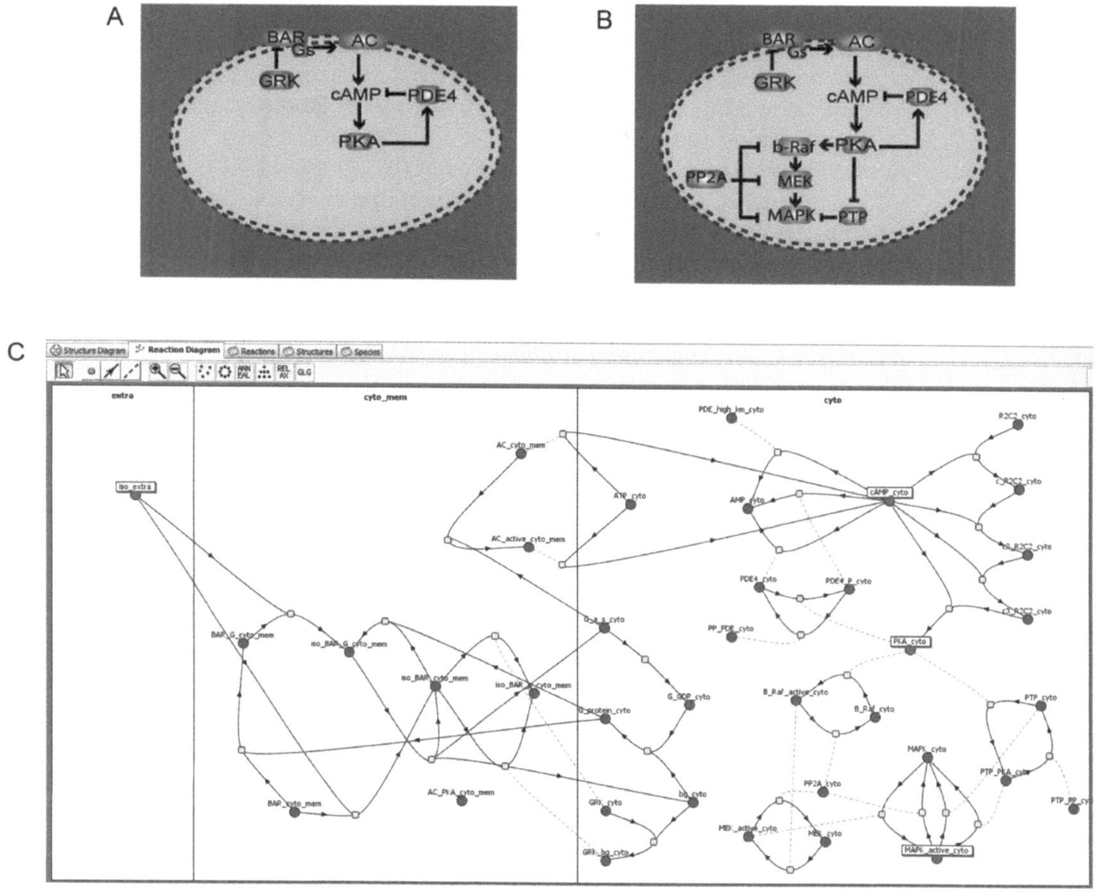

Fig. 2 VCell model of a signaling pathway. The top panels show a schematic diagram of signaling through the beta-adrenergic receptor at the membrane level (A) and extended to the mitogen-activated protein kinase effector (B). The bottom (C) is a screenshot of the detailed reaction diagram corresponding to the summary diagram from panel (B) in the "Physiology" component of a VCell BioModel. (A and B are from Neves *et al.*, 2008; reprinted by permission of Cell Press). (For interpretation of the references to color in this figure legend, the reader is referred to the web version of this book.)

more readily modeled as a continuous average density. For most models of eukaryotic cells, spatially resolving the plasma membrane and the nuclear envelope provide the basic encapsulation, whereas the finer structures are represented in a mean field manner by considering their effects to be continuously distributed within these compartments. Thus, a spatial model can be a hybrid where some features are represented with an explicit geometry, whereas fine structures are represented as compartments within the geometry that occupy a continuous volume fraction with a continuous surface-to-volume ratio. Importantly, these continuously distributed compartments may be nonuniform.

To build a spatial model in VCell, a cellular geometry must be defined using raw experimental images, segmented images, or shapes described using mathematical expressions (Analytic Geometry). The analytic geometry capability of VCell allows arbitrary inequalities in x, y, and z combined with Boolean operators to identify volume domains (e.g., "$(x^2 + y^2 + z^2 < 3^2)$ OR $((x - 2)^2 + y^2 + z^2 < 3^2)$" is the union of two spheres of radius 3, centered at (0,0,0) and (2,0,0)).

Through the use of a universal image format library (Linkert *et al.*, 2010), various native microscopy formats (e.g., Zeiss LSM) can be imported into VCell as raw image data and additional microscope data information describing the size of the field of view and space between image planes. Thus, the imported image data can be treated as samples from a scalar field, which can be used to describe protein distributions. The experimental images can also be used to define the geometry of the cell or subcellular structures.

To assist in this process, VCell provides tools for image processing and segmentation (Fig. 3). For example, if the cell interior is fluorescently labeled, then the cell membrane can often be determined by using an isosurface of pixel intensities and the set of all pixels that are brighter than a specified threshold in the cell interior. VCell provides a tool for 2D or 3D intensity histogram segmentation, which can apply a low-pass filter to accommodate noise and punctate staining. Thin or narrow processes (e.g., dendrites, lamellipodia, and filopodia) have a relatively weaker observed fluorescence and are often underrepresented by threshold segmentations. Therefore, manual editing tools are provided as well as tools for merging numerous small objects due to uneven staining. This process is repeated for each resolved cellular feature (e.g., nucleus and cytosol) and the resulting geometric domains are given appropriate labels. In Fig. 3, one cell is chosen with the cropping tool and segmented by this combined process of histogram thresholding and manual editing. The resultant 3D surface rendered geometry is shown in the inset on the upper right of Fig. 4.

Once the geometry is complete, one must map the compartments from the *Physiology* to the geometric domains defined in the geometry so that the reaction and transport mechanisms can be distributed spatially. Fig. 4 shows how the "extra" and "cyto" compartments in the Physiology (Fig. 2) are mapped, respectively, to the geometry domains labeled "background" and "cell" in the segmented geometry produced from Fig. 3; the software simply requires the user to draw a line connecting the compartment to the color assigned to the corresponding domain in the geometry. The same geometry can be reused in multiple models as it is accessible through the VCell database. In this example, the compartment to domain mapping is one-to-one. Often, however, multiple physiological compartments may be mapped to the same geometric domain, with specification of the volume fraction of each compartment within the domain. For example, the "cytosol" and "ER lumen" compartments can be mapped to the same "cellular" geometric domain with volume fractions of 0.85 and 0.15, respectively. This approximation assumes that the ER structure is fine enough to be not resolvable on the spatial scale of the model and avoids having to represent the difficult geometry and expensive computation of a spatially resolved ER.

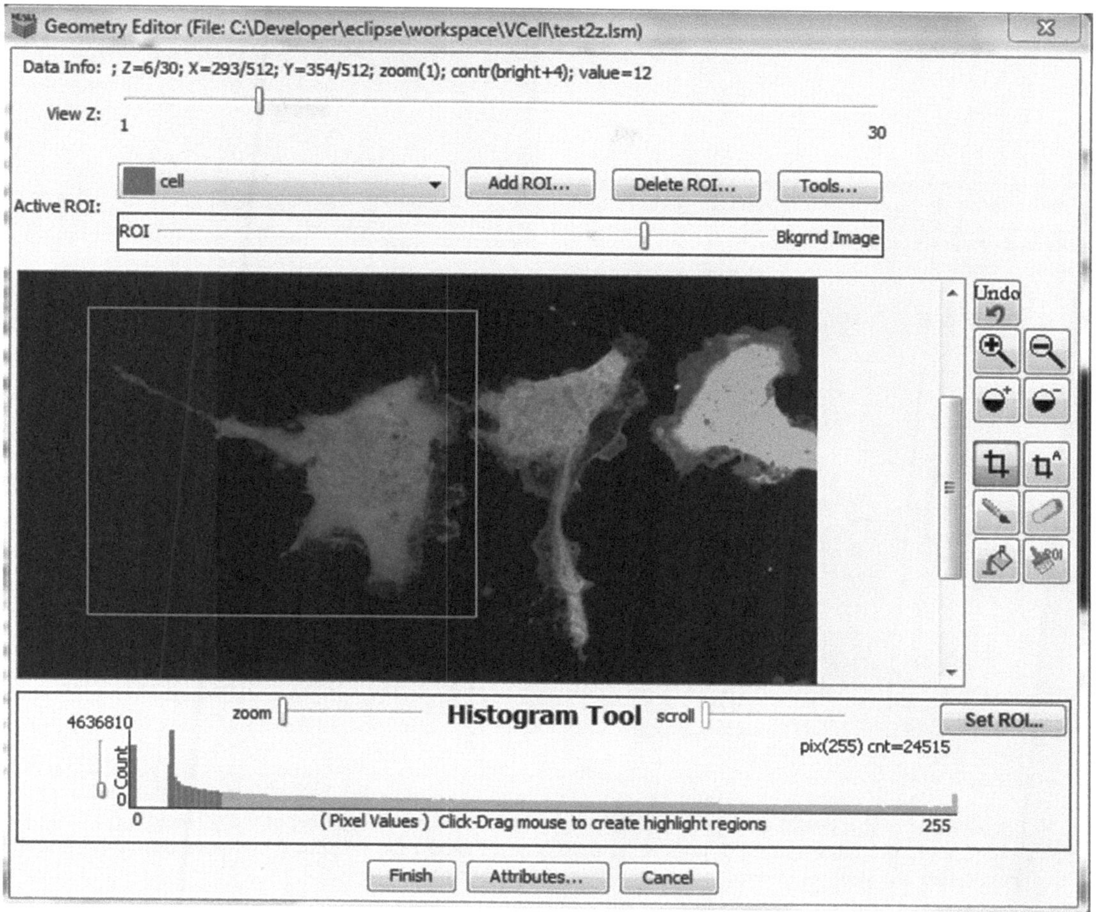

Fig. 3 Image segmentation screen in VCell. VCell allows geometries for spatial models to be derived from 2D or 3D experimental images. The figure shows utilities for editing and segmenting an image to define the regions of a cell. (For interpretation of the references to color in this figure legend, the reader is referred to the web version of this book.)

To complete a spatial model, initial concentrations, diffusion coefficients, and velocities need to be specified (Fig. 5). In this example, we highlight the species iso_extra, the only species in the "extra" compartment in the reaction network of Fig. 2. Its initial concentration is set to 1 μM. Diffusive transport is specified by the diffusion coefficient that defaults to 10 $\mu m^2 s^{-1}$ for volumetric species and 0.1 $\mu m^2 s^{-1}$ for membrane species; iso_extra represents a small drug molecule with a much higher diffusion coefficient than a protein, so it is set at 300 $\mu m^2 s^{-1}$. The flux due to advection of a molecular species can be specified by the *x*, *y*, and *z* components of the velocity for each molecular species; these are all set to 0 for our

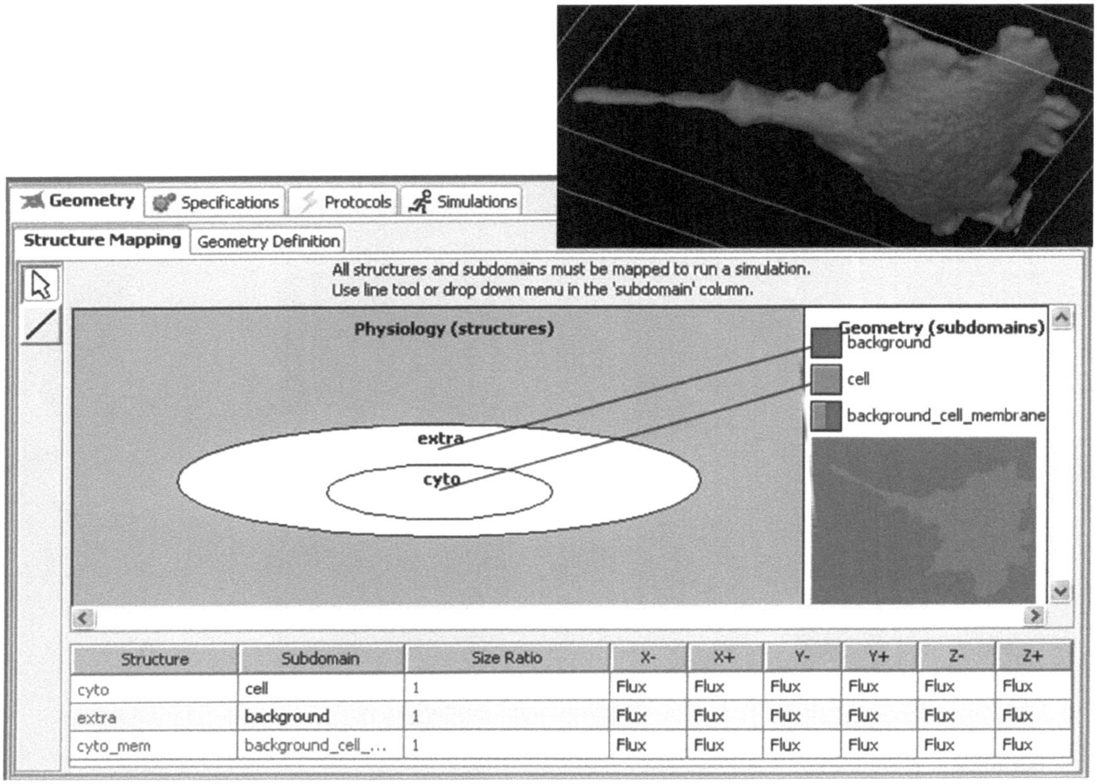

Fig. 4 Screen shot of the Virtual Cell geometry mapping window. A surface rendering of the geometry produced by the cropping and segmentation utilities of Fig. 3, is shown in the inset on the upper right. (For interpretation of the references to color in this figure legend, the reader is referred to the web version of this book.)

example, but if the *Application* is meant to simulate a microfluidics experiment, it could be set to a value representing the perfusion rate. Spatially invariant initial concentration, diffusion coefficient, or the velocity components are specified with numerical constants. The initial concentration, diffusion, and advection specifications need not be constants, however, but can be expressions of spatial coordinates, time, or variables in the model. An initial concentration may be specified as an explicit function of spatial coordinates. It can also be based on an image of a protein distribution as a *fieldData* object, which is imported as an image file. This object is a named dataset that may be multivariate for multichannel recordings. Diffusion and advection coefficients may also be explicit functions of coordinates or of fieldData. If other model parameters must be specified as a nonuniform distribution, then one may define a dummy species that can then be referenced anywhere in the model, and can be given a spatial or spatiotemporal profile using explicit functions or fieldData.

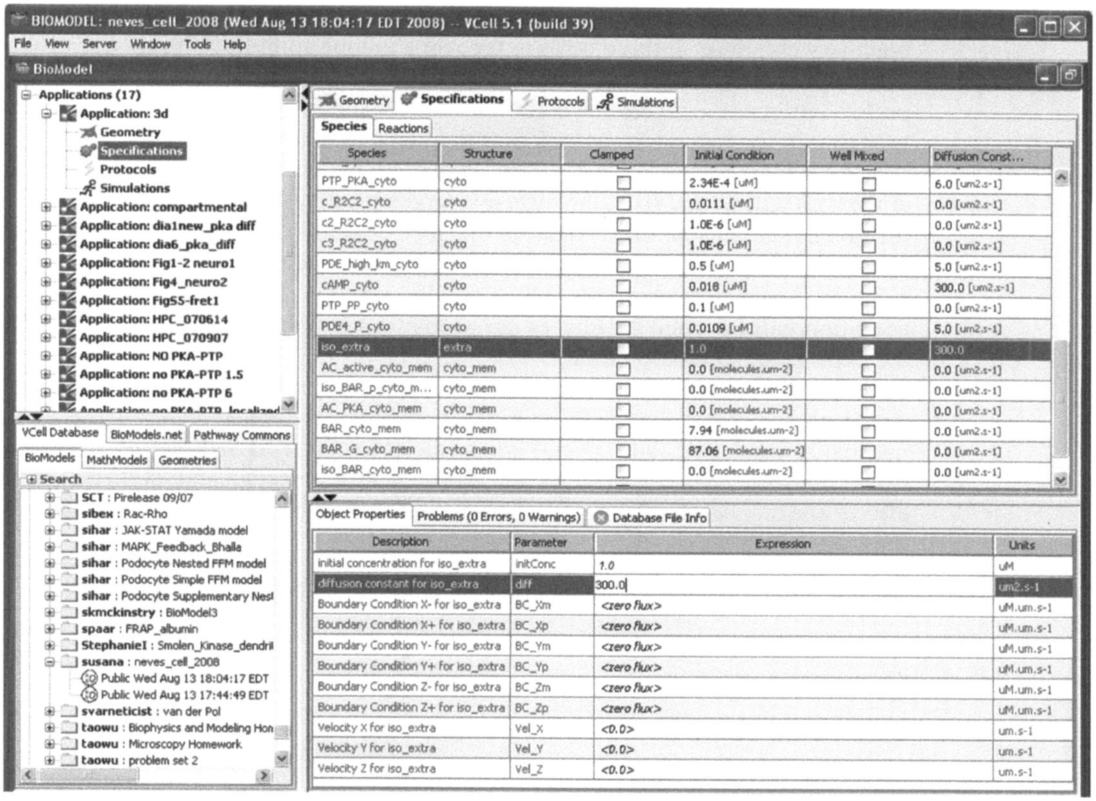

Figure 5 Specification of initial concentrations, diffusion coefficients, velocities and boundary conditions within a VCell *Application*. The species iso_extra is highlighted in the upper table and its editable properties are listed in the lower panel. (For interpretation of the references to color in this figure legend, the reader is referred to the web version of this book.)

Finally, in models that aim to directly fit experimental spatiotemporal fluorescence image data, the simulated fluorescence and the experimental measurements may be directly compared to evaluate the fitting error.

The flux density of a molecular species at a membrane results from the sum of the transmembrane transport and adsorption and desorption due to surface-binding reactions. This flux density is constrained to equal the flux due to diffusion and advection at the surface of the membrane and results in the generation of gradients near the membrane for nonzero flux. Thus, VCell handles boundary conditions of domains bounded by membranes automatically, based on the specified membrane-associated reactions and transport mechanisms. Boundary conditions at the edges of the simulations geometry, however, must be specified for any species that exists in a domain that intersects with one or more of these edges, as is the case for iso_extra in Fig. 5. The portion of the domain boundary that coincides with the walls of the geometry's bounding box represents an artificial truncation of a larger geometric domain and so either the concentration or the flux density must be given as a boundary condition at these walls to recover a unique solution. If the *concentration* is specified at a wall (i.e., "value" boundary condition), then this wall acts as a perfect buffer at the given concentration and will supply the required molecular flux to maintain the given concentration at the boundary. If instead the molecular flux density is given at the wall (i.e., "flux" boundary condition), then the concentration at the wall will be such that the diffusive (and advective) flux at the wall is equal to the given flux density. This "flux" boundary condition is often specified with zero flux, as is the case for the example in Fig. 5; 0 flux is equivalent to either a plane of symmetry or an impermeable wall. In all cases, the influence of the boundary condition will be made smaller as the geometry's bounding box is made larger (at the expense of simulation time).

To summarize, the VCell BioModel workspace provides the flexibility to map a single physiological model to multiple mathematical frameworks, ODEs, PDEs, nonspatial stochastic, and spatial stochastic (Fig. 1). This rich set of simulation capabilities enables the modeler to independently consider and evaluate spatial effects and stochastic effects as illustrated in Fig. 6, which summarizes the process

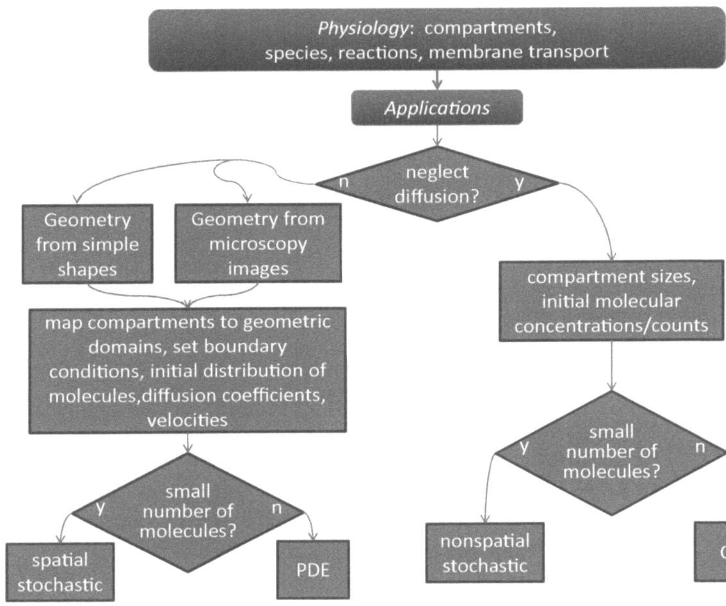

Fig. 6 Choices and required information when building a Virtual Cell BioModel. A single *Physiology* can spawn multiple *Applications* (varying geometries, initial conditions, boundary conditions, ODE, PDE, nonspatial stochastic, spatial stochastic, and so on). Here, the "y" choice for "neglect diffusion?" implies that the diffusion is fast enough to uniformly distribute all molecules within a compartment. The "small number of molecules?" decision reflects the desire to explore random variation as well as rare events. (For interpretation of the references to color in this figure legend, the reader is referred to the web version of this book.)

for specifying multiple *Applications* based a single *Physiology*. Of note, even within a single physical formulation, say a deterministic spatial model, there can be many *Applications*, each with different initial concentrations of species, clamped species concentrations, different geometries for different cell types, and so on. In other words, any combination of the specifications described in Fig. 6 can be chosen to exercise the parent *Physiology*. A VCell *Application* is thus akin to a "virtual experiment." After an *Application* is completely specified, VCell automatically generates a full mathematical description, which contains all the constants, variables, functions, ODEs, PDEs, or stochastic processes of the system. This can be viewed and even copied into an editable "MathModel" workspace. This mathematical description (whether generated automatically in the BioModel workspace or edited manually in the MathModel workspace) is directly translated into the input to our various simulation solvers. This separation of biological model construction from the mathematics of numerical simulation enables independent development and verification of modeling and simulation capabilities. The considerations required to set up and run spatial simulations are the subject of Section IV.

IV. Running Spatial Simulations with VCell: Numerical Method and Simulation Parameters

Mathematically, spatial aspects of cell signaling are deterministically modeled by systems of PDEs that involve rates of change of concentrations of signaling molecules in both space and time (time and spatial derivatives), as described in Section II, Eqs. (1) and (2). Most spatial mathematical models, particularly models with realistic geometries and complex nonlinear behaviors, do not have analytical solutions, and the problems need to be solved numerically.

Numerical solution of a PDE entails discrete sampling of both spatial domains and a time interval of interest. The spatial sampling should be fine enough to capture the essential geometric features of both the volumetric and surface domains of the geometry. The temporal sampling should be fine enough to capture any fast events in the mathematical system. This spatiotemporal sampling produces a solution in the form of tables of floating point numbers. It is important to realize that the numerical solution of a PDE is approximate: what the computer actually solves is not the original PDE but rather a system of algebraic equations that approximates the original PDE. This algebraic system of equations is obtained through discretization of time and spatial derivatives included in a PDE. Although this can be done using different discretization approaches (i.e., a PDE can be approximated by different algebraic systems), the numerical solution of the algebraic system converges to the exact solution of the original PDE with increasing sampling density, that is, the numerical solution can in principle be made as close to the exact solution of a PDE as desired, by refining the spatial grid and decreasing the integration time step.

Still, various discretization schemes have different characteristics with respect to order of convergence, numerical stability, conservation of mass, and other parameters. In VCell, for example, spatial discretization of PDEs is performed using a finite volume scheme (Ferziger and Peric, 2002), a conservative method with a built-in mass balancing, a feature that is particularly important in biological applications (Novak *et al.*, 2007; Schaff *et al.*, 1997, 2000; Slepchenko and Loew, 2010; Slepchenko *et al.*, 2000, 2003). More details about handling geometry in VCell can be found in (Novak *et al.*, 2007; Resasco *et al.*, 2011; Schaff *et al.*, 2001; Slepchenko and Loew, 2010).

Time discretization methods can differ by how they advance the solution from one time point to the next. The methods that advance the system based on the rates evaluated at the "old" time point (explicit solvers) require a sufficiently small integration step to ensure numerical stability (numerical instability can manifest itself as qualitatively wrong behaviors, such as unphysical oscillations or negative concentrations, or an exponential growth of numerical error resulting in machine infinity). In contrast, the implicit methods that propagate the system on the basis of rates corresponding to the "new" time point are unconditionally stable but they result in a system of nonlinear algebraic equations that must be solved iteratively.

Efficient numerically stable solvers have long been provided for simulating temporal behaviors of cell signals (Alves *et al.*, 2006). In particular, systems biologists have come to rely on so-called stiff solvers that retain their numerical stability and good performance in the presence of vastly disparate time scales – a common situation in biological applications. However, for the case of spatially resolved systems described by PDEs, solvers that meet such requirements and apply to a general class of problems are less common. A relatively new addition to the list of VCell spatial algorithms is a fully implicit spatial simulator (Resasco *et al.*, 2011; Slepchenko and Loew, 2010), which meets requirements of numerical stability and efficiency that modelers are used to in nonspatial simulators. With a built-in automatic time-step control and in combination with automatic meshing, the new spatial integrator in VCell is easy to use. It is freely accessible through the VCell user interface (www.vcell.org).

The fully implicit spatial solver in VCell is based on the well-known method of lines (Schiesser, 1991): after applying spatial discretization, a system of PDEs is first replaced with a large system of ODEs, which is then solved using a stiff ODE solver. The stiff solver advances the solution in time using implicit differentiation formulas and adaptive time step control. The latter allowed us to relieve the user of the burden of specifying the integration time step, which is generally a nontrivial task. Based on model parameters and tolerances, the solver automatically determines the initial integration time step and adjusts it along the way: the required accuracy is maintained by applying small time steps during periods of rapid change in the solution, whereas the time step is allowed to safely grow outside of these periods. The adaptive time step control essentially eliminates the time discretization error and significantly enhances efficiency of the solver.

The method, however, results in a large coupled nonlinear system of algebraic equations that must be solved at each step in time. A commonly used alternative called operator-splitting (Sportisse, 2000) avoids solving large, coupled nonlinear systems of algebraic equations, but it can carry more error and is not always applicable, especially in situations when stiffness originates from binding of molecules to membranes, interactions that are represented in terms of fast, nonlinear boundary conditions. The fully implicit approach, although sometimes dismissed as inefficient because of the large size and complexity of the nonlinear solves, can in fact be efficient when implemented with the use of effective technologies designed to optimize storage requirements and computation time. In particular, the adaptive control of the time step and order of the integration method, efficient iterative approaches to solving large-scale sparse nonlinear systems (see e.g., Knoll and Keyes (2004) and references within), and the application of effective physics-based preconditioners (Saad, 2003) are the main ingredients that contribute to the robustness and good performance of this solver. More details of the implementation of the fully implicit solver in VCell are provided in (Resasco *et al.*, 2011).

The VCell fully implicit simulator is essentially turnkey. The solver is accessed through a Solver dialog box from the Simulation Editor (Fig. 7). Using this panel, one can switch between the solvers using a dropdown menu. This, plus simulation start and end times and the desired output time interval, is all the information needed to run the fully implicit solver. The user may also choose to adjust the relative and absolute tolerances for local (time-discretization) errors and the maximum integration time-step allowed. These simulation parameters are initially set at default values: 1e-7 and 1e-9, for the relative and absolute tolerances respectively, and 0.1 for the maximum integration time step.

The other two tabs of the Simulation Editor reveal panels for Parameter and Mesh specification. The Mesh panel allows a user to refine or coarsen the mesh by changing the mesh size for each Cartesian direction. In this way, the user can increase or decrease the number of grid points for which the solution is computed. In the Parameter panel, the user can vary model parameters, such as rate constants or initial concentrations, for a given run. This can be done individually or through the option of parameter scanning. The latter allows a user to run a batch of simulations for a selected set of combinations of parameter values. For this, the user specifies parameter ranges and the number of values within a range, which will be selected, either uniformly or logarithmically, for scanning. This is done by checking boxes in the "Scan" column, under the "Edit > Parameters" tab. VCell then automatically initiates simulations for all combinations of selected parameters. The results for individual parameter combinations can be viewed by selecting a set of parameter values from the table at the bottom of the "Results" window, in which simulation results are displayed.

The continuous description in terms of PDEs becomes inadequate when concentrations of signaling molecules are relatively low and stochastic fluctuations need to be taken into account. Instead, spatial stochastic approaches should be applied. Stochastic modeling of signaling events seeks to predict dynamics of a probability distribution over states of signaling molecules and their spatial location. Realistic

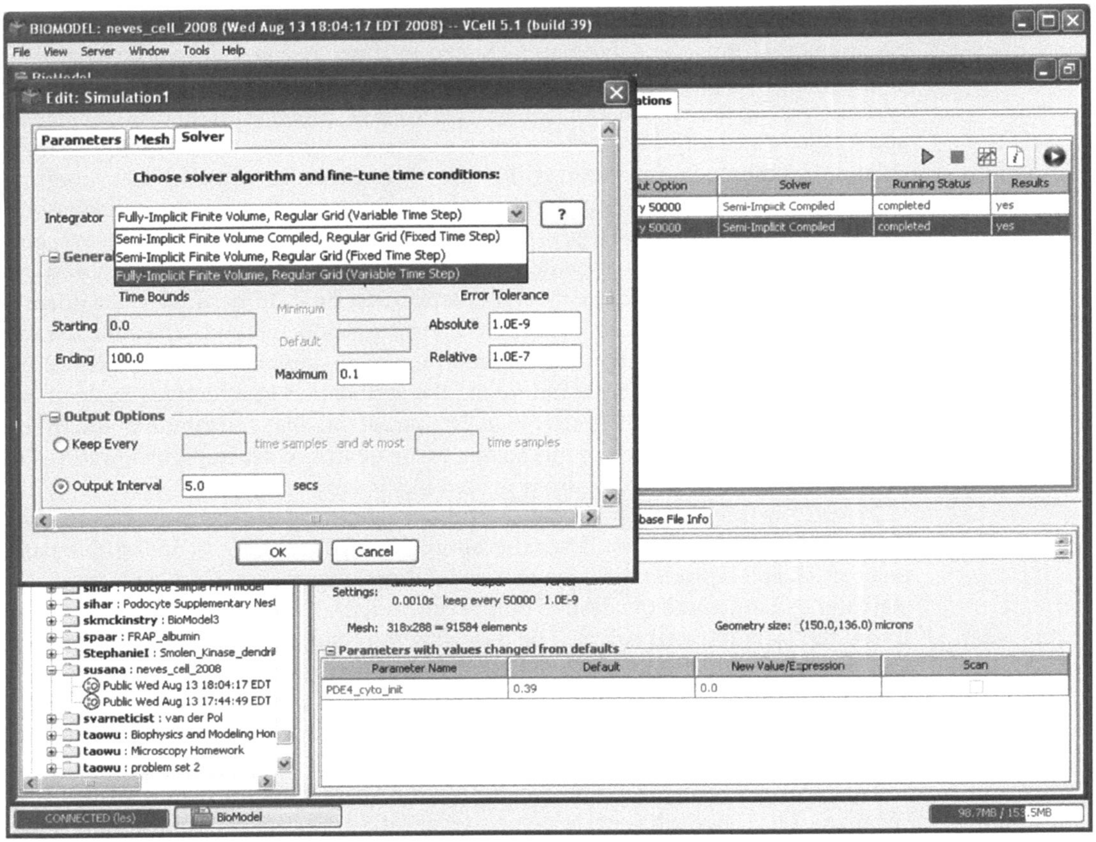

Fig. 7 Specifying and running simulations in VCell. The background shows the overall simulations interface and the foreground shows the Solver tab in the "Edit Simulation" dialog box. (For interpretation of the references to color in this figure legend, the reader is referred to the web version of this book.)

models formulated in terms of stochastic processes rarely have exact solutions (Gardiner, 2004; Kim and Shin, 1999; van Kampen, 1992), and computer simulations have inevitably become a method of choice for stochastic applications to cell biology.

Numerical methods for modeling stochastic processes often rely on random number generators, which are nowadays built in every computer; such algorithms are called Monte Carlo methods. In recent years, there has been significant progress in developing Monte Carlo algorithms designed for applications in cell biology (Andrews *et al.*, 2010, Elf *et al.*, 2003; Isaacson and Peskin, 2006; Kerr *et al.*, 2008; Morelli and ten Wolde, 2008; Morton-Firth and Bray, 1998; Plimpton and Slepoy, 2005; Saxton, 2007; van Zon and ten Wolde, 2005). Some of these methods have been implemented in software packages such as MCell (http://www.mcell.cnl.salk.edu), StochSim (http://www.pdn.cam.ac.uk/groups/comp-cell/StochSim.html), Smoldyn (http://www.smoldyn.org/), ChemCell (http://ipal.sandia.gov/ip_details.php?ip=8030), and MesoRD (http://mesord.sourceforge.net/).

Stochastic simulators utilize different approaches and yield results with a varying degree of detail. A method based on a reaction-diffusion master equation (Elf *et al.*, 2003; Isaacson and Peskin, 2006; Stundzia and Lumsden, 1996) approximates volume elements as well-stirred compartments and describes a system in terms of a number of copies of each signaling molecule in each volume element. Diffusion fluxes between neighboring locations are treated as a subset of unimolecular reaction steps, and Gillespie-type (event-driven) algorithms (Gibson and Bruck, 2000a; Gillespie, 1976, 1977) optimized for spatial simulations (Fange and Elf, 2006) are used to generate stochastic trajectories of the system. Although conceptually appealing, the method does not have a well-defined scope of applicability and diverges as the mesh is refined.

In a more detailed description, the state of the system is defined in terms of locations and states of every signaling molecule (velocities of the molecules are considered to be in equilibrium at any time because of low Reynolds numbers). Particle-based Brownian dynamics algorithms that advance a system in time with a fixed time step are widely used to simulate dynamics of interacting molecules in cells (Andrews and Bray, 2004; Kerr *et al.*, 2008; Plimpton and Slepoy, 2003). In one of most recent developments, VCell has incorporated Smoldyn (Andrews *et al.*, 2010) (Fig. 1D), which is among the most efficient particle-based spatial stochastic simulators. Interpolation techniques implemented in Smoldyn become exact in the diffusion-controlled limit. The algorithm is designed to reproduce observable on- and off- reaction rates, and therefore results are expected to be sufficiently accurate even when obtained with relatively large steps. Still, numerical error cannot be regularly estimated, other than for the case of diffusion-limited reactions. In this regard, more accurate, but likely less efficient, methods were recently proposed in (Morelli and ten Wolde, 2008; van Zon and ten Wolde, 2005).

V. Application to a Specific Example: cAMP Signaling in Neuronal Cells

Intracellular signal transduction often leads to localized regulation of specific cellular processes. Such spatial heterogeneity can be experimentally observed by *in vivo* high-resolution subcellular imaging and can involve localized production of small molecule messengers and/or localized activation of protein kinases, protein phosphatases and other signaling components. Concentration gradients of key signaling components can form dynamic subcellular regions, called signaling microdomains, where selected molecules have elevated (or decreased) levels compared to other contiguous areas. It has long been thought that some of the specificity observed in the effects of signals that use similar or identical signaling pathways arise from such spatial domains of signaling components within the cells. Signaling microdomains involving cAMP and protein kinase cascades have been recently studied in a number of different cell types. These studies revealed a surprising complexity that is not easily explained by simple reaction mechanisms and the geometrical shape of the particular cells. Performing quantitative simulations of the biochemistry in spatially resolved geometries using software such as VCell can be of critical help to guide experiments and unravel this complexity. This is illustrated in this Section by example, using a recent study of beta-adrenergic receptor signaling in hippocampal neurons (Neves *et al.*, 2008).

A. Initial Hypothesis and Overall Approach

The overall goal was to study the role of morphology and spatial anisotropy in regulating intracellular signaling in neurons. The original hypothesis was the following: due to the shape of hippocampal neurons, microdomains of cAMP may form in response to global beta-adrenergic receptor stimulation, and this would then lead to localized activation of the downstream effector, mitogen-activated protein kinase (MAPK; active in its phosphorylated state, P-MAPK). The approach used was to resolve the information flow within the cell using spatial specifications from realistic cell shapes and locations of relevant components. Models were developed using VCell and simulations were used to analyze how the various factors (signaling network connectivity map, individual reaction kinetics, diffusion constraints, shape, etc.) could affect the dynamics of the signaling microdomains. The predictions generated from these simulations were tested experimentally in an iterative cycle of model building/refinement – simulation predictions – new experiments. This work has resulted in exciting new insights into the interplay between signaling network topology, biochemical kinetics, and spatial anisotropy in the regulation of the cellular response to receptor stimulation.

B. Creating and Testing the Model and the Initial Hypothesis

The first step was to build a simple, but quantitatively accurate model of the signal transduction at the membrane that controls the concentration of the cytosolic second messenger, cAMP. This initial model included the extracellular ligand (isoproterenol), the beta-adrenergic membrane receptor, the G-protein cascade leading to adenylyl cyclase activation, cAMP generation, the activation of protein kinase A (PKA), and the cAMP degradation by PKA activation of phosphodiesterase (PDE4) (see Fig. 2A). This simple stimulatory pathway with one negative feedback loop had many parameters known from experimental measurements. The unknown parameters were constrained by running simulations as a compartmental (i.e., ODE) model to fit time curves and dose–response curves of a few selected components whose average concentration was measured in brain slice experiments.

The second step was to use microscope images of cultured hippocampal neurons as geometries for simulating the spatially resolved activity of this signaling network (an example is shown in Fig. 8D). All molecules involved were assumed to be evenly

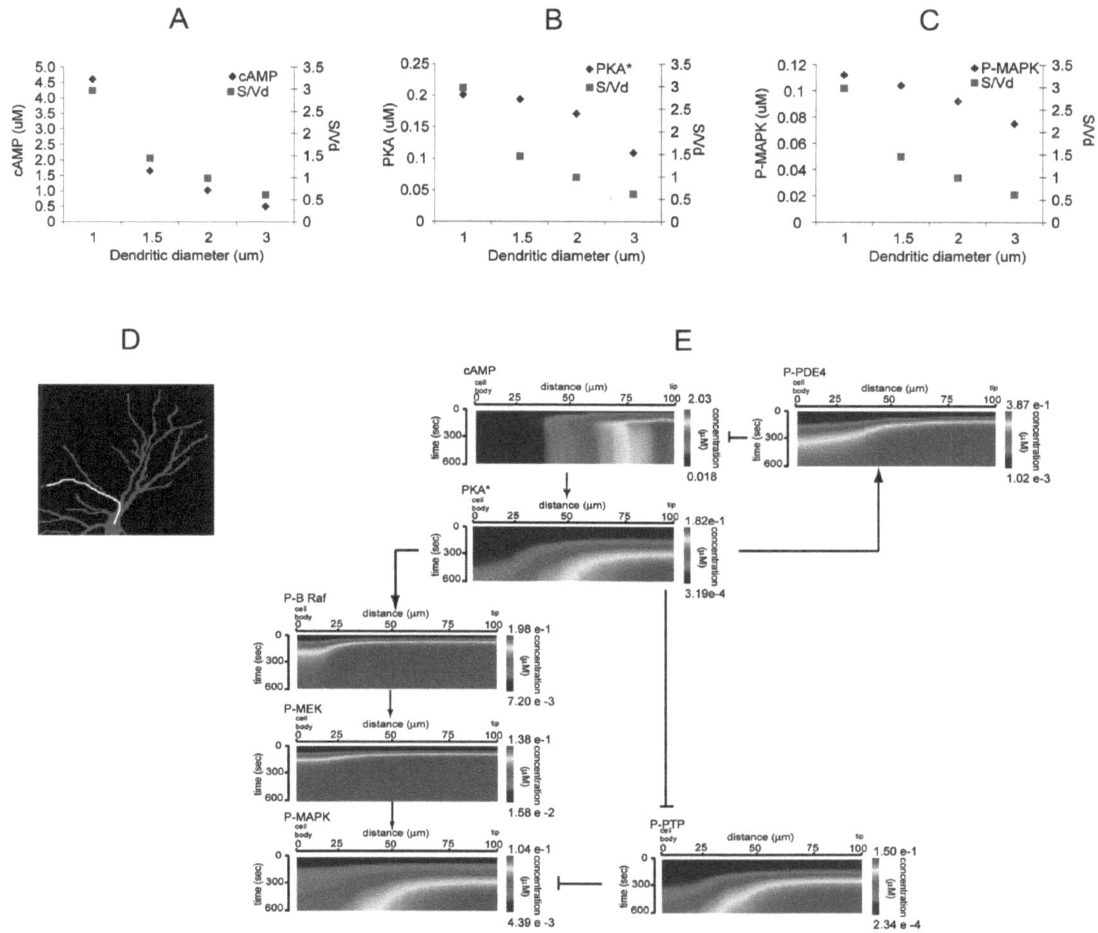

Fig. 8 Spatiotemporal distribution of key signaling molecules. The top three panels show the simulation results analyzing the dependence of local concentration of cAMP (panel A), activated PKA (panel B) and activated MAPK (panel C) on dendrite diameter. Maximum concentrations at 600 s poststimulus are plotted (blue diamonds) together with dendritic surface/body ratio (red squares). Note how cAMP concentration strongly depends on S/V values as opposed to PKA and MAPK. Panel D shows a typical image of the cultured hippocampal neuron geometries that were used for spatially resolved simulations. Panel E shows kymographs (produced in the VCell Results Viewer) depicting the activation profile of key signaling components downstream of cAMP. A line scan was done on the dendrite highlighted in white in panel D (*X*-axis) over a 10-min time course (*Y*-axis); each image has the respective pseudocolor key to concentration values on the right. Note how the information flow for conservation of microdomain downstream of PKA appears to occur through the inhibitory path (modulation of PTP by PKA). This Figure is adapted from Neves *et al.* (2008), and is reproduced by permission of Cell Press. (See color plate.)

distributed at the initial steady state. The time-course simulations after ligand stimulus predicted that cAMP microdomains would form, with relatively steep concentration gradients (strong activation in the dendrites and essentially no activation in the cell body). This was then confirmed by live cell measurements of cAMP concentration in slice experiments (using a cAMP FRET sensor), which were compared to further simulations that used the actual geometries of the neurons from the slice experiments' microscope images. Theoretical analysis of this (still relatively simple) model had shown that a critical parameter that influences the cAMP gradients is dendrite diameter. This mathematical analysis was confirmed by simulation results on idealized geometries where the dendrite diameter was varied and a strong correlation was seen with cAMP concentrations (Fig. 8A).

C. First Unexpected Model Prediction

The model was then extended to include the signaling elements connecting the second messenger production to effector activation: a stimulatory (PKA → b-Raf → MAPK) and an inhibitory (PKA ¬ PTP ¬ MAPK) feed-forward link, both with negative regulation by phosphatases (PPP2A and PP1, respectively). The schematic overview of the extended signaling network is shown in Fig. 2B and the detailed reaction diagram of the corresponding VCell model is shown in Fig. 2C. Spatially resolved simulations of the extended model were started by activating the receptors with saturating concentrations of isoproterenol, and they showed that, similar to the case of cAMP, distinct microdomains for P-MAPK were also formed in the distal dendrites, as originally hypothesized. However, when analyzing the dependency of the P-MAPK microdomains on dendrite diameter, the results were very different than for the cAMP microdomain, in that the P-MAPK microdomain was far more robust (compare Fig. 8A and C). Thus, the model predicted that localized cAMP signaling is directly dependent on cellular geometry, but other factors must contribute to the control of the spatial distribution of MAPK activation. One possible hypothesis was that this was due to differences in diffusion constants of the different molecules involved, for example, cytosolic proteins diffuse much slower than the small molecule cAMP. This hypothesis was tested *in silico*, by performing parameter scan simulations where the diffusion constants of several of the molecules were changed to higher values. (It is worth noting that this was a quite important model analysis step independent of the hypothesis because many of these diffusion constants were estimated based on molecular radius and not experimentally measured values). These simulations showed, however, that the characteristics of the P-MAPK microdomains were not significantly affected when any of the diffusion constants were increased (within physically reasonable bounds).

So what could be the cause? When looking at the simulation data for other cytosolic proteins, the very first protein in the signaling path, PKA* (the cAMP-activated PKA), also exhibited a microdomain formation that was more robust to changes in dendrite size and shape, similar to the case of P-MAPK (Fig. 8B). A natural hypothesis was that the cause must reside in the upstream part of the signaling – for example, the phosphodiesterase-mediated upper negative feedback loop. An *in silico* "virtual knockout" experiment was performed where simulations where run with the PDE4 activity set to zero. These simulations showed a complete loss of microdomain formation for both PKA* and P-MAPK. This important prediction was then confirmed experimentally: hippocampal tissue slices were treated with the PDE4 inhibitor rolipram before beta-adrenergic stimulation, and as a result, global (body + dendrite) activation of P-MAPK was seen, perfectly matching simulation predictions.

D. More Unexpected Model Predictions

While doing the above model analysis to compare microdomain characteristics for the second messenger cAMP versus the downstream effectors PKA* and P-MAPK, a second surprising feature was observed. Looking at the spatial dynamics of all of the diffusible components involved, the proteins that form the intermediary link between them in the stimulatory pathway, P-b-Raf and P-MEK, not only do not have robust spatial microdomains, they essentially do not form any microdomain, and show little difference in activation between different areas of the cell (Fig. 8E). Given the previous insight gained into the importance of the upper negative feedback loop in propagating spatial information, a new round of *in silico* and *in vivo* experiments was done, focused on the negative feedback component of the downstream activation path: phosphatases PP2A and P1. Simulations of inhibited PP2A/PP1 activity showed a disappearance of the downstream P-MAPK microdomain, while the upstream PKA* microdomain was unaffected. This was then confirmed experimentally by treatment with the phosphatase inhibitor, okadaic acid.

Given the apparent critical role of negative regulators, further efforts were made to understand the role of the parallel inhibitory feed-forward loop (PTP inhibition by PKA followed by MAPK inhibition by PTP). Indeed, P-PTP (currently not measurable experimentally) also showed a similar robust microdomain like PKA* and P-MAPK (Fig. 7E). A PTP "virtual knockout" experiment *in silico* (where phosphatase activities were kept normal) resulted in simulations where only shallow P-MAPK microdomains were formed, which were also not at all robust to changes in cellular shape characteristics such as dendrite versus cell body surface-to-volume ratios. Experimental testing of this prediction was far more complicated because no good specific direct inhibitor of PTP was available. Gene knockdown experiments were performed, where animals were treated with antisense oligonucleotides that were designed to reduce the expression of the *PTPRR* gene products, prior to sacrificing them for tissue slice experiments. Comparison of P-MAPK imaging after isoproterenol stimulation in tissue slices from the antinsense oligonucleotide-treated animals versus those from scrambled oligonucleotide-treated controls matched the simulation predictions. This also prompted further more detailed analyses to study the relative role of kinetic parameters of the reactions in the stimulatory and inhibitory feed-forward paths, as well as of their respective reverse loops via phosphatases.

E. Conclusions

This study achieved a new understanding of how cell morphology characteristics, biochemical parameters, and network topology combine in subtle ways to control the propagation of spatial information through the signaling networks (in the particular case of beta-adrenergic stimulation of neurons). To put in perspective the complexity of the study, it should be noted that the presentation here only briefly summarizes the stepwise progression from an initial hypothesis through many cycles of *in silico* and *in vivo* experiments. Considerable more data is provided and discussed in the original paper by Neves *et al.* (2008): the six multi-paneled results figures are accompanied by a 50-page supplement with more simulation and experimental data (34 more figures and tables) as well as additional theoretical mathematical analyses. Overall, this study is an excellent example of how geometry can control cell signaling and how modeling and experiment can interact to solve a complex cell biological problem.

Acknowledgments

We are grateful to the Ed Boyce and Frank Morgan for their help in visualizing the volumetric simulations presented in Fig. 1. This work is supported by grant # P41RR013186 from the NIH National Center for Research Resources.

References

Alves, R., Antunes, F., and Salvador, A. (2006b). Tools for kinetic modeling of biochemical networks. *Nat. Biotechnol.* **24**, 667–672.

Andrews, S. S., Addy, N. J., Brent, R., and Arkin, A. P. (2010). Detailed simulations of cell biology with Smoldyn 2.1. *PLoS Comput. Biol.* **6**, e1000705.

Andrews, S. S., and Bray, D. (2004). Stochastic simulation of chemical reactions with spatial resolution and single molecule detail. *Phys. Biol.* **1**, 137–151.

Brown, S. A., Morgan, F., Watras, J., and Loew, L. M. (2008). Analysis of phosphatidylinositol-4,5-bisphosphate signaling in cerebellar Purkinje spines. *Biophys J.* **95**, 1795–1812.

Elf, J., Doncic, A., and Ehrenberg, M. (2003). Mesoscopic reaction-diffusion in intracellular signalling. *SPIE* **5110**, 114–124.

Fange, D., and Elf, J. (2006). Noise-induced min phenotypes in *E. coli*. *PLoS Comput. Biol.* 2.

Ferziger, J. H., and Peric, M. (2002). *Computational Methods for Fluid Dynamics.* Springer, .

Fink, C. C., Slepchenko, B., Moraru, I. I., Watras, J., Schaff, J. C., and Loew, L. M. (2000). An image-based model of calcium waves in differentiated neuroblastoma cells. *Biophys J.* **79**, 163–183.

Gardiner, C. (2004). Handbook of Stochastic Methods for Physics, Chemistry and the Natural Sciences. Springer-Verlag, New York.

Gibson, M. A., and Bruck, J. (2000a). Efficient exact stochastic simulation of chemical system with many species and many channels. *J. Phys. Chem. A* **104**, 1876–1889.

Gibson, M. A., and Bruck, J. (2000b). Efficient exact stochastic simulation of chemical systems with many species and many channels. *J. Phys. Chem. A.* **104**, 1876–1889.

Gillespie, D. T. (1976). A general method for numerically simulating the stochastic time evolution of coupled chemical reactions. *J. Comput. Phys.* **22**, 403–434.

Gillespie, D. T. (1977). Exact stochastic simulation of coupled chemical reactions. *J. Phys. Chem.* **81**, 2340–2361.

Gillespie, D. T. (2001). Approximate accelerated stochastic simulation of chemically reacting systems. *J. Chem. Phys.* **115**, 1715–1733.

Goryachev, A. B., and Pokhilko, A. V. (2008). Dynamics of Cdc42 network embodies a Turing-type mechanism of yeast cell polarity. *FEBS Lett.* **582**, 1437–1443.

Hernjak, N., Slepchenko, B. M., Fernald, K., Fink, C. C., Fortin, D., Moraru, I. I., Watras, J., and Loew, L. M. (2005). Modeling and analysis of calcium signaling events leading to long-term depression in cerebellar Purkinje cells. *Biophys J.* **89**, 3790–3806.

Holt, M., Cooke, A., Neef, A., and Lagnado, L. (2004). High mobility of vesicles supports continuous exocytosis at a ribbon synapse. *Curr. Biol.* **14**, 173–183.

Isaacson, S. A., and Peskin, C. S. (2006). Incorporating diffusion in complex geometries into stochastic chemical kinetics simulations. *SIAM J. Sci. Comput.* **28**, 47–74.

Kapustina, M., Vitriol, E., Elston, T. C., Loew, L. M., and Jacobson, K. (2010). Modeling capping protein FRAP and CALI experiments reveals in vivo regulation of actin dynamics. *Cytoskeleton (Hoboken)* **67**, 519–534.

Kerr, R. A., Bartol, T. M., Kaminsky, B., Dittrich, M., Chang, J. -C., Baden, S. B., Sejnowski, T. J., and Stiles, J. R. (2008). Fast Monte Carlo Simulations Methods for Biological Reaction-Diffusion Systems in Solution and on Surfaces. *SIAM J. Sci. Comput.* **30**, 3126–3149.

Kholodenko, B. N., Hancock, J. F., and Kolch, W. (2010). Signalling ballet in space and time. *Nat. Rev. Mol. Cell. Biol.* **11**, 414–426.

Kim, H., and Shin, K. J. (1999). Exact solution of the reversible diffusion-influenced reaction for an isolated pair in three dimensions. *Phys. Rev. Lett.* **82**, 1578–1581.

Knoll, D. A., and Keyes, D. E. (2004). Jacobian-free Newton-Krylov methods: a survey of approaches and applications. *J. Comput. Phys.* **193**, 357–397.

Linkert, M., Rueden, C. T., Allan, C., Burel, J. M., Moore, W., Patterson, A., Loranger, B., Moore, J., Neves, C., Macdonald, D., Tarkowska, A., Sticco, C., Hill, E., Rossner, M., Eliceiri, K. W., and Swedlow, J. R. (2010). Metadata matters: access to image data in the real world. *J. Cell. Biol.* **189**, 777–782.

Ma, L., Janetopoulos, C., Yang, L., Devreotes, P. N., and Iglesias, P. A. (2004). Two complementary, local excitation, global inhibition mechanisms acting in parallel can explain the chemoattractant-induced regulation of PI(3,4,5)P3 response in dictyostelium cells. *Biophys J.* **87**, 3764–3774.

Moissoglu, K., Slepchenko, B. M., Meller, N., Horwitz, A. F., and Schwartz, M. A. (2006). In vivo dynamics of Rac-membrane interactions. *Mol. Biol. Cell.* **17**, 2770–2779.

Morelli, M. J., and ten Wolde, P. R. (2008). Reaction Brownian dynamics and the effect of spatial fluctuations on the gain of a push-pull network. *J. Chem. Phys.* **129**, 054112.

Morton-Firth, C. J., and Bray, D. (1998). Predicting temporal fluctuations in an intracellular signalling pathway. *J. Theor. Biol.* **192**, 117–128.

Neves, S. R., Tsokas, P., Sarkar, A., Grace, E. A., Rangamani, P., Taubenfeld, S. M., Alberini, C. M., Schaff, J. C., Blitzer, R. D., Moraru, I. I., and Iyengar, R. (2008). Cell shape and negative links in regulatory motifs together control spatial information flow in signaling networks. *Cell* **133**, 666–680.

Novak, I. L., Gao, F., Choi, Y. -S., Resasco, D., Schaff, J. C., and Slepchenko, B. M. (2007). Diffusion on a curved surface coupled to diffusion in the volume: application to cell biology. *J. Comput. Phys.* **226**, 1271–1290.

Plimpton, S., Slepoy, A., ChemCell: A Particle-Based Model of Protein Chemistry and Diffusion in Microbial Cells. Sandia Technical Report SAND2003-4509, 2003.

Plimpton, S. J., and Slepoy, A. (2005). Microbial cell modeling via reacting diffusive particles. *J. Phys. Conf. Ser.* **16**, 305–309.

Resasco, D. C., Gao, F., Morgan, F., Novak, I. L., Schaff, J. C., and Slepchenko, B. M. (2011 Dec 2). Virtual Cell: computational tools for modeling in cell biology. *Wiley Interdiscip Rev Syst Biol Med.* 2011 Dec 2. doi: 10.1002/wsbm.165 [Epub ahead of print].

Roy, P., Rajfur, Z., Jones, D., Marriott, G., Loew, L., and Jacobson K. (2001). Local photorelease of caged thymosin beta4 in locomoting keratocytes causes cell turning. *J. Cell. Biol.* **153**, 1035–1048.

Saad, Y. (2003). Iterative Methods for Sparse Linear Systems. SIAM, Philadelphia, PA.

Saucerman, J. J., Zhang, J., Martin, J. C., Peng, L. X., Stenbit, A. E., Tsien, R. Y., and McCulloch, A. D. (2006). Systems analysis of PKA-mediated phosphorylation gradients in live cardiac myocytes. *Proc. Natl. Acad. Sci. U S A* **103**, 12923–12928.

Saxton, M. J. (2007). Modeling 2D and 3D diffusion. *Methods Mol. Biol.* **400**, 295–321.

Schaff, J., Fink, C. C., Slepchenko, B., Carson, J. H., and Loew, L. M. (1997). A general computational framework for modeling cellular structure and function. *Biophys J.* **73**, 1135–1146.

Schaff, J. C., Slepchenko, B. M., Choi, Y. S., Wagner, J., Resasco, D., and Loew, L. M. (2001). Analysis of nonlinear dynamics on arbitrary geometries with the Virtual Cell. *Chaos* **11**, 115–131.

Schaff, J. C., Slepchenko, B. M., and Loew, L. M. (2000). Physiological modeling with virtual cell framework. *Methods Enzymol.* **321**, 1–23.

Schiesser, W. E. (1991). The Numerical Method of Lines: Integration of Partial Differential Equations. Academic Press, San Diego.

Shen, L., Weber, C. R., and Turner, J. R. (2008). The tight junction protein complex undergoes rapid and continuous molecular remodeling at steady state. *J. Cell. Biol.* **181**, 683–695.

Slepchenko, B. M., and Loew, L. M. (2010). Use of Virtual Cell in studies of cellular dynamics. *Int. Rev. Cell. Mol. Biol.* **283**, 1–56.

Slepchenko, B. M., Schaff, J. C., and Choi, Y. S. (2000). Numerical approach to fast reactions in reaction-diffusion systems: application to buffered calcium waves in bistable models. *J. Comput. Phys.* **162**, 186–218.

Slepchenko, B. M., Schaff, J. C., Macara, I., and Loew, L. M. (2003). Quantitative cell biology with the Virtual Cell. *Trends Cell Biol.* **13**, 570–576.

Sportisse, B. (2000). An analysis of operating splitting techniques in the stiff case. *J. Comput. Phys.* **161**, 140–168.

Stundzia, A. B., and Lumsden, C. J. (1996). Stochastic simulation of coupled reaction-diffusion processes. *J. Comput. Phys.* **127**, 196–207.

van Kampen, N. G. (1992). Stochastic Processes in Physics and Chemistry. North-Holland, Amsterdam.

van Zon, J. S., and ten Wolde, P. R. (2005). Simulating biochemical networks at the particle level and in time and space. *Phys. Rev. Lett.* **94**, 128103.

Vilela, M., Morgan, J. J., and Lindahl, P. A. (2010). Mathematical model of a cell size checkpoint. *PLoS Comput. Biol.* **6**, e1001036.

Zhong, H., Sia, G. -M., Sato, T. R., Gray, N. W., Mao, T., Khuchua, Z., Huganir, R. L., and Svoboda, K. (2009). Subcellular dynamics of type II PKA in neurons. *Neuron.* **62**, 363–374.

Further Reading

Alves, R., Antunes, F., and Salvador, A. (2006a). Tools for kinetic modeling of biochemical networks. *Nat. Biotech.* **24**, 667–672.

Andrews, S. S., Addy, N. J., Brent, R., and Arkin, A. P. (2010b). Detailed simulations of cell biology with Smoldyn 2.1. *PLOS Comput. Biol.* **6**, e1000705.

Czech, J., Dittrich, M., and Stiles, J. R. (2009). Rapid creation, Monte Carlo simulation, and visualization of realistic 3D cell models. *Methods Mol. Biol.* **500**, 237–287.

Hattne, J., Fange, D., and Elf, J. (2005). Stochastic reaction-diffusion simulation with MesoRD. *Bioinformatics.* **21**, 2923–2924.

Moraru, I. I., Schaff, J. C., Slepchenko, B. M., Blinov, M. L., Morgan, F., Lakshminarayana, A., Gao, F., Li, Y., and Loew, L. M. (2008). Virtual Cell modelling and simulation software environment. *IET Syst, Biol.* **2**, 352–362.

Slepchenko, B. M., Schaff, J. C., Carson, J. H., and Loew, L. M. (2002). Computational cell biology: spatiotemporal simulation of cellular events. *Annu. Rev. Biophys. Biomol. Struct.* **31**, 423–441.

CHAPTER 9

Stochastic Models of Cell Protrusion Arising From Spatiotemporal Signaling and Adhesion Dynamics

Erik S. Welf and Jason M. Haugh

Department of Chemical and Biomolecular Engineering, North Carolina State University, Raleigh, North Carolina, USA

Abstract

During cell migration, local protrusion events are regulated by biochemical and physical processes that are in turn coordinated with the dynamic properties of cell-substratum adhesion structures. In this chapter, we present a modeling approach for integrating the apparent stochasticity and spatial dependence of signal transduction pathways that promote protrusion in tandem with adhesion dynamics. We describe our modeling framework, as well as its abstraction, parameterization, and validation against experimental data. Analytical techniques for identifying and evaluating the

0091-679X/10 $35.00
DOI 10.1016/B978-0-12-394620-1.00009-6

effects of model bistability on stochastic simulation results are shown, and implications of this analysis for understanding cell protrusion behavior are offered.

I. Introduction

Cell crawling over an adhesive surface is a mechanical phenomenon marked by coordinated protrusion at the cell front and retraction at the rear (Parsons *et al.*, 2010). Fibroblasts and other migratory cells of mesenchymal lineage exhibit discrete, micron-sized adhesive contacts that are nucleated by transmembrane adhesion receptors and which form as the leading edge of a migrating cell protrudes over the surface. These adhesions are dynamic, responsive to external and intracellular forces, and conducive to assembly of molecularly diverse protein complexes. Although protrusion, adhesion, and retraction are clearly mechanical processes, they are apparently organized by the timing and partitioning of biochemical signaling pathways. This handoff from chemical regulation to mechanical actuation, together with the ability of cells to sense and respond to mechanical forces, creates a bidirectional feedback mechanism that is thought to play a critical role in controlling cell migration (Welf and Haugh, 2011). In this chapter, we discuss development and simulation of a computational model representing adhesion dynamics and adhesion-mediated signaling both as a cause and a consequence of localized protrusion.

How do diverse cellular protrusion behaviors arise from the interplay among physical and biochemical subprocesses? As with any complex system, the variety of interacting molecular components and processes at work during cellular protrusion demand analytical approaches for parsing their influence on cellular behaviors, and the apparently random nature of those events suggests that stochastic computational modeling is well-suited for representing them. The key challenges include how to model the components that are not fully understood at a mechanistic level, and, for those components that are better understood, deciding how much detail to include (Mogilner, 2009). As we discuss, our approach for dealing with these challenges has been to simplify (coarse-grain) certain aspects of the system while employing phenomenological assumptions to balance the scope and desired detail of the model with computational tractability and physical understanding.

II. Model Synthesis

A. Adhesion Dynamics

During active membrane protrusion in cells of mesenchymal origin, actin polymerization at the leading edge of the cell applies force on the membrane, which is balanced by immobile adhesion structures that couple with the actin network and transmit force to the substrate. These cell–matrix adhesions thus serve as mechanical linkages that enable the cell to pull itself along, but they also mediate the localization of numerous intracellular signaling proteins. The signaling properties of adhesion structures differ according to their sizes and intracellular locations. The small nascent adhesions that form at the leading edge of a cell protrusion facilitate actin polymerization through activation of Rac and certain other signaling intermediates (Cox *et al.*, 2001); however, when these nascent adhesions mature to form larger, more stable focal adhesions, they accumulate actomyosin activity and inhibit protrusion and cell shape change either by biochemical means or by acting as firm anchors for the actin cytoskeleton (Vicente-Manzanares *et al.*, 2007, 2011). Thus, within a local region at the cell periphery, spontaneous transitions between predominantly protrusive and adhesive phenotypes are observed.

Observation of adhesion dynamics by live-cell microscopy directly illustrates why a stochastic framework, in which adhesions are treated as discrete entities, is well-suited for modeling adhesion and migration (Fig. 1a). In tandem with adhesion formation and turnover, protrusion is rarely smooth with respect to time and space; it is most common to see leading edges bulge in transient, localized bursts (Machacek and Danuser, 2006; Tsukada *et al.*, 2008). Likewise, the process of adhesion maturation occurs infrequently, and as a result there are relatively low numbers of stable adhesions that nonetheless have dramatic phenotypic effects.

To model the spatial aspects of local adhesion formation, turnover, maturation, and signaling, we consider a control volume comprising a region starting at the leading edge of a protruding region of the cell and extending rearward toward the cell center, terminating just following the boundary between the lamellipodium (LP) and the lamella (LM), as shown in Fig. 1b. The LP is the region of dense, dynamic actin

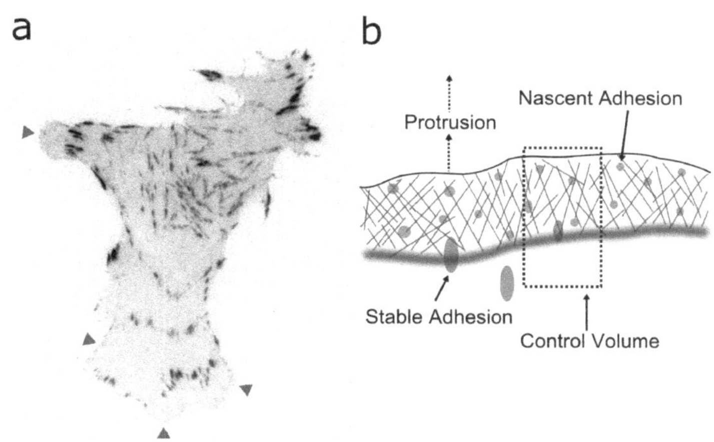

Fig. 1 Stochastic nature and spatial representation of adhesions. (a) Inverted grayscale image of a CHO.K1 cell expressing GFP-paxillin, monitored by TIRF microscopy, showing regions of nascent adhesion formation/turnover (red arrowheads) and the discrete nature of the larger, mature adhesions (adapted from Cirit *et al.* (2010)). (b) Diagram illustrating the control volume for the model system and the locations of adhesions therein. Nascent adhesions are formed and move rearward relative to the leading edge as the cell protrudes, and they either mature or turn over when they reach the back edge of the lamellipodium. (For color version of this figure, the reader is referred to the web version of this book.)

starting at the leading edge of the cell and extending several microns into the cell body, terminating at the location where actin depolymerization thins the dense actin network of the LP to form the LM. As the boundary between the LP and the LM moves forward relative to immobile adhesions, nascent adhesions that reach the LM-LP boundary either turn over (i.e., disintegrate) or mature to form stable adhesions (Nayal *et al.*, 2006). As the front of a protruding region moves forward, adhesions move rearward relative to the leading edge, which is the frame of reference for the model; increases in protrusion velocity directly affect adhesion turnover by increasing the rate at which nascent adhesions reach the LM-LP boundary (Choi *et al.*, 2008). Likewise, the effects of stable adhesions are assumed to fade with increasing distance from the cell front.

Nascent adhesions form at a rate proportional to the rate of local protrusion (Choi *et al.*, 2008), thus placing them within the positive feedback loop, protrusion $\rightarrow$ nascent adhesion formation $\rightarrow$ Rac signaling $\rightarrow$ protrusion, which we term the *core protrusion cycle* (Fig. 2). As explained below, nascent adhesions are assumed to mediate localized activation of Rac by one of two mechanisms, which differ in mathematical form (rate law). It is further assumed that the velocity of leading-edge protrusion is a monotonically increasing function of the local Rac concentration.

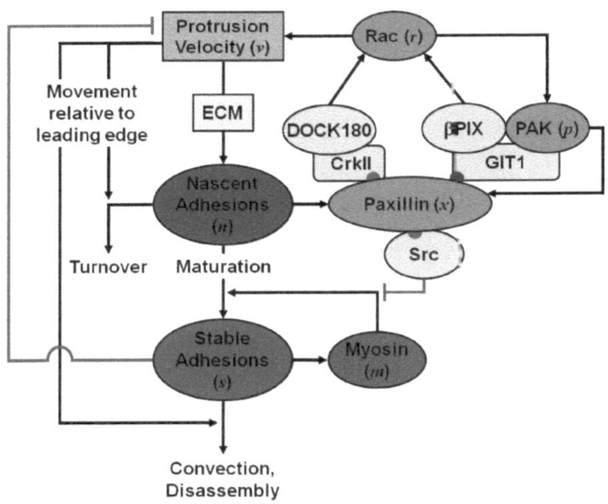

Fig. 2 Conceptual model framework. The rate of formation of nascent adhesions depends on the ECM concentration, and the rates of nascent adhesion formation and turnover depend on the velocity of membrane protrusion. Nascent adhesions promote protrusion via Rac activation, either via a pathway utilizing βPix that is reinforced by positive feedback through PAK, or through a pathway involving DOCK180. Those nascent adhesions that are not turned over mature to form stable adhesions, a process that is reinforced by myosin-mediated feedback and attenuated by Src. Stable adhesions directly antagonize protrusion, disassemble over a relatively long time scale, and have a diminishing influence on processes at the leading edge as a function of their growing distance from the leading edge during protrusion. (For color version of this figure, the reader is referred to the web version of this book.)

Experimental observations indicate that formation of stable adhesions coincides with local pauses in protrusion, and the myosin-dependent contractile processes stimulated by stable adhesions encourage adhesion maturation (Choi *et al.*, 2008). Thus, the core protrusion cycle is subject to an opposing feedback loop whereby stable adhesions reduce the rate of nascent adhesion formation while enhancing the probability of nascent adhesion maturation (Fig. 2). Although the effects of stable adhesions on local protrusion have not been characterized in mechanistic detail, their importance demands that these effects be included at least phenomenologically in our model.

B. Adhesion-Mediated Signaling

A host of scaffolding proteins and kinases are recruited to cell–matrix adhesions, and our focus here is on adhesion-associated signaling pathways that promote local protrusion. Paxillin is a scaffold protein recruited to nascent adhesions shortly after their formation, and once phosphorylated on specific sites, paxillin mediates binding of guanine exchange factors (GEFs) that activate Rac, which in turn enhances actin polymerization (Deakin and Turner, 2008). The Rac effector p21-activated kinase (PAK) phosphorylates paxillin on serine 273, providing a binding site for the recruitment of the scaffold protein GIT1, which forms a complex with both the Rac-GEF βPIX and PAK; this positive feedback loop involving local Rac activation, embedded within the core protrusion cycle, is apparently required for maintenance of protrusion, at least in certain cell contexts (Nayal *et al.*, 2006). In parallel, paxillin phosphorylated on tyrosine residues 31 and 118 by focal adhesion kinase (FAK) recruits the CrkII adapter and the unconventional Rac-GEF DOCK180, further amplifying Rac activation in response to paxillin localization and phosphorylation (Smith *et al.*, 2008; Kiyokawa and Matsuda, 2009). Paxillin phosphorylated on tyrosines 31 and 118 also mediates binding of the tyrosine kinase Src, which opposes myosin function and may thus attenuate adhesion maturation (Tsubouchi *et al.*, 2002).

C. Model Formulation

We are concerned with both biochemical signaling and physical processes governing adhesion dynamics and extension of the cell membrane, and how stochastic fluctuations in these processes are coupled. The appropriate level of detail therefore involves biochemical interactions and reactions at the molecular level; however, as with many biochemical systems we employ simplifying assumptions to reduce the number of adjustable model parameters and decrease the computational burden. A compromise in the degree of coarse-graining was reached by including abbreviated descriptions of biochemical mechanisms that are relatively well-characterized while employing phenomenological descriptions of other important processes. For example, we model the phosphorylation of different amino acid residues on paxillin as distinct events but assume that the subsequent binding and modifications of adapter proteins and GEFs are implicit in the activation of Rac (for a discussion of kinetic model simplification, see Cirit and Haugh, 2011). Adhesion maturation is an example of a process that is less well-understood, and our phenomenological approach was to treat nascent and mature adhesions as discrete entities and cast the transition between the two in terms of a probability that increases according to the local myosin activity.

The locations of adhesions relative to the leading edge and laterally along the cell contour determine the degree to which the two adhesion types influence protrusion or adhesion maturation (Fig. 1b). The mechanical effects of adhesion formation and maturation are widely speculated to involve force-responsive proteins (i.e., mechanotransduction) and propagation of stress within the heterogeneous actin network (Anderson *et al.*, 2008; Gardel *et al.*, 2010; Parsons *et al.*, 2010). Although the molecular and physical details involved in these mechanical processes form the basis of continuing theoretical and experimental work, it seems reasonable to focus the details of a coarse-grained model on either the signaling or mechanical aspects of the system. Whereas other studies have dealt primarily with the mechanical side (Chan and Odde, 2008; Li *et al.*, 2010; Sabass and Schwarz, 2010; Zimmermann *et al.*, 2010; Barnhart *et al.*, 2011), we chose to emphasize the properties of adhesion-mediated signaling.

As described above, our approach for dealing with the spatial relationships between model variables is to define a control volume that moves along with the leading edge of a cell; within a control volume, molecular species are assumed to be well-mixed, and the width of the control volume is set so that the validity of this approximation is ensured. We investigated the possibility of spatial propagation along the one-dimensional leading-edge contour by performing spatially extended simulations (described in detail under *Computational Methods*).

III. Model Analysis

The performance of the model was evaluated in part by comparing the qualitative behavior of model simulations at different values of the parameter representing the effect of extracellular matrix (ECM) to experiments assessing protrusion of CHO.K1 cells on different densities of the ECM protein fibronectin ([Fn]). An intermediate [Fn] (2 μg/mL coating concentration) fosters optimal cell migration speed of this cell line (Palecek *et al.*, 1997), and the relative abundance of nascent and stable adhesions at different [Fn] apparently contributes to this optimality. As shown in both experiments and simulations, intermediate [Fn] supports maximal protrusion in conjunction with a high abundance of nascent adhesions, whereas high [Fn] supports mostly stable adhesions, and low [Fn] does not support many adhesions of either type (Cirit *et al.*, 2010).

Although values for some of the model parameters were chosen based on experimental evidence, other parameters representing phenomenological relationships were varied systematically (Table I). Fig. 3a shows stochastic simulation

Table I
Model parameters

Parameter	Description	Comment
$k_{a,n}^{ECM}$	Rate constant, N assembly	Model input; varied from 0.01–100 min^{-1}
E_n	Rac → protrusion coupling	Set to 100 (>> 1)
K_v	Saturation of protrusion velocity	Set to 1; moderate saturation
I_n	S → protrusion inhibition	Varied from 0–10
I_s	Src → maturation inhibition	Varied from 0–100
$k_{d,n}$	Rate constant, basal N turnover	Set to 0.1 min^{-1}; same value as $k_{d,s}$
C_n	Protrusion → N turnover coupling	Set to 20 (Nayal *et al.*, 2006)
$k_{a,s}$	Rate constant, basal S growth	Set to 0.01 min^{-1} (<< $k_{d,s}$)
E_s	Myosin → S growth coupling	Varied from 0 – 100
$k_{d,s}$	Rate constant, S disassembly	Set to 0.1 min^{-1} (Nayal *et al.*, 2006)
C_s	Protrusion → S convection	Varied from 1–100
k_{d,x_i}	Rate constant, X dephosphorylation†	Set to 10 min^{-1} (arbitrarily fast)
K_{x_i}	Saturation of phospho-paxillin†	Set to 1; moderate saturation
p_o	Basal paxillin phosphorylation	Set at 0.01 (<< 1)
$k_{d,r}$	Rate constant, Rac deactivation	Set at 4 min^{-1} (Moissoglu *et al.*, 2006)
K_p	Saturation of Pak activation	Set to 1; moderate saturation
$k_{d,m}$	Rate constant, myosin deactivation	Set at 4 min^{-1}; same value as $k_{d,r}$
N^*	Scaling factor, N	Varied from 1 – 10
K_m	Amplification factor, S → Myosin	Set to 10
K_r	Amplification factor, Paxillin → Rac	Set to 10
D_r	Mobility coefficient, Rac	Next subvolume model; set to 15 μm^2/min (Moissoglu *et al.*, 2006)

† *i* denotes paxillin phosphorylation on serine 273 or tyrosines 31 and 118.

results for the model with βPix/no DOCK180 signaling and different values of the parameters I_n and E_s, which characterize the phenomenological effects of protrusion inhibition by stable adhesions and enhancement of adhesion maturation by myosin, respectively. Qualitative characterization of the stochastic simulation behaviors, as shown in Fig. 3b, facilitates comparison of simulation results across different combinations of parameter values. The effect of myosin-mediated adhesion strengthening, modeled by the E_s parameter, can be seen clearly in the spatially extended simulations shown in Fig. 4 – when E_s is low, protrusion dominates, but when E_s is higher, the formation of stable adhesions inhibits protrusion yet allows for stochastic protrusion bursts that propagate laterally as active Rac diffuses. Another approach for characterizing stochastic simulation results is to calculate the mean lifetimes of protrusion and adhesion events occurring during an extended simulation period (1000 min was used). Protrusion events were identified as periods of time during which the dimensionless protrusion velocity $v > 0.5$, and adhesion events were identified as periods of time during which the number of stable adhesions was nonzero ($S \geq 1$).

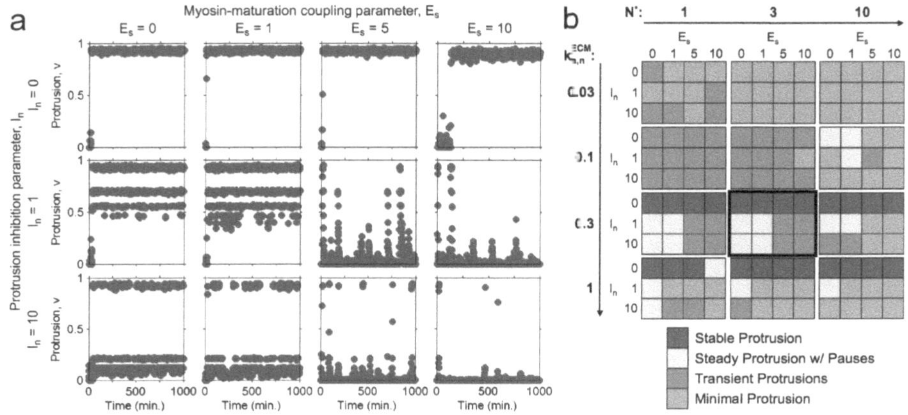

Fig. 3 Characterization of protrusion/adhesion phenotypes via stochastic simulation. The model system was simulated starting with all species numbers initialized at zero. (a) Protrusion velocity v is plotted as a function of time for $k_{a,n}^{ECM} = 0.3$ min^{-1}, $N^* = 3$, and a matrix of E_s and I_n values as indicated. (b) The same (E_s, I_n) matrix was repeated for different values of $k_{a,n}^{ECM}$ and N^* as indicated, and the apparent phenotype of each simulation is categorized qualitatively. The matrix framed with a thicker border corresponds to the simulations shown in a. (See color plate.)

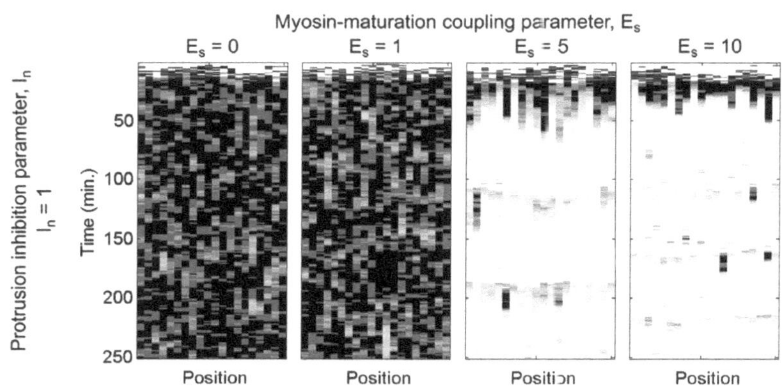

Fig. 4 Spatially extended simulation results. Spatially extended simulations were performed using the Next Subvolume Method to account for lateral diffusion of active Rac; the virtual leading edge is subdivided into 20 subvolumes, each 1.94 μm in size. Protrusion velocity is indicated in grayscale (white: $v = 0$; black: $v = 1$) as a function of time and position for a range of E_s values. Adapted from Cirit *et al.* (2010).

Fig. 5 shows the effect of changes to E_n and C_s for different values of $k_{a,n}$ and E_s, respectively, in the model with DOCK180/no βPix signaling.

A deterministic analysis of the model equations, treating the molecular species as continua rather than as discrete entities, was also performed. Although a

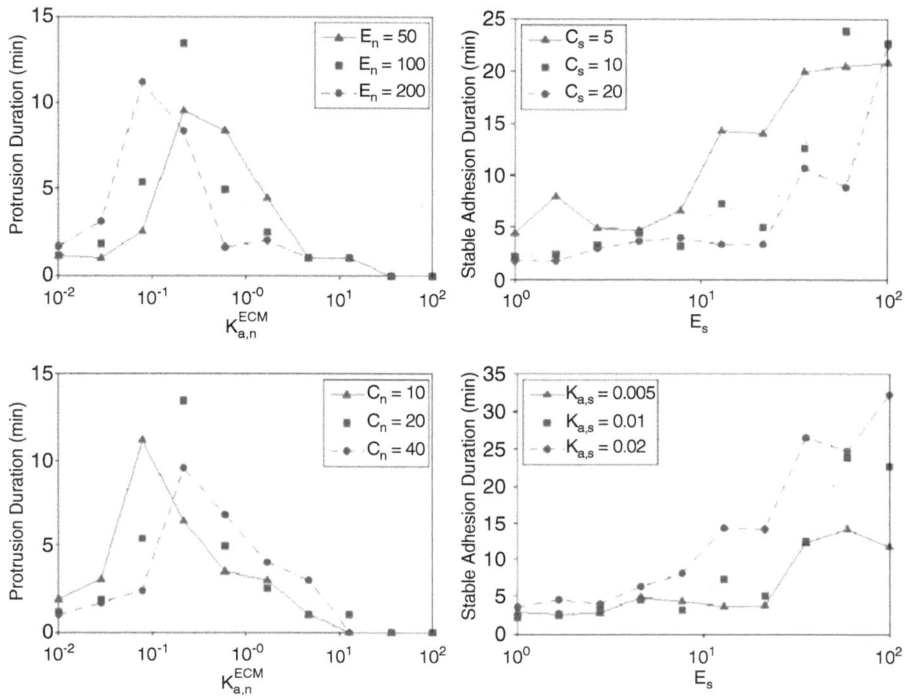

Fig. 5 Characterization of stochastic simulation results. Mean lifetimes of protrusion and adhesion events at various values for $k_{a,n}^{ECM}$, E_s, E_n, C_n, C_s, and $k_{a,s}$ were calculated. Protrusion events were identified as periods of time during which the dimensionless protrusion velocity $v > 0.5$, and adhesion events were identified as periods of time during which the number of stable adhesions was nonzero ($S \geq 1$). Adapted from Welf and Haugh (2010). (For color version of this figure, the reader is referred to the web version of this book.)

purely deterministic treatment was not able to produce switching between protrusive and adhesive states, the calculations proved useful in identifying conditions where the model exhibits bistability, which is related to the existence of multiple steady states (Cirit *et al.*, 2010). In the context of our models, bistability is a condition in which both the protrusive and adhesive phenotypes are stable. Fig. 6 shows identification of regions of model bistability via phase plane analysis, with the nullclines for nascent adhesions and stable adhesions plotted in (v,s) space. Intersections of the n and s nullclines indicate fixed points in the system, and regions of bistability are shown as functions of model parameters in Fig. 6b. A method for comparing regions of model bistability with stochastic model simulation results (in terms of mean protrusion or stable adhesion lifetimes) is shown in Fig. 7.

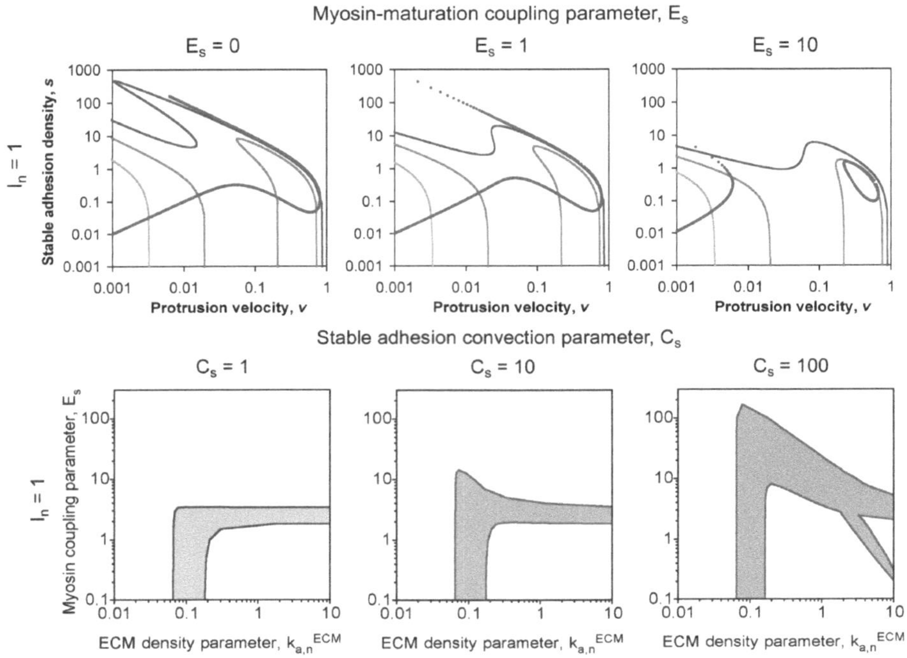

Fig. 6 Determination of regions of bistability by phase plane analysis. In the upper panels, the nullclines for *n* (green) and *s* (magenta) are plotted in (*v*, *s*) space. For the *n*-nullclines, the values of the ECM parameter are 0.03 (light green), 0.1 (green), and 0.3 (dark green) min^{-1}. Intersections of the *n*- and *s*-nullclines are fixed points of the system. In the lower panels, the shaded region of ($k_{a,n}^{ECM}$, E_s) parameter space indicates where there are multiple fixed points ($k_{a,n}^{ECM}$ values given in units of min^{-1}). Adapted from Cirit *et al.* (2010). (See color plate.)

IV. Biological Insights from the Modeling Approach

In many cell signaling systems, the coupling of multiple feedback mechanisms complicates the mapping of stimuli to cell responses. Feedback loops can give rise to nonlinear effects such as amplification, oscillation, and hysteresis (Besser and Schwarz, 2010; López, 2010). In the context of cell migration, it was of interest to investigate how feedback loops might amplify or attenuate signaling events to affect the observed stochastic switching between protrusion and adhesion phenotypes. For example, measurements of the leading-edge protrusion velocity in migrating CHO.K1 cells clearly show isolated bursts in protrusion that appear to arise randomly; monitoring the localization of adhesions in these cells confirms that a lack of protrusion is accompanied by formation of large stable adhesions (Cirit *et al.*, 2010).

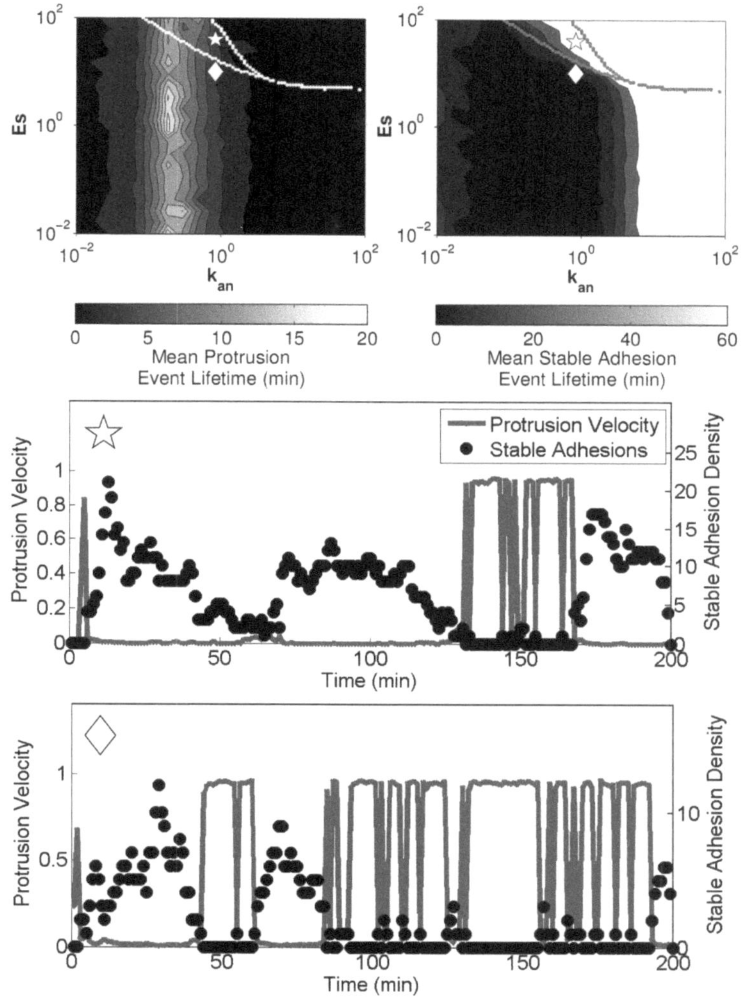

Fig. 7 Regions of bistability overlaid on stochastic simulation results. Mean lifetimes of protrusion and adhesion events were calculated as described in the caption for Fig. 5, and regions of bistability were identified by finding the steady state(s) of the deterministic model equations numerically using different combinations of initial conditions (upper panels). Stochastic simulation results corresponding to the parameter values indicated by the symbols in the upper panels are shown in the lower panels. Results are adapted from Welf and Haugh (2010) with $I_n = 10$, $I_s = 1$, and $C_s = 100$.

Based on our modeling studies, we can propose biochemical mechanisms that generate, through amplification of stochastic fluctuations, transient yet dramatic excursions from a particular stable state (Cirit *et al.*, 2010; Welf and Haugh, 2010). Such transient behavior takes the form of accelerations from an otherwise low protrusion state or decelerations (pauses) from an otherwise persistent protrusion state. Positive feedback amplification via Rac/PAK signaling and negative feedback attenuation via Src-mediated inhibition of adhesion maturation are capable of mediating these respective behaviors. If both signaling mechanisms are in play as we would propose, the same cell could employ one or the other mechanism at different times and/or at different subcellular locations.

Simulation results show that in order to achieve protrusion under high ECM density or high myosin activity conditions, the magnitude of the Src-mediated inhibition of maturation must be of a certain magnitude relative to the inhibition of protrusion by stable adhesions. Because Src-mediated buffering of adhesion maturation does not prevent adhesion formation under low ECM/low myosin conditions, this hypothetical mode of regulation presents an attractive means for maintaining sensitivity to changes in ECM density or myosin activity across wide ranges of these variables (Welf and Haugh, 2010).

Our original hypothesis held that model bistability would be important for stochastic phenotype switching. Although such behavior is likely to occur in regions of parameter space that are close to the bistability envelope, we found that model bistability is not required for the model to produce switching behavior. Bistable regions of parameter space usually lie between those regions that give monostable low and monostable high protrusion, and in the vicinity of the interface between the two, the stochastic model readily produces transient departures from the stable state.

V. Open Challenges

A central issue in formulating increasingly useful models of cellular processes is how best to rectify the increasing molecular-level detail of the biology knowledge base with a desire to create holistic models encompassing a large set of regulatory interactions. In general, the granularity of a model should be determined by how well the constituent mechanisms are understood, balanced by the need to specify values of their corresponding rate parameters and tempered by the availability of quantitative data (Cirit and Haugh, 2011; Mogilner *et al.*, 2006). In many biological systems, biochemical complexity is combined with the need to describe mechanical effects and account for spatial concentration and stress gradients. Particularly in systems where spatial considerations are clearly important, as in cell migration, inclusion of all known biochemical interactions is computationally intractable. Further, many important cellular phenomena, such as those mediating mechanotransduction, remain to be characterized mechanistically (Bershadsky *et al.*, 2006). Although mechanical and biochemical models of cell migration have been independently proposed, and integration of biochemical and mechanical phenomena has been achieved recently in the context of leukocyte rolling and firm adhesion (Caputo and Hammer, 2009), these two fundamental modes of regulation have yet to be combined in a satisfactory way in a single model of cell migration.

In this chapter we have presented one approach for integrating the spatial and mechanical processes mediated by stable adhesion formation and myosin contractility into the biochemical framework that regulates cell protrusion. Our treatment of these processes represents only the most basic relationships between model variables, and these relationships should be refined as new data, especially those of a quantitative nature, become available. The recent development of new experimental approaches for perturbing and analyzing the spatial, temporal, and mechanical aspects of cell signaling will enable collection of such data (Grashoff *et al.*, 2010; Toomre and Bewersdorf, 2010; Wu *et al.*, 2009). Hence, as increasingly detailed descriptions of the underlying network are developed, it will be necessary to evaluate and compare their emergent properties, mapped to the behaviors encoded by the more coarse-grained or phenomenological treatments used to construct necessarily less detailed, holistic models.

VI. Computational Methods

A. Parameter Nomenclature

Certain model parameters are dimensionless and phenomenological; these are classified by whether they characterize enhancement of species i formation (E_i), inhibition of species i formation (I_i), or augmentation of species i consumption rate (C_i). Other parameters have dimensions and include first-order rate constants with units of inverse time, characterizing assembly/activation or disassembly/deactivation of species i ($k_{a,i}$ or $k_{d,i}$, respectively), and diffusion coefficients with units of area/time (D_i). Dimensionless parameters K_i denote ratios of rate constants, characterizing the rate of assembly or activation relative to that of disassembly or deactivation for species i ($K_i = k_{a,i}/k_{d,i}$). Definitions of all model parameters are listed in Table 1.

B. Model Equations

We constructed model equations considering conservation of molecular and adhesion-based species based on the conceptual model shown in Fig. 2. We have explored two variations of the model, each corresponding to the scaffolding effect of a different phosphorylation site or sites on paxillin: serine 273 (Cirit *et al.*, 2010) or tyrosines 31 and 118 (Welf and Haugh, 2010). The equations for each instance of the model were identical, except as indicated. The dimensionless densities of nascent adhesions (n), stable adhesions (s), and recruited myosin (m) are generally written as follows.

$$\frac{dn}{dt} = k_{a,n}^{ECM}(1 + E_n v) - k_{d,n}(1 + C_n v)n - k_{a,s} f\left(m, x_{31/118}\right)n \tag{1}$$

$$\frac{ds}{dt} = k_{a,s} f\left(m, x_{31/118}\right)n - k_{d,s}(1 + C_s v)s \tag{2}$$

and

$$\frac{dm}{dt} = k_{d,m}(s - m) \tag{3}$$

The value of the parameter $k_{a,n}^{ECM}$ maps in some way to the density and character of the ECM, and v is the dimensionless protrusion velocity. The algebraic function $f(m,x_{31/118})$ describes the enhancement of adhesion maturation by myosin and its inhibition by paxillin phosphorylated on tyrosines 31 and 118, which directs Src recruitment; in the model considering serine 273 only, the dependence on the $x_{31/118}$ variable is absent.

Previous theoretical studies have analyzed in detail how the kinetics of actin polymerization might affect local membrane protrusion (Barnhart *et al.*, 2011; Gov, 2006; Zimmermann *et al.*, 2010); in this work we employ a simple functional relationship between Rac activity (r) and membrane protrusion, such that the protrusion velocity increases in response to Rac signaling until a saturation limit is reached.

$$v = \frac{K_v r}{(1 + K_v r)g(s)} \tag{4}$$

The function $g(s)$ specifies the relationship between stable adhesion density and inhibition of protrusion.

Although various phenomenological forms of the f and g functions may be proposed, we adopted simple, linear forms as follows.

$$f\left(m, x_{31/118}\right) = 1 + \frac{E_s m}{1 + I_s x_{31/118}} \tag{5}$$

$$g(s) = 1 + I_n s \tag{6}$$

Again, the variable $x_{31/118}$ in the f function is effectively fixed at zero in the βPix/no DOCK180 model (Cirit *et al.*, 2010).

The equations for the signaling circuit variables are as follows. The variable x_i (i = 273 or 31/118) represents the subset of n harboring phosphorylated paxillin (and, implicitly, GIT1/βPIX/PAK or CrkII/DOCK180 complexes), r is the density of active Rac (activated by βPIX or DOCK180), and p is the subset of x_{273} harboring Rac-activated PAK. For the case where we consider phosphorylation of serine 273 (Cirit *et al.*, 2010), we write:

$$\frac{dx_{273}}{dt} \approx k_{d,x_{273}}\left[K_{x_{273}}(p_o + p)(n - x_{273}) - x_{273}\right] \tag{7}$$

The small basal paxillin phosphorylation activity, p_0, is included so that x_{273}, r, and p can evolve in time when all initial values are zero. Likewise, the fraction of nascent adhesions harboring paxillin phosphorylated on tyrosine 31/118 (Welf and Haugh, 2010) is written

$$\frac{dx_{31/118}}{dt} \approx k_{d,x_{31/118}}\left[K_{x_{31/118}}\left(n - x_{31/118}\right) - x_{31/118}\right] \tag{8}$$

The equation for the activation of Rac is written

$$\frac{dr}{dt} = k_{d,r}(x_{273} - r) \tag{9}$$

or

$$\frac{dr}{dt} = k_{d,r}(x_{31/118} - r) \tag{10}$$

In spatially extended simulations, the conservation of active Rac also includes lateral diffusion. For the βPix/no DOCK180 model, an additional equation describes the activation of PAK on paxillin/GIT1/PAK complexes.

$$\frac{dp}{dt} \approx k_{d,p}[K_p r(x_{273} - p) - p] \tag{11}$$

C. Specification of Stochastic Models

To specify the stochastic model, we convert dimensionless model variables to numbers of molecules via scaling factors, indicated with an asterisk, for example, $N = N^*n$, where N is the absolute number of nascent adhesions in the control volume and n is the corresponding dimensionless variable. Based on the scaling of the conservation equations listed in the previous section, the other scaling factors are related to N^* as follows.

$$S^* = X^* = P^* = N^* \tag{12}$$

$$M^* = K_m N^* \tag{13}$$

$$R^* = K_r N^* \tag{14}$$

Because our model contains certain phenomenological rate laws, the stochastic formulation is not automatically specified as in the case of a mass action model. Our reaction propensity functions, in units of number of molecules per minute, are specified as follows.

Nascent adhesion assembly (Ø → N):

$$k_{a,n}^{ECM}(1 + E_n v)N^* \tag{15}$$

Nascent adhesion turnover (N → Ø):

$$k_{d,n}(1 + C_n v)N \tag{16}$$

Adhesion maturation (N → S):

$$k_{a,s} f\left(m, x_{31/118}\right)N \tag{17}$$

Disappearance of stable adhesions (S → Ø):

$$k_{d,s}(1 + C_s v)S \tag{18}$$

Myosin activation (Ø → M):

$$k_{d,m}K_m S \tag{19}$$

Myosin deactivation (M → Ø):

$$k_{d,m}M \tag{20}$$

Paxillin phosphorylation on serine 273 (Ø → X_{273}):

$$k_{d,x_{273}}K_{x_{273}}(p_o + p)(N - X_{273}) \tag{21}$$

Paxillin dephosphorylation of serine 273 (X_{273} → Ø):

$$k_{d,x_{273}}X_{273} \tag{22}$$

Paxillin phosphorylation on tyrosine 31/118 (Ø → $X_{31/118}$):

$$k_{d,x_{31/118}}K_{x_{31/118}}(N - X_{31/118}) \tag{23}$$

Paxillin dephosphorylation on tyrosine 31/118 ($X_{31/118}$ → Ø):

$$k_{d,x_{31/118}}X_{31/118} \tag{24}$$

Rac activation by βPix (Ø → R):

$$k_{d,r}K_r(X_{273}) \tag{25}$$

Rac activation by DOCK180 (Ø → R):

$$k_{d,r}K_r(X_{31/118}) \tag{26}$$

Rac deactivation (R → Ø):

$$k_{d,r}R \tag{27}$$

PAK activation (Ø → P):

$$k_{d,p}K_p r(X_{273} - P) \tag{28}$$

PAK deactivation (P → Ø):

$$k_{d,p}P \tag{29}$$

Stochastic simulations were performed using the Next Reaction Method (Gibson and Bruck, 2000), a modification of the Gillespie algorithm (Gillespie, 1977), implemented in MATLAB (MathWorks, Natick, MA). These methods simulate trajectories of the chemical master equation describing discrete stochastic systems, such as those encountered in cells where small numbers of reacting species or rare reaction events dominate system dynamics.

D. Spatially Extended Simulations

Spatially extended stochastic simulations were performed using the Next Subvolume Method (Elf and Ehrenberg, 2004), whereby diffusion of species i between adjacent compartments is modeled as a "hopping" reaction with first-order rate constant D_i/L^2, where D_i is the diffusivity of species i, and L is the node spacing between adjacent compartments. A compromise between numerical accuracy and computational expense is achieved by setting the node spacing L equal to the smallest of the dynamic length scales, $L_i = \sqrt{D_i\tau_i}$, where τ_i is the mean lifetime of diffusible species i. In most of our spatially extended simulations (Cirit *et al.*, 2010), we assumed that only active Rac is diffusible, with $\tau_r = 1/k_{d,r}$. Estimates of D_r and $k_{d,r}$ were obtained from the literature (Moissoglu *et al.*, 2006), yielding $L = L_r \approx 2\ \mu m$. We assumed a one-dimensional geometry, corresponding to the contour of a leading edge, with periodic boundary conditions.

Acknowledgments

This work was supported by NSF grant CBET-0828936, and E.S.W. was supported by the Cell Migration Consortium under National Institutes of Health grant U54-GM064346.

References

Anderson, T. W., Vaughan, A. N., and Cramer, L. P. (2008). Retrograde flow and myosin II activity within the leading cell edge deliver F-actin to the lamella to seed the formation of graded polarity actomyosin II filament bundles in migrating fibroblasts. *Mol. Biol. Cell.* **19**, 5006–5018.

Barnhart, E. L., Lee, K. -C., Keren, K., Mogilner, A., and Theriot, J. A. (2011). An adhesion-dependent switch between mechanisms that determine motile cell shape. *PLoS Biol.* **9**, e1001059.

Bershadsky, A., Kozlov, M., and Geiger, B. (2006). Adhesion-mediated mechanosensitivity: a time to experiment, and a time to theorize. *Curr. Opin. Cell. Biol.* **18**, 472–481.

Besser, A., and Schwarz, U. S. (2010). Hysteresis in the cell response to time-dependent substrate stiffness. *Biophys. J* **99**, L10–L12.

Caputo, K. E., and Hammer, D. A. (2009). Adhesive dynamics simulation of G-protein-mediated chemokine-activated neutrophil adhesion. *Biophys. J* **96**, 2989–3004.

Chan, C. E., and Odde, D. J. (2008). Traction dynamics of filopodia on compliant substrates. *Science* **322**, 1687–1691.

Choi, C. K., Vicente-Manzanares, M., Zareno, J., Whitmore, L. A., Mogilner, A., and Horwitz, A. R. (2008). Actin and alpha-actinin orchestrate the assembly and maturation of nascent adhesions in a myosin II motor-independent manner. *Nat. Cell. Biol.* **10**, 1039–1050.

Cirit, M., and Haugh, J. M. (2011). Quantitative models of signal transduction networks: how detailed should they be? *Commun. Integr. Biol.* **4**, 353–356.

Cirit, M., Krajcovic, M., Choi, C. K., Welf, E. S., Horwitz., Alan, F., and Haugh, J. M. (2010). Stochastic model of integrin-mediated signaling and adhesion dynamics at the leading edges of migrating cells. *PLoS Comput. Biol.* **6**, e1000688.

Cox, E. A., Sastry, S. K., and Huttenlocher, A. (2001). Integrin-mediated adhesion regulates cell polarity and membrane protrusion through the Rho family of GTPases. *Mol. Biol. Cell.* **12**, 265–277.

Deakin, N. O., and Turner, C. E. (2008). Paxillin comes of age. *J. Cell. Sci.* **121**, 2435–2444.

Elf, J., and Ehrenberg, M. (2004). Spontaneous separation of bi-stable biochemical systems into spatial domains of opposite phases. *Syst. Biol. (Stevenage)* **1**, 230–236.

Gardel, M. L., Schneider, I. C., Aratyn-Schaus, Y., and Waterman, C. M. (2010). Mechanical integration of actin and adhesion dynamics in cell migration. *Annu. Rev. Cell. Dev. Biol.* **26**, 315–333.

Gibson, M. A., and Bruck, J. (2000). Efficient exact stochastic simulation of chemical systems with many species and many channels. *J. Phys. Chem. A* **104**, 1876–1889.

Gillespie, D. T. (1977). Exact stochastic simulation of coupled chemical reactions. *J. Phys. Chem.* **81**, 2340–2361.

Gov, N. S. (2006). Dynamics and morphology of microvilli driven by actin polymerization. *Phys. Rev. Lett.* **97**, 018101.

Grashoff, C., Hoffman, B. D., Brenner, M. D., Zhou, R., Parsons, M., Yang, M. T., McLean, M. A., Sligar, S. G., Chen, C. S., and Ha, T., *et al.* (2010). Measuring mechanical tension across vinculin reveals regulation of focal adhesion dynamics. *Nature* **466**, 263–266.

Kiyokawa, E., and Matsuda, M. (2009). Regulation of focal adhesion and cell migration by ANKRD28-DOCK180 interaction. *Cell. Adh. Migr.* **3**, 281–284.

Li, Y., Bhimalapuram, P., and Dinner, A. R. (2010). Model for how retrograde actin flow regulates adhesion traction stresses. *J. Phys. Condens. Matter* **22**, 194113.

Lo'pez, J. M. (2010). Digital kinases: a cell model for sensing, integrating and making choices. *Commun. Integr. Biol.* **3**, 146–150.

Machacek, M., and Danuser, G. (2006). Morphodynamic profiling of protrusion phenotypes. *Biophys. J.* **90**, 1439–1452.

Mogilner, A. (2009). Mathematics of cell motility: have we got its number? *J. Math. Biol.* **58**, 105–134.

Mogilner, A., Wollman, R., and Marshall, W. F. (2006). Quantitative modeling in cell biology: what is it good for? *Dev. Cell.* **11**, 279–287.

Moissoglu, K., Slepchenko, B. M., Meller, N., Horwitz., Alan, F., and Schwartz, M. A. (2006). In vivo dynamics of Rac-membrane interactions. *Mol. Biol. Cell.* **17**, 2770–2779.

Nayal, A., Webb, D. J., Brown, C. M., Schaefer, E. M., Vicente-Manzanares, M., and Horwitz, A. R. (2006). Paxillin phosphorylation at Ser273 localizes a GIT1-PIX-PAK complex and regulates adhesion and protrusion dynamics. *J. Cell. Biol.* **173**, 587–589.

Palecek, S. P., Loftus, J. C., Ginsberg, M. H., Lauffenburger, D. A., and Horwitz, A. F. (1997). Integrin-ligand binding properties govern cell migration speed through cell-substratum adhesiveness. *Nature* **385**, 537–540.

Parsons, J. T., Horwitz, A. R., and Schwartz, M. A. (2010). Cell adhesion: integrating cytoskeletal dynamics and cellular tension. *Nat. Rev. Mol. Cell. Biol.* **11**, 633–643.

Sabass, B., and Schwarz, U. S. (2010). Modeling cytoskeletal flow over adhesion sites: competition between stochastic bond dynamics and intracellular relaxation. *J. Phys. Condens. Matter* **22**, 194112.

Smith, H. W., Marra, P., and Marshall, C. J. (2008). uPAR promotes formation of the p130Cas-Crk complex to activate Rac through DOCK180. *J. Cell. Biol.* **182**, 777–790.

Toomre, D., and Bewersdorf, J. (2010). A new wave of cellular imaging. *Annu. Rev. Cell. Dev. Biol.* **26**, 285–314.

Tsubouchi, A., Sakakura, J., Yagi, R., Mazaki, Y., Schaefer, E., Yano, H., and Sabe, H. (2002). Localized suppression of RhoA activity by Tyr31/118-phosphorylated paxillin in cell adhesion and migration. *J. Cell. Biol.* **159**, 673–683.

Tsukada, Y., Aoki, K., Nakamura, T., Sakumura, Y., Matsuda, M., and Ishii, S. (2008). Quantification of local morphodynamics and local GTPase activity by edge evolution tracking. *PLoS Comput. Biol.* **4**, e1000223.

Vicente-Manzanares, M., Newell-Litwa, K., Bachir, A. I., Whitmore, L. A., and Horwitz, A. R. (2011). Myosin IIA/IIB restrict adhesive and protrusive signaling to generate front-back polarity in migrating cells. *J. Cell. Biol.* **193**, 381–396.

Vicente-Manzanares, M., Zareno, J., Whitmore, L., Choi, C. K., Horwitz., &., and Alan, F. (2007). Regulation of protrusion, adhesion dynamics, and polarity by myosins IIA and IIB in migrating cells. *J. Cell. Biol.* **176**, 573–580.

Welf, E. S., and Haugh, J. M. (2011). Signaling pathways that control cell migration: models and analysis. *Wiley Interdiscip. Rev. Syst. Biol. Med.* **3**, 231–240.

Welf, E. S., and Haugh, J. M. (2010). Stochastic dynamics of membrane protrusion mediated by the DOCK180/Rac pathway in migrating cells. *Cell. Mol. Bioeng.* **3**, 30–39.

Wu, Y. I., Frey, D., Lungu, O. I., Jaehrig, A., Schlichting, I., Kuhlman, B., and Hahn, K. M. (2009). A genetically encoded photoactivatable Rac controls the motility of living cells. *Nature* **461**, 104–108.

Zimmermann, J., Enculescu, M., and Falcke, M. (2010). Leading-edge-gel coupling in lamellipodium motion. *Phys. Rev. E Stat. Nonlin. Soft Matter Phys.* **82**, 051925.

CHAPTER 10

Nonparametric Variable Selection and Modeling for Spatial and Temporal Regulatory Networks

Anil Aswani*, Mark D. Biggin†, Peter Bickel‡ and Claire Tomlin*

*Department of Electrical Engineering and Computer Sciences, University of California, Berkeley, California, USA

†Genomics Division, Lawrence Berkeley National Laboratory, Berkeley, California, USA

‡Department of Statistics, University of California, Berkeley, California, USA

0091-679X/10 $35.00
DOI 10.1016/B978-0-12-388403-9.00010-2

Abstract

Because of the increasing diversity of data sets and measurement techniques in biology, a growing spectrum of modeling methods is being developed. It is generally recognized that it is critical to pick the appropriate method to exploit the amount and type of biological data available for a given system. Here, we describe a method for use in situations where temporal data from a network is collected over multiple time points, and in which little prior information is available about the interactions, mathematical structure, and statistical distribution of the network. Our method results in models that we term Nonparametric exterior derivative estimation Ordinary Differential Equation (NODE) model's. We illustrate the method's utility using spatiotemporal gene expression data from *Drosophila melanogaster* embryos. We demonstrate that the NODE model's use of the temporal characteristics of the network leads to quantifiable improvements in its predictive ability over nontemporal models that only rely on the spatial characteristics of the data. The NODE model provides exploratory visualizations of network behavior and structure, which can identify features that suggest additional experiments. A new extension is also presented that uses the NODE model to generate a comb diagram, a figure that presents a list of possible network structures ranked by plausibility. By being able to quantify a continuum of interaction likelihoods, this helps to direct future experiments.

I. Introduction

Understanding gene expression is important because it will provide greater insights into the generation of phenotype from genetic information and enable a better comprehension of many disease processes. But studying gene expression is challenging because it is the end result of complex interactions between many quantitative inputs, including protein–DNA binding and regulatory protein expression. These inputs are often experimentally measured, leading to large data sets that are difficult to understand without the help of computational methods. Modeling is a useful approach for studying gene expression and its associated data sets because it handles data in an automated manner. When correctly done, modeling is capable of using the data to quickly invalidate many hypotheses, while also generating new hypotheses that are consistent with the data and suggest novel biology.

Considerable work has been done on modeling regulatory networks at the level of protein and mRNA expression (Aswani *et al.*, 2009b; Bansal *et al.*, 2007; Bonneau *et al.*, 2006; Cinquemani *et al.*, 2009; D'haeseleer *et al.*, 1999; Eisen *et al.*, 1998; Fakhouri *et al.*, 2010; Friedman *et al.*, 2000; Jong and Ropers, 2006; Markowetz and Spang, 2007; Porreca *et al.*, 2008; Rao *et al.*, 2008; Stuart *et al.*, 2003; Werhli *et al.*, 2006). These models span a spectrum of detail ranging from heuristic Boolean networks to mechanistic dynamical models. The particular modeling technique that is chosen is dependent on the biological data available; and this choice requires implicitly or explicitly making strong assumptions about the biological behavior, mathematical structure, or statistical distribution of the network. Making these assumptions is arguably reasonable when the purpose of modeling a regulatory network is to reverse engineer the network structure. However, such assumptions may be inaccurate when studying the biological processes of regulation itself. For this reason, alternative modeling techniques are needed.

In this chapter, we present an exploratory modeling technique for regulatory networks where temporal data is collected over a large number of time points or experiments. We propose what we call a Nonparametric exterior derivative estimate Ordinary Differential Equation (NODE) model (Aswani *et al.*, 2010). The method was originally designed to study a data set of gene expression from *Drosophila melanogaster* embryogenesis that has been collected (Fowlkes *et al.*, 2008; Luengo Hendriks *et al.*, 2006) by the Berkeley *Drosophila* Transcription Network Project (BDTNP). This data set provides relative levels of transcription factor protein and target gene mRNA expression at cellular resolution for multiple time points. The NODE model does not make any prior assumptions on the structure of the underlying network or on the biological modes of regulation and so is well suited for studying the process of regulation itself. Specifically, the NODE model has been able to identify novel modes of regulation, suggesting future directions of experimental study.

Transcriptional regulation in all animals occurs through a complicated set of interactions between transcription factor proteins and the *cis*-control regions (CCRs) of DNA that regulate target genes (Davidson, 2006; Young, 2011). Different combinations of transcription factors in a nucleus influence diverse protein–protein interactions on the CCR, and in this way generate complex spatial and temporal patterns of expression. However, while it is clear that the levels of occupancy of factors on CCRs and the spatial arrangement of their recognition sites within the CCR are important (Fakhouri *et al.*, 2010), we are not yet able to correctly predict how particular combinations of binding lead to specific spatiotemporal patterns of transcription. This has important implications for parametric modeling where a mathematical form of the network is assumed.

An advantage of the NODE model in the context of the *Drosophila* gene expression data set is that it enables the methodical study of the modes of regulation of a transcription factor when CCR-architecture is not considered. The classical picture is to categorize factors into activators/promoters and inhibitors/repressors. When applied to the early *Drosophila* embryo data, the NODE model suggests that this classification is too coarse, and finer notions of regulation should be considered and modeled. Furthermore, the NODE model explicitly considers the temporal features of biological networks in order to generate quantitatively more accurate models of the gene expression, in comparison to a (quasi-)static model. The temporal nature of the NODE model brings it closer to the biology.

This chapter provides an overview of the NODE method, and describes some resulting biological insights into modes of regulation in the early *Drosophila* system. A new extension of the NODE model is also described that generates a list of possible network structures ranked by likelihood. This approach recognizes that it may be difficult to identify the "best" model for a system, and that by instead generating a list of "good" models experiments can then be designed to distinguish between them. The chapter concludes by providing a detailed description of NODE and discussing some of the open computational challenges for nonparametric approaches that incorporate temporal characteristics.

II. Overview of the NODE Model

The NODE model seeks to capture the total net effect of direct and indirect influence of each transcription factor on a target gene. The model is generated by looking at the correlation between factor protein concentrations and the *change* in target mRNA concentration over time. By looking at the change in target mRNA over time, we are able to generate a dynamic equation model that describes each transcription factor's influence on each gene. This model describes the regulatory network at cellular resolution using the concentrations of gene products like protein and mRNA from each cell. In addition, the model can be extended to generate a comb diagram, which shows a variety of network structures that are ranked by their plausibility. Some technical assumptions are made by NODE and its extension, but they are quite general and apply to many biological systems.

A. Experimental Data

We apply our technique to experimental data that has been collected and processed by the BDTNP (Fowlkes *et al.*, 2008; Luengo Hendriks *et al.*, 2006), where measurements of protein and mRNA concentrations are taken by analyzing images of many *Drosophila* embryos to create a virtual embryo. The virtual embryo consists of 6078 cells and is a computational, spatial decomposition that is determined by averaging the geometry and number of cells of different embryos. The virtual embryo has measurements of the concentration (averaged over multiple embryos at a fixed time point) of various transcription factor protein and target mRNAs at the cellular level for six different time points during Stage 5 of the *Drosophila* embryo. For example, Fig. 1 shows the experimentally measured pattern of even-skipped (*eve*) mRNA on the virtual embryo during late Stage 5 of embryogenesis.

B. Assumptions

Three assumptions are made by NODE to enable the use of an ordinary differential equation (ODE) model. First, it assumes that the rate-limiting species (e.g., transcription factor protein concentrations) that drive the behavior of the network have been measured; actions on faster timescales (e.g., the dynamics of factors binding to target genes) are not considered by the model. Second, transcription factor protein

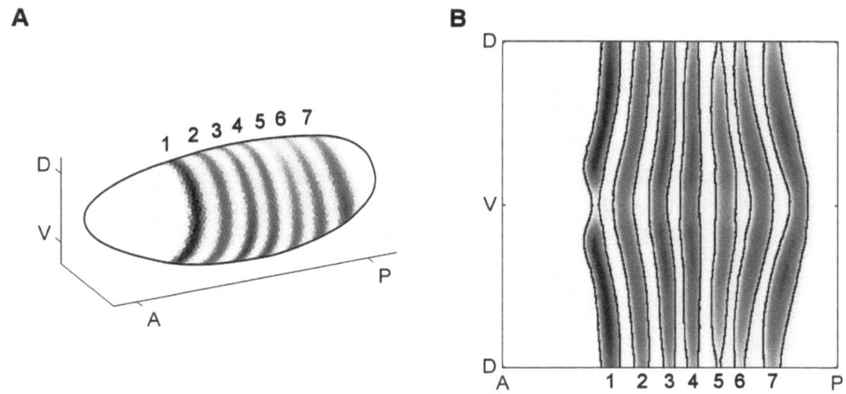

Fig. 1 Quantitative cellular resolution 3D gene expression. (A) A three-dimensional plot of the *Drosophila* embryo showing the experimentally measured pattern of *eve* mRNA as it appears in late Stage 5. There are seven distinct expression stripes located along the anterior–posterior axis (AP) of the embryo, with the intensity of each stripe also varying moderately along the dorsal–ventral axis (DV). (B) A two-dimensional cylindrical projection of a Stage 5 *Drosophila* embryo provides an easier visualization of the details of the *eve* mRNA patterns, showing that expression of each stripe is similar on either side of the ventral mid line (V). (For color version of this figure, the reader is referred to the web version of this book.)

concentrations in nuclei are assumed to be large enough for the levels of occupancy on CCRs to be deterministic. Note that a small degree of randomness is introduced into the modeling procedure due to measurement noise present in the data. Third, there is an assumption that spatial processes such as diffusion of gene products between cells are a negligible portion of the system behavior.

For statistical reasons, there are a few other assumptions that are made about the biological network. There are no assumptions made on whether a factor for a particular target CCR is always a repressor or always an activator, in contrast to many other modeling methods that often make this assumption. However, the extension of the NODE model for the generation of a comb diagram assumes that a particular transcription factor on a particular gene has a greater tendency toward activation or repression. Whether or not this is biologically valid is not clear, but it seems intuitively reasonable. This extension also assumes that only a small subset of transcription factors actually regulates each individual target gene to a significant degree.

Under these assumptions, the system can be reasonably described by an ODE

$$\frac{dx}{dt} = f(x), \tag{1}$$

where x is a vector whose elements are the concentrations of the rate-limiting species. Nonlinear regression techniques (Bansal *et al.*, 2007; Markowetz and Spang, 2007) start by making additional assumptions about the network in order to hypothesize a function with unknown coefficients, and then they regress the data onto this function. This can be problematic if the prior knowledge is incorrect. In contrast, our NODE method does not make any assumptions on the functional form of $f(x)$. Nevertheless, both approaches require the use of statistical regularization to protect against overfitting.

Note that no assumptions are made regarding the presence or absence of feed-forward or feedback loops or other cross-regulatory interactions in the network. The NODE model ignores the complexity of such loops because it does not attempt to predict which interactions are direct and which are indirect, but only seeks to determine the net effect that a given transcription factor has on a target, including all direct and indirect influences.

C. Interpretation

Instead of using a single ODE model to describe the regulatory network, the NODE model uses a group of ODE models consisting of the first-order Taylor expansion (i.e., linearization) of the ODE given in Eq. (1). Each equation of the NODE model describes how the behavior of the regulatory network changes if protein concentrations of the transcription factors in the cell of a particular experiment are slightly perturbed. This approach requires fewer assumptions because it does not require knowing the mathematical structure of the single ODE model in Eq. (1). The disadvantage of this flexibility is that it is more difficult to interpret the NODE model.

To further understand the intuition of the NODE model, consider Fig. 2. The target gene is *eve* mRNA, and there are five protein transcription factors: bicoid (*bcd*), giant (*gt*), hunchback (*hb*), knirps (*kni*), Krüppel (*Kr*) (Arnosti *et al.*, 1996; Fujioka *et al.*, 1999; Small *et al.*, 1996). The horizontal axes give the measured transcription factor protein concentrations for different time points and cells of the embryo, and the vertical axis gives the measured change in *eve* mRNA concentration for the corresponding time point and cell. Generating a model of the form of Eq. (1) means identifying a function that takes protein factor concentrations as an input and then gives change in target gene mRNA concentration as an output. The idea of the NODE model is to have a separate submodel for each time point and cell.

Each submodel is valid for only when the factor concentrations are moderately perturbed, and it technically corresponds to the linearization of Eq. (1). In the case of Fig. 2A, the nth submodel corresponding to a vector of transcription factor concentrations $\xi[n]$ is given by

$$\frac{dx_{eve,\mathrm{mRNA}}[n]}{dt} = a_{bcd}[n](x_{bcd} - \xi_1[n]) + \cdots + a_{Kr}[n](x_{Kr} - \xi_5[n]) + b[n], \qquad (2)$$

and it is valid for when the factor concentrations are perturbed within the gray circle. Because of this formulation, the NODE model has many submodels and requires the use of statistical tools that protect against overfitting.

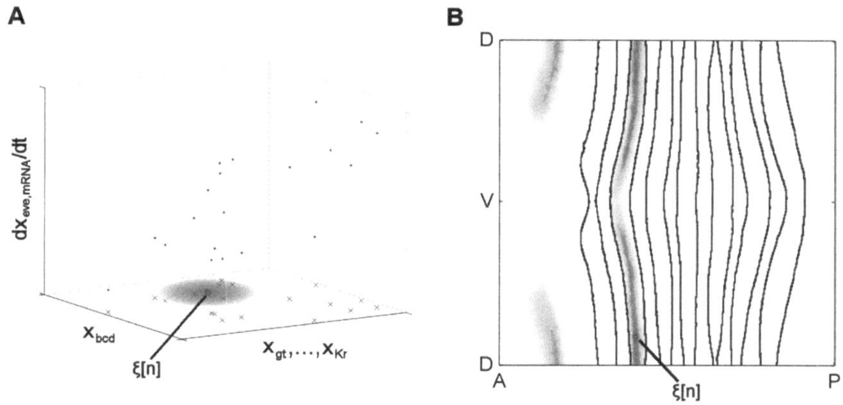

Fig. 2 Neighborhood of cells with similar factor concentrations. (A) Identifying the model in Eq. (1) is mathematically equivalent to computing a function that takes transcription factor protein concentrations as an input and then gives change in target gene mRNA concentration as an output. The NODE model has a separate submodel for each time point and cell, and each submodel is computed by looking at a window of cells with similar concentrations of regulatory factors. An example of such a window is given by the gray circle centered about the cell with factor concentrations $\xi[n]$, which is labeled with a black line. Cells with concentrations similar to $\xi[n]$ are those points that fall within the gray circle, and darker shades of gray indicate more similar concentrations. Cells with dissimilar concentrations are those points that are not within the gray circle. (B) The same window of cells is indicated on the virtual embryo by the set of cells colored gray. Darker shades of gray indicate higher similarity in concentration to $\xi[n]$, and white indicates large dissimilarity. The black lines show the boundaries of the experimental *eve* pattern. (For interpretation of the references to color in this figure legend, the reader is referred to the web version of this book.)

For a particular vector of transcription factor protein concentrations $\xi[n]$, the gray circle in Fig. 2A represents measurements of factor concentrations that are similar in value to $\xi[n]$. When the parameters of the submodel in Eq. (2) are fit, the points within the gray circle are used to do the regression. In the case of the *Drosophila* data, the gray circle can be visualized as a group of cells with similar concentrations of factors. Such a group of cells is shown in gray in Fig. 2B. As expected, cells nearby often have similar concentrations of factors. More surprisingly, cells far away can sometimes have similar concentrations of factors.

D. Statistical Improvements

To improve the accuracy of the model, our method uses a novel regression technique (Aswani *et al.*, 2011) known as the nonparametric exterior derivative estimator (NEDE) that makes local estimates of the ODE in Eq. (1) and can scale to networks with hundreds of species. It is able to utilize measurements from multiple experiments while protecting against overfitting. The NEDE estimator adds constraints to the identification problem by learning correlations between factors, and these constraints protect the model from overfitting and erroneously identifying

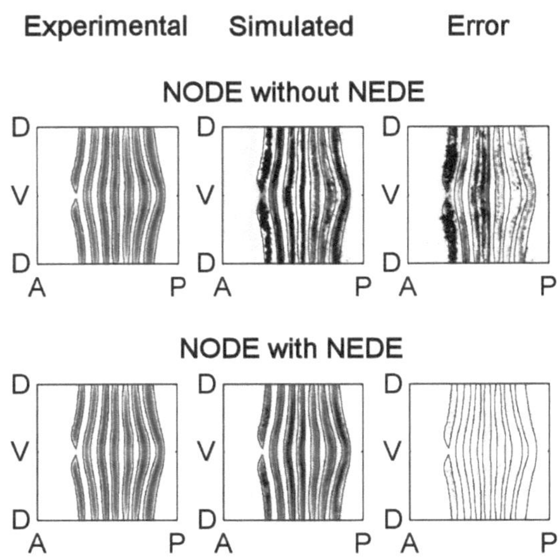

Fig. 3 Comparison of NODE model without and with the NEDE estimator. (A) A NODE model was generated without using the NEDE method: Ordinary least squares was used instead. The model was fit with data from the first two measured time points of Stage 5, and then the model was simulated using the first time point as an initial condition. The experimental and simulated *eve* mRNA patterns, as well as simulation error, are shown for Stage 5:51–75 of development, which corresponds to the fifth time step. (B) A NODE model was generated using the NEDE method. The model was fit and simulated in the same manner as before, and visual examination shows that the NEDE statistical method results in a model that makes significantly better predictions. (For color version of this figure, the reader is referred to the web version of this book.)

weak biological effects as strong effects. Tuning parameters for our method are selected in a data-driven manner using cross-validation. Details on its theoretical properties can be found in (Aswani *et al.*, 2011).

It is the use of the NEDE estimator that distinguishes the NODE model from similar models that build a set of local models (Cinquemani *et al.*, 2009; Jong and Ropers, 2006; Porreca *et al.*, 2008). The use of this statistical tool is necessary for reducing overfitting, and Fig. 3 shows a striking example of its importance. The figure compares a simulation of the NODE model for the *eve* mRNA stripes when the NEDE statistical tool is and is not used. The initial condition of the simulation in both cases is the experimentally measured *eve* concentration, and the models are fit using data from only the first two time measured points. Fig. 3 shows the concentration predicted by each simulation at Stage 5:51–75 (i.e., the portion of Stage 5 in which cell membranes are 51–75% formed) of embryo development, which corresponds to the fifth time point. Examining it shows that the NODE model with NEDE leads to a significantly better simulation.

E. Extension for Network Selection

One of the original aims in developing the NODE model was to create a new technique for learning the network structure of a regulatory network that could exploit the mathematical features of temporal data. The original NODE model is not able to do this, but it can be easily extended for this purpose by using a variant of the NEDE regression tool that is able to select relevant regulators. This regression technique (Aswani *et al.*, 2011) is known as the nonparametric adaptive lasso exterior derivative estimator (NALEDE), and it uses lasso regularization to select a sparse set of regulators of the target gene for each submodel. Next, our extension combines the sparsity structure of these submodels in order to generate a list of possible network structures. Cross-validation is used to select the network topology that best explains the experimentally measured data, and it is in this way that the NODE model can be extended to identify network topologies.

The NODE model and its extension generate distinct network topologies from each other where the edges have different biological interpretations, and so it is important to explain the differences. In both the original and extended NODE models, an edge from a regulator to a target indicates that on the timescale of the measurements that make up the data set, the net effect of the regulator through direct and indirect interactions is to control the expression of the target. For the original NODE model, each submodel has a corresponding network topology; and the edge between a regulator and a target can be classified into one of four potential categories:

1. Type I Repression – At current factor concentrations, the target mRNA will decrease in concentration over time. An increase in factor concentration will lead to a faster rate of decrease in target mRNA amounts over time.
2. Type II Repression – At current factor concentrations, the target mRNA will increase in concentration over time. An increase in factor concentration will lead to a slower rate of increase in target mRNA amounts over time.
3. Type I Activation – At current factor concentrations, the target mRNA will increase in concentration over time. An increase in factor concentration will lead to a faster rate of increase in target mRNA amounts over time.
4. Type II Activation – At current factor concentrations, the target mRNA will decrease in concentration over time. An increase in factor concentration will lead to a slower rate of decrease in target mRNA amounts over time.

In contrast, the extension of the NODE model generates a single network topology that describes the aggregate behavior of regulators and their influence on a target. For the network topology generated by the extension, the edge does not categorize the nature of the interaction (e.g., activation or inhibition), and in fact biological insights from our modeling results suggest that such a cumulative classification is too restrictive. Rather, the absence of an edge indicates that any potential interactions are not statistically significant according to the extension of the NODE model. The opposite holds as well.

III. Biological Insights

We identified a NODE model for *eve* mRNA pattern formation in the *Drosophila* blastoderm that assumed that the only protein regulatory factors were bcd, gt, hb, Kr, kni (Arnosti *et al.*, 1996; Fujioka *et al.*, 1999; Small *et al.*, 1996). The model was validated by fitting the model to only the first two time points and then measuring how well a simulation could predict the remaining time points. This shows that the NODE model fits the experimental *eve* data well, and that it has some predictive ability when the network and the transcription factor concentrations are slightly perturbed from wild type (Aswani *et al.*, 2010). The NODE model can also be analyzed and extended to suggest hypothesis for further study through additional biological experiments and mathematical modeling.

In general, biological validation is done by fitting a NODE model to one set of data and then quantifying how well it can predict another set of data. The most basic comparison is when both sets of data occur under the same biological conditions, and this is the one used here to evaluate the NODE model for the *eve* stripes. The disadvantage of this approach is that it does not examine the applicability of the model to new biological conditions. Genetic point mutation experiments are arguably the most powerful method for evaluating this, but they are too drastic because the generated NODE models will in general not apply such situations. Gene over- and underexpression studies are better suited for validation, because NODE models apply to situations in which the biological conditions are slightly perturbed from those of the data set that was used to generate the model.

A. Importance of Temporal Dynamics

To evaluate the importance of considering the temporal features of the network when building the mathematical model, we also generated a nontemporal, spatial-correlation model

$$x_{eve,\mathrm{mRNA}}[n] = a_{bcd}[n](x_{bcd} - \xi_1[n]) + \cdots + a_{Kr}[n](x_{Kr} - \xi_5[n]) + b[n], \quad (3)$$

where the left-hand side of the equation is simply the *eve* mRNA concentration (cf., Eq. (2) that has the change in *eve* on the left-hand side). The spatial-correlation and NODE models were generated using only the first two time points of data, and then their simulations were compared to the experimentally measured patterns. This is a nonbiased comparison between spatial and temporal models for this system, because the mathematical form and statistical methods used for both are identical.

The NODE model gives 59% better agreement than the spatial-correlation model to the experimentally measured *eve* pattern (Aswani *et al.*, 2010). For comparison, Fig. 4 shows the concentration predicted by each simulation at Stage 5:51–75 of embryo development, which corresponds to the fifth time point. This result supports the intuitive assumption that the NODE model is intrinsically more biologically realistic than a spatial-correlation model. As stated earlier, biological networks are marked by temporal effects. For instance, a protein binds to DNA that initiates

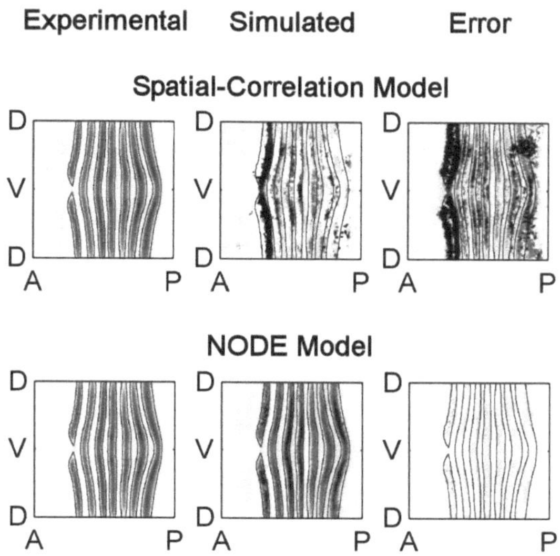

Fig. 4 Comparison of spatial-correlation model and NODE model. (A) A spatial-correlation model was generated using the NEDE method. The model was fit with data from the first two measured time points of Stage 5, and then the model was simulated using the first time point as an initial condition. The experimental and simulated *eve* mRNA patterns, as well as simulation error, are shown for Stage 5:51–75 of development, which corresponds to the fifth time step. (B) A NODE model was generated using the NEDE method. The model was fit and simulated in the same manner as before, and visual examination shows that the NODE model makes significantly better predictions. (For color version of this figure, the reader is referred to the web version of this book.)

transcription. This is not an instantaneous process, and there is some delay between when a regulatory factor initiates transcription and when the target mRNA is expressed. The spatial-correlation model does not model this notion of temporal effects, whereas the NODE model does.

This is not to say spatial-correlation models are incorrect. Visual comparison of the spatial-correlation model to the NODE model shows many similarities, and there are many matches between the interactions predicted by the two models (Aswani *et al.*, 2010). This is encouraging because many experimentally validated regulatory interactions have been implicitly interpreted using a spatial-correlation model (Arnosti *et al.*, 1996; Fowlkes *et al.*, 2008; Fujioka *et al.*, 1999; Marco *et al.*, 2009; Small *et al.*, 1996), and this agreement provides mutual support both for our NODE model and the previously determined interactions.

B. Concentration-Dependent Effects of Transcription Factors

The NODE (and the spatial-correlation) model can be visualized as spatiotemporal maps of transcription factor activities. These factor activity plots show the intensity and variation of predicted effects of factors at different locations on the embryo and at different time points. An example of such a map for our NODE model for Stage 5:9–25 can be found in (Aswani *et al.*, 2010). This plot shows how the five transcription factors (directly or indirectly) affect *eve* mRNA pattern formation, indicating the predicted degree of repression (i.e., an anticorrelation between factor expression and the rate of change of target expression) and the predicted degree of activation (i.e., a positive correlation between factor and the change in target).

The NODE model is not a mechanistic model because it cannot capture the various mechanisms involved in the regulation of *eve* mRNA. However, this is a strength because of the flexibility gained by not having to make *a priori* assumptions on regulatory mechanisms. This comes at the cost of not being able to identify which interactions are direct or indirect, however.

In many cases it is known that individual gene expression stripes can be controlled via a single CCR and current computational models generally assume that a given factor acts only as an activator or a repressor on a given CCR (Bansal *et al.*, 2007; Markowetz and Spang, 2007). However, the NODE (and the spatial-correlation) model frequently predicts concentration-dependent effects whereby, on and around the same expression stripe, a transcription factor has both repressing and activating effects (Aswani *et al.*, 2010). For example, while the NODE model for *eve* stripe 2 formation is consistent with previous molecular genetic evidence that Kr is a repressor at the posterior margin of this expression stripe (Arnosti *et al.*, 1996), the model also implies that Kr is an activator just anterior of this in cells where Kr concentrations are lower.

This and the many other similar cases could represent spurious correlations, perhaps due to other transcription factors having dominant effects on targets in cells where the transcription factor under study is expressed at lower levels. However, there are a number of cases where factors, including Kr, have been shown to switch from activating to repressing the same target as their concentrations increase (Ptashne *et al.*, 1980; Sauer and Jäckle, 1991). Thus, the predictions of our NODE (and spatial-correlation) model make it more obvious that gene regulation can involve multiple mechanisms of transcription factor action that should be considered.

C. Network Structure Hypotheses

Analysis of *in vivo* binding data (Li *et al.*, 2008; MacArthur *et al.*, 2009) shows that there is a continuous spectrum of transcription factor binding levels to CCRs, in which much of the lowest levels of binding do not result in functionally significant regulation of transcription. There are many reasons for this, including that transcription factors are expressed at high enough concentrations in cells that they will be driven thermodynamically to bind to fortuitous occurrences of their recognition sites in any parts of the genome that are accessible within chromatin (Kaplan *et al.*, 2011; Li *et al.*, 2011). Conceptually, this is important because it suggests that many regulatory networks should be thought of in terms of potential models with varying levels of resolution that can be validated through experimentation, rather than in terms of an exact model that encompasses all of the important biological behaviors.

Here, different models mean different collections of transcription factors that enter into Eq. (1) that describes the mRNA expression of a target gene, and this notion of modeling is related to reverse engineering the network structure.

This conceptual framework of a set of models, rather than an exact model, can be applied to an extension of the NODE model. We applied the NODE model with the NALEDE estimator to expression data of the second *eve* expression stripe, and we only considered the portions of the embryo where (a) the stripe is expressed, (b) immediately anterior of the stripe, or (c) immediately posterior of the stripe. Here, we used 16 protein regulatory factors in the model. The results are summarized in Fig. 5, which is called a comb diagram and individually ranks each factor by its likelihood of being a regulator. A horizontal line on the comb diagram represents a likelihood threshold, and the network structure corresponding to this threshold value is the set of factors that extend below this line. By varying the likelihood threshold, different sets of models can be generated.

An example of a likelihood threshold is given by the horizontal line in Fig. 5 that is labeled "cv," which stands for cross-validation. This is a special likelihood threshold because it denotes the threshold value that corresponds to the model with the best predictive ability, given the experimentally measured data. In other words, the model that has bcd, gt, hb, hkb, Kr, run, and tll as regulators of *eve* stripe 2 is the best model for the data set out of all the possible network structures given by the comb diagram. This shows that this extension of the NODE model is promising because the comb diagram is able to accurately select the most well characterized regulators of the second *eve* stripe (i.e., bcd, gt, hb, and Kr) (Arnosti *et al.*, 1996) and several additional likely regulators – hkb, run, and tll – that have been found to bind to stripe 2 at high levels *in vivo* (MacArthur *et al.*, 2009). Additionally, the results of the extension conclude that

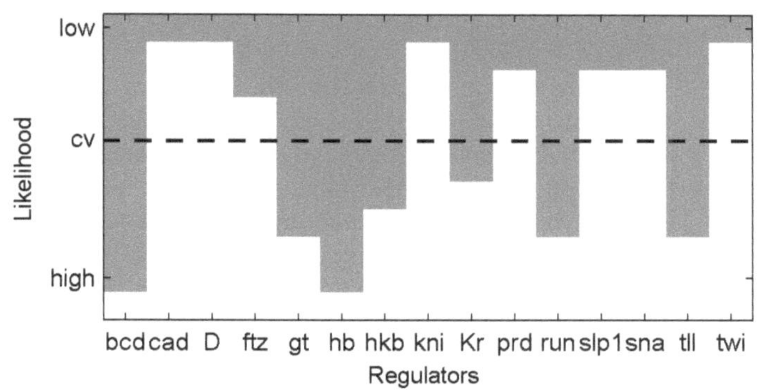

Fig. 5 Comb diagram for regulators of *eve* stripe 2 at Stage 5:0–3. Sixteen protein regulatory factors are individually ranked based on their likelihood of being a regulator of *eve* stripe 2, as determined by the extension of the NODE model that uses the NALEDE estimator. A horizontal line on the comb diagram represents a likelihood threshold, and the network structure corresponding to this threshold value is the set of factors that extend below this line. The likelihood threshold that corresponds to the model with the best predictive ability as measured by cross-validation error is labeled "cv." This threshold generates a set of factors that has good agreement with what is biologically known or hypothesized.

ftz, prd, slp1, sna, and twi are not significant regulators, which is also consistent with *in vivo* DNA binding data (MacArthur *et al.*, 2009). On the other hand, *in vivo* DNA binding data suggests that cad, kni, and D may also be regulators (MacArthur *et al.*, 2009), but the comb diagram does not provide strong evidence for this. There are several possible explanations: These transcription factors may be redundant with respect to other regulators, or more likely their role occurs in portions of the embryo that are not covered by our model. These explanations show the importance of considering multiple data sets when building models of regulatory networks.

IV. Open Challenges

There are still some unanswered, theoretical questions regarding the NODE model and its extensions. Cross-validation is an approach for selecting the tuning parameters of the method, in a data-driven manner; however, there are many specific implementations of cross-validation, and the most accurate methods are computationally slow. The current implementation of the NODE model uses a less accurate version of cross-validation that is guided by theory (Aswani *et al.*, 2011; Shao, 1993; Yang, 2007), but there are no results on which cross-validation procedure provides the best tradeoff between computational complexity and statistical performance. Furthermore, the theoretical properties on the consistency of variable selection when using the NODE model with the NALEDE estimator are unexplored; intuition suggests that combining the theorems of (Aswani *et al.*, 2011; Bertin and Lecué, 2008) will lead to this result, but it needs to be rigorously checked.

V. Computational Methods

In this section, we describe the technical details of the NODE model and its extension. We denote the vector of transcription factor protein concentrations as $x[t, e]$ and the vector of target gene mRNA concentrations as $y[t, e]$, where $t \in T$ is the time of the measurement and $e \in E$ is an index, which uniquely identifies each experiment. For the virtual embryo data set, there are six time points $T = \{1, \ldots, 6\}$ and each cell in the virtual embryo is considered to be a separate experiment $E = \{1, \ldots, 6078\}$. Note that notation like $x_{bcd}[t, e]$ denotes the bcd protein concentration in cell e at time t.

VI. Building a NODE Model

The NODE technique is summarized in the following algorithm. Any tuning parameters are chosen in a data-driven manner using cross-validation (Aswani *et al.*, 2011; Shao, 1993; Yang, 2007). Without loss of generality, this section describes the algorithm for the specific case of the NODE model for the *eve* mRNA stripes with five regulatory factors: bcd, gt, hb, kni, and Kr.

Inputs: Transcription factor protein concentrations $x[t, e]$, target gene mRNA concentrations $y[t, e]$

Outputs: NODE model

(1) Presmooth the transcription factor protein concentrations $x[t, e]$ and then compute time derivatives of the target gene mRNA concentrations $y[t, e]$.

(a) For each $e \in E$

(i) Do a least-squares fit of the polynomial $\hat{x}[t, e] = c_0 + c_1 t + \cdots + c_r t^r$ (where $c_0, c_1, \ldots, c_r$ are coefficients and r is a tuning parameter) with the data points: $x[t, e]$ for each $t \in T$.

(ii) Do a least-squares fit of the polynomial $\hat{y}[t, e] = k_0 + k_1 t + \cdots + k_r t^r$ (where $k_0, k_1, \ldots, k_r$ are coefficients and r is a tuning parameter) with the data points: $y[t, e]$, for each $t \in T$.

(b) Presmoothed factor protein concentration data is given by $\hat{x}[t, e]$, and time derivative of target gene mRNA data is given by $d\hat{y}[t, e]/dt = k_1 + k_2 t + \cdots + r k_r t^{r-1}$.

(2) Define matrix Y with rows given by $(d\hat{y}[t, e]/dt)$, for each $t \in T$ and $e \in E$.

(3) Calculate the NODE model.

(a) For each $t \in T$ and $e \in E$

(i) Define matrix $X_{[t,e]} = [\,1 \;\; \Xi_{[t,e]}\,]$, where first column is all ones and $\Xi_{[t,e]}$ is the matrix with rows given by $(\hat{x}[u, v] - \hat{x}[t, e])$, for each $u \in T$ and $v \in E$.

(ii) Define weighting matrix $W_{[t,e]}$ to be the diagonal matrix with entries along diagonal given by

$$w[u, v] = \begin{cases} \frac{3}{4}\left(1 - \left(\frac{n[u, v]}{h}\right)^2\right), & \text{if } n[u, v] \leq h \\ 0, & \text{otherwise} \end{cases} \tag{4}$$

for each $u \in T$ and $v \in E$, where $n[u, v] = \|\hat{x}[u, v] - \hat{x}[t, e]\|$ is the Euclidean distance and h is a tunable parameter.

(iii) Define matrix $P_{[t,e]}$ by making its columns be the ($p-d$) principal components of $\Xi_{[t,e]}{}^T W_{[t,e]} \Xi_{[t,e]}$ with smallest eigenvalues, where p is number of factors ($p = 5$ for the NODE model of target *eve* mRNA) and d is a tuning parameter.

(iv) Coefficients of NODE model, for eth cell at tth time point, are given by NEDE estimator

$$\begin{aligned}[\,b_{[t,e]} \; a_{bcd,[t,e]} \; \cdots \; a_{Kr,[t,e]}\,]^T = arg\,\min_{\beta} \left\|W_{[t,e]}{}^{1/2}(Y - X_{[t,e]}\beta)\right\|_2^2 \\ +\lambda\left\|P_{[t,e]}\beta\right\|_2^2,\end{aligned} \tag{5}$$

where

$$\frac{dx_{eve,\text{mRNA}[t,e]}}{dt} = a_{bcd,[t,e]}(x_{bcd} - \hat{x}_{bcd}[t,e]) + \cdots + a_{Kr,[t,e]}(x_{Kr} - \hat{x}_{Kr}[t,e]) + b_{[t,e]}. \quad (6)$$

Step 1 presmoothes the experimental data and computes its time derivative. If this were not done, the resulting NODE model would be statistically biased (Schneeweiß, 1976). Step 1.a describes polynomial regression (PR) for the purpose of simplifying the presentation, though the actual NODE method uses local polynomial regression (LPR) for this step; LPR is a variant of PR which protects against oversmoothing the data, and it can be quickly computed by doing a weighted linear regression.

Caution must be used when deciding to presmooth certain data sets in which the measurements are very noisy and taken at a sparse grid of points in time. In such cases, there is a risk of smoothing out biologically relevant, temporal trends in the data because of the sparsity of the temporal grid. There is another implementation issue that needs to be considered. The NODE model uses a moderate to large amount of data, and so it may be that another modeling technique is better suited to the data in such cases.

A NODE model is computed in Step 3 using the NEDE estimator, which is an advanced statistical tool that protects against overfitting (Aswani *et al.*, 2011). The NEDE estimator can be calculated quickly on a computer because it is a convex optimization problem, and it is statistically well behaved. Step 3.a.ii determines a window of data points that have measured concentrations similar to experiment e at time t, and the size of this window is selected by the parameter h. Data points with highly (weakly) similar concentrations are weighted highly (weakly) in the estimation of the coefficients of the NODE model. Eq. (4) uses the Epanechnikov kernel to do this weighting.

The coefficients of the NODE model are computed in Step 3.a.iv using the NEDE estimator in Eq. (5). It protects against overfitting by learning constraints that the data obeys (Step 3.a.iii), and then using these constraints to reduce the degrees of freedom in the regression. In general, the data points form a manifold, and the projection matrix in Eq. (5) enforces that the regression coefficients lie close to the manifold. This methodology is motivated by differential geometry, which says that the exterior derivative of a function on an embedded submanifold lies in the cotangent space (Lee, 2003).

VII. Building a Comb Diagram with the NALEDE Extension

The procedure for computing a NODE model with the NALEDE estimator (Aswani *et al.*, 2011) is nearly identical to that for the original NODE model. The only difference is Eq. (5), which is replaced by the NALEDE estimator

$$[b_{[t,e]}\ a_{bcd,[t,e]}\ \cdots\ a_{tll,[t,e]}]^T = \arg\min_{\beta} \left\| W_{[t,e]}{}^{1/2}(Y - X_{[t,e]}\beta) \right\|_2^2 + \lambda \left\| P_{[t,e]}\beta \right\|_2^2 + \mu \sum_{j=1}^{p} \omega_i |\beta_i|, \tag{7}$$

where $\omega_i = \left| a_{i,[t,e]} \right|^{\gamma}$ and $a_{i,[t,e]}$ are the estimated coefficients from Eq. (5). The idea is that this estimator adds a lasso regularization that promotes sparsity of the coefficient values and leads to local variable selection. The weighting values ω_i help to ensure that the NALEDE estimator has good theoretical properties on its ability to select the correct variables.

Generating a comb diagram from a NODE model with the NALEDE estimator is a simple process. For each transcription factor, we compute its inclusion frequency that is defined as

$$f_i = \frac{1}{\#T \cdot \#E} \sum_{t \in T, e \in E} [(1(a_{i,[t,e]} \neq 0))]. \tag{8}$$

This counts the fraction of submodels in which a factor is included. If a transcription factor is included in more (fewer) submodels, it is individually more (less) likely to be a regulator of the target gene. The comb diagram plots the inclusion frequency for each transcription factor, with bars for each factor that go from 0 (at the top of the plot) to f_i (at the bottom of the plot).

For a likelihood threshold of value L, the transcription factors that are deemed significant regulators are those for which $f_i > L$. From a modeling perspective, this helps to select different network structures based on whether or not a transcription factor is a significant regulator. Furthermore, statistics can be used to select the likelihood threshold, which gives the set of regulatory factors that generate a NODE model that best fits the experimentally measured data set. This is done by computing a NODE model and its cross-validation error for different likelihood thresholds and then selecting the threshold with the lowest cross-validation error.

Further Reading

More details on the NODE model and its biological insights for *Drosophila* embryogenesis can be found in (Aswani, 2010; Aswani *et al.*, 2010). The NODE model has been used in engineering domains as well, including modeling for a helicopter system (Aswani, 2010; Aswani *et al.*, 2009a). Theoretical properties of the statistical methods NEDE and NALEDE that underlie the NODE model and its extensions were studied in (Aswani, 2010; Aswani *et al.*, 2011).

Acknowledgments

This work is part of a broader collaboration by the BDTNP. We are grateful for the frequent advice, support, criticisms, and enthusiasm of its members. The gene expression data were funded by the U.S.

National Institutes of Health (NIH) under grants GM704403 (to MDB). Computational analyses were funded by National Science Foundation (NSF) award #CCR-0225610 (ITR) and #CNS-0931843 (CPS-ActionWebs) (to CT) and NSF awards #DMS-0605236 and #DMS-0906808 (to PB). Work at Lawrence Berkeley National Laboratory was conducted under Department of Energy contract DE-AC02-05CH11231.

References

Arnosti, D. N., Barolo, S., Levine, M., and Small, S. (1996). The eve stripe 2 enhancer employs multiple modes of transcriptional synergy. *Development* **122**, 205–214.

Aswani, A. (2010). Systems Theory for Pharmaceutical Drug Discovery. Available at the following link: http://www.eecs.berkeley.edu/Pubs/TechRpts/2010/EECS-2010-68.html

Aswani, A., Bickel, P., and Tomlin, C. (2009a). Statistics for sparse, high-dimensional, and nonparametric system identification. Proceedings of the IEEE International Conference on Robotics and Automation, 2133–2138.

Aswani, A., Bickel, P., and Tomlin, C. (2011). Regression on manifolds: estimation of the exterior derivative. *Ann. Stat.* **39**, 48–81.

Aswani, A., Guturu, H., and Tomlin, C. (2009b). *System identification of hunchback protein patterning in early Drosophila embryogenesis* 7723–7728.

Aswani, A., Keranen, S., Brown, J., Fowlkes, C., Knowles, D., Biggin, M., Bickel, P., and Tomlin, C. (2010). Nonparametric identification of regulatory interactions from spatial and temporal gene expression data. *BMC Bioinform.* **11**, 413.

Bansal, M., Belcastro, V., Ambesi-Impiombato, A., and di Bernardo, D. (2007). How to infer gene networks from expression profiles. *Mol. Syst. Biol.* 3.

Bertin, K., and Lecué, G. (2008). Selection of variables and dimension reduction in high-dimensional nonparametric regression. *Electron. J. Stat.* **2**, 1224–1241.

Bonneau, R., Reiss, D., Shannon, P., Facciotti, M., Hood, L., Baliga, N., and Thorsson, V. (2006). The inferelator: an algorithm for learning parsimonious regulatory networks from systems-biology data sets *de novo*. *Genome Biol.* **7**, R36.

Cinquemani, E., Milias-Argeitis, A., Summers, S., and Lygeros, J. (2009). Local identification of piecewise deterministic models of genetic networks. In Hybrid Systems: Computation and Control, pp. 105–119.

Davidson, E. H. (2006). *The Regulatory Genome: Gene Regulatory Networks In Development And Evolution, 1st ed. Academic Press,* .

D'haeseleer, P., Wen, X., Fuhrman, S., and Somogyi, R. (1999). Linear modeling of mRNA expression levels during CNS development and injury. In Pacific Symposium on Biocomputing, pp. 41–52.

Eisen, M., Spellman, P., Brown, P., and Botstein, D. (1998). Cluster analysis and display of genome-wide expression patterns. *PNAS* **98**, 14863–14868.

Fakhouri, W. D., Ay, A., Sayal, R., Dresch, J., Dayringer, E., and Arnosti, D. N. (2010). Deciphering a transcriptional regulatory code: modeling short-range repression in the *Drosophila* embryo. *Mol. Syst. Biol.* **6**, 341.

Fowlkes, C., Hendriks, C. L., Keränen, S., Weber, G., Rübel, O., Huang, M. -Y., Chatoor, S., Simirenko, L., DePace, A., Henriquez, C., Beaton, A., Weiszmann, R., Celniker, S., Hamann, B., Knowles, D., Biggin, M., Eisen, M., and Malik, J. (2008). Constructing a quantitative spatio-temporal atlas of gene expression in the *Drosophila* blastoderm. *Cell* **133**, 364–374.

Friedman, N., Linial, M., Nachman, I., and Pe'er, D. (2000). Using Bayesian networks to analyze expression data. *J. Comput. Biol.* **7**, 601–620.

Fujioka, M., Emi-Sarker, Y., Yusibova, G. L., Goto, T., and Jaynes, J. B. (1999). Analysis of an even-skipped rescue transgene reveals both composite and discrete neuronal and early blastoderm enhancers, and multi-stripe positioning by gap gene repressor gradients. *Development* **126**, 2527–2538.

Jong, H de., and Ropers, D. (2006). Qualitative approaches towards the analysis of genetic regulatory networks. *In* "System Modeling in Cellular Biology: From Concepts to Nuts and Bolts," (Z. Szallasi, V.

Periwal, J. Stelling, eds.), System Modeling in Cellular Biology: From Concepts to Nuts and Boltspp. 125–148. MIT Press.
Kaplan, T., Li, X. -Y., Sabo, P. J., Thomas, S., Stamatoyannopoulos, J. A., Biggin, M. D., and Eisen, M. B. (2011). Quantitative models of the mechanisms that control genome-wide patterns of transcription factor binding during early *Drosophila* development. *PLoS Genet.* **7**, e1001290.
Lee, J. (2003). *Introduction to Smooth Manifolds.* Springer.
Li, X-yong., MacArthur, S., Bourgon, R., Nix, D., Pollard, D. A., Iyer, V. N., Hechmer, A., Simirenko, L., Stapleton, M., Hendriks, C. L. L., Chu, H. C., Ogawa, N., Inwood, W., Sementchenko, V., Beaton, A., Weiszmann, R., Celniker, S. E., Knowles, D. W., Gingeras, T., Speed, T. P., Eisen, M. B., and Biggin, M. D. (2008). Transcription factors bind thousands of active and inactive regions in the *Drosophila* blastoderm. *PLoS Biol.* **6**, e27.
Li, X. -Y., Thomas, S., Sabo, P. J., Eisen, M. B., Stamatoyannopoulos, J. A., and Biggin, M. D. (2011). The role of chromatin accessibility in directing the widespread, overlapping patterns of *Drosophila* transcription factor binding. *Genome Biol.* **12**, R34.
Luengo Hendriks, C., Keränen, S., Fowlkes, C., Simirenko, L., Weber, G., DePace, A., Henriquez, C., Kaszuba, D., Hamann, B., Eisen, M., Malik, J., Sudar, D., Biggin, M., and Knowles, D. (2006). Three-dimensional morphology and gene expression in the *Drosophila* blastoderm at cellular resolution I: data acquisition pipeline. *Genome Biol.* **7**, R123.
MacArthur, S., Li, X. -Y., Li, J., Brown, J., Chu, H. C., Zeng, L., Grondona, B., Hechmer, A., Simirenko, L., Keranen, S., Knowles, D., Stapleton, M., Bickel, P., Biggin, M., and Eisen, M. (2009). Developmental roles of 21 *Drosophila* transcription factors are determined by quantitative differences in binding to an overlapping set of thousands of genomic regions. *Genome Biol.* **10**, R80.
Marco, A., Konikoff, C., Karr, T., and Kumar, S. (2009). Relationship between gene co-expression and sharing of transcription factor binding sites in *Drosophila melanogaster. Bioinformatics* **25**, 2473–2477.
Markowetz, F., and Spang, R. (2007). Inferring cellular networks – a review. *BMC Bioinform.* **8**, S5.
Porreca, R., Drulhe, S., Jong, H de., and Ferrari-Trecate, G. (2008). Structural identification of piecewise-linear models of genetic regulatory networks. *J. Computat. Biol.* **15**, 1365–1380.
Ptashne, M., Jeffrey, A., Johnson, A., Maurer, R., Meyer, B., Pabo, C., Roberts, T., and Sauer, T. (1980). How the lambda repressor and cro work. *Cell* **19**, 1–11.
Rao, A., Hero, A., States, D., and Engel, J. (2008). Using directed information to build biologically relevant influence networks. *J. Bioinform. Computat. Biol.* **6**, 493–519.
Sauer, F., and Jäckle, H. (1991). Concentration-dependent transcriptional activation or repression by Krüppel from a single binding site. *Nature* **353**, 563–566.
Schneeweiß, H. (1976). Consistent estimation of a regression with errors in variables. *Metrika* **23**, 101–115.
Shao, J. (1993). Linear model selection by cross-validation. *J. Am. Stat. Assoc.* **88**, 486–494.
Small, S., Blair, A., and Levine, M. (1996). Regulation of two pair-rule stripes by a single enhancer in the *Drosophila* embryo. *Dev. Biol.* **175**, 314–324.
Stuart, J., Segal, E., Koller, D., and Kim, S. (2003). A gene-coexpression network for global discovery of conserved genetic modules. *Science* **302**, 249–255.
Werhli, A., Grzegorczyk, M., and Husmeier, D. (2006). Comparative evaluation of reverse engineering gene regulatory networks with relevance networks, graphical Gaussian models and Bayesian networks. *Bioinformatics* **22**, 2523–2531.
Yang, Y. (2007). Consistency of cross validation for comparing regression procedures. *Ann. Stat.* **35**, 2450–2473.
Young, R. A. (2011). Control of the embryonic stem cell state. *Cell* **144**, 940–954.

CHAPTER 11

Quantitative Models of the Mechanisms that Control Genome-Wide Patterns of Animal Transcription Factor Binding

Tommy Kaplan[*,†] **and Mark D. Biggin**[‡]

[*]Department of Molecular and Cell Biology, California Institute of Quantitative Biosciences, University of California, Berkeley, California, USA

[†]School of Computer Science and Engineering, The Hebrew University, Jerusalem, Israel

[‡]Genomics Division, Lawrence Berkeley National Laboratory, Berkeley, California, USA

0091-679X/10 $35.00
DOI 10.1016/B978-0-12-388403-9.00011-4

Abstract

Animal transcription factors drive complex spatial and temporal patterns of gene expression during development by binding to a wide array of genomic regions. While the *in vivo* DNA binding landscape and *in vitro* DNA binding affinities of many such proteins have been characterized, our understanding of the forces that determine where, when, and the extent to which these transcription factors bind DNA in cells remains primitive.

In this chapter, we describe computational thermodynamic models that predict the genome-wide DNA binding landscape of transcription factors *in vivo* and evaluate the contribution of biophysical determinants, such as protein–protein interactions and chromatin accessibility, on DNA occupancy. We show that predictions based only on DNA sequence and *in vitro* DNA affinity data achieve a mild correlation ($r = 0.4$) with experimental measurements of *in vivo* DNA binding. However, by incorporating direct measurements of DNA accessibility in chromatin, it is possible to obtain much higher accuracy ($r = 0.6$–0.9) for various transcription factors across known target genes. Thus, a combination of experimental DNA accessibility data and computational modeling of transcription factor DNA binding may be sufficient to predict the binding landscape of any animal transcription factor with reasonable accuracy.

I. Introduction

Animal transcription factors each bind to many thousands of DNA regions throughout the genome in cells (Boyer *et al.*, 2005; Georlette *et al.*, 2007; MacArthur *et al.*, 2009; Robertson *et al.*, 2007; Zeitlinger *et al.*, 2007; reviewed by Biggin, 2011). While many of the most highly occupied regions are functional *cis*-regulatory regions and are evolutionarily conserved, many thousands of other genomic regions that are bound at lower levels *in vivo* do not appear to be functional targets (Carr and Biggin, 1999; MacArthur *et al.*, 2009). It is, therefore, a critical challenge to quantitatively predict the DNA binding levels of regulatory transcription factors in cells and to determine the biochemical mechanisms that direct these complex patterns of factor occupancy.

Animal transcription factors recognize short (5–12 bp) sequences of DNA that occur with high frequency throughout the genome (Wunderlich and Mirny, 2009), yet most occurrences of these recognition sites are not detectably bound *in vivo* (Carr and

Biggin, 1999; Li *et al.*, 2008; Liu *et al.*, 2006). There are several mechanisms that could account for this discrepancy between predicted and observed transcription factor DNA binding in cells. Competitive inhibition of binding at those DNA recognition sites that overlap sequences occupied either by other sequence-specific factors (Stanojevic *et al.*, 1991) or nucleosomes (Agalioti *et al.*, 2000; Cosma *et al.*, 1999; Narlikar *et al.*, 2002) could selectively inhibit DNA occupancy at these sites. In addition, direct or indirect cooperative interactions between transcription factors bound at close by recognition sites could increase their occupancy at other genomic locations (Buck and Lieb, 2006; Mann *et al.*, 2009; Miller and Widom, 2003; Zeitlinger *et al.*, 2003). Here we describe a computational modeling strategy that can analyze the relative influence of each of these biochemical mechanisms on the overall pattern of transcription factor DNA binding *in vivo* (Kaplan *et al.*, 2011). A glossary is provided to explain key technical terms used in describing the computational modeling (Section VII).

II. Overview of Model/Algorithm

A. Alternate Modeling Strategies

Many computational algorithms have been developed for predicting *in vivo* DNA binding. Crudely, these studies fall into two categories:

Qualitative models aim at identifying statistically significant occurrences of DNA binding sites or *cis*-regulatory regions (Agius *et al.*, 2010; Ernst *et al.*, 2010; Frith *et al.*, 2001; Granek and Clarke, 2005; Narlikar *et al.*, 2007; Narlikar and Ovcharenko, 2009; Rajewsky *et al.*, 2002; Ramsey *et al.*, 2010; Schroeder *et al.*, 2004; Sinha, 2006; Sinha *et al.*, 2003; Ward and Bussemaker, 2008; Whitington *et al.*, 2009; Won *et al.*, 2010). These computational methods usually rely on modeling the background distribution of transcription factor DNA recognition sites and focus on identifying significant p-values, that is, sites where the background hypothesis is rejected. These algorithms can identify a subset of *cis*-regulatory binding sites and provide a putative transcriptional regulatory architecture for an organism by connecting regulators to a set of putative target genes. They are less adequate, however, for predicting the levels of transcription factor DNA occupancy, which has been shown to be critical for relating DNA binding patterns to biological function (Carr and Biggin, 1999; MacArthur *et al.*, 2009).

Quantitative models, on the other hand, estimate the occupancy of a factor along the genome. Statistically, they aim to calculate the binding probability (hence, the percent of time or cells) at which the protein binds a specific DNA locus. These methods are, thus, more suitable for modeling the continuous quantitative landscape of transcription factor DNA occupancy as measured by genome-wide chromatin immunoprecipitation (ChIP) studies. An additional advantage of the quantitative approach is its natural generative probabilistic settings, which allow for easy integration of external data, such as chromatin state, the concentration of transcription factors in cells, or interactions between neighboring proteins (He *et al.*, 2009; He *et al.*, 2010; Roider *et al.*, 2007; Wasson and Hartemink, 2009).

In addition to direct quantitative models of transcription factor DNA binding, a related set of models have focused on predicting the gene transcription patterns driven by predefined DNA *cis*-regulatory regions. These studies generally use thermodynamic models to predict transcription factor DNA binding within known *cis*-regulatory regions as well as the resulting expression patterns driven by these target regions in animal embryos (He *et al.*, 2010; Kazemian *et al.*, 2010; Raveh-Sadka *et al.*, 2009; Segal *et al.*, 2008; Zinzen *et al.*, 2006). Three-dimensional changes in the concentration of regulatory transcription factors result in differential occupancy at the same DNA locus over different nuclei, which in turn results in different expression outputs in each cell. Unfortunately, these models do not explicitly train their models (or test them) on experimental *in vivo* DNA binding data, and limit their scope to predict the expression levels driven by specific target genes or *cis*-regulatory regions across the embryo. Therefore, their success in predicting *in vivo* DNA occupancy cannot be directly assessed.

B. Generalized Hidden Markov Models

Most direct quantitative algorithms for predicting transcription factor DNA binding rely on a probabilistic framework based on generalized hidden Markov models (gHMMs). These models use inference algorithms to estimate the occupancies (or DNA binding probability) of one or more transcription factors across any DNA sequence given their concentrations and protein–DNA binding specificities (Frith *et al.*, 2001; Granek and Clarke, 2005; Kulp *et al.*, 1996; Raveh-Sadka *et al.*, 2009; Segal *et al.*, 2008; Sinha *et al.*, 2003; Wasson and Hartemink, 2009).

We have adopted a form of gHMM for modeling transcription factor-DNA binding *in vivo,* as this class of model offers several advantages. These models have very few parameters and are therefore straightforward to optimize. Unlike most probabilistic graphical models, they offer exact inference of posterior probabilities in linear time, using a forward-backward dynamic programming algorithm (Durbin *et al.*, 1998; Rabiner, 1989). Finally, gHMMs are related to thermodynamic equilibrium models: they view the ensemble of all possible configurations of bound factors along the DNA as a Boltzmann distribution in which each configuration is assigned a weight (or probability) depending on its energetic state; the probability that a factor is bound at a specific location is calculated by summing the probabilities of all configurations in which it is bound (Ackers *et al.*, 1982; Buchler *et al.*, 2003; Granek and Clarke, 2005; Rajewsky *et al.*, 2002; Schroeder *et al.*, 2004; Segal *et al.*, 2008; Sinha, 2006; Wasson and Hartemink, 2009).

On the other hand, gHMMs are limited in their modeling power due to their Markovian property: these models lack any memory for past states, and so when estimating the probability of binding at a certain position the model is agnostic of other (nonoverlapping) DNA binding sites. This prevents this class of models from considering the full context in which DNA binding occurs. We will address this limitation below and offer an approximation to allow a full thermodynamic model using sampling procedures.

C. Experimental Datasets

We demonstrated our approach by modeling the genomic binding of five regulators of early embryonic anterior-posterior (A-P) patterning in *Drosophila melanogaster*: Bicoid (BCD), Caudal (CAD), Hunchback (HB), Giant (GT), and Kruppel (KR).

ChIP-seq data for the five factors in stage 5 blastoderm embryos were used to provide the measure of *in vivo* DNA occupancy (Bradley *et al.*, 2010). 20-bp-long sequence reads were mapped to the genome (Apr. 2006 assembly, BDGP Release 5). To minimize mapping noise, we only considered reads uniquely mapped to the genome with up to one mismatch. The mapped reads were then extended according to their orientation to a length of 150 bp, and binned (down-sampled) to a 10 bp resolution. Finally, the genomic binding landscape of each factor was smoothed using a running window of 10 bins (or 100 bp), to account for sampling noise.

DNA binding affinities of the five factors (expressed as position weight matrices – PWMs) were derived from previous *in vitro* measurements that used SELEX-Seq (Berkeley *Drosophila* Transcription Network Project, unpublished data; (MacArthur *et al.*, 2009) (Fig. 1). The PWM counts were normalized to probabilities,

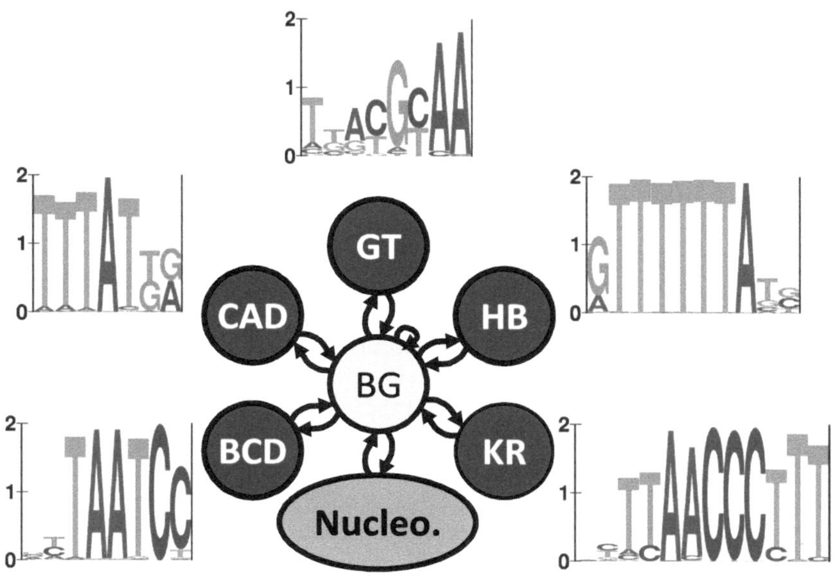

Fig. 1 The generalized hidden Markov model. Diagram of the model states, including the unbound background (BG) state, five states corresponding to the five transcription factors in the model (BCD, CAD, GT, HB, and KR), and a 141-bp-long nucleosomal binding state (Nucleo.). The emission probabilities of each transcription factor state are visualized using sequence logos that are based on position weight matrices (PWMs). (See color plate.)

after adding a pseudo-count of 0.01 to avoid zero probabilities (available at http://bdtnp.lbl.gov/gHMM). Additional sources of PWMs (Noyes *et al.*, 2008; Segal *et al.*, 2008) were also tested in the model, yielding similar results (Kaplan *et al.*, 2011). In all cases, the DNA binding specificities defined by these various experiments (i.e., the PWMs) were maintained and were not optimized as parameters in the model. We found that the tradeoff between having potentially more accurate PWMs (and a better fit to experimental *in vivo* DNA binding data) versus the cost of additional parameters to optimize was not beneficial for most factors. Moreover, using fixed PWMs from external studies prevents overfitting or drift toward additional motifs that are often present near developmental regions, such as the CAGGTAG sequence known to be bound by the transcription factor Zelda (Bradley *et al.*, 2010).

The accessibility of DNA in chromatin was obtained from DNase-seq data resulting from the DNase I digestion of isolated stage 5 blastoderm embryo nuclei (Li *et al.*, 2011; Thomas *et al.*, 2011). 34-bp-long reads were mapped to the genome by requiring unique matches with no more than two mismatches, then these were extended to a length of 150 bp, binned (down-sampled) to a 10 bp resolution, and smoothed using a running window of 10 bins.

Estimates of transcription factor protein concentrations in each nucleus were derived from three-dimensional fluorescence microscopy of *D. melanogaster* embryos at early stage 5 (Fowlkes *et al.*, 2008).

All of the above experimental datasets are available from the supplemental website for Kaplan *et al.* (2011) at http://bdtnp.lbl.gov/gHMM.

D. Model Overview

Hidden Markov models are probabilistic frameworks where the observed data (such as, in our case the DNA sequence) are modeled as a series of outputs (or emissions) generated by one of several (hidden) internal states. The model then uses inference algorithms to estimate the probability of each state along every position along the observed data. In our case, the model is composed of the various states that the DNA could be in: unbound (the background state), bound by transcription factor t_1, bound by transcription factor t_2, etc., or wrapped around a nucleosome (Fig. 1). Each state holds some probability distribution of the DNA sequences it favors (and emits according to the HMM). In our case, the background state is derived using the simple mononucleotide (single base) probability (frequency) in the genome to model the A/T distribution along the noncoding parts of the genome. The "bound" states hold a probabilistic DNA model that represents the sequences that each protein prefers to bind (its recognition sites). Additional parameters of the gHMM include the prior probabilities of entering each state, which are modeled using the transition probabilities between states. For example, a highly expressed protein that is more likely to be in the bound state along the DNA will have a higher transition probability than a protein present at lower concentrations in cells. Once the parameters of the gHMM are optimized (using a held-out set of training sequences) and given a new DNA sequence, it is straightforward to infer the probability of each state (unbound,

bound by factor t_1, bound by factor t_2, etc.) at each position along the sequence. See Section V for further details of these models.

All our computational models estimate the DNA binding probability of each transcription factor at a single-nucleotide resolution. A model-based algorithm is then used to transform these predictions into smoothed ChIP-like landscapes so they can be compared to the *in vivo* ChIP-seq measurements of protein–DNA binding (Fig. 2). For this, the length distribution of DNA fragments recovered by the ChIP process is used to simulate the overall shape of one peak, corresponding to a single DNA binding event measured by ChIP-seq. For a length distribution *c(l)*, the estimated shape F of a peak is described as:

$$F(\Delta_x) \propto \sum_{l=\Delta_x}^{\infty} c(l)$$

where Δx denotes the relative distance from the binding locus or peak center. In other words, the probability of obtaining a read Δx bp away from the binding event is proportional to the total number of reads at least Δx bp long (Capaldi *et al.*, 2008; Kaplan *et al.*, 2011).

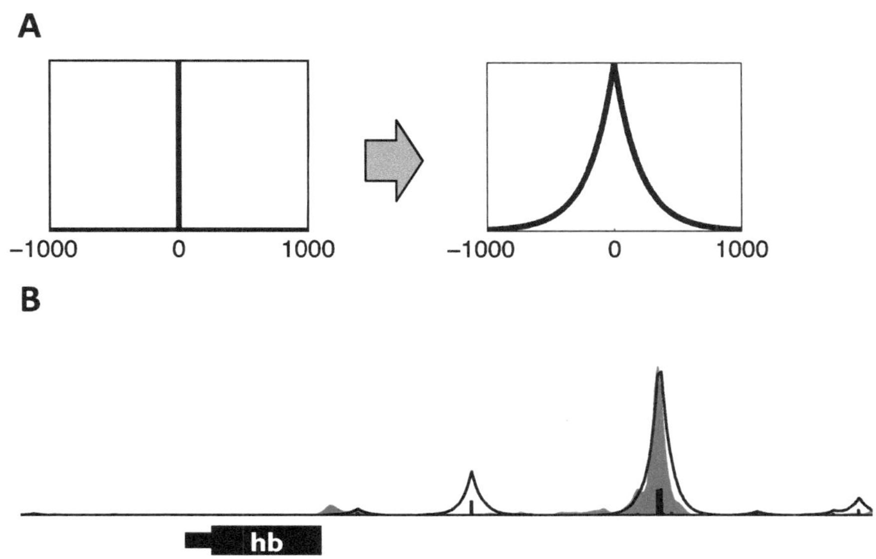

Fig. 2 From DNA binding probabilities to ChIP landscape. **(A)** Each DNA binding event (left) was transformed to a model-based estimation of expected ChIP peak shape based on the average length of the DNA fragments immunoprecipitated in the ChIP experiment (right) (Kaplan *et al.*, 2011). **(B)** This model was then used to convolve the model's binding predictions (vertical black bars) to the expected landscape of ChIP sequencing assay (thin black line), which was then compared to the measured *in vivo* DNA binding landscape (gray shaded landscape).

III. Biological Insights

A. A Simple Model is Mildly Successful

We began with the simplest model – a single transcription factor binding to DNA. This required optimizing only a single parameter, $P(t)$, for each transcription factor that corresponds to its effective concentration in nuclei and assuming, for this first simple case at least, that the protein is expressed at the same concentration in all embryo cells. We used standard optimization techniques (based on a combination of genetic algorithms and gradient ascent-based algorithm) to optimize these parameters (see section V). For each tested value of $P(t)$, we used the generalize hidden Markov model to estimate the binding probability per position, and then convoluted these predictions into the predicted DNA binding landscape.

To analyze our predictions, we compiled a list of 21 known target loci of the A-P patterning system. Each target gene was expanded by ~10 Kb upstream and downstream of the transcription unit to capture its *cis*-regulatory regions. In each analyses presented in this chapter, we trained the model parameters to optimize the fit between the predicted and the observed ChIP-seq landscapes at a set of six loci, which spanned ~87 Kb, and evaluated the trained model on the remaining set of 15 loci, which spanned ~280 Kb. To account for long genomic regions where no DNA binding is observed *in vivo* by ChIP, the training and test sets were enhanced by addition of three or five control regions, spanning a total of 100 and 221 Kb, respectively (Kaplan *et al.*, 2011).

After parameter optimization using the training set, the model was applied to the test set. The predicted DNA binding landscape around one gene in the test set is shown in Fig. 3A. The total correlation between the model predictions and measured data was quite weak when averaged over all ~500 kb of the test set ($r = 0.36$), with specific factors varying from $r = 0.15$ (GT) to $r = 0.66$ (BCD) (Kaplan *et al.*, 2011).

In addition to estimating accuracy using the correlation between the model's predictions and experimentally measured *in vivo* DNA binding, we also tried two alternatives. In one, we used distance-based measures such as the root mean square deviation (RMSD) between the predicted and measured genomic landscapes. In the second, we tried a peak-centric comparison method, where a peak calling algorithm was used to identify "bound regions" in both the predicted and the measured data and then the overlap between called peaks was compared. These alternate scoring methods resulted in qualitatively similar results to the correlation coefficients given in the rest of the text and in Fig. 4.

B. Allowing Transcription Factor Competition Does Not Improve the Predictions' Accuracy

Encouraged by the results with each transcription factor considered singly, we examined the effect of DNA binding site competition between the five factors on our ability to predict *in vivo* DNA occupancy. Overlapping DNA recognition sites can

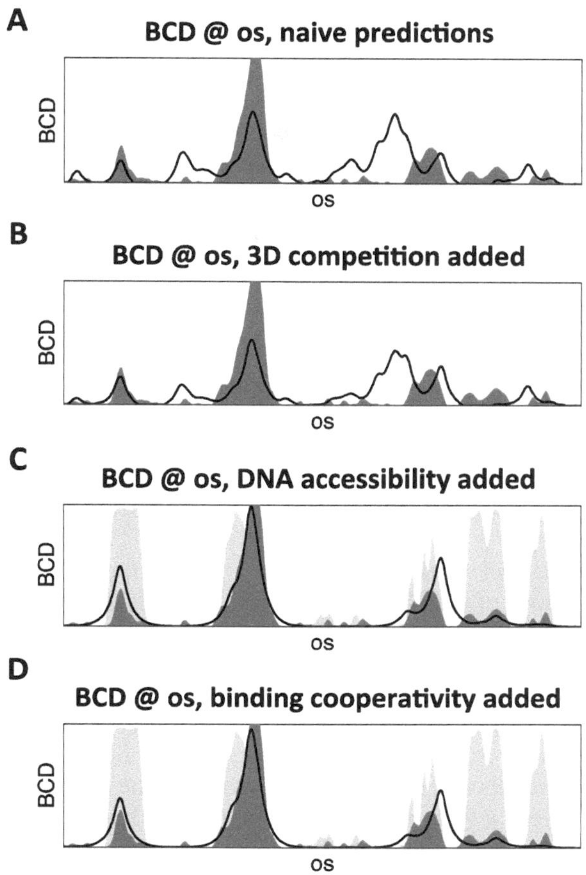

Fig. 3 High-resolution predictions of protein–DNA binding landscape. **(A)** The model's DNA binding predictions (thin black line) for BCD are compared to the measured *in vivo* DNA binding landscape (dark shaded landscape) across the 15 Kb around the *os* locus. In this example, the BCD binding landscape was predicted without considering the other transcription factors. **(B)** Same as (A), except that direct DNA binding competition between the five factors and with nucleosomes was allowed, and BCD binding was modeled independently in each of 6,078 nuclei of the stage 5 blastoderm embryo. **(C)** Same as (B), but also incorporating a nonuniform DNase I hypersensitivity-based prior on transcription factor binding to account for variations in DNA accessibility (shown as light shaded landscape). **(D)** Same as (C), but further adding cooperative interactions between adjacently bound transcription factor molecules in a thermodynamic setting.

allow direct competition between transcription factors (Stanojevic *et al.*, 1991). Moreover, overlapping sites are often conserved at long evolutionary distances, suggesting an important role for inter-factor competition (Hare *et al.*, 2008). Therefore, we expanded the gHMM in our model to consider all five transcription

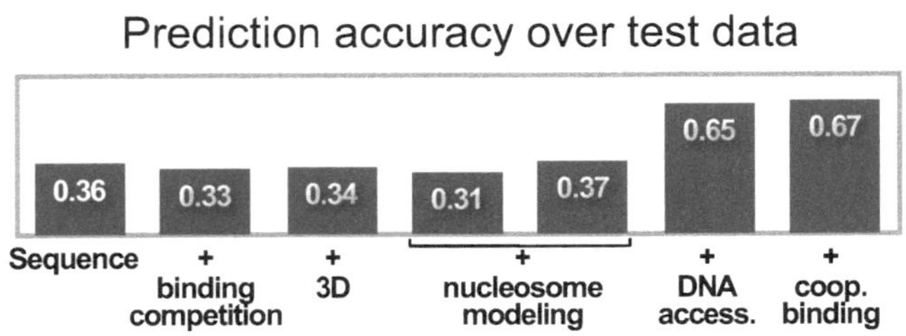

Fig. 4 Prediction accuracy at increasing degrees of model complexity. Accuracy of DNA binding predictions for the test set of 15 known A-P targets and five control loci. Shown are the correlation coefficients between model prediction and measured *in vivo* DNA binding landscape for increasing degrees of model complexity. These are, from left to right: independent predictions per transcription factor using our simplest model; allowing DNA binding site competition between transcription factors; making predictions at a single-nucleus resolution; including nucleosomes using a sequence-specific or a sequence-independent model of nucleosome binding; adding a nonuniform prior on transcription factor binding using DNA accessibility measurements; and adding cooperative DNA binding interactions in a thermodynamic setting.

factors simultaneously in a probabilistic framework (Fig. 1), where the concentrations of each factor t is modeled by an additional probabilistic term $P(t)$. In the single transcription factor model, binding of one protein to a recognition site did not affect the DNA occupancy of a different transcription factor at an overlapping site. In this new model, however, because the total occupancy at a site cannot exceed 1, transcription factors effectively compete for DNA binding to overlapping recognition sites. Surprisingly, this competitive model gave slightly less accurate predictions than its single factor counterpart. On the test data, the model's predictions decreased from a total correlation of 0.36 to 0.33 (see Fig. 4).

C. Expanding the Model to Three Dimensions With Single Nucleus Resolution Has Only Slight Effects

One reason why the model did not improve when competition was allowed could have been that, because we treated the embryo as a homogenous entity, the model allowed competition between transcription factors that are not expressed together at high levels in the same cells. We therefore expanded our algorithm to model the DNA binding of all transcription factors in each of the ~6000 nuclei of the embryo separately. To scale the optimized concentration parameters of the five transcription factors for each nuclei, we further scaled the prior probability $P(t)$ of every transcription factor t proportionally to its protein expression level, as measured at a single-cell resolution (Fowlkes *et al.*, 2008). We then averaged the predicted DNA binding landscape of all nuclei to obtain whole-embryo genomic predictions, which were then compared to the (whole-embryo average) *in vivo* DNA binding measurements from ChIP-seq (Kaplan *et al.*, 2011). This slightly improved the predictions relatively to the whole-embryo predictions (Figs. 3B and 4). However, combining

DNA binding site competition and 3D expression data yields a model that is only about as effective as the simplest model. Thus, while competition between transcription factors is likely important at a subset of recognition sites, it does not appear to be a principal determinant of the overall distribution of transcription factor DNA occupancy *in vivo*.

D. Predicting Nucleosome Location Does Not Improve the Model's Predictive Power

To test if chromatin state influences the accuracy of our model, we first attempted to predict the locations of nucleosomes to enable modeling of the competition between transcription factors and nucleosomes in binding to DNA (Narlikar *et al.*, 2007; Raveh-Sadka *et al.*, 2009; Wasson and Hartemink, 2009). As there are no direct measurements of nucleosome positions from early *Drosophila* embryos, we modeled these computationally. We extended our Markov model to represent the sequence bound by a single nucleosome. This was done by including an additional state in the gHMM that comprised a sequence-independent model of nucleosome DNA binding in which nucleosomes are viewed as long "space-fillers" that, when present, prevent regulators from binding to DNA. We used a 141-bp long model of nucleosome binding, based on a fixed distribution of nucleotides as in the background state P_B of the Markov model (0.32 for A/T, 0.18 for G/C). Similarly to the transcription factor states, the nucleosomal state was assigned a prior probability term $P(t)$ to reflect a fixed nucleosomal concentration along the embryo. $P(t)$ was optimized together with other concentration-related parameters $P(t)$ for all transcription factors. Alternatively, due to uncertainty in the literature about the contribution of DNA sequence specificity to *in vivo* nucleosome positioning, we also tested a sequence-specific model of nucleosome binding (Segal *et al.*, 2006). Neither of these nucleosomal models dramatically improved the DNA binding predictions for the five transcription factors (Fig. 4).

E. DNA Accessibility Data Greatly Improve DNA binding Predictions

A weakness of the above strategies to predict nucleosome location is that only one constitutive model is derived for all cells of the organism for all stages of development. Yet it is known that chromatin accessibility varies dramatically over time and between cells (Kharchenko *et al.*, 2011; Thomas *et al.*, 2011). Therefore, we sought to exploit direct genome-wide measurements of DNA accessibility for the same developmental stage from which the ChIP-seq data were derived (Li *et al.*, 2011; Thomas *et al.*, 2011). Interestingly, when we compared these DNA accessibility data to the predictions of the original, simple version of our gHMM, we found that the model correctly predicts DNA binding on the most highly accessible genomic regions but tended to predict stronger DNA binding than was actually measured on less accessibility regions (Kaplan *et al.*, 2011). We therefore leveraged the statistical framework of generalized hidden Markov models and incorporated

DNA accessibility data into the model as a nonuniform prior probability of regulatory binding along the genome – with regions of low accessibility being given a greatly reduced probability of binding.

The incorporation of differential DNA accessibility in this way dramatically boosted the model's accuracy by almost twofold to a correlation of $r = 0.67$ with the measured *in vivo* occupancy data when averaged over all the ~500 kb test set, with the factor-specific correlation varying from 0.58 (HB) to 0.79 (BCD) (Figs. 3C and 4).

In addition to the sigmoidal prior described in Section V, we investigated additional methods to transform the DNA accessibility data DD_x into probabilities PD_x. First, we tried to linearly scale the accessibility data DD_x and limit the maximal PD_x values at one. This resulted with slightly less accurate predictions ($r = 0.66$ on test data). Also, we tried an even simpler model using a step function, namely modeling PD_x as one value below some minimal value of DD_x, and another value above it. Even this naive model achieved comparable accuracy, at $r = 0.64$. This slightly reduced correlation suggests that the effect of DNA accessibility on transcription factor binding may be almost binary – low accessibility regions show almost no regulatory binding, while binding at accessible regions is modeled quite accurately by DNA sequence alone (Kaplan *et al.*, 2011).

F. Modeling Direct Cooperative DNA Binding Does Not Affect Model Performance

Although our predictions that included DNA accessibility data were reasonable, we sought to further refine our model by considering factor-factor interactions other than the simple direct competition (via overlapping recognition sites) described earlier. For example, direct physical interactions between transcription factors bound at neighboring recognition sites have often been found to increase the occupancy of one or both proteins on DNA, for both homomeric and heteromeric cooperative interactions, and to sharpen the regulatory response to changes in transcription factor concentration (Arnosti *et al.*, 1996; Small *et al.*, 1992).

Generalized hidden Markov models, however, have limited ability to model the broader context of DNA binding events, including cooperative interactions between neighboring sites. We therefore added a second, sampling-based phase to our computational model. In this phase, a large ensemble of DNA binding configurations is sampled, each with a different set of protein–DNA interactions. The probability of each configuration is then estimated based on all pairs of nearby occupied sites (up to 95 bp apart) and the parameterized energetic gain of each pair. Finally, the overall DNA binding probability at each position is quantified as a weighted sum of all sampled configurations.

By adopting a statistical mechanics perspective, the exponential space of protein–DNA binding configurations can be viewed as a canonical ensemble in a thermodynamic equilibrium. Here, the probability of each configuration is directly linked to its energetic state, including direct protein–DNA interactions, steric hindrance constraints, and cooperative interactions with neighboring factors (Ackers *et al.*, 1982; Segal *et al.*, 2008). We extended our model to capture cooperative interactions between transcription factors using a novel set of 15 parameters (one for each nonredundant pair of the five factors), modeling the energy gain for the nearby binding of every possible pair of the five transcriptional regulators in our model.

The optimized set of cooperative DNA binding parameters includes predictions of interactions between many homomeric and heteromeric pairs (Kaplan *et al.*, 2011). These cooperativity parameters improved the predictive power of the model to a correlation of $r = 0.67$ on the test data, ranging from $r = 0.58$ (HB) to $r = 0.79$ (BCD), a marginal improvement over the Markovian approach (Figs. 3D and 4). Thus, our model suggests that cooperative interactions between transcription factor molecules have a rather limited contribution in shaping the genomic landscape of *in vivo* DNA binding (Kaplan *et al.*, 2011).

G. Implications for Determining Transcription Factor DNA Occupancy *in vivo*

The increasing availability of genome-wide *in vivo* measurements of DNA accessibility (via DNase I, FAIRE) for a variety of cell types, developmental stages, and environmental conditions, together with the laborious nature of direct ChIP measurements, suggests a mixed computational-experimental streamlined strategy for estimating the genome-wide binding landscape of proteins. While we often fail to predict transcription factor DNA binding levels from DNA sequence and *in vitro* DNA affinity measurements alone, by incorporating DNA accessibility data into a thermodynamic model, a reasonable job of quantitatively predicting the occupancy of transcription factors can be made. While such an approach should not be viewed as a substitute for systematic experimental measurement of transcription factor DNA binding *in vivo*, we believe our predictions are good enough to be useful when such experimental data are unavailable or impractical to obtain.

IV. Open Challenges

Quantitative computational models of sequence-specific protein–DNA interactions offer a fast approximation of the genomic landscape of protein–DNA binding. Nonetheless, these predictions are still far from being reliable enough to fully replace experimental *in vivo* measurements.

One of the greatest challenges for improving future models is in modeling locus-specific DNA accessibility using genomic DNase I hypersensitivity data. Our current models rely on a probabilistic platform, in which we tested various ways to transform read coverage into *a priori* DNA binding probabilities, with a sigmoid function being the most useful. While this approach worked well on relatively accessible regions (*cis*-regulatory regions and the regions flanking actively expressed genes), it was not as accurate on a full genomic scale, giving a correlation coefficient of only 0.33 for an entire chromosome arm (Kaplan *et al.*, 2011). Most false predictions arose from *bona fide* sites predicted to be strongly bound, but which show limited or no binding *in vivo* due to limited accessibility. In addition, we observed some highly accessible regions bound by several transcription factors, even in the absence of cognate sequence recognition sites. We believe that optimizing the transformation from DNase I read densities into binding probabilities at very low and very high DNase-seq read densities could strengthen the model.

A second challenge is to improve the modeling of cooperative DNA binding (both direct and indirect). In our work to date, we applied a somewhat simple approach, where two nearby transcription factor molecules contribute some constant energetic value only if they bind in close proximity (<95 bp). It seems probable that more sophisticated methods, with a greater number of parameters, could model the biological/physical effect with greater accuracy.

Wasson and Hartemink (2009) recently used hidden Markov models to analyze transcription factor DNA binding in yeast and showed that their predictions improve as more sequence-specific transcription factors are added to the model. While we did not observe this trend with our data, possibly because we only analyzed five transcription factors, revisiting this approach with a greater number of transcription factors could be revealing.

Finally, while the direct goal of the work described in this chapter was to predict *in vivo* DNA binding from DNA sequence, *in vitro* affinity, and chromatin accessibility data, a more challenging question is to understand and predict *de novo* how dynamic patterns of DNA accessibility are themselves generated in cells. This may require correctly modeling the activities of hundreds of sequence-specific transcription factors, the chromatin remodeling proteins that they recruit, nucleosomes, and other chromatin proteins. We doubt that sufficient data or knowledge is available to yet take up this task.

V. Computational Methods

A. Generalized Hidden Markov Models

Generalized hidden Markov models were used to predict transcription factor DNA binding based on the factor concentration and the DNA sequence. We followed a thermodynamic rationale, and considered the space of all valid DNA binding configurations as a Boltzmann distribution. Under this statistical framework, the probability of each configuration, P_i, is proportional to its energetic state E_i

$$P_i \propto e^{-\beta E_i}$$

where β equals $1/k_B T$, with k_B being the Boltzmann constant and T the temperature (25 °C).

The energetic state of each configuration could therefore be calculated from its binding probability. Under this model, bound nucleotides are generated according to the protein–DNA binding preference, or PWM, of the transcription factor. The probability of a subsequence S_i to be bound by transcription factor t equals

$$P_t(S_i) = P(t) \prod_{j=0}^{l_t - 1} P_j(S_{i+j} | \theta_t)$$

with $P(t)$ being the *a priori* binding probability of transcription factor t, l_t the length of the binding site for factor t, and $P_j(S_{i+j}|\theta_t)$ corresponds to the probability of the nucleotide S_{i+j}, at the j position of a binding site for factor t, as modeled by its recognition parameters θ_t. Unbound nucleotides are generated from a mononucleotide background distribution P_B (0.32 for A/T, 0.18 for G/C).

It is useful to visualize this family of models as a series of probabilistic transitions between the internal states of the model (Fig. 1). The different types of DNA sequence (unbound DNA; DNA bound by factor t, etc.) are the nodes, and the allowed transitions between states are shown as arrows in the figure. The parameters of the model correspond to the probabilities of transition between states. Each state is associated with one transcription factor; the probability of the corresponding DNA subsequence is calculated using its binding site model $P_j(S_i)$. Each configuration is viewed as one path along the internal states of the model, starting in one state at the beginning of the DNA sequence, and transitioning among the states until the end of the sequence. The full binding configuration of DNA sequence S, with multiple factors $t_1,\ldots,t_k$ bound at positions $x_1\ldots x_k$, respectively, is viewed as a path that loops into the unbound state along most of the DNA sequence except for positions $x_1\ldots x_k$ where it enters the states corresponding to the transcription factors $t_1,\ldots,t_k$. We can then write the probability of this path as:

$$P(S) = P_B(S)\prod_{i=1}^{k} P(t)\frac{P_{t_i}(S_{x_i})}{P_B(S_{x_i})}$$

Note that no overlapping binding sites are allowed in each configuration. To further account for steric hindrance, each PWM was extended by two flanking regions of 3 bp. These were modeled by a nonspecific background distribution P_B (0.32 for A/T, 0.18 for G/C). The minimal distance between two occupied sites in one binding configuration is therefore 7 bp (two 3 bp flanks plus a 1 bp transition through the unbound state).

To infer the overall binding probability of each transcription factor at each DNA position, one must account for the exponentially large number of possible configurations, while weighting each configuration based on its probability. While this task seems difficult at first, it can be solved in a linear time using the dynamic programming inference algorithm (Durbin *et al.*, 1998; Rabiner, 1989). Specifically, we use the forward-backward algorithm. First, we calculate the local probabilities of each transcription factor t to bind DNA at each position i, $U_{t,i} = P(t) * P_t(S_i)$. We then calculate the Forward Potentials $F_{t,i}$ and the Backward Potentials $B_{t,i}$ by summing the probabilities of all configurations (paths) that end (for Forward Potentials) or begin (for Backward Potentials) at position i with a binding site of t. Finally, we calculate the exact *a posteriori* probability of transcription factor t bound at position i by multiplying the forward and backward potentials. This calculates the binding probability of factor t at position i, given all possible combinations of other transcription factors along the entire sequence S.

B. Model-based Simulation of Chromatin

We used a sigmoid transformation to convert the genomic landscape of DNase I hypersensitivity data DD_x (density of sequenced reads along the genome) into the *a priori* probability PD_x of entering a bound state at position x:

$$PD_x = \frac{1}{1 + e^{-\beta\, DD_x + a}}$$

The parameters of this equation, $\alpha = 6.008$ and $\beta = 0.207$, were optimized over the training data, separately from the concentration parameters in an iterative manner (piecewise optimization). Those probabilities PD_x are then multiplied by the prior probability of binding *P(t)* for each transcription factor t in order to calculate the actual transition probability into the bound state of transcription factor t at position x along the genome.

C. Thermodynamic Modeling of Protein–DNA Interactions Using Boltzmann Ensembles

To predict transcription factor DNA binding in a full thermodynamic setting we first used the generalized hidden Markov model to analyze the underlying sequence and predict proteins' DNA binding according to the different protein concentrations within each nucleus in the *Drosophila* blastoderm stage embryo. This was used to calculate an approximate map of DNA binding. To allow for cooperative interactions between the transcription factor molecules, we then used the DNA binding probabilities described above to sample 10,000 binding configurations per sequence/run and reweighted them to account for the energetic gain due to cooperative DNA binding interactions. This was done in a thermodynamic setting, where every configuration i was reweighted by W_i

$$W_i = \exp\left(- \sum_{|x_j - x_k| < 95} C_{j,k} \right)$$

where x_j and x_k are the binding locations of factors j and k, while $C_{j,k}$ corresponds to their optimized cooperativity parameter. The reweighted samples are then averaged, and the binding probability of every factor at every position is calculated. This combination of direct gHMM calculations followed by importance-weighted sampling allows us to approximate the full thermodynamic landscape of binding using a fast framework with few parameters.

D. Optimization of Model Parameters

We optimized all the parameters in our models by focusing on the train set loci and maximizing the correlation among the model predictions and the *in vivo* measurements of transcription factor DNA binding. The prior probabilities *P(t)* of entering into the bound state for each transcription factor, which reflect the nuclear protein concentration of each factor, were first optimized by a genetic optimization algorithm (Goldberg and Holland, 1988) with 25 generations and a population size of 15. We then further optimized the *P(t)* variables using a gradient-based trust-region algorithm (Steihaug, 1983).

VI. Glossary

Qualitative Models of DNA Binding Sites: A family of computational models aimed at identifying transcription factor binding sites along a given DNA sequence.

Quantitative Models of DNA Binding: A family of computational models aimed at estimating the occupancy of DNA-binding proteins along the positions of a given DNA sequence. For example, given an input sequence, a qualitative model may identify two putative recognition sites, while a quantitative model may predict that one of these sites is occupied twice as often (i.e., for longer periods of time) as the other.

Position Weight Matrix (PWM): A statistical representation of a DNA motif. Commonly used to model the DNA recognition element of a transcription factor, a PWM is a table of 4-by-N that records the probability of observing each of the four nucleotides at every position of the motif. These models assume independence between the N positions of the motif such that each nucleotide position is represented as a single column with the estimated probabilities for each of the four nucleotides. To calculate the probability of transcription factor binding at a DNA word of size N given the PWM, the probabilities given in the cells of the table that correspond to the nucleotide at each of the N positions of the word are multiplied.

Background Model of DNA: A statistical representation of DNA sequences, typically used as a negative control when scanning DNA for sequence motifs. These models typically model only the general nucleotide (A-T content) of the DNA and as a result are too weak to model the entire length of a sequence-specific binding site for transcription factors.

Thermodynamic Model: *According* to statistical thermodynamics, the relative amount of time a complex system with multiple states would spend in each state is related to its energetic states. Using a Boltzmann distribution, the energetic state of each configuration is used to estimate the probability of the system being in each state. For example, every position along a DNA sequence could be bound by many transcription factors, but it is more likely the system is usually in a "stable" state – such as no binding at all or binding of one or more proteins at their higher-affinity recognition sites.

Hidden Markov Model (HMM): A probabilistic framework for modeling a series of observations (in our case a DNA sequence) using a series of unobserved transitions between the internal states of the model. The parameters of the model include the probabilities of transition between the various states (the transition probabilities), and the probabilities for each of the possible outputs of each state (the emission probabilities). Using inference algorithms, HMMs are used to efficiently find the most probable explanation (path over the states of the model) of the data, or to infer the posterior probability of a given state at a given position.

Generalized Hidden Markov Model (gHMM): An extended class of HMMs that allow states with longer outputs as well as mute states with no output at all. We employ a gHMM with "bound" states that use PWM to model sequence-specific binding sites, and a "not-bound" state that uses a background model of DNA nucleotide distribution. Given a DNA sequence, we use the gHMM to infer which positions along the sequence are likely to correspond to the "bound" states and to what extent.

Prior and Posterior Probabilities: In Bayesian statistics, the prior and posterior probabilities estimate the likelihood of an event before or after we take evidence into account, respectively. For example, the prior probability of a given state in model corresponds to how often we believe a given transcription factor binds DNA in general, while the posterior probability of the protein's binding depends of the actual sequence of the DNA.

Dynamic Programming: A class of algorithms in computer science that solve certain problem by breaking them down into simpler overlapping subproblems.

Forward-Backward Algorithm: A dynamic programming inference algorithm for calculating the posterior probability of all states at all the positions of an input series of observations. Here, we use the algorithm to estimate the posterior binding probability of each transcription factor along a sequence of DNA. First, the algorithm calculates the probabilities of each state at any position given the DNA sequence from the start until that point (forward probabilities). It then calculates the probabilities of all states given the remaining part of the DNA sequence (background probabilities). Finally, these are combined to produce the posterior probability given the full sequence.

Further Reading

More details on our model and the implications of our analysis for transcription factor DNA binding can be found in Kaplan *et al.* (2011). A companion paper providing additional biochemical arguments suggesting that chromatin accessibility plays a more important role than direct heteromeric cooperative association between transcription factors in directing factor binding in cells can be found in Li *et al.* (2011). Finally, Biggin (2011) comprehensively reviews the relationship between the continuum of transcription factor DNA occupancy levels seen in animal cells and biological function.

References

Ackers, G. K., Johnson, A. D., and Shea, M. A. (1982). Quantitative model for gene regulation by lambda phage repressor. *Proc. Natl. Acad. Sci. USA.* **79**, 1129–1133.

Agalioti, T., Lomvardas, S., Parekh, B., Yie, J., Maniatis, T., and Thanos, D. (2000). Ordered recruitment of chromatin modifying and general transcription factors to the IFN-beta promoter. *Cell* **103**, 667–678.

Agius, P., Arvey, A., Chang, W., Noble, W. S., and Leslie, C. (2010). High resolution models of transcription factor-DNA affinities improve in vitro and in vivo binding predictions. *PLoS Comput. Biol.* 6.

Arnosti, D. N., Barolo, S., Levine, M., and Small, S. (1996). The eve stripe 2 enhancer employs multiple modes of transcriptional synergy. *Development* **122**, 205–214.

Biggin, M. D. (2011). Animal transcription networks as highly connected, quantitative continua. *Dev. Cell* **21**, 611–626.

Boyer, L. A., Lee, T. I., Cole, M. F., Johnstone, S. E., Levine, S. S., Zucker, J. P., Guenther, M. G., Kumar, R. M., Murray, H. L., Jenner, R. G., Gifford, D. K., Melton, D. A., Jaenisch, R., and Young, R. A. (2005). Core transcriptional regulatory circuitry in human embryonic stem cells. *Cell* **122**, 947–956.

Bradley, R. K., Li, X. -Y., Trapnell, C., Davidson, S., Pachter, L., Chu, H. C., Tonkin, L. A., Biggin, M. D., and Eisen, M. B. (2010). Binding site turnover produces pervasive quantitative changes in transcription factor binding between closely related Drosophila species. *Plos Biol.* **8**, e1000343.

Buchler, N. E., Gerland, U., and Hwa, T. (2003). On schemes of combinatorial transcription logic. *Proc. Natl. Acad. Sci. USA.* **100**, 5136–5141.

Buck, M. J., and Lieb, J. D. (2006). A chromatin-mediated mechanism for specification of conditional transcription factor targets. *Nat. Genet.* **38**, 1446–1451.

Capaldi, A., Kaplan, T., Liu, Y., Habib, N., Regev, A., Friedman, N., and O'Shea, E. (2008). Structure and function of a transcriptional network activated by the MAPK Hog1. *Nat. Genet.* **40**, 1300–1306.

Carr, A., and Biggin, M. D. (1999). A comparison of in vivo and in vitro DNA-binding specificities suggests a new model for homeoprotein DNA binding in Drosophila embryos. *EMBO J.* **18**, 1598–1608.

Cosma, M. P., Tanaka, T., and Nasmyth, K. (1999). Ordered recruitment of transcription and chromatin remodeling factors to a cell cycle- and developmentally regulated promoter. *Cell* **97**, 299–311.

Durbin, R., Eddy, S. R., Krogh, A., and Mitchison, G. (1998). *Biological Sequence Analysis: Probabilistic Models of Proteins and Nucleic Acids.* Cambridge University Press, Cambridge, UK New York.

Ernst, J., Plasterer, H. L., Simon, I., and Bar-Joseph, Z. (2010). Integrating multiple evidence sources to predict transcription factor binding in the human genome. *Genome Res.* **20**, 526–536.

Fowlkes, C. C., Hendriks, C. L. L., Keränen, S. V. E., Weber, G. H., Rübel, O., Huang, M. -Y., Chatoor, S., Depace, A. H., Simirenko, L., Henriquez, C., Beaton, A., Weiszmann, R., Celniker, S., Hamann, B., Knowles, D. W., Biggin, M. D., Eisen, M. B., and Malik, J. (2008). A quantitative spatiotemporal atlas of gene expression in the Drosophila blastoderm. *Cell.* **133**, 364–374.

Frith, M. C., Hansen, U., and Weng, Z. (2001). Detection of cis-element clusters in higher eukaryotic DNA. *Bioinformatics* **17**, 878–889.

Georlette, D., Ahn, S., MacAlpine, D. M., Cheung, E., Lewis, P. W., Beall, E. L., Bell, S. P., Speed, T., Manak, J. R., and Botchan, M. R. (2007). Genomic profiling and expression studies reveal both positive and negative activities for the Drosophila Myb MuvB/dREAM complex in proliferating cells. *Genes Dev.* **21**, 2880–2896.

Goldberg, D., and Holland, J. (1988). Genetic algorithms and machine learning. *Machine Learning* **3**, 95–99.

Granek, J. A., and Clarke, N. D. (2005). Explicit equilibrium modeling of transcription-factor binding and gene regulation. *Genome Biol.* **6**, R87.

Hare, E. E., Peterson, B. K., Iyer, V. N., Meier, R., and Eisen, M. B. (2008). Sepsid even-skipped enhancers are functionally conserved in Drosophila despite lack of sequence conservation. *PLoS Genet.* **4**, e1000106.

He, X., Chen, C. C., Hong, F., Fang, F., Sinha, S., Ng, H. H., and Zhong, S. (2009). A biophysical model for analysis of transcription factor interaction and binding site arrangement from genome-wide binding data. *PLoS One* **4**, e8155.

He, X., Samee, M. A., Blatti, C., and Sinha, S. (2010). Thermodynamics-based models of transcriptional regulation by enhancers: the roles of synergistic activation, cooperative binding and short-range repression. *PLoS Comput. Biol.* 6.

Kaplan, T., Li, X. Y., Sabo, P. J., Thomas, S., Stamatoyannopoulos, J. A., Biggin, M. D., and Eisen, M. B. (2011). Quantitative models of the mechanisms that control genome-wide patterns of transcription factor binding during early Drosophila development. *PLoS Genet.* **7**, e1001290.

Kazemian, M., Blatti, C., Richards, A., McCutchan, M., Wakabayashi-Ito, N., Hammonds, A. S., Celniker, S. E., Kumar, S., Wolfe, S. A., Brodsky, M. H., and Sinha, S. (2010). Quantitative analysis of the Drosophila segmentation regulatory network using pattern generating potentials. *PLoS Biol.* 8.

Kharchenko, P. V., Alekseyenko, A. A., Schwartz, Y. B., Minoda, A., Riddle, N. C., Ernst, J., Sabo, P. J., Larschan, E., Gorchakov, A. A., Gu, T., Linder-Basso, D., Plachetka, A., Shanower, G., Tolstorukov, M. Y., Luquette, L. J., Xi, R., Jung, Y. L., Park, R. W., Bishop, E. P., Canfield, T. K., Sandstrom, R., Thurman, R. E., MacAlpine, D. M., Stamatoyannopoulos, J. A., Kellis, M., Elgin, S. C., Kuroda, M. I., Pirrotta, V., Karpen, G. H., and Park, P. J. (2011). Comprehensive analysis of the chromatin landscape in *Drosophila melanogaster. Nature* **471**, 480–485.

Kulp, D., Haussler, D., Reese, M. G., and Eeckman, F. H. (1996). A generalized hidden Markov model for the recognition of human genes in DNA. *Proc. Int. Conf. Intell. Syst. Mol. Biol.* **4**, 134–142.

Li, X. -Y., MacArthur, S., Bourgon, R., Nix, D., Pollard, D. A., Iyer, V. N., Hechmer, A., Simirenko, L., Stapleton, M., Hendriks, C. L. L., Chu, H. C., Ogawa, N., Inwood, W., Sementchenko, V., Beaton, A., Weiszmann, R., Celniker, S. E., Knowles, D. W., Gingeras, T., Speed, T. P., Eisen, M. B., and Biggin, M. D. (2008). Transcription factors bind thousands of active and inactive regions in the Drosophila blastoderm. *PLoS Biol.* **6**, e27.

Li, X. Y., Thomas, S., Sabo, P. J., Eisen, M. B., Stamatoyannopoulos, J. A., and Biggin, M. D. (2011). The role of chromatin accessibility in directing the widespread, overlapping patterns of Drosophila transcription factor binding. *Genome Biol.* **12**, R34.

Liu, X., Lee, C. K., Granek, J. A., Clarke, N. D., and Lieb, J. D. (2006). Whole-genome comparison of Leu3 binding in vitro and in vivo reveals the importance of nucleosome occupancy in target site selection. *Genome Res.* **16**, 1517–1528.

MacArthur, S., Li, X. -Y., Li, J., Brown, J. B., Chu, H. C., Zeng, L., Grondona, B. P., Hechmer, A., Simirenko, L., Keränen, S. V. E., Knowles, D. W., Stapleton, M., Bickel, P., Biggin, M. D., and Eisen, M. B. (2009). Developmental roles of 21 Drosophila transcription factors are determined by quantitative differences in binding to an overlapping set of thousands of genomic regions. *Genome Biol.* **10**, R80.

Mann, R. S., Lelli, K. M., and Joshi, R. (2009). Hox specificity unique roles for cofactors and collaborators. *Curr. Top. Dev. Biol.* **88**, 63–101.

Miller, J. A., and Widom, J. (2003). Collaborative competition mechanism for gene activation in vivo. *Mol. Cell Biol.* **23**, 1623–1632.

Narlikar, G. J., Fan, H. -Y., and Kingston, R. E. (2002). Cooperation between complexes that regulate chromatin structure and transcription. *Cell.* **108**, 475–487.

Narlikar, L., Gordan, R., and Hartemink, A. J. (2007). A nucleosome-guided map of transcription factor binding sites in yeast. *PLoS Comput. Biol.* **3**, e215.

Narlikar, L., and Ovcharenko, I. (2009). Identifying regulatory elements in eukaryotic genomes. *Brief Funct. Genomic. Proteomic.* **8**, 215–230.

Noyes, M. B., Meng, X., Wakabayashi, A., Sinha, S., Brodsky, M. H., and Wolfe, S. A. (2008). A systematic characterization of factors that regulate Drosophila segmentation via a bacterial one-hybrid system. *Nucleic Acids Res.* **36**, 2547–2560.

Rabiner, L. (1989). A Tutorial on hidden Markov models and selected applications in speech recognition. *P IEEE.* **77**, 257–286.

Rajewsky, N., Vergassola, M., Gaul, U., and Siggia, E. D. (2002). Computational detection of genomic cis-regulatory modules applied to body patterning in the early Drosophila embryo. *BMC Bioinformat.* **3**, 30.

Ramsey, S. A., Knijnenburg, T. A., Kennedy, K. A., Zak, D. E., Gilchrist, M., Gold, E. S., Johnson, C. D., Lampano, A. E., Litvak, V., Navarro, G., Stolyar, T., Aderem, A., and Shmulevich, I. (2010). Genome-wide histone acetylation data improve prediction of mammalian transcription factor binding sites. *Bioinformatics* **26**, 2071–2075.

Raveh-Sadka, T., Levo, M., and Segal, E. (2009). Incorporating nucleosomes into thermodynamic models of transcription regulation. *Genome Res.* **19**, 1480–1496.

Robertson, G., Hirst, M., Bainbridge, M., Bilenky, M., Zhao, Y., Zeng, T., Euskirchen, G., Bernier, B., Varhol, R., Delaney, A., Thiessen, N., Griffith, O. L., He, A., Marra, M., Snyder, M., and Jones, S. (2007). Genome-wide profiles of STAT1 DNA association using chromatin immunoprecipitation and massively parallel sequencing. *Nat. Meth..* **4**, 651–657.

Roider, H. G., Kanhere, A., Manke, T., and Vingron, M. (2007). Predicting transcription factor affinities to DNA from a biophysical model. *Bioinformatics* **23**, 134–141.

Schroeder, M. D., Pearce, M., Fak, J., Fan, H., Unnerstall, U., Emberly, E., Rajewsky, N., Siggia, E. D., and Gaul, U. (2004). Transcriptional control in the segmentation gene network of Drosophila. *Plos Biol.* **2**, E271.

Segal, E., Fondufe-Mittendorf, Y., Chen, L., Thåström, A., Field, Y., Moore, I. K., Wang, J. -P. Z., and Widom, J. (2006). A genomic code for nucleosome positioning. *Nature* **442**, 772–778.

Segal, E., Raveh-Sadka, T., Schroeder, M., Unnerstall, U., and Gaul, U. (2008). Predicting expression patterns from regulatory sequence in Drosophila segmentation. *Nature* **451**, 535–540.

Sinha, S. (2006). On counting position weight matrix matches in a sequence, with application to discriminative motif finding. *Bioinformatics* **22**, e454–e463.

Sinha, S., van Nimwegen, E., and Siggia, E. D. (2003). A probabilistic method to detect regulatory modules. *Bioinformatics* **19**(Suppl 1), i292–i301.

Small, S., Blair, A., and Levine, M. (1992). Regulation of even-skipped stripe 2 in the Drosophila embryo. *EMBO J.* **11**, 4047–4057.

Stanojevic, D., Small, S., and Levine, M. (1991). Regulation of a segmentation stripe by overlapping activators and repressors in the Drosophila embryo. *Science (New York, NY)* **254**, 1385–1387.

Steihaug, T. (1983). The conjugate gradient method and trust regions in large scale optimization. *SIAM J. Numerical Analysis* **20**, 626–637.

Thomas, S., Li, X. Y., Sabo, P. J., Sandstrom, R. B., Thurman, R. E., Canfield, T. D., Giste, E., Fisher, W., Hammonds, A., Celniker, S. E., Biggin, M. D., and Stamatoyannopoulos, J. A. (2011). Dynamic reprogramming of chromatin accessibility during Drosophila embryo development. *Genome Biol.* **12**, R43.

Ward, L. D., and Bussemaker, H. J. (2008). Predicting functional transcription factor binding through alignment-free and affinity-based analysis of orthologous promoter sequences. *Bioinformatics* **24**, i165–i171.

Wasson, T., and Hartemink, A. J. (2009). An ensemble model of competitive multi-factor binding of the genome. *Genome Res.* **19**, 2101–2112.

Whitington, T., Perkins, A. C., and Bailey, T. L. (2009). High-throughput chromatin information enables accurate tissue-specific prediction of transcription factor binding sites. *Nucleic Acids Res.* **37**, 14–25.

Won, K. -J., Ren, B., and Wang, W. (2010). Genome-wide prediction of transcription factor binding sites using an integrated model. *Genome Biol.* **11**, R7.

Wunderlich, Z., and Mirny, L. A. (2009). Different gene regulation strategies revealed by analysis of binding motifs. *Trends Genet.* **25**, 434–440.

Zeitlinger, J., Simon, I., Harbison, C. T., Hannett, N. M., Volkert, T. L., Fink, G. R., and Young, R. A. (2003). Program-specific distribution of a transcription factor dependent on partner transcription factor and MAPK signaling. *Cell* **113**, 395–404.

Zeitlinger, J., Zinzen, R. P., Stark, A., Kellis, M., Zhang, H., Young, R. A., and Levine, M. (2007). Whole-genome ChIP-chip analysis of Dorsal, Twist, and Snail suggests integration of diverse patterning processes in the Drosophila embryo. *Genes Dev.* **21**, 385–390.

Zinzen, R. P., Senger, K., Levine, M., and Papatsenko, D. (2006). Computational models for neurogenic gene expression in the Drosophila embryo. *Curr. Biol.* **16**, 1358–1365.

CHAPTER 12

Computational Analysis of Live Cell Images of the *Arabidopsis thaliana* Plant

Alexandre Cunha[*,†], **Paul T. Tarr**[‡], **Adrienne H.K. Roeder**[†,‡], **Alphan Altinok**[‡], **Eric Mjolsness**[¶] **and Elliot M. Meyerowitz**[‡]

[*]Center for Advanced Computing Research, Division of Engineering and Applied Science, California Institute of Technology, Pasadena, California, USA

[†]Center for Integrative Study of Cell Regulation, California Institute of Technology, Pasadena, California, USA

[‡]Division of Biology, California Institute of Technology, Pasadena, California, USA

[¶]Department of Computer Science, University of California, Irvine, California, USA

0091-679X/10 $35.00
DOI 10.1016/B978-0-12-388403-9.00012-6

Abstract

Quantitative studies in plant developmental biology require monitoring and measuring the changes in cells and tissues as growth gives rise to intricate patterns. The success of these studies has been amplified by the combined strengths of two complementary techniques, namely live imaging and computational image analysis. Live imaging records time-lapse images showing the spatial-temporal progress of tissue growth with cells dividing and changing shape under controlled laboratory experiments. Image processing and analysis make sense of these data by providing computational ways to extract and interpret quantitative developmental information present in the acquired images. Manual labeling and qualitative interpretation of images are limited as they don't scale well to large data sets and cannot provide field measurements to feed into mathematical and computational models of growth and patterning. Computational analysis, when it can be made sufficiently accurate, is more efficient, complete, repeatable, and less biased.

In this chapter, we present some guidelines for the acquisition and processing of images of sepals and the shoot apical meristem of *Arabidopsis thaliana* to serve as a basis for modeling. We discuss fluorescent markers and imaging using confocal laser scanning microscopy as well as present protocols for doing time-lapse live imaging and static imaging of living tissue. Image segmentation and tracking are discussed. Algorithms are presented and demonstrated together with low-level image processing methods that have proven to be essential in the detection of cell contours. We illustrate the application of these procedures in investigations aiming to unravel the mechanical and biochemical signaling mechanisms responsible for the coordinated growth and patterning in plants.

I. Introduction

One of the great challenges in developmental biology is to understand how the growth, regulation, and division of individual cells result in the overall morphogenesis of the organ. Understanding morphogenesis requires time-lapse live imaging of the cells in the tissue to observe their growth and their division pattern. Extracting information from this complex four-dimensional imaging data requires image processing to automatically quantify features of interest and changes in developing tissues.

Current methods in imaging and image analysis unfortunately require a trial and error approach to optimize the technique for each individual situation. Here, we give general recommendations and specific details about methods we have used, with the intention that these methods will be modified to fit each specific individual situation. In this chapter, we illustrate and detail the methodologies of live imaging and image processing of two specific tissues, the sepal epidermis and the shoot apical meristem (SAM), both of which relate to understanding the roles of growth and cell division in morphogenesis of these tissues. One reason to use the model plant

Arabidopsis thaliana, in addition to its well-known properties for genetics and genomics, is that plant cells cannot move or migrate relative to their neighbors. These properties make *Arabidopsis* a tractable model system for image analysis, simplifying the study of growth and cell division in morphogenesis.

First, we examine the sepal, which is the outermost, green, leaf-like floral organ that encloses and protects the developing bud (Fig. 1A). The sepal is a representative example of a plant lateral organ that, unlike many other candidates including leaves and petals, is easily accessible for imaging. The outer epidermal cell layer of *Arabidopsis* sepals has a characteristic pattern consisting of highly elongated giant cells interspersed between a diversity of smaller cells (Roeder *et al.*, 2010). We asked how the growth and division of these cells simultaneously generated the pattern and the organ.

In our second example, we discuss live imaging and computational techniques to study the cell division and growth patterns of one of the two stem cell niches of *Arabidopsis*, the SAM. The indeterminate growth of shoots is accompanied by the continual production of new organ primordia on the flanks of the SAM. Elucidation of the underlying genetic and biochemical networks that maintain this stem cell population at the growing shoot and the molecular signals required for organ initiation are fundamental questions in plant developmental biology. Understanding how these networks operate in real time to regulate cell division and growth patterns in the SAM is approached by time-lapse live imaging using laser scanning confocal microscopy (LSCM) to follow changes in gene expression patterns, hormone synthesis or signaling, and cell–cell communication.

The combined strengths of live imaging and image processing give us ways to generate quantitative data necessary to assist in the design and validation of mathematical and computational models of growth and patterning. *In silico* simulation allows one to quickly explore diverse hypotheses that are otherwise time-consuming and maybe difficult to set up in the wet lab. A key component of this computational endeavor is the ability to routinely turn the objects seen in images into faithful geometrical models amenable to computer manipulation. Once we are able to determine the real geometry (volume, surface area, distances) of cells and their topology (the neighboring cells of every cell), we are then one step closer to computationally investigate cell–cell signaling on entire tissues and make simulations to assess and predict the importance of hormones and proteins, mechanics, geometry, and network organization in organ development with digital cells closely resembling their living counterparts.

Describing plant cells and tissues in all their geometrical and topological complexity has intrigued researchers for many decades. Lewis systematically investigated almost a century ago the actual three-dimensional shape of cells in plant and animal tissues using serial sections and showed that cells rarely have regular shapes and sizes (a hypothesis defended by many but not verified until then); however, the great majority of cells are surrounded by an average of 14 neighboring cells (Lewis, 1923). Matzke and Duffy (1955) performed a similar study of meristem cells and concluded after observations that cells deeper in the meristem do not exhibit a dominant shape but have all kinds of forms, mostly convex. The primary goal of these investigations and others from the same period was to precisely reveal what shape cells have in tissues, how they are connected to each other, and what was the prevalent polygonal shape of their faceted walls. They were very quantitative in nature, characterizing the three-dimensional cell shapes in terms of total number of facets per cell, the number of same shape facets per cell, and providing average numbers over thousands of cells from different species. Both investigations had difficulties in image interpretation that are still very contemporary. For example, Lewis stated that "*the chief difficulty will be in interpreting the cell walls which fall nearly in the plane of section and consequently appear as hazy films*". We are faced with exactly the same problem when using confocal microscopy: cell walls, perpendicular to the focal plane, are poorly resolved, hazy, and sometimes invisible in the acquired images. By improving image acquisition strategies and considering such limitations during the development of algorithms, we should be able to indirectly localize with some precision such cell walls and determine cell shapes and connectivity. We hope the work presented here sheds some light on how to successfully image sepals and meristems and obtain accurate geometrical reconstructions of plant cells and tissues from confocal images.

II. Overview of Systems and Methods

We give below a brief description of the two plant systems we introduced in the previous section together with recommendations for imaging and image processing that can be freely modified to suit specific needs. They have been successfully applied in our laboratory experiments and they continue to be further refined and expanded to accommodate our reconstruction needs. We believe a close collaboration and constant interaction between imaging and image-processing specialists can produce superior quantitative results and thus we address both areas in this chapter. We stress the importance of having a well-planned image acquisition step to promote a successful computational analysis. Determining the requirements and difficulties of each allows one to identify adequate methods on both fronts to cope with physical and computational limitations.

Before we cover the general methodologies of time-lapse live imaging of sepal and meristem tissues, we will first point out several basic considerations to keep in mind before beginning any time-lapse live imaging study. Similarly, we present a few low-level image processing methods that we suggest apply to treat the acquired microscope images before performing the segmentation and tracking of cells.

A. Sepal Epidermis

The sepal is a defensive organ that encloses and protects the developing reproductive structures. At maturity, the sepal opens when the flower blooms. The outer sepal epidermis (see Fig. 1) contains a characteristic pattern of diverse cell sizes

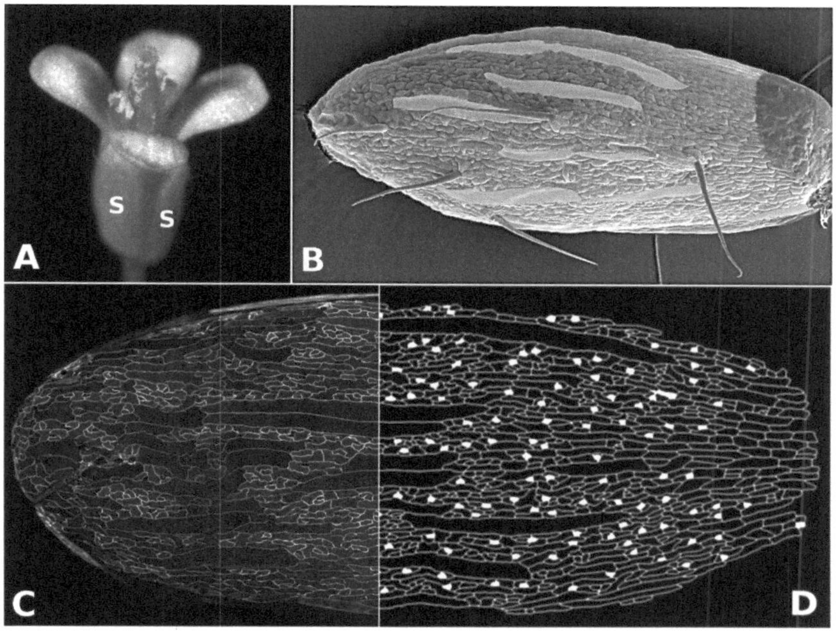

Fig. 1 Sepals and giant cells (refer to color figure). (A) The sepal is a defensive organ that opens when the flower blooms. Sepals are marked with letter S. (B) A scanning electron micrograph shows the pattern of diverse cell sizes in sepal epidermis. Giant cells (shown in pink online) are cells that undergo endoreduplication and grow much larger than other cells. Measuring cell size distribution and cell network topology helps in the investigation of a fundamental question in biology: how a pattern of different cell types develops from a field of relatively uniform cells. We developed algorithms (Cunha *et al.*, 2010) to segment sepal images and provide cell size and shape information enabling the endoreduplication study of sepals in Roeder *et al.* (2010). (C-D) We show a montage of a confocal maximum intensity projection of the plasma membrane of sepal cells (C) and its segmentation (D) where stomata and unrecognized regions (due to image aberrations) are manually filled with white color after segmentation. Segmented plasma membranes (D) are slightly blurred to improve visualization. (See color plate.)

ranging from giant cells one-fifth the length of the sepal to small cells one-hundredth the length of the sepal (Roeder *et al.*, 2010). To determine how growth and cell division are coordinated to create this pattern, we used live imaging to track the cell division pattern and static imaging with image processing to measure the cell sizes in wild-type and mutant sepals (Roeder *et al.*, 2010).

B. Shoot Apical Meristem Anatomy

The SAM sits atop the most apical part of the growing shoot, which can be identified grossly under a dissecting microscope as the dark green region between the developing floral primordia. The SAM is defined as the first 4–5 cell layers in the center of the shoot between the developing organ primordial (see Fig. 2). The SAM is further subdivided into three zones based on gene expression patterns, cellular behavior, and a cell's differentiation status/potential. Fluorescent reporter constructs are available for each of these genes and can serve as a starting point for anyone learning live imaging techniques for the SAM.

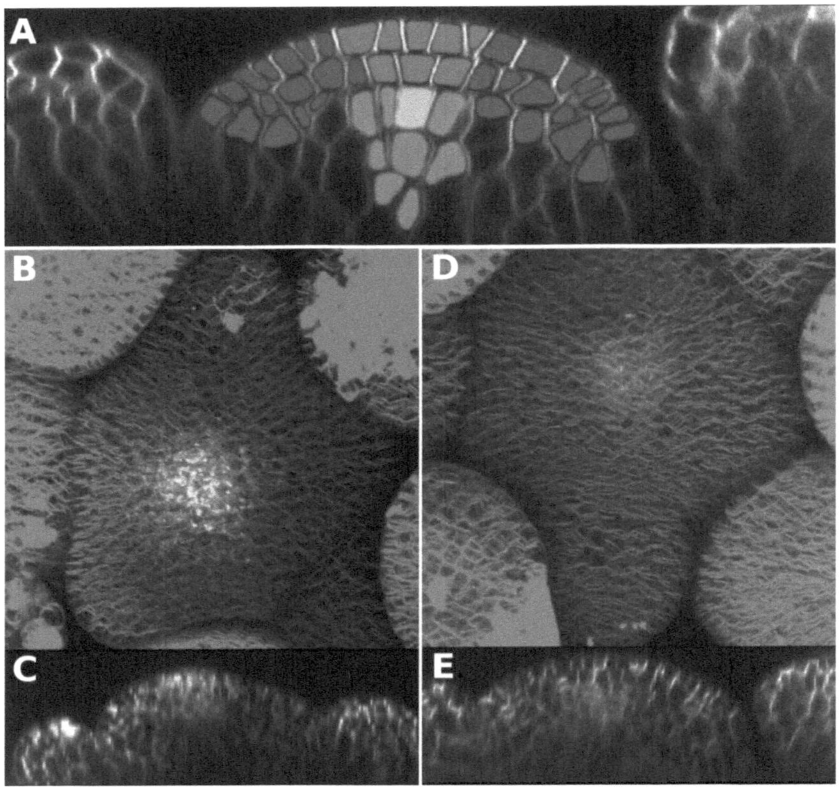

Fig. 2 Anatomy of the SAM (refer to color figure). (A) Image of a transverse section of the SAM in which the SAM is subdivided into three primary domains where the central zone (CZ) is labeled in medium grey (green online), the rib meristem (RM) is labeled in light grey (yellow online), and the peripheral zone is labeled in dark gray (blue online). Note that a few cells in the middle of the L3 may express both the *CLV3* and *WUS* genes (pink cell online) as visualized by the expression of *CLV3* and *WUS* in the middle of the L3 (C). (B-E) Maximum intensity projection (B and D) and transverse sections (C and E) of SAMs showing the cellular architecture together with gene expression patterns. (B, C) The CZ is marked by the expression of the *pCLV3::mTFP-ER* reporter in the 20–30 pluripotent stem cells in the first three central layers of the SAM. (D, E) Below the CZ is the RM, which is delineated by the expression domain for the homeodomain transcription factor *WUSCHEL* (*pWUS::mTFP-ER*). Surrounding the CZ and the RM is the peripheral zone (PZ), which is roughly marked by the expression domain for several genes, including *UNUSAL FLORAL ORGANS* (*UFO*) and *FILAMENTOUS FLOWER* (*FIL*) (not shown, blue region in A). (See color plate.)

C. Time-Lapse Live Imaging Versus Static Imaging of Living Tissue

The first choice to make is whether to undertake time-lapse live imaging of the organ while it is attached to the living plant or to remove it from the plant and take a single static image. Both protocols are detailed here and both procedures are often used in concert. We recommend starting with static imaging of individual organs at different time points to gain a sense of the relevant time frame. Then live imaging of a few samples should give detailed information about the changes occurring. The computer can assist with tracking these cells over time. Finally, many static images of a specific time point can be measured after image segmentation to produce large datasets that can be analyzed with statistics.

For example, we first imaged sepals from various flowers and found that giant cells were established very soon after the sepal primordium formed. We observed the development of sepals throughout this period using live imaging and found that giant cells stopped dividing and entered the specialized endoreduplication cell cycle in which they replicated their DNA and grew (Roeder *et al.*, 2010). Concurrently, neighboring cells divided, thereby degreasing their size. Finally, we took static images of mature sepals, segmented them, and measured the final cell sizes of nearly all the epidermal cells in the sepal.

D. Considerations for Engineering Transgenic Lines and Use of Fluorescent Probes for Live Imaging

The selection of the correct combinations of fluorescent proteins (FPs) or probes is very important for a successful live imaging study. Generation of suitable transgenic reporters is a labor-intensive process; therefore, careful consideration of the goals of the live imaging study can save substantial time. The following are a set of recommendations pertaining to the engineering of T-DNA vectors for making transgenic plant lines.

The first step in the production of any transgenic fluorescent reporter line requires deciding whether to engineer a transcriptional or translational reporter. There are advantages and disadvantages to both. A transcriptional reporter is constructed by replacing the gene of interest's (GOI) open reading frame (ORF) with a FP such that the expression of FP is now controlled by the GOI genetic regulatory elements. A transcriptional reporter is useful for visualizing the expression pattern of the GOI and level of promoter activity in different cell types. However, a transcriptional reporter may lack additional transcriptional regulatory elements present outside the gene's proximal promoter and 3′UTR.

A translational reporter requires fusing the ORF of a FP in frame with the GOI ORF. This can be done to produce either an N-terminal or C-terminal fusion protein. A translational reporter can faithfully reproduce a gene's subcellular localization and post-transcriptional/translation regulation. However, care must be taken to ensure that an N-terminal or C-terminal FP tag does not alter the normal subcellular targeting of the fusion protein. Furthermore, some subcellular compartments can be difficult to visualize by confocal microscopy. In general, the larger the cellular organelle, the more pronounced it is in the image. As a general guide, transcription factors and stable membrane-targeted proteins are the easiest to visualize as translational fusions. Proteins targeted to smaller cellular compartments (such peroxisomes and mitochondria) can be difficult to visualize in the image. To correctly identify these subcellular compartments requires co-visualization with known marker dyes or transgenes known to localize to these cellular compartments. When confronted with this issue it is generally wise to make a transcriptional reporter of the GOI.

The production of cytoplasmic transgene FP reporters, whether transcriptional or translational, should be avoided when reasonable due to the difficulty these markers present in downstream image processing. Many subcellular localization tags are available to avoid this complication when making transcriptional reporters (Cutler *et al.*, 2000). For nuclear localization, the FP can be fused to Histone2B (H2B) such that it is incorporated into the nucleosomes (Boisnard-Lorig *et al.*, 2001), or N7 such that it is in the nuclear envelope (Cutler *et al.*, 2000). The advantage of H2B is that the chromosome dynamics of mitosis can be observed. For the plasma membrane, the 29-1 tag is common (Cutler *et al.*, 2000) and the low-temperature-induced transmembrane protein RCI2A is also used (Thompson and Wolniak, 2008).

Standard molecular cloning techniques are used to engineer fluorescent reporter transgenes. We routinely clone approximately 5 Kb of genomic sequence upstream of the GOIs initiating ATG as the promoter. If another gene is present within this region, we will clone the entire sequence up to the adjacent gene. As a general rule, we also clone approximately 1.5 Kb of the GOI's 3′UTR, which is also determined by the proximity of an adjacent gene. To aid in the generation of new reporters we have adopted the gateway recombination-based technology (Invitrogen Life Technologies Corporation, Grand Island, New York, USA) as the preferred method for introduction of FPs or FP translational fusions into the GOIs regulatory region. There are several published gateway compatible plasmids available to accomplish this (Karimi *et al.*, 2007a, b). While these plasmids are useful, we have found engineering our own gateway destination vectors by inserting a gateway recombination cassette between the GOI's cloned regulatory regions for transgene expression gives us additional flexibility. In addition to the available gateway recombination technology, a BAC clone-based recombineering method for generating transgene reporters is gaining traction (Venken *et al.*, 2009). One advantage of this system is that it allows a larger region of the chromosomal DNA to be present, which may contain distal enhancer elements (>5 Kb from the GOIs ATG).

Cloning the genetic control elements is the first step. The next step is deciding whether to construct a transcriptional or translational reporter. The main advantage of the gateway recombination-based system is that it makes this step versatile. Once the gateway recombination cassette (that contains the *attR1* and *attR2* sites) is inserted in between the GOIs genetic control elements transcriptional or/and translational reporters can are easily constructed. Our lab has created a standard set of FP constructs in pENTR-D/TOPO that allows us to introduce the FP into the converted

GOI destination vector. To increase the expression levels of these FP reporters, we have introduced the Tobacco Mosaic Virus translation enhancer (Ω translational enhancer) upstream of the FP ORF (Sleat *et al.*, 1987). The inclusion of the Ω translational enhancer results in a consistent level of transgene expression across independent insertion clonal lines. It also enhances the FP signal from weak promoters due to increased translation of the FP mRNA. We have also included an idealized Kozak sequence (a stretch of six adenines) found in the most highly expressed *Arabidopsis* genes (Nakagawa *et al.*, 2008). In addition, we have multimerized the FP ORF such that each of the FP reporters are 2X or 3X versions separated by a Pro-Ala(9)-Pro linker sequence to allow for proper folding. The signal from these engineered FPs is further enhanced by their targeting to subcellular compartments (as discussed above). One note on making an FP targeted to the endoplasmic reticulum (ER). We have noticed that the signal from ER-targeted transcriptional FP reporters has a major drawback of perdurance, where the reporter signal is maintained after cell division in cells that may not express the GOI. Care should be taken when looking at the ER-localized proteins for the first time before making a conclusion about the expression pattern and domain size of your GOI.

Research on the biochemistry and biophysics of FPs continues to yield an expanding spectrum of FP colors that are brighter and more photostable (Shaner *et al.*, 2005, 2008). One should select a set of FPs that are compatible with the laser lines and filter sets on the available microscope and the other FP reporters that will be used in your study. Some of our favorite FPs for expression in *Arabidopsis* include YPet (yellow), mGFP (green), mTFP (blue), and tdTomato (red) (Nguyen and Daugherty, 2005; Shaner *et al.*, 2008). Other FPs that have been used with success include mYFP (yellow), mCitrine (yellow), and Venus (yellow). Note the signal for Venus bleaches quickly and should be used sparingly. When choosing combinations of FPs to use, consider the excitation wavelengths available and the emission filters on your microscope, such that optimal imaging conditions can be achieved. We have found that the following FPs sets offer a nice combination of photostability, brightness, and compatible emission spectra, which can be separated in double and triple marker lines; mTFP, YPET, and mCherry; mGFP, YPET, and mCherry; mGFP, YPET, and tdTomato; mGFP and tdTomato. For the combination of YPET and tdTomato, the emission spectra for the two FPs partially overlap due to the long emission tail of YFP. This can be overcome if the two FPs are targeted to different subcellular compartments and bright enough. Also if using more than one marker, imaging of separable colors greatly reduces the difficulty of automatic segmentation. Markers of the same colors with different subcellular localizations can be segmented manually.

The previous discussion has outlined the steps involved and issues to be considered when engineering new transgene constructs. However, some applications require the ubiquitous expression of a translational reporter gene. Choosing the appropriate promoter to drive expression of the fusion protein is essential. Many promoters considered to be ubiquitous are not expressed well in certain tissues. For example, the cauliflower mosaic virus 35S RNA promoter (35S) is not expressed well in the SAM or embryos but it is highly expressed in lateral organs such as sepals. Furthermore, transgenes made with the 35S promoter are silenced at a high frequency. A better choice we have found is the *UBIQUITIN10* (*UBQ10*) gene promoter (Nimchuck *et al.*, 2011. While this promoter is not as strong as the 35S promoter, it gives a nice middle level of transgene expression without the problem of silencing. We now use the *UBQ10* promoter as our primary driver for transgenes in the SAM and tissues outside the SAM and embryo.

E. Important Considerations Regarding Selection of the Microscope and Microscopy Settings for Live Imaging

Live imaging of plant development described in the methods herein involves the use of a LSCM. While several variations of confocal microscope technology exist, such as a spinning disk confocal microscope or two-photon or two-photon light sheet microscopy, it is not clear if any of these microscopes offer any improvements over the standard LSCM microscope when referring to time-lapse live imaging of sepals or the SAM over long periods. What is of upmost importance, however, are the parameters used during a time-lapse live imaging or static live imaging study. The methods described here are based on imaging with a Zeiss 510 LSCM microscope, as used by our lab, and the general guidelines outlined may need to be altered if you are using a different microscope, such as a Leica or Nikon LSCM.

The total working distance available should be the first thing checked to see if it will suit your needs, before proceeding with any new live imaging experiment of the SAM. Without sufficient working distance it may not be possible to image all the time points desired for your time-lapse live imaging experiment due to plant growth. The working distance on an upright Zeiss 510 LSCM generally allows you to conduct a 72 h time course.

The next two important considerations are scan speed versus resolution. Increasing the frame size (e.g., 1024 instead of 512) will increase the time required to scan each image in the *z*-stack. Although this will improve the final quality of the image (i.e., pixel or voxel number), it increases the total scan time required to capture the *z*-stack. One of the greatest challenges in time-lapse or static live imaging is to acquire the total image stack before tissue growth pushes the desired region of interest (ROI) outside the specified imaging distance in the *z*-plane. Cell expansion in the growing stem occurs at a significant rate, such that during the acquisition of the average SAM *z*-stacks (30–60 μm) the cell expansion in the stem will have pushed your ROI out of the specified scan region. Ideally, downstream image processing applications will benefit from a pixel resolution close to 1:1:1 in the *x*–*y*–*z* planes. To achieve a 1:1:1 pixel resolution for the acquisition of a 30 μm image stack at a 1024 resolution with the maximum scan speed of 1.94 s per image (this is the maximum scan speed of the Zeiss 510 with the 63X/0.95NA water-dipping objective commonly used in live imaging SAM tissue) will take approximately 6 min and 54 s (see Fig. 3). The best way to solve this dilemma of growth

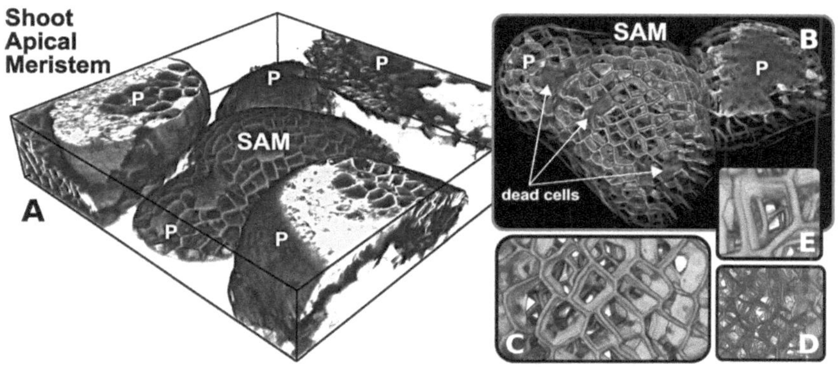

Fig. 3 High-resolution shoot apical meristem. A large, high-resolution z-stack with 307 slices of a shoot apical meristem, SAM is shown in (A) together with primordia (marked with letter P). The stack was acquired with x:y:z resolution of 0.11:0.11:0.13 μm using a slice spacing of 0.13 μm. Each slice has 1024 × 1024 pixels in the xy plane. We manually dissect the digital central SAM from the other parts to obtain the image in (B) after denoising. Note in (A) and (B) the presence of dead cells which are sometimes caused while preparing the plant for imaging. Magnified details are shown in (C–E). In (C) and (E) we can visualize many layers within the SAM, starting from the epidermal layer L1. Note that the top and bottom walls of cells are not shown because they were not captured in the raw z-stack, thus allowing one to see through the cells. (D) Close-up view of the SAM looking from the bottom. (For color version of this figure, the reader is referred to the web version of this book.)

during the imaging session is to determine what the minimum image quality required is for any downstream image processing application and adjusting scan speed, image resolution, and the z-interval distance parameters accordingly to capture an image that is sufficiently good for your specific downstream image processing applications. However, if your downstream image processing needs do require a 1:1:1 pixel resolution placing your plants at 4°C for approximately 15–30 min and then using cold water for the imaging session will slow growth enough during that time period to allow you to image your ROI without significant altering of the image quality from growth.

The final consideration for any time-lapse live imaging study is the selection of a reporter line that is sufficiently strong such that the laser light input can be kept to a minimum to reduce the potential for damaging the tissue. In fact, the major complication for any time-lapse live imaging experiment is tissue damage/death caused by laser light radiation. This is an unavoidable side effect but can be minimized by engineering transgene constructs to maximize brightness (discussed above) and selecting transgenic lines with the highest level of expression while retaining the fidelity of its endogenous expression pattern. This will allow for the reduction of tissue damage, increasing the possibility of achieving the maximum desired time course of your experiment. However, cellular damage from laser light radiation and free radicals generated by the excitation of the FP fluorophore will make damage to the tissue an unavoidable outcome of any time-lapse live imaging experiment. Therefore, it is of utmost importance that the viability of the SAM or sepal tissue is monitored, especially at later time points (>24 h). This is most easily accomplished by monitoring primordia outgrowth and development in the SAM or monitoring cell division and expansion in the sepal. If you notice that the position of the primordia is unchanged or outgrowth has ceased after successive image stacks taken 16–24 h apart, the SAM tissue has been severely damaged and cellular growth has all but stopped, and further (and perhaps even earlier) time points will yield no useful data.

F. Noise Reduction

Images acquired using a LSCM are inevitably corrupted by shot noise, a type of noise due to detection of photons or other particles, that follows a Poisson distribution and is accentuated in low light emission conditions (here, low photon count by the CCD detector). The consequence of reducing light intensity is an amplification of noise and hence an increase in the uncertainty in the location of labeled regions. It is worth remembering that the labeled regions in a fluorescent image are the result of a probabilistic measurement of photon count at a certain location in space and time recorded by the microscope camera. Therefore, the spatiotemporal location of a labeled region is a probability measure with an associated error. Any computational contouring detection scheme can only offer, at best, an approximate location for the true contours of the labeled regions.

Other important factors, in no particular order, regulating noise and image resolution in fluorescence LSCM for live imaging include scanning speed (faster/slower scans increase/decrease noise), size of raster line and zoom magnification (for the same amount of light, the larger the size, the noisier the image will be), light wavelength–dye combination (proper matching reduces noise), instability of laser (heavily used equipment tends to be less stable and noisier), and numerical aperture (affects amount of collected light). Explaining the mechanisms of influence of each of these factors is beyond the scope of this chapter, so we refer the reader to Pawley (2006a) for a contemporary account on the subject.

Reducing noise in fluorescent images is a natural first step prior to attempting segmentation. The hope is that by reducing noise one can improve the chances of restoring contour locations in low-contrast images and detecting cell boundaries with better accuracy. In fact, we show that combining robust computational methods for noise reduction and contrast enhancement does reveal cell contours even in deep tissue locations. Since we want to reduce Poisson noise, it is natural that we explore noise-reduction methods adopting a Poisson distribution model for the noise component. But, perhaps surprisingly, we do not have to restrict ourselves to such a case. As we will show through examples, white noise models represented by a Gaussian distribution are also adequate in removing Poisson noise toward accentuating plasma membrane location. There is an abundance of algorithms designed for additive white Gaussian noise reduction and recent methods have been shown to be extremely effective (e.g., Buades *et al.*, 2005; Dabov *et al.*, 2007). While these are developed with natural, mesoscopic images in mind, mostly for digital-camera-like devices, we have experimented with them on confocal images and successfully generated denoised images with quality suitable for segmentation. Recent work by Makitalo and Foi (2011) for Poisson noise removal has suggested transforming image data to match the constant variance property of white Gaussian noise and then use these powerful Gaussian-based denoising schemes followed by an inverse transformation. But we have yet to verify any practical application suggesting that approaches of this nature outperform the white Gaussian noise removal methods for our problems. We have favored the nonlocal means method (Buades *et al.*, 2005) to denoize our confocal images and will present details about the method and results in Section V.B.

G. Segmentation

Segmentation refers to the partitioning of an image into homogeneous regions each representing a single object or parts of the same object. This is a classical image processing problem with decades of development and a vast literature. Methods range from simple thresholding (classify pixels in an image as either foreground or background based on a global cutoff value) to more elaborate energy-minimization methods that require numerical optimization to cluster pixels into uniform regions. Unfortunately, most methods are designed to work well for a particular set of images that conform to prior knowledge built into the method. In our experience, algorithms that claim to be general cannot cope with all possible variations in image quality and content. It is thus common practice to rework and tune an already developed segmentation method to fit specific needs or build from scratch. In all cases, results are most often not complete, missing details, even if minor, that are sometimes important, and a post-processing manual intervention is necessary to augment the automatic results. Given these practical limitations, our goal is to develop a semi-automatic segmentation strategy requiring the least amount of manual editing to achieve high-quality results.

We have experimented with much-used segmentation methods (e.g., Chan and Vese, 2001) and customized pipelines and have shown that good-quality semi-automatic segmentation can be achieved for sepals and meristems (Cunha *et al.*, 2010). Because our live confocal images can exhibit poor quality, we start our processing by enhancing the images prior to segmentation – denoising and contrast enhancement. It is much easier to segment a high-contrast, clean image where the separation of regions is more evident than a noisy, low-contrast image where it is difficult to distinguish where boundaries begin and end. All our processing is done for slices of a *z*-stack (SAM) and projections obtained from *z*-stacks (sepals). Our strategy for the SAM is to build up from the segmented slices a volumetric segmentation. We assume, based on observations from our own work and from others, that the SAM is composed of convex or almost convex cells. This property has advantageous implications when considering reconstructing from slices. If slices are acquired with sufficient resolution in all Cartesian directions and are properly segmented, then it is possible to merge them and obtain full three-dimensional segmentation of cells. In this chapter, we present the first part of our work, segmenting SAM slices.

There are a multitude of image segmentation approaches for two-dimensional images. We have opted to work with the active contours without edges method (Chan and Vese, 2001) and build our own segmentation pipeline from a collection of low-level operations. The former is a well-known method in the image processing literature and it was chosen because it is designed for separating an image into only two regions, matching our needs (our two regions are plasma membrane and cell interior which matches the dark background). Given the good quality of the denoized images we obtained with nonlocal means, we decided to invest in an alternative method, a segmentation pipeline combining mathematical morphology (Soille, 2004) and other low-level image processing operations (Gonzales and Woods, 2008). The pipeline is tuned to our own sepal and meristem images but we believe it is applicable to other images presenting similar content. The results have been proven to greatly reduce the manual labor required to obtain good-quality segmentation of sepal and meristem images.

There are few publications presenting computational methods to segment plant tissues. We are not aware of any method for sepals besides our own (Cunha *et al.*, 2010) and only a few have been developed for root meristems (Marcuzzo *et al.*, 2008) and shoot meristems (Cunha *et al.*, 2010; Fernandez *et al.*, 2010; Liu *et al.*, 2010; de Reuille *et al.*, 2005). The work of Reuille *et al.* is mostly manual where users manually select junction points in the image and specify which sets of junctions comprise each cell in the L1 layer of the SAM. The work of Liu *et al.* (2010) is similar to ours as it segments slices of the SAM but they use the watershed transform (Soille, 2004) to segment cell boundaries. Fernandez *et al.* (2010) also uses the watershed transform to segment cells but now directly in three dimensions. They use more than one *z*-stack image of the same meristem to improve the signal of plasma membranes in the overall floral meristem tissue thus increasing the number of correctly segmented cells when compared to using a single *z*-stack. Although highly computational, these proposed methods still need some manual intervention in their segmentation pipeline to correct faulty segmentation results and to align the *z*-stacks prior to segmentation. The work of Cunha *et al.* (2010) also uses manual editing to augment the automatic segmentation but the segmentation methods differ from those mentioned above. We present the details of this work in Section V.

H. Tracking

From the computer vision perspective, cell tracking presents a combination of specific challenges: (1) Low frame rates: Images are often acquired in multiple independent channels, such as alternate fluorescence colors. The need to avoid photobleaching and phototoxicity, and other natural constraints such as developmental cycles can result in very long intervals between successive frames. For example, *Arabidopsis* movies considered in this work have a 6 hours interval between consecutive frames. The resulting abruptness of motion, as well as morphological changes may render many classical motion estimation and tracking algorithms ineffective (Arulampalam *et al.*, 2002; Comaniciu *et al.*, 2003; Peterfreund, 1999; Veenman, 2001). (2) Cell division and cell death: The number of cells often changes during tracking because of cell division and death (cell fusion is also possible). In our particular plant tissues, neither cell death nor cell fusion occurs, so the actual number of cells in the tissue changes only through cell division. But in addition, cells can disappear and reappear as they transiently leave the imaging surface or field of view. Some algorithms are robust to missing data points in time (Veenman, 2001); however, object integrity is assumed by most algorithms (Broida and Chellappa, 1986; Veenman, 2001; Yilmaz *et al.*, 2006). (3) Change in cell morphology and appearance: Cell shape and appearance can change significantly over time. For example, differentiation involves rapid morphological and gene expression changes. (4) Tight clustering: Many cell types, including ES cells, tend to literally stick together, forming tight cell clusters, making individual cells more difficult to identify and track. (5) Quantitation: Due to fluctuations in illumination, changes in expression levels, and photobleaching, cell appearance may vary greatly over time, even between successive frames. For example, boundaries may blend into background or may overlap with a neighboring cell.

To tackle the issues mentioned above, various cell-tracking algorithms have been developed (Song *et al.*, 2000). These include commercial microscopy analysis packages (MediaCybernetics, Bethesda, MD, USA); academic cell-tracking softwares, such as CellTracker (Shen *et al.*, 2006), CellTrack (Sacan *et al.*, 2008), DynamiK (Jaeger *et al.*, 2009), and Soft-Assign (Gold *et al.*, 1998; Gor *et al.*, 2005; see also Section V.E); and plug-in modules in more general analysis frameworks such as CellProfiler (Jones *et al.*, 2008) and ImageJ (Rasband, 1997). These methods range in sophistication and generality, from manual clicking on object positions (ImagePro Software - Media Cybernetics, Inc., Bethesda, Maryland, USA; Metamorph Software, Molecular Devices, Inc., Sunnyvale, California, USA) to standalone implementations (Shen *et al.*, 2006) and to cell lineages (Mosig *et al.*, 2009; Wang *et al.*, 2010).

Assessing the performance of a tracking algorithm is less characterized. It can be evaluated by several domain-specific metrics. Intensity coherencies along tracks can be evaluated (Black *et al.*, 2003). An event-based metric, which is analogous to identifying cell division and death events, was suggested in Roth *et al.* (2008) and a tracking-difficulty metric was introduced in Pan *et al.* (2009), which considers local tracking failures over time. Paired with temporal locations of cell events, this type of evaluation can measure the effectiveness of tracking algorithms on capturing cell events. We illustrate this on the dataset considered in this work.

III. Biological Insights

The ability to track cells or follow changes in gene expression patterns during plant development has already yielded significant advances in our understanding of morphogenesis and cell–cell communication in both lateral organ, such as the sepal, and stem cell maintenance and morphogenesis of organ outgrowth in the SAM

(Reddy and Meyerowitz, 2005; Roeder *et al.*, 2010). For example, following the change in the size of the pluripotent stem cell pool in the central zone (CZ) of the SAM following transient reduction in the expression of the *CLAVATA3* gene product suggested that cells that exit the CZ and enter the PZ can be respecified as *CLAVATA3* expressing CZ cells without undergoing cell division (Reddy and Meyerowitz, 2005).

Roeder *et al.* (2010) combined time-lapse live imaging and image processing with computational modeling to conclude that variations in cell cycle decisions to divide or endoreduplicate are responsible for generating the characteristic pattern of cells in sepals. Live imaging and computational modeling have also contributed to our understanding of the relationship between the hormone-mediated lateral organ growth and the mechanical properties of the SAM epidermis (Hamant *et al.*, 2008; Heisler *et al.*, 2010). Each of these studies reached conclusions that would have been very difficult without time-lapse live imaging, image processing, and computational modeling. As the time-lapse live imaging technique combined with image processing and computational modeling is extended to new applications, even more will be learned about cell growth and morphogenesis in plant development.

IV. Open Computational Challenges

Despite many decades of research, there are enduring challenges in the image processing and analysis area that directly affect our reconstruction developments of plant tissues. We will address a few important ones below and offer some thoughts of what might lie ahead.

One computational challenge is the proper stitching of the sub-images forming the larger whole sepal tissue image. Sepal images presented in this chapter are composed of up to six overlapping images each acquired by carefully repositioning the sepal on the microscope to avoid shifts and distortions. Large shifts and distortions might render a poorly formed composite image, which will inevitably lead to contouring errors during the image segmentation stage. We used specialized commercial software (Photoshop Photomerge, Adobe Systems Inc., San Jose, California, USA) followed by manual verification and repositioning to mosaic the sub-images, but solutions were suboptimal: cell walls on the overlapping areas were not fully matched in some images leaving cells with broken, unaligned contours. We believe this is mostly due to nonlinear warping of the tissue during acquisition, a feature that is not detected by the photo merger software and difficult to realize manually (rotations and translations alone are not sufficient to align walls). We have experimented with tools developed for aligning medical images but to no avail. A specialized nonlinear image registration procedure is necessary to unwarp and bring all cell walls in the overlapping regions to a full alignment.

A fully three-dimensional geometrical reconstruction of a living SAM of *Arabidopsis* continues to be a daunting task. A few groups have proposed semi-automatic solutions (see Section II.G) with different levels of manual intervention and computational complexity. There are presently quite a few challenges to computationally generate a faithful reconstruction in three dimensions. The poor signal-to-noise ratio typical of live confocal images prevents obtaining an accurate location of cell contours especially in deep parts of the tissue. A more limiting factor is the signal absence of partial or entire cell walls (as occurs when walls in the X–Y plane are skipped in the z-axis because of the finite distance between z-stack slices), generating gaps that are challenging to be automatically detected especially in three dimensions. Using the fusion of three or more image stacks of the same meristem, each taken from a different viewing angle, helps reduce the number of gaps and it improves cell wall localization (Fernandez *et al.*, 2010). This comes at the expense of submitting the plant to a much higher dosage of laser radiation (45–60 min to acquire three z-stacks), which in turn might limit the total amount of light that can deeply penetrate the tissue, thus preventing resolving the cell network architecture in interior parts of the meristem. The z-stack fusion process requires manually collocating and establishing the correspondence of fiducial points to assist in the alignment of all stacks after acquisition, a task not easily accomplished in three dimensions. We believe, the development of new markers and image acquisition techniques will allow the generation of high-resolution images in all directions with better quality thus facilitating the three-dimensional reconstruction of meristems from single stacks.

Tuning of algorithm and software parameters is usually necessary to achieve robust image processing results. Repeated unsuccessful trials can lead to frustrations when end-users are not familiar with the underlying methods and their limitations, when such limitations are not clearly presented in the software, or simply due to unpreparedness. For example, not fully knowing what to expect, a biologist might try several times an open source general image segmentation software to process data before realizing it only produces partially acceptable results. The process is repeated a few more times with new promising software packages producing similar outcomes. User frustration builds up and the tendency is to abandon the automatic route and solely rely on the manual labeling where success is likely, though not necessarily immediate or reproducible by peers. When presented later on with a similar problem and slightly different data set, the entire process might be repeated. This is not an uncommon scenario. Image processing software is specialized, and general algorithms rarely produce complete results (though for simple, high-quality images, most methods work pretty well). In general, building good software is an art in itself and constructing effective processing pipelines requires substantial knowledge of all of its components. In our experience, the most successful users of bioimage processing are those who take the time to understand the methods, at least in a high-level fashion, partner with image processing specialists, accept the current limitations of algorithms, and are willing to manually correct what results from the automatic processing. The challenge here is to make this knowledge widespread, foster partnership between specialists and nonspecialists, and to build the necessary interactive and friendly software tools easily adaptable to multiple scenarios.

Our processing pipeline is not different from the scenario pictured above: it requires tuning of parameters to achieve suitable results, it was designed for a specific class of images and with a specific problem in mind, and it usually produces results requiring some manual intervention for completeness. On the other hand, it does produce good results much faster than manual labeling alone, it is reproducible, and by providing easy ways to manually remedy results it engages and, surprisingly, empowers the biologist in the solution-seeking process, giving more confidence with the achieved results. The crowdsourcing paradigm (demonstrated recently, e.g., in the Galaxy Zoo and the RNA folding game) has shown that even nonspecialists can beat the best algorithms when the problems are well explained, the interactive software is straightforward to use, and the efforts of crowds are properly harvested. What might lie ahead is thus a future where computers diligently propose a set of best possible solutions for challenging image processing tasks and human crowds help refine them to a desirable result. Such refinement might in turn feed back into the algorithms that learn and incorporate new rules to avoid repeating prior mistakes.

V. Imaging and Computational Methods

We present in this section protocols for image acquisition and processing. First, we describe four imaging protocols (Methods 1–4) we have developed in our lab to acquire images of sepals and meristems suitable for visual inspection and image processing. See Fig. 4 for a hands-on illustration of sample preparation and image acquisition of a shoot meristem. We then outline the image processing algorithms and methods used to do segmentation and tracking of cells.

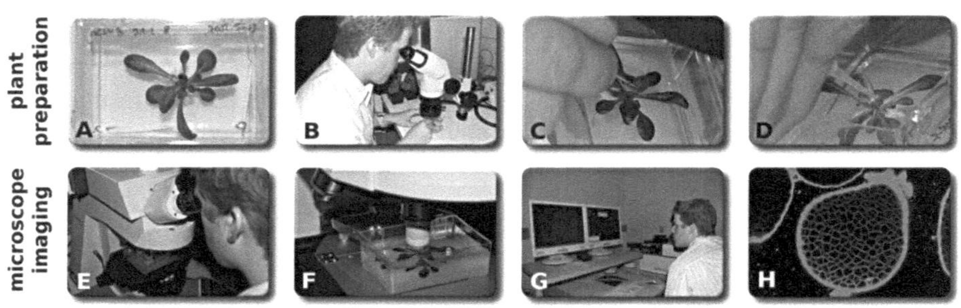

Fig. 4 SAM imaging session (refer to color figure). Well-developed protocols for sample preparation and microscope imaging of the shoot meristem are essential to obtain good-quality images for processing. Plant preparation for imaging at the lab is illustrated on the top row. An *Arabidopsis* sample (shown in A) is dissected with the help of a dissecting microscope (B, C), to expose and clear the way to the minuscule SAM. It is very important at this stage not to touch the shoot with the fine forceps to avoid damaging and killing the stem cells. This can only be verified later on during the imaging session. The shoot can then be stained with a drop of the selected fluorescent dye at its tip (D). After resting a few minutes, the plant is ready for imaging (bottom row). Adjusting the microscope (E) and positioning the stage at the right distance (F) will prepare the SAM for image acquisition. At the computer station (G) one can fine-tune the acquisition parameters (filters, scan speed, aperture, image bit depth, image size, and resolution, etc.) to produce high-quality images (H) suitable for image processing. (See color plate.)

A. Imaging Protocols

1. Method 1: Time-lapse live imaging of sepal growth

   a. The plants must be planted in special pots to prevent the soil from spilling during imaging. First, cut fiberglass window screen in roughly 15 cm $\times$ 15 cm squares and soak them in water with a few squirts of Simple Green all-purpose cleaner overnight. Rinse with water. Fill a small square pot (about 6.5 cm $\times$ 6.5 cm) with moist soil and cover it tightly with a square of washed window screen. Use a rubber band to pinch the window screen firmly around the top of the pot. Remove remaining window screen below the rubber band with scissors. Plant the transgenic seeds expressing both a plasma membrane marker and a nuclear marker and grow as is standard. A total of 10–15 plants is about the maximum that can be imaged in one session.
   b. When the plants have bolted about 5 cm and are actively flowering, start the time-lapse live imaging.
   c. Dissect the inflorescence to reveal a single stage 3 flower under a dissecting microscope at about 50$\times$ magnification. Use fine forceps (Dumont #5) and 23-gauge needles. First, remove all of the open flowers. Starting on one side, remove all the overlying flowers until the stage 3 flower is revealed between two older flowers which can serve to protect it. Never touch the stage 3 flower. Remove the older flowers from the other side of the inflorescence so that it will lie flat on the slide. Some stage 12 flowers should remain on the lateral sides of the inflorescence because these improve the health of the plant. Watch for these flowers to further develop during the whole time series if the plant is healthy, and if not, discard the sample. Dissecting is the most difficult step and it is typical for many samples to be thrown out due to damage when the youngest flowers need to be observed.
   d. Tape the stem of the plant to the frosted part of a slide with lab tape about 0.5 cm below the inflorescence such that the stage 3 flower of choice faces up. The slide will remain attached to the plant throughout and serve as a guide for returning the plant to the same orientation for each imaging session.
   e. To return the plant to the growth room, cut a small hole in the mesh behind the plant, place a small plant stake through it (bamboo shish kabob skewers work well). Tape the back of the slide to the stake so that it is oriented vertically with the meristem upward. If it is not vertical, the meristem will grow away from the slide making imaging difficult. Allow the plant to recover several hours before the first imaging time point. This greatly improves the plant survival rate.
   f. To image, tip the plant on its side, detach the slide from the stake. Mount the inflorescence in 0.02% silwet (a surfactant that does not affect plant viability) and 0.1 mg/ml propidium iodide (Sigma P4170-10 mg - Sigma-Aldrich Co. LLC, St. Louis, Missouri, USA). Especially for the first time point, pipette the liquid up and down around the inflorescence to remove bubbles. Make sure the flower of interest is completely covered with solution and not obscured by any bubbles. Cover with a cover slip.

g. Carefully transport the prepared plants to the upright confocal microscope. We use a Zeiss 510 Meta. To minimize damage to the plant, it is best to either push the plants on a cart or use a tray that holds the entire plant on its side instead of carrying the base in one hand and the slide in the other. Use a small box of same height as the stage placed adjacent to the stage to hold the plant pot while inserting the slide in the holder on the microscope stage. First, visualize the inflorescence under the 10× objective to find the center where the flower of interest is located. Switch to a 40× water-dipping objective and create a column of water between the objective and the cover slip. Water-dipping objectives provide greater working distances and good optics, while not damaging the plant. In our experience, oil immersion objectives do not provide sufficient working distances and oil can be quite toxic to the plant.

h. On the confocal microscope, find the flower of interest and make sure it is not damaged. Damaged tissue will stain strongly with propidium iodide. In healthy tissue, only the cell walls will stain with propidium iodide. If the flower is damaged, either select another flower to image, or discard the sample and try the next plant.

i. Take a confocal *z*-stack of the flower while exposing the plant to as little laser light as possible. Reduce laser transmission to around 10%. Maximize scan speed and use the zoom to scan as small a region around the flower as possible to minimize the scan time. Scan the image only once. Only image one flower per plant.

j. Remove the plant from the microscope and remove the cover slip. Gently blot the inflorescence dry with a kimwipe. Replace the stake and tape the slide to it so the plant sits directly upright. Return the plant to the growth room.

k. For the next time point, repeat the procedure of mounting the plant and imaging the flower. Be sure to find the same flower by visual identification. Compare to the previous image if necessary. Giant cells in the sepals are good landmarks. The interval between time points should be short enough to capture the dynamics of the process. For cell division 6 h is reasonable.

l. Continue taking time point images as long as the plant maintains viability and the process of interest continues. As the organ grows, change zoom and eventually the objective to maintain a field of view encompassing the whole organ. When cells start to stain brightly with propidium iodide, viability is compromised, but the preceding image sequence can be used. Sometimes plants will arrest, growth and cell division will stop, but no damage is evident. Again the time points while the plant is actively growing can be analyzed and those after it stops are excluded.

m. After the whole time series has been acquired, the images are aligned using the Affine Registration function in Amira (Visage Imaging Gmbh, Berlin, Germany). Open the confocal stack in Amira, and use the Voltex function to display the volume rendering. Use Volume edit to crop away any other flowers. Use the hand tools to pull the image into approximate alignment with the image of the preceding time point. Then set the Affine Registration parameters such that the computer produces a good alignment of the second time point to the first. Continue by aligning the third time point to the second and so on. From Amira, the aligned stacks can be exported for further three-dimensional analysis of the cells. In addition, a series of two-dimensional snapshots of the volume rendering can be produced. Choose a level of zoom and angle in which the first time point and the last are visible and take snapshots of each time point with the camera button. Finally, movies rotating the three-dimensional image and showing the time points can be scripted within Amira. Alternatively, the snapshots can be combined into a movie using QuickTime Pro (Apple Inc., Cupertino, California, USA).

2. Method 2: Static imaging of living sepals for quantitative image analysis
   a. Use a transgenic plant expressing a fluorescent plasma membrane marker, such as *ATML1p::mCitrine-RCI2A*. The plant should be healthy and actively flowering.
   b. Under a dissecting scope at about 32× times magnification, gently open the flower using fine forceps (Dumont #5). Use a 23-gauge, 1-inch needle to cut downward along the inner side of the sepal to remove the base of the sepal from the flower.
   c. Wet the sepal by placing it in 50 μl of 0.01% triton X-100 on a precleaned Gold Seal microslide (Cat No 3010). Cover with an 18 mm square cover slip (Corning Cat. No. 2865-18). Tap the side of the slide to displace air bubbles from the sepal. Remove the cover slip.
   d. Mount the sepal by placing it on a new slide in fresh 0.01% triton X-100. Carefully turn the sepal such that desired side faces up for imaging. In this case, the outer abaxial side was imaged. Note that different brands of slide have different properties that make orientation of the sepal easier or more difficult. Again carefully lower a cover slip over the sepal and tap the slide to remove air bubbles. For mosaic images, removing excess mounting solution by placing the corner of a Kimwipe against the edge of the cover slip is essential. Otherwise, the sepal flattens as the liquid evaporates causing shifts between parts of the mosaic.
   e. Examine the sepal with epifluorescence on the confocal microscope to ensure that it is properly mounted, is not obscured by air bubbles, and was not damaged in the dissecting process.
   f. Set the light path of the microscope such that the proper excitation is used and the proper emission is captured. Make sure to exclude chlorophyll (>635 nm). We used a Zeiss 510 Meta upright confocal microscope. For mCitrine, 514 excitation was used together with a dichroic mirror reflecting only light less than 545 nm and a 530–600 nm band pass filter, such that only 530–545 nm wavelength light reached the photomultiplier tube (PMT).
   g. Optimize the brightness of the signal and decrease the background noise as much as possible through adjusting the laser output, transmission, pinhole, detector gain, and amplifier offset. For segmentation, compromising signal to achieve less noise is often better than increasing signal with more noise. In our example, laser output = 50%, transmission = 28.1%, pinhole = 100 μm, detector gain = 711, amplifier offset = 0, and amplifier gain = 1.

h. Take multiple confocal stacks such that adjacent images tile (cover) the whole sepal with some overlap between images. The mature sepal is larger than the field of view with either the 10× or 20×. Two images can be used at 10×, whereas six are generally required at 20×. The increased resolution at 20× is important for subsequent image processing.
i. To create a complete image, make projections of each stack using the confocal software. Use Adobe Photoshop Photomerge function to make a single complete image. There are often slight alignment problems at the junctions between images. Carefully crop away background that will interfere with automated segmentation. The image is now ready for segmentation.

3. Method 3: Live imaging of the SAM

a. The live imaging time course is typically over 24–72 h but can be extended if the plants remain viable. Depending on the goal of the live imaging experiment, you will need to image your samples every 6–24 h. You will have to do a pilot experiment to determine what the optimal time points between imaging sessions are before beginning the experiment. The longer you can go between imaging sessions and still visualize the cellular process you wish to observe the better, as you will do less damage to the tissue and prolong the viability of the tissue thereby extending the possible time course duration.
b. To begin the experiment, germinate seeds on plates containing the appropriate selective growth media.
c. About 7–10 days post-germination the seedlings are transferred onto one of the following growth media: (1) The seedlings are transferred into plastic boxes containing B5 or MS growth media. It is of upmost importance that this step is carried out following aseptic practices in a tissue culture hood and all instruments be sterilized prior to use. Failure to do so will result in contamination of the boxes with mold or bacteria, which will impact the viability of your plants. If you notice mold before the beginning of the live imaging session, the experiment should be stopped and a new batch of seeds should be germinated and transferred to the boxes. To prevent contamination of the boxes, prior to removal for the sterile tissue culture hood, tape the boxes with cloth tape. Place the sealed boxes into the growth room under constant light. (2) Alternatively, if the seeds are a homozygous stock they can be germinated on soil. The plants should be watered from the top to prevent the formation of taproots that will complicate their transfer into the shallow containers for imaging. After 7–10 days of germination, the seedlings are transferred into small round containers making sure to fill them with enough soil from the pot to fill the container. We have found that this method practically eliminates the complication of mold contamination producing healthier plants for imaging.
d. Check the plants every 1–3 days watching for the first sign of shoot emergence. As soon as the bolting shoot is visible and the first floral buds can be seen, begin dissection of the cauline leaves and floral buds to expose the SAM. Every care

should be taken to remove as many of the developing flowers as possible without touching or damaging the SAM. Remove all floral primordia that are growing over the edges of the SAM, as these will cause shadowing of the SAM during image acquisition.

e. Once the developing flowers have been dissected and the SAM is exposed, the imaging session can begin. If you are using the boxes, remove the tape such that the tape can be replaced after the imaging is complete. If you do not have a plasma membrane marker in your sample stain the SAM with FM4-64. Otherwise place the plants at 4 °C for 15 min.
f. Dilute the FM4-64 stock solution (1 mg/ml) 1:10 in ddH_2O (final concentration of 100 μg/ml). Add ~20 μl drop to the tip of the shoot and transfer the plants into a cold room for ~10–15 min. If you are using a plasma membrane marker line (i.e., *UBQ10::Ω29-1tdTomato*) staining with FM4-64 is not necessary as the cells of plasma membrane will be marked. Once staining is complete begin imaging.
g. Fill the box with ice-cold water or place a drop of ice-cold water on the objective lens and the tip of the shoot to form a water column for the water immersion lens (40× or 63×). Lower the stage to a point that the plant will fit under an upright microscope. Raise the stage such that the objective is over the SAM under epifluorescent light and find the SAM.
h. Once the SAM has been visualized in the eyepieces, switch over to using the LSM mode. Note: We use a Zeiss 510 LSCM so the following will be a description of how to proceed using the Zeiss LSM software and interface. If you use a different microscope you will have to modify the following protocol based on your particular microscope software and interface.
i. Set the light path for the appropriate filter sets for the laser line/s of interest. When imaging samples where the plasma membrane will be image with FM4-64 or a RFP, it is best to change the first secondary dichroic beam splitter to 635 Vis setting (NFT 635 Vis). This will remove fluorescent emmission from chloroplasts that will interfere with downstream image processing steps.
j. We set our Argon/2 laser to 5.7A (50% transmission) and attenuate the output down to 10–20%. The lower the laser transmission the healthier your samples will remain throughout the duration of your time-lapse live imaging session.
k. Click on the fast XY to visualize the SAM. Adjust the detector gain, amplifier offset, and amplifier gain. Typical settings for us are in the following ranges depending on the samples being imaged: detector gain – 650–850, amplifier offset – 0.0–0.1, amplifier gain – 1.0–1.5.
l. Once all the settings are set, switch to the *z*-settings and find the top and bottom of your image stack, using the fast XY scan. If you are performing image segmentation following the collection of data points, it is best to scan at a *x*:*y*:*z* ratio of 1:1:1. In the Zeiss *z*-settings, there is a button labeled 1:1:1, click on this and it will automatically set the *z*-scan setting such that the image resolution is 1:1:1.

m. In the Scan control window under the MODE tab set the FRAME SIZE to 1024 and hit the MAX scan speed button. It is best to change the DATA DEPTH to 8-bit and the SCAN DIRECTION as return. Make sure the MODE is LINE, METHOD is MEAN, NUMBER is 1. This will ensure that the scan speed is as fast as possible.
n. Once everything is set, click the start the button and begin imaging.
o. Once the image stack is acquired, proceed to the next sample.
p. After the last sample has been imaged, place plants back into the growth chamber. If using boxes, try to make sure as much of the water is removed and replace the cloth tape around the upper and lower seam to prevent mold contamination.
q. Repeat steps e–o. At certain time points, you may have to dissect away newly emerging primordia that begin to grow over the SAM.
r. Once all time points are collected proceed to image registration and image processing.

4. Method 4: Static live imaging of the SAM

a. To begin the experiment seeds are germinated on plates containing the appropriate selective growth media.
b. After 7–10 days of germination, the seedlings are transferred onto soil.
c. Check the plants for ~10 days watching for the first sign of shoot emergence. Once the shoots have bolted about 5 cm they are ready to image (Note: you can image early or later depending on the experiment).
d. Fill 2–3 60 × 20 mm deep Petri dishes half way with 1% molten agarose in water. Set to the side and allow the agarose to cool.
e. Cut approximately 2 cm of the apical shoot. Dissect away all cauline leaves, flowers, and siliques. Under a dissecting microscope remove the developing floral buds to expose the SAM. Every care should be taken to remove as many of the developing flowers as possible without touching or damaging the SAM. Remove all floral primordia that are growing over the edges of the SAM, as these will cause shadowing of the SAM during image acquisition.
f. Once the developing flowers have been dissected and the SAM is exposed, the imaging session can begin. If you do not have a plasma membrane marker in your sample stain the SAM with FM4-64. Dilute the FM4-64 stock solution (1 mg/ml) 1:10 in ddH_2O (final concentration of 100 μg/ml). Place 200 μl of FM4-64 staining solution into a 1.7 ml eppendrof tube. Invert the dissected shoot apex and insert it into eppendrof tube containing the staining solution and tap the tube to remove air bubbles. Transfer the dissected shoot apex into a cold room for ~10–15 min (Note: If imaging a fluorophore with a temperature-sensitive maturation time you may lose the signal). If you are using a plasma membrane marker line (i.e., *UBQ10::Ω29-1tdTomato*) staining with FM4-64 is not necessary as the cells plasma membrane will be marked. Once staining is complete begin imaging.
g. After staining is complete, make a small hole in the hardened agarose in the Petri dish. Insert the stem into the agarose such that 0.5–1 mm of the shot apex is above the agarose surface. Fill the dish with water such that the shoot apex is covered.

h. Place sample on the stage and center the objective over the SAM under epifluorescent light to find the SAM. Once the SAM has been located switch over to using the LSM mode.
i. Refer to steps i–r in the Method 3: Live Imaging of the SAM section to complete imaging experiment.

B. Denoising with Nonlocal Means

The nonlocal means method for white Gaussian noise reduction (Buades *et al.*, 2005) introduced a new paradigm in image denoising. Its success can be attributed to its originality, simplicity, and ability to greatly reduce noise while sharply resolving the boundaries (edges) of objects. Like many of its predecessors, nonlocal means is a neighborhood filtering method: the noiseless signal value at a pixel location is the weighted average intensity value of its closest neighbors (we consider the pixel itself to be part of its neighborhood). If the signal data has a normal distribution with zero mean and constant variance, this approach is mathematically grounded and it works well in practice, that is, it does reduce noise. Methods based on this framework basically differ in two essential aspects: the averaging scheme and the selection of neighboring pixels. When considering all neighboring pixels to have the same exact importance, the noiseless pixel value is simply the mean value of its neighbors. This old, classical method guarantees noise reduction throughout but unfortunately the final image is severely blurred with sharp edges destroyed. This is not a recommended strategy prior to image segmentation as we should always prefer to have sharp edges in the image, which facilitates the separation of objects of interest.

By distinguishing how much each pixel contributes to the noiseless value of its neighbors, denoising results can be significantly improved. Instead of giving the same weight to all its neighbors, nonlocal means uses a weighted average of pixel values,

$$u_i^{k+1} = \sum_{j \in N_i} w_{ij}^k u_j^k$$

where weight w_{ij}^k measures the contribution of neighboring pixel value u_j^k to the noise reduced value u_i^{k+1} of pixel i at iteration k. The sum above is over all pixels j belonging to the neighborhood N_i of pixel i. The superscript index k refers to the current denoising iteration – we can repeat the average scheme as many times as necessary – starting with the original, $k = 0$, noisy image (for simplicity, we drop this index in the expressions below). In practice, we use up to four iterations to denoize confocal images of sepals and meristem. We mostly work with images having an 8-bit depth for each color channel, so $u \in [0, 255]$. We linearly map Zeiss LSM images quantized using 12 or 16 bit values to this interval. Although one might rightfully suspect that such shortening of range values can lead to a loss of information, it has not proven detrimental at all in our processing of the *Arabidopsis* images.

The neighborhood N_i may comprise all pixels in the image but this is neither efficient nor justifiable. In practice, we chose as neighbors only those other pixels belonging to a square region centered at a given pixel. This is called the search window, shown in Fig. 5B, and its size plays an important role: large search windows tend to over smooth the image while very small windows might leave the image almost unchanged. We usually compute with search windows ranging from 11×11 to 21×21 in size – in order to have a unique center they need to be odd sized. It should be clear that the computational cost increases with the size of the search window as we need to perform more arithmetic calculations

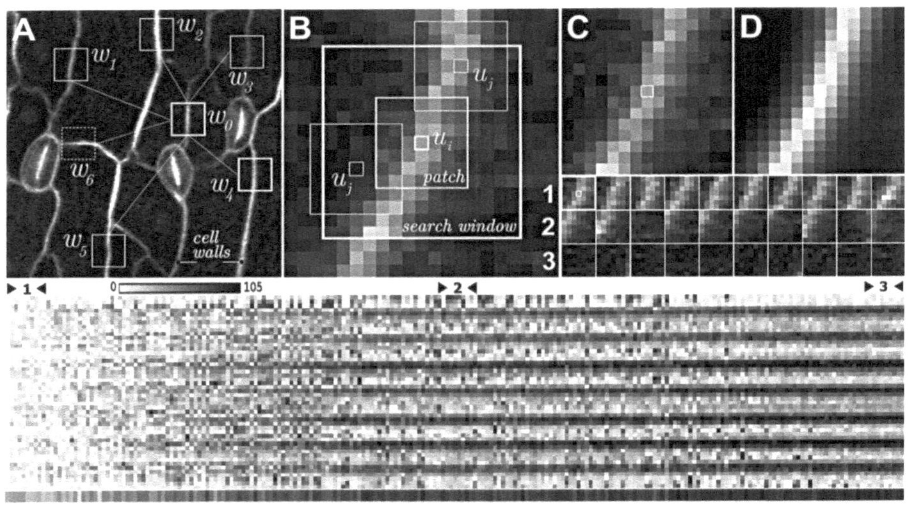

Fig. 5 Nonlocal means scheme (refer to color version). The nonlocal means method succeeds by denoising patches that are structurally similar. (A) We have a collection of square patches each containing a piece of a plasma membrane in a sepal. Averaging pixels whose patches are similar decreases noise while keeping the original image structures intact. (B) How the method works for every pixel in the image. The pixel at the center (small central red square) is denoized using all its neighbors present in the search window (large thick white square). Each pixel in the search window entails a patch (middle-sized squares) containing structural information used in the patch similarity computation. (C, D) A small portion of a plasma membrane before (C) and after (D) denoising. The noise is greatly reduced leaving a sharp contrast-enhanced edge suitable for segmentation. In each of the rows (1-2-3), we show 10 patches which are the closest (1), halfway closest (2), and farthest (3) from the central patch of the target pixel (large central red box in B and repeated as the first patch in row 1), classified by distance between patches. One can visually recognize that patches along the plasma membrane are most similar (row 1) and those away from the wall (row 3) are the most dissimilar to the central patch and should contribute little to the denoising of the central pixel. The heatmap (see color version) on the bottom of the picture contains 225 columns (15×15 search window), a column for each gray pixel which is shown below the heatmap, and it contains 49 rows referring to the number of pixels in 7×7 patches. Each column is color coded to show the difference vector $|v_i - v_j|$ for every patch P_j in the search window shown in (B). They are classified, from left to right, from closest to farthest in similarity to the central patch shown as the first block in row 1. The numbers 1,2, and 3 in the heatmap correspond to the rows 1-2-3 below panel (C). (See color plate.)

for larger windows. Depending on the algorithm implementing the nonlocal means scheme the computational cost can increase from linearly (preferable) to quadratically.

For the classical denoising method described above, we have constant weight values, that is, $w_{ij} = 1/|N_i|$, where $|N_i|$ gives the total number of pixels present in the square search window N_i. Note that this constant weight expression does not account for any structural information that could help during denoising; that is, regardless of the position of a pixel, in an edge or not, it contributes equally as any other pixel to the denoized value of its neighbors and itself. Intuitively, it should not be that way. By incorporating structural information on the weight computation, the nonlocal means method is able to differentiate the contribution of pixels according to their structural similarity. Similar pixels should have a mutually positive contribution on their denoized values while dissimilar pixels should have little or no influence at all. As an example, the square boxes in Fig. 5A are all structurally similar, each containing a piece of and centered at a plasma membrane, except that w_6 contains a rotated version. When we averaged their center pixels, we expect to obtain intensity values consistent with those found exclusively in a plasma membrane, not something else. It is as if we are averaging only those pixels along the plasma membranes without knowing *a priori* their location.

Mathematically, we can express the weight between any two pixels i and j as

$$w_{ij}^k = f^k(d_{ij}, N_i, h)$$

where d_{ij} measures the similarity between pixels i and j, h is a filtering parameter controlling the amount of smoothing, and f is a function that returns high/low values for highly similar/dissimilar pixels. It is customary to use a normalized weighting function $w_{ij} \in [0, 1]$, such as

$$w_{ij} = C\,e^{-d_{ij}/h^2}, \quad C = \frac{1}{\sum_{j \in N_i} e^{-d_{ij}/h^2}}$$

or

$$w_{ij} = C\frac{1}{1 + h\,d_{ij}^2}, \quad C = \frac{1}{\sum_{j \in N_i} 1/(1 + h\,d_{ij}^2)}$$

where C is a normalization constant so we have $\sum_j w_{ij} = 1$. To compute the similarity value d_{ij} one should use a measurement reflecting the structural properties of the square *patches* P_i and P_j surrounding, respectively, pixels i and j (see Fig. 5B). We consider the squared norm of the difference vector $v_{ij} = v_i - v_j$ where v_i is the vector listing all pixel values in patch P_i (likewise for v_j) and have $d_{ij} = |v_{ij}|^2 = v_{ij} \dot{s} v_{ij}$. In practice, this is a cheap and effective way to measure the structural similarity between patches. Since $v_{ii} = 0$, which will always give the maximum possible value for w_{ii} ($w_{ii} = 1$), we use a trick to compute w_{ii} : $w_{ii} = \max_{j \neq i}\{w_{ij}\}$, that is, the weight of the central pixel is the maximum weight of all its neighbors excluding itself. This removes a strong bias toward the central pixel value.

The algorithm is thus very simple. For each pixel in the image, compute the weighted average of neighboring pixels as given above using a suitable expression for weights and similarity values. Repeat for every pixel in the image. And if one needs to apply more than one denoising pass to reduce noise, repeat the process as many times as needed. Anyone with basic programming skills can quickly implement the algorithm and experiment to find which combination of parameter values will give the best denoized results for their particular images.

Once we choose the functions to measure similarity and weight, there are only three parameters that will control denoising: the size of the search window N_i, the size of the patch P_i, and the filtering parameter h. Sizes of patches and search windows are usually kept constant for all pixels but this is not necessary. Adaptive schemes exist to control these sizes benefitting final results but at an increased computational cost. The filtering parameter h gives much control on how smooth is the final result: a value too large might produce a very smooth denoized image and destroy fine details present in the noisy image. If using a very small value, we practically have no real denoising. One has to experiment a few times to find the sweet spot. The patch size should be large enough to capture structural information but not too large to avoid including mixed information (we do not want to compare apples to oranges). We use the fast implementation of nonlocal means from Darbon *et al.* (2008).

The images in Fig. 5C and 5D show a portion of plasma membrane before and after denoising. Note the sharpness of the membrane after denoising and contrast enhancement. It is this recovered sharpness that facilitates segmenting the plasma membrane of cells in sepals and the meristem.

C. Contrast Enhancement

Averaging schemes used by neighborhood filtering methods usually produce images with reduced contrast. This is due to averaging which brings intensities of neighboring pixels closer to a central value thus reducing the difference of their intensities and hence sharpness. The lower the gradient the less contrast we have in an image and hence the more difficult it will be to segment it. In areas where we already have low contrast, something we might commonly find in some slices of a *z*-stack, filtering might completely destroy edge information. The denoising scheme presented above is not immune to this effect and we have to be careful when selecting the denoising parameters for low-contrast images. After denoising, we apply a contrast-enhancement strategy to try to recover as much as possible the initial gradient prior to denoising on the plasma membranes or even locally boost the local intensity gradient to greater values, something that will tremendously help segmenting the image. The main goal here is to accentuate faint plasma membranes so we can easily distinguish regions of plasma membrane and cell interiors.

The nature of our denoized images – thin light edges (plasma membranes) on dark background – have led us to adopt a proven contrast-enhancement technique, namely high-boost filtering. This is a technique available in many image processing packages and can be easily controlled via three parameters. If u gives the intensity of a pixel in the image, its high-boosted version u' is

$$u' = u + a(u_\sigma - u)$$

where u_σ is the Gaussian blurred version of the original pixel using a variance of σ^2 and a is a weight value (amplitude) that gives the strength of boosting. The higher its value, the larger the gradient between the pixel and its neighbors; but a very high value might saturate the image in areas interior to the cell and close to the plasma membrane where the intensity might not be uniform even after denoising. We use amplitude values in the [2,10] interval. The Gaussian blur uses a square kernel whose size ranges from 3×3 to 5×5 – we mostly use the later. How much blurring to use is image dependent and thus we tune σ to suitable values, typically something in the [0.5, 2.5] range. When boosted values are negative or surpass the image quantization value they are truncated to, respectively, 0 and 255 (we use 8-bit quantized images). See Fig. 8C for a contrast-enhanced meristem slice.

D. Segmentation

Our basic strategy to segment sepals and meristems is to manipulate their original raw images such that the plasma membrane of cells is strongly accentuated throughout. This should turn the image into a promising segmentation candidate because high contrast is a key feature to a successful segmentation.

1. Active Contours

We find the nonlocal means method for image denoising straightforward and easily understandable as it involves very simple mathematics. The same is not true for the active contours without edges segmentation method of Chan and Vese (2001) as one would need to understand level set ideas, methods for solving partial differential equations, and numerical optimization, to name a few, which would take more space to fully explain than we can afford in this chapter. We will present the main ideas of the method and refer the reader to the original article (Chan and Vese, 2001) and our own work with it on *Arabidopsis* (Cunha *et al.*, 2010) for extended details.

The active contours without edges segmentation model is attractive due to its simplicity and because it does not require computing image derivatives typically employed to detect region boundaries. It is suitable for our problem because we need to separate only two regions, plasma membranes and cell interiors whose average color matches the dark background. One can view the method as a set of rubber bands that are allowed to bend, break, and join such that regions enclosed by a rubber band (e.g., plasma membranes) have similar intensity values. Likewise, regions external to any rubber band (e.g., cell interior and background) also have similar intensity values that differ from internal regions. The algorithm works by modifying the rubber band such that it adds pixels to its interior when their color is closer to the interior color rather than the exterior color. The same reasoning applies to pixels whose color is closer to the exterior color. The interior and exterior colors are given by the average color of all pixels comprising, respectively, the current interior and exterior regions. So, the game here is to move pixels either inside or outside the current boundary configuration according to their color and the average color of interior and exterior regions. As pixels are exchanged between regions, the regions themselves change their average color. The process is repeated until there is no more change in the colors of interior and exterior regions or until the number of pixels jumping from inside to out, or vice versa, is too small to make any significant change in the boundary configuration. When this happens, we have achieved a final segmentation with one region representing the plasma membranes and the other region representing everything else (see far right columns in Fig. 6). A key rule of the game is that we cannot allow the rubber band to wiggle indefinitely and have an infinite length so we constrain how much it is allowed to extend. Otherwise, we risk every single pixel having a rubber band around it, which is not what we want – we want to form geometrically plausible regions with similar average pixel intensities.

Mathematically, and for completeness, the above can be translated in minimizing the following energy model:

$$E(\phi) = \int_{\Omega} (u - c_1)^2 H(\phi) + (u - c_2)^2 (1 - H(\phi)) + \mu \int_{\Omega} |\nabla \phi|$$

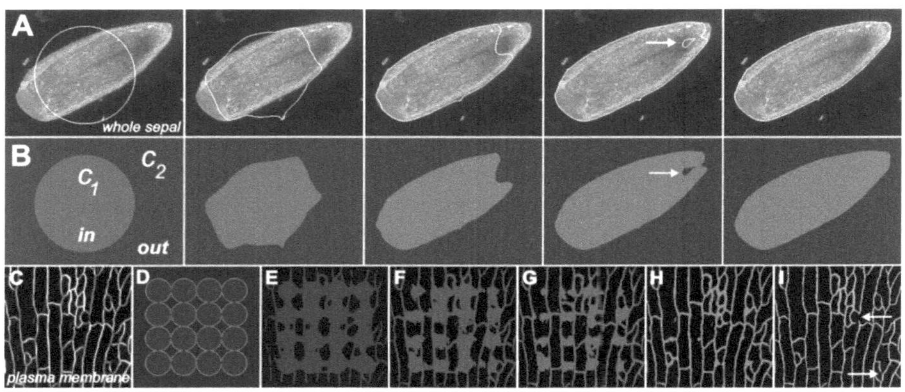

Fig. 6 Segmenting with active contours. In row (A) we show in sequence a few of the transformations a rubber band (white line) takes, starting from a circle, to segment an entire sepal using the active contours without edges approach ($\mu = 0.50$). Images in row (B) show the respective inside and outside regions of the rubber band with their respective, different average colors. Note the subtle color change in and out as the rubber evolves to a final solution. The arrows in (A) and (B) highlight when the rubber band breaks up (new topology) and then rejoins. A small portion of the plasma membranes of a sepal (shown in C) is segmented using active contours. The initial guess for the position of the rubber band is a collection of 16 evenly spaced circles (D), which quickly evolve ($\mu = 0.01$) to capture the plasma membranes (E-I). (I) Arrows show where the method failed, mostly due to the weak signal on the cell walls. (For interpretation of the references to color in this figure legend, the reader is referred to the web version of this book.)

where u gives the pixel intensity in the image, c_1 is the average intensity in the interior region (e.g., plasma membranes), and c_2 is the average intensity in the outside region (see Fig. 6B). The last integral of the energy gives the rubber band length that can be controlled using the model parameter μ. The ϕ is the level set function whose zero iso-contour gives the position of the rubber band and $H(\phi)$ is a Heaviside function that tell us if we are inside, $H(\phi) = 1$, or outside, $H(\phi) = 0$, the region delineated by the boundary. Presenting methods to minimize this energy function is beyond the scope of this chapter so we refer the reader to Chan and Vese (2001) and Cunha *et al.* (2010) for further information including numerical methods to solve the corresponding Euler–Lagrange partial differential equation. We would like to note that μ is an important parameter that must be experimented with to achieve a good final segmentation. If it is a very large number (corresponding to a very stiff rubber) it will be difficult to modify the band and it might not evolve enough to capture all plasma membrane positions. If we use a very small number (corresponding to very soft rubber) we might end up with a very wiggly rubber band with many disjoint regions representing the plasma membranes.

2. Low-Level Pipeline

As an alternative to active contours, we developed a faster segmentation pipeline. The good quality of the nonlocal means denoized images encouraged us to pursue a low-level segmentation pipeline composed of a few mathematical morphology operations combined with standard image processing procedures. The basic idea to obtain cell contours is to locally or globally threshold the enhanced image followed by the formation of a skeleton, removal of dangling edges and other tiny, insignificant regions. Editing is necessary at the end when automatic results are not complete. The steps of this pipeline are shown below, where we start with a denoized image. We refer the reader to Gonzales and Woods (2008) for further details on how to perform the operations below on an image except editing.

- *Edge detection*. We used a simple first derivative method (e.g., Prewitt) to detect the edges.
- *Threshold*. The current image is then thresholded to remove as much as possible regions away from the edges, where the gradient is low (within cells).
- *Hole closing*. The ridges of the original wide edges have zero derivative and they might need to be closed after thresholding is applied.
- *Thinning*. We thin edges to form single pixel-wide skeleton lines.
- *Pruning*. We prune dangling lines and open contours and remove isolated pixels and tiny areas.
- *Editing*. After a visual inspection, the user can adjust the segmentation results whenever necessary by editing the binary image obtained after hole closing and then reapply thinning and pruning to obtain a new set of complete cells. Repeat these steps if needed.

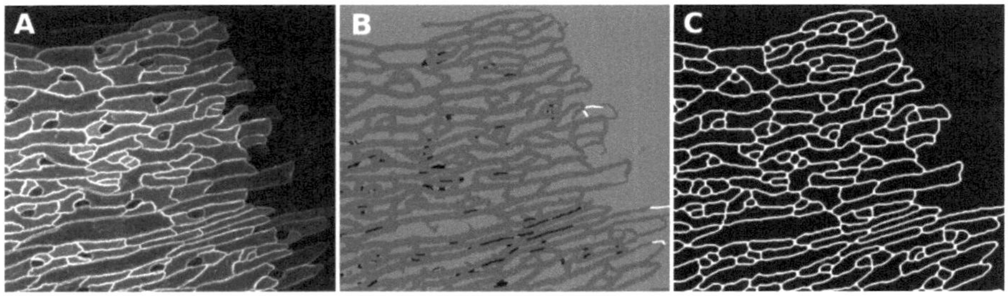

Fig. 7 Low-level segmentation (see color version). (A) A portion of a denoised sepal image. We have enhanced the contrast of the labeled plasma membranes. (B) Automatically constructed thick edges representing the plasma membranes from which we will extract the final cell contours shown in (C). Green and magenta are, respectively, the areas manually removed and added to binary image in order to fix the missing and extra walls. (See color plate.)

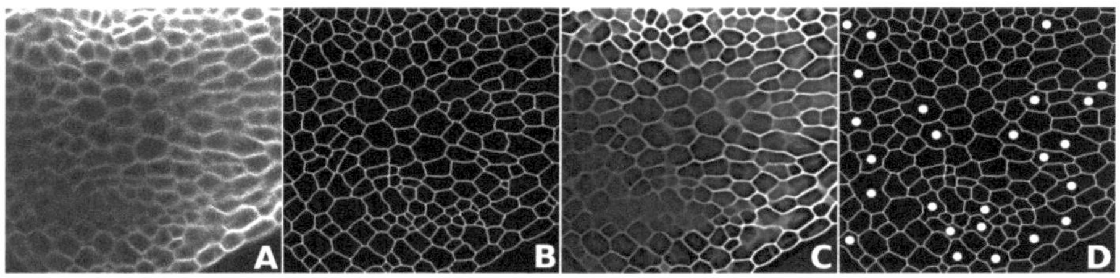

Fig. 8 Segmentation of a SAM slice. (A) Last slice of a 187 slices, uniform x:y:z resolution (pixel size 0.11 μm) floral meristem z-stack; (B) its segmented cell walls. One can note (in A) the different luminosity across the slice including dark regions where it is difficult to visually distinguishing where plasma membranes are located and how they are connected. After four iterations of nonlocal means denoising, contrast enhancement, and the application of mathematical morphology filters to slightly reduce edge thickness, we obtain the image in (C). To obtain the preliminary segmented image (in D) showing one pixel-wide edges (blurred here for visualization), we first apply a local normalization step to even the colors in the entire slice and then local threshold it using mean values of 25 × 25 patches. A pruned skeleton gives then the final edges. Disks (red online) in (D) mark some non-convex regions that potentially need to be corrected to obtain the final segmentation shown in (B). Correction is done in a post-processing editing stage where a few edges are manually added and removed. (For interpretation of the references to color in this figure legend, the reader is referred to the web version of this book.)

In Fig. 7, we show results of using this pipeline in sepals. Fig. 8 shows the segmentation of the last slice of a SAM z-stack. Note in this case that even in regions of low contrast and apparently missing information, we can automatically recover most of the cell walls with very few edits needed to correct the automatic results.

E. Tracking with Soft-Assign

We describe here our tracking method based on the Soft-Assign algorithm. Consider the following problem: given two sets of points X_i and Y_j find the affine transformation $y = Ax + b$ and the correspondence matrix $M = \{M_{ij}\}$ that provides the best match between the points. The correspondence matrix M has 1 at position ij if cell x in frame i matches to cell y in frame j, and 0 otherwise. A general framework for this problem was proposed in Gold *et al.* (1998) and adapted to cell division in Gor *et al.* (2005).

The overall energy function to be minimized is a total error (Euclidian distance between cell positions, defined as the cell centroids, in consecutive frames) of all matches scaled by the expected variance of this error σ^2:

$$E_p = \sum_{i,j} M_{ij}\left(\frac{||Ax_i - y_j^2||^2}{\sigma^2}\right)$$

The mapping may be assumed to affine, with geometric parameters $\{A_{ij}\}$. Alternatively, one may use image warping models such as thin plate splines (Chui and Rangarajan, 2000; Gor *et al.*, 2005) to allow for image warping as discussed in Section IV and also for nonuniform tissue growth. When not known, the geometric (affine or warping) parameters $\{A_{ij}\}$ are jointly estimated with correspondence (Gold *et al.*, 1998). To incorporate possible cell divisions in this objective, this function is modified to minimize total error between current cells and their siblings, as well as their common parents (Gor *et al.*, 2005):

$$E_s = \sum_{k,l} L_{kl}\left(\frac{||y_k - y_l||^2}{\sigma_1^2} + \frac{||y_k - y_l^p||^2}{\sigma_2^2}\right)$$

where the first term stands for the Euclidean distance between two siblings, y_k and y_l, on the same frame, scaled by the expected variance σ_1^2. The second term stands for the Euclidean distance between y_k and the parent of its sibling y_l, denoted by y_l^p, similarly scaled by the expected variance σ_2^2. The correspondence matrix $\{L_{kl}\}$ includes a slack row and column for possible non-matches, to allow for disappearing cells as discussed above in Section II.H. Entries of $\{L_{kl}\}$ will have a value of 1 if cells y_k and y_l are siblings with common parent, and 0 otherwise.

To represent cell and sibling correspondences simultaneously, a full objective function is created from the E_p and E_s terms above (Gor *et al.*, 2005). In addition, to account for affine transformations, the full objective function is jointly optimized using Thin Plate Spline transformations with Soft-Assign algorithm embedded in a deterministic annealing loop (Chui and Rangarajan, 2000; Gor *et al.*, 2005).

To avoid getting trapped in local minima, this framework uses deterministic annealing (Kosowsky and Yuille, 1994) to turn a discrete assignment problem into a continuous one. Initially, it allows the $\{M_{ij}\}$ to take on fractional values, eventually converging to binary values. It works by minimizing the objective function described above, indexed by a control parameter (inverse temperature), which is gradually increased. Within each iteration the iterative optimization algorithm returns an optimized (usually globally optimal, always at least locally optimal), doubly

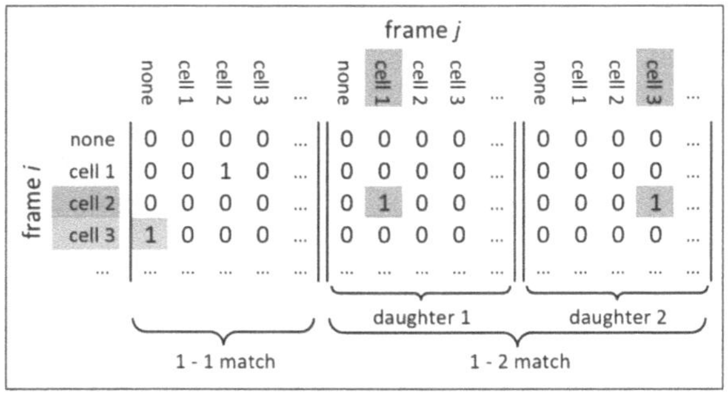

Fig. 9 Handling of 1-1 and 1-2 matching in Soft-Assign. In this example, cell 1 in frame i matches to cell 2 in frame j. Cell 2 in frame i divides to produce cell 1 and cell 3 in frame j. Cell 3 in frame i disappears, that is, moves out of the field of view. (For color version of this figure, the reader is referred to the web version of this book.)

stochastic matrix of point correspondences for the current value of the control parameter. Since the interim assignments are "soft" rather than binary and thus able to smoothly improve, the algorithm is called Soft-Assign.

Cell tracking involves not just matches between individual points, but also cell division and death events. To represent 1-1 (cell to cell), 1-2 (e.g., cell division), and 1-0 (e.g., cell death) matches between cells, the algorithm was generalized to use multiple correspondence $M = \{M_{ij}\}$'s padded with a slack column (denoted by *none*) (Gor *et al.*, 2005). One matrix represents the usual 1-1 matches, and the other two matrices encode possible cell divisions as shown in Fig. 9. For the optimization problem, the property of double stochasticity (1-1 or 1-2 matchings only) must be preserved. However, slack variables permit also 1-0 and 0-1 matchings. To ensure that there is no more than one match for 1, a related normalization is applied to the rows:

$$\sum_j M_{i,j}^{\alpha=1} = \sum_j M_{i,j}^{\alpha=2} = 1 - \sum_j M_{i,j}^{\alpha=0}$$

where α indexes the three sub-matrices (see Fig. 9). This procedure ensures that annealing preserves either 1-1 or 1-2 correspondences, but not both simultaneously.

For example in one sepal live imaging sequence, there are a total of 1491 individual nuclei and 106 division events on 17 frames (Fig. 10). The Soft-Assign algorithm correctly tracked 1403 nuclei as they belong with the correct lineage. Thus, a 94% of the nuclei were tracked in correct lineages. However, out of 106 division events, 42 of them (40%) were identified successfully where both daughters were assigned correctly to the same parent. All other division events assigned one of the daughters to the parent, while the other daughter was not assigned to any parent (Fig. 11). All subsequent division events may still be captured correctly, following an erroneously detected division event.

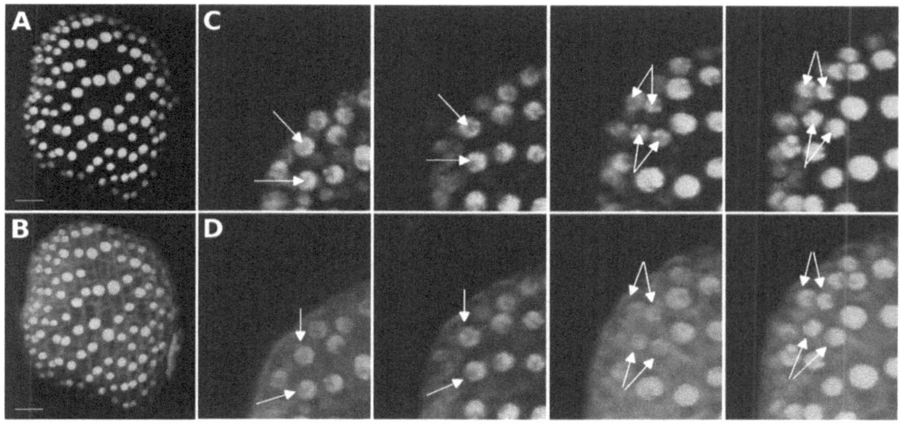

Fig. 10 Sepal lineage. We apply our tracking method in a time-lapse series of sepals consisting of a total of 1491 individual nuclei and 106 division events on 17 frames similar to the ones shown in (A) and (B) (scale bar = 20 μm). In row (C) we have enlarged top-left regions with only nuclei shown in each frame; arrows indicate before and after tracked division events. The Soft-Assign algorithm correctly tracked 1403 nuclei as they belong to the correct lineage (see text). Two cells, indicated by the arrows in the first frame of row (C) divide between frames 2 and 3 in the same row. In (D) we have enlarged top-left regions from original frames (6 hours interval) showing labeled nuclei and cell walls and the same tracked division events. (For color version of this figure, the reader is referred to the web version of this book.)

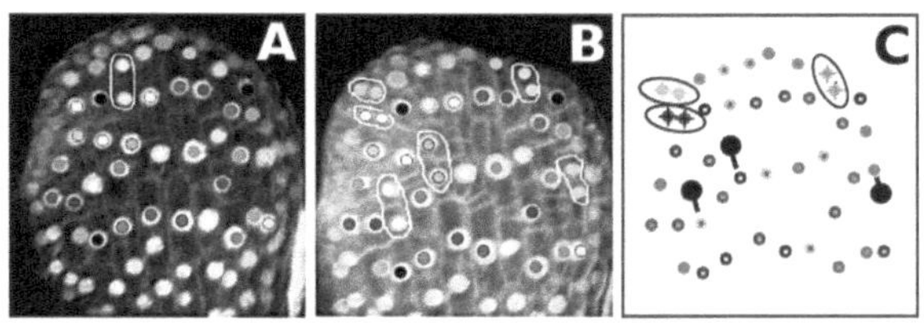

Fig. 11 Tracking in sepals (see color version). (A, B) Two consecutive frames of a sepal time-lapse image. (B) Six division events manually labeled. Out of these six events, three of them were correctly detected: the daughters were assigned to the same parent (shown with ellipses in C). Three other division events (marked using black disks with tails in C) were not assigned to any lineage, they wrongly appear as the first nuclei on a new lineage. The tails point to the missed daughter. (See color plate.)

VI. Further Reading

For those interested in basic and advanced concepts in image processing including image enhancement, segmentation, and mathematical morphology the textbook by Gonzales and Woods (2008) offers a comprehensive account. It is widely adopted in the classroom and the explanations are easy to follow with many practical examples. A more advanced text in computer vision that we recommend is the recent book by Szeliski (2010), which gives a good account of why computer vision is difficult and presents modern methods including machine learning approaches. The book of Soille (2004) offers basic and advanced concepts on mathematical morphology with illustrations on many practical problems including cell segmentation.

The *Handbook of Biological Confocal Microscopy* (Pawley, 2006a) is a rich source of material explaining the image acquisition process in confocal microscopy including live imaging. Chapter 4 (Pawley, 2006b), in particular, address many of the issues related to image acquisition tradeoffs and the nature of noise present in fluorescent laser scanning images. Pawley (2000) offers some practical suggestions to prevent difficulties during a confocal session.

The most recent articles by Fernandez *et al.* (2010) and Liu *et al.* (2010) present advanced developments for segmenting and tracking in the floral and SAM. Their audience is biologists with some knowledge of computing and algorithm development, and they provide software with which one can readily experiment.

Our recent review articles address the challenges and propose strategies to integrate imaging, image processing and analysis, and mathematical and computational modeling as an enabling paradigm in the morphodynamics studies of plant development (Chickarmane *et al.*, 2010; Roeder *et al.*, 2011).

Acknowledgments

We acknowledge funding support from the Gordon and Betty Moore Foundation Cell Center (http://www.cellcenter.caltech.edu/) (AC and AHKR), the Department of Energy Office of Basic Energy Sciences, Division of Chemical Sciences, Geosciences and Biosciences grant DE-FG02-88ER13873 (EMM), National Science Foundation grant IOS-0846192 (EMM), and NIH NRSA Postdoctoral Fellowship F32-GM090534 (PTT). Work of EM was partially supported by NIH R01 GM086883, and NSF's Frontiers in Biological Research (FIBR) program Award No. EF-0330786.

References

Arulampalam, M. S., Maskell, S., and Gordon, N. (2002). A tutorial on particle filters for online nonlinear/non-gaussian bayesian tracking. *IEEE Transact. Signal Process.* **50**, 174–188.

Black, J., Ellis, T., and Rosin, P. (2003). A novel method for video tracking performance evaluation. In *Joint IEEE Int. Workshop on Visual Surveillance and Performance Evaluation of Tracking and Surveillance* (pp. 125–132).

Boisnard-Lorig, C., Colon-Carmona, A., Bauch, M., Hodge, S., Doerner, P., Bancharel, E., Dumas, C., Haseloff, J., and Berger, F. (2001). Dynamic analyses of the expression of the HISTONE::YFP fusion protein in Arabidopsis show that syncytial endosperm is divided in mitotic domains. *The Plant Cell* **13**, 495–509.

Broida, T. J., and Chellappa, R. (1986). Estimation of object motion parameters from noisy images. *IEEE Transact. PAMI* **8**, 90–99.

Buades, A., Coll, B., and Morel, J. M. (2005). A review of denoising algorithm, with a new one. *SIAM J. Multiscale Model Simul.* **4**, 490–530.

Chan, T. F., and Vese, L. A. (2001). Active contours without edges *IEEE Transact. Image Process.* **10**, 266–277.

Chickarmane, V., Roeder, A. H. K., Tarr, P. T., Cunha, A., Tobin, C., and Meyerowitz, E. M. (2010). Computational morphodynamics: a modeling framework to understand plant growth. *Annu. Rev. Plant Biol.* **61**, 65–87.

Chui, H., and Rangarajan, A. (2000). A new algorithm for non-rigid point matching. In *IEEE Conference on Computer Vision and Pattern Recognition* (pp. 44–51).

Comaniciu, D., Ramesh, V., and Meer, P. (2003). Kernel-based object tracking. *IEEE Trans. Pattern Anal. Mach. Intell.* **25**, 564–575.

Cunha, A. L., Roeder, A. H. K., and Meyerowitz, E. M. (2010). Segmenting the sepal and shoot apical meristem of Arabidopsis thaliana. In *Annual International Conference of the IEEE Engineering in Medicine and Biology Society (EMBC)* (pp. 5338–5342).

Cutler, S. R., Ehrhardt, D. W., Griffitts, J. S., and Somerville, C. R. (2000). Random GFP::cDNA fusions enable visualization of subcellular structures in cells of Arabidopsis at a high frequency. *Proc. Natl. Acad. Sci.* **97**, 3718–3723.

Dabov, K., Foi, A., Katkovnik, V., and Egiazarian, K. (2007). Image denoising by sparse 3D transform-domain collaborative filtering. *IEEE Transact. Image Process.* **16**, 2080–2095.

Darbon, J., Cunha, A., Chan, T. F., Osher, S., and Jensen, G. J. (2008). Fast nonlocal filtering applied to electron cryomicroscopy. In *IEEE International Symposium on Biomedical Imaging* (pp. 1331–1334).

Fernandez, R., Das, P., Mirabet, V., Moscardi, E., Traas, J., Verdeil, J. -L., Malandain, G., and Godin, C. (2010). Imaging plant growth in 4D: robust tissue reconstruction and lineaging at cell resolution. *Nat. Meth.* **7**, 547–553.

Gold, S., Rangarajan, A., Lu, C. -P., Pappu, S., and Mjolsness, E. (1998). New algorithms for 2D and 3D point matching: pose estimation and correspondence. *Pattern Recog.* **31**, 1019–1031.

Gonzales, R. C., and Woods, R. E. (2008). *Digital Image Processing, 3rd ed. Prentice Hall, .*

Gor, V., Elowitz, M., Bacarian, T., and Mjolsness, E. (2005). Tracking cell signals in fluorescent images. In *IEEE Computer Society Conference on Computer Vision and Pattern Recognition* (pp. 142–148).

Hamant, O., Heisler, M. G., Jönsson, H., Krupinski, P., Uyttewaal, M., Bokov, P., Corson, F., Sahlin, P., Boudaoud, A., Meyerowitz, E. M., Couder, Y., and Traas, J. (2008). Developmental patterning by mechanical signals in Arabidopsis. *Science* **322**, 1650–1655.

Heisler, M. G., Hamant, O., Krupinski, P., Uyttewaal, M., Ohno, C., Jönsson, H., Traas, J., and Meyerowitz, E. M. (2010). Alignment between PIN1 polarity and microtubule orientation in the shoot apical meristem reveals a tight coupling between morphogenesis and auxin transport. *PLoS Biol.* **8**, e1000516.

Jaeger, S., Song, Q., and Chen, S. -S. (2009). Dynamik: a software environment for cell dynamics, motility, and information tracking, with an application to Ras pathways. *Bioinformatics* **25**, 2383–2388.

Jones, T., Kang, I., Wheeler, D., Lindquist, R., Papallo, A., Sabatini, D., Golland, P., and Carpenter, A. (2008). Cellprofiler analyst: data exploration and analysis software for complex image-based screens. *BMC Bioinformat.* **9**, 482.

Karimi, M., Bleys, A., Vanderhaeghen, R., and Hilson, P. (2007a). Building blocks for plant gene assembly. *Plant Physiol.* **145**, 1183–1191.

Karimi, M., Depicker, A., and Hilson, P. (2007b). Recombinational cloning with plant gateway vectors. *Plant Physiol.* **145**, 1144–1154.

Kosowsky, J. J., and Yuille, A. L. (1994). The invisible hand algorithm: solving the assignment problem with statistical physics. *Neural Networks* **7**, 477–490.

Lewis, F. T. (1923). The typical shape of polyhedral cells in vegetable Parenchyma and the restoration of that shape following cell division. *Proc. Amer. Acad. Arts Sci.* **58**, 537–554.

Liu, M., Yadav, R., Roy-Chowdhury, A., and Reddy, G. V. (2010). Automated tracking of stem cell lineages of Arabidopsis shoot apex using local graph matching. *Plant J.* **62**, 135–147.

Makitalo, M., and Foi, A. (2011). Optimal inversion of the Anscombe transformation in low-count Poisson image denoising. *IEEE Trans. Image Process.* **20**, 99–109.

Marcuzzo, M., Quelhas, P., Campilho, A., Mendonça, A. M., and Campilho, A. C. (2008). Automatic cell segmentation from confocal microscopy images of the Arabidopsis root. In *IEEE International Symposium on Biomedical Imaging* (pp. 712–715).

Matzke, E. B., and Duffy, R. M. (1955). The three-dimensional shape of interphase cells within the apical meristem of Anacharis densa. *Amer. J. Bot.* **42**, 937–945.

Mosig, A., Jager, S., Wang, C., Nath, S., Ersoy, I., Palaniappan, K., and Chen, S. -S. (2009). Tracking cells in life cell imaging videos using topological alignments. *Algor Mol. Biol.* **4**, 10.
Nakagawa, S., Niimura, Y., Gojobori, T., Tanaka, H., and Miura, K. (2008). Diversity of preferred nucleotide sequences around the translation initiation codon in eukaryote genomes. *Nucleic Acids Res.* **36**, 861–871.
Nguyen, A. W., and Daugherty, P. S. (2005). Evolutionary optimization of fluorescent proteins for intracellular FRET. *Nat. Biotechnol.* **23**, 355–360.
Nimchuck, Z. L., Tarr, P. T., Ohno, C., Qu, X., and Meyerowitz, E. M. (2011). Plant stem cell signaling involves ligand-dependent trafficking of the clavata1 receptor kinase. *Curr. Biol.* **21**, 345–352.
Pan, P., Porikli, F., and Schonfeld, D. (2009). A new method for tracking performance evaluation based on a reflective model and perturbation analysis. In *IEEE International Conference on Acoustics, Speech and Signal Processing* (pp. 3529–3532).
Pawley, J. B. (2000). The 39 steps: a cautionary tale of quantitative 3-D fluorescence microscopy. *BioTechniques* **28**, 884–886.
Pawley, J. B. (2006a). *Handbook of Biological Confocal Microscopy, 3rd ed. Springer, .*
Pawley, J. B. (2006b). Points, pixels, and gray levels: digitizing imaging data. In "*Handbook of Biological Confocal Microscopy*," (J. Pawley, ed.), (pp. 59–79). Springer, 3rd ed..
Peterfreund, N. (1999). Robust tracking of position and velocity with Kalman snakes. *IEEE Transactions on Pattern Analysis and Machine Intelligence* **21**, 564–569.
Rasband, W. S. (1997). *ImageJ - image analysis software.* U.S. National Institutes of Health, Bethesda Maryland.
Reddy, G. V., and Meyerowitz, E. M. (2005). Stem-cell homeostasis and growth dynamics can be uncoupled in the Arabidopsis shoot apex. *Science* **28**, 663–667.
de Reuille, P. B., Bohn-Courseau, I., Godin, C., and Traas, J. (2005). A protocol to analyse cellular dynamics during plant development. *Plant J.* **44**, 1045–1053.
Roeder, A. H. K., Chickarmane, V., Cunha, A., Obara, B., Manjunath, B. S., and Meyerowitz, E. M. (2010). Variability in the control of cell division underlies sepal epidermal patterning in Arabidopsis thaliana. *PLoS Biol.* **8**, .
Roeder, A. H. K., Tarr, P. T., Tobin, C., Zhang, X., Chickarmane, V., Cunha, A., and Meyerowitz, E. M. (2011). Computational morphodynamics of plants: integrating development over space and time. *Nat. Rev. Mol. Cell Biol.* **12**, 265–273.
Roth, D., Koller-Meier, E., Rowe, D., Moeslund, T., and Van Gool, L. (2008). Event-based tracking evaluation metric. In *IEEE Workshop on Motion and video Computing* (pp. 1–8).
Sacan, A., Ferhatosmanoglu, H., and Coskun, H. (2008). Celltrack: an opensource software for cell tracking and motility analysis. *Bioinformatics* **24**, 1647–1649.
Shaner, N. C., Lin, M. Z., McKeown, M. R., Steinbach, P. A., Hazelwood, K. L., Davidson, M. W., and Tsien, R. Y. (2008). Improving the photostability of bright monomeric orange and red fluorescent proteins. *Nat. Meth.* **5**, 545–551.
Shaner, N. C., Steinbach, P. A., and Tsien, R. Y. (2005). A guide to choosing fluorescent proteins. *Nat. Meth.* **2**, 905–909.
Shen, H., Nelson, G., Nelson, D. E., Kennedy, S., Spiller, D. G., Griffiths, T., Paton, N., Oliver, S. G., White, M. R. H., and Kell, D. B. (2006). Automated tracking of gene expression in individual cells and cell compartments. *J. Royal Soc. Interface* **3**, 787–794.
Sleat, D. E., Gallie, D. R., Jefferson, R. A., Bevan, M. W., Turner, P. C., and Wilson, T. M. (1987). Characterisation of the 5′-leader sequence of tobacco mosaic virus RNA as a general enhancer of translation in vitro. *Gene* **60**, 217–225.
Soille, P. (2004). *Morphological Image Analysis: Principles and Applications, 2nd ed. Springer-Verlag, Inc, New York.*
Song, Y., Feng, X., and Perona, P. (2000). Towards detection of human motion. In *IEEE Computer Society Conference on Computer Vision and Pattern Recognition* (pp. 810–817).
Szeliski, R. (2010). *Computer Vision: Algorithms and Applications.* Springer, .
Thompson, M. V., and Wolniak, S. M. (2008). A plasma membrane-anchored fluorescent protein fusion illuminates sieve element plasma membranes in arabidopsis and tobacco. *Plant Physiol.* **146**, 1599–1610.

Veenman, C. J. (2001). Resolving motion correspondence for densely moving points. *IEEE Transactions on Pattern Analysis and Machine Intelligence* **23**, 54–72.
Venken, K. J. T., Carlson, J. W., Schulze, K. L., Pan, H., He, Y., Spokony, R., Wan, K. H., Koriabine, M., de Jong, P. J., White, K. P., Bellen, H. J., and Hoskins, R. A. (2009). Versatile P[acman] BAC libraries for transgenesis studies in Drosophila melanogaster. *Nat. Meth.* **6**, 431–434.
Wang, Q., Niemi, J., Tan, C. -M., You, L., and West, M. (2010). Image segmentation and dynamic lineage analysis in single-cell fluorescence microscopy. *Cytometry Part A* **77A**, 101–110.
Yilmaz, A., Javed, O., and Shah, M. (2006). Object tracking: A survey. *ACM Computing Surveys* **38**.

CHAPTER 13

Multi-Scale Modeling of Tissues Using CompuCell3D

Maciej H. Swat[*], **Gilberto L. Thomas**[*,†], **Julio M. Belmonte**[*], **Abbas Shirinifard**[*], **Dimitrij Hmeljak**[*] **and James A. Glazier**[*]

[*]Department of Physics, Biocomplexity Institute, Indiana University, Bloomington, Indiana, USA

[†]Instituto de Física, Universidade Federal do Rio Grande do Sul, C.P. 15051, Porto Alegre, Brazil

Abstract

The study of how cells interact to produce tissue development, homeostasis, or diseases was, until recently, almost purely experimental. Now, multi-cell computer simulation methods, ranging from relatively simple cellular automata to complex immersed-boundary and finite-element mechanistic models, allow *in silico* study of multi-cell phenomena at the tissue scale based on biologically observed cell behaviors and interactions such as movement, adhesion, growth, death, mitosis, secretion of chemicals, chemotaxis, etc. This tutorial introduces the lattice-based Glazier–Graner–Hogeweg (GGH) Monte Carlo multi-cell modeling and the open-source GGH-based CompuCell3D simulation environment that allows rapid and intuitive modeling and simulation of cellular and multi-cellular behaviors in the context of

0091-679X/10 $35.00
DOI 10.1016/B978-0-12-388403-9.00013-8

tissue formation and subsequent dynamics. We also present a walkthrough of four biological models and their associated simulations that demonstrate the capabilities of the GGH and CompuCell3D.

I. Introduction

A key challenge in modern biology is to understand how molecular-scale machinery leads to complex functional structures at the scale of tissues, organs, and organisms. While experiments provide the ultimate verification of biological hypotheses, models and subsequent computer simulations are increasingly useful in suggesting both hypotheses and experiments to test them. Identifying and quantifying the cell-level interactions that play vital roles in pattern formation will aid the search for treatments for developmental diseases like cancer and for techniques to develop novel cellular structures.

Unlike experiments, models are fast to develop, do not require costly apparatus, and are easy to modify. However, abstracting the complexity of living cells or tissues into a relatively simple mathematical/computational formalism is difficult. Creating mathematical models of cells and cell–cell interactions that can be implemented efficiently in software requires drastic simplifications: no complete model could be solved within a reasonable time period.

Consequently, the quality and reliability of mathematical models depend on how well complex cell behaviors can be represented using simplified mathematical approaches.

Tissue-scale models explain how local interactions within and between cells lead to complex biological patterning. The two main approaches to tissue modeling are (1) *Continuum* models, which use cell-density fields and partial differential equations (PDEs) to model cell interactions without explicit representations of cells, and (2) *Agent-based* models, which represent individual cells and interactions explicitly. Agent-based *in silico* experiments are gaining popularity because they allow control of the level of detail with which individual cells are represented.

II. Glazier-Graner-Hogeweg (GGH)Modeling

The GGH model (Glazier and Graner, 1992; Graner and Glazier, 1993) provides an intuitive mathematical formalism to map observed cell behaviors and interactions onto a relatively small set of model parameters – making it attractive both to wet-lab and computational biologists.

Like all models, the GGH technique has a typical application domain: modeling soft tissues with motile cells at single-cell resolution. The GGH has been continuously and successfully applied to model biological and biomedical processes, including *Tumor growth* (Dormann *et al.*, 2001; dos Reis *et al.*, 2003; Drasdo *et al.*, 2003; Holm *et al.*, 1991; Turner and Sherratt, 2002), *Gastrulation* (Drasdo and Forgacs, 2000;

Drasdo *et al.*, 1995; Longo *et al.*, 2004), *Skin pigmentation* (Collier *et al.*, 1996; Honda *et al.*, 2002; Wearing *et al.*, 2000), *Neurospheres* (Zhdanov and Kasemo, 2004a,b), *Angiogenesis* (Ambrosi *et al.*, 2004; Ambrosi *et al.*, 2005; Gamba *et al.*, 2003; Merks *et al.*, 2008; Merks and Glazier, 2006; Murray, 2003; Pierce *et al.*, 2004; Serini *et al.*, 2003), the *Immune system* (Kesmir and de Boer, 2003; Meyer-Hermann *et al.*, 2001), *Yeast colony growth* (Nguyen *et al.*, 2004; Walther *et al.*, 2004), *Myxobacteria* (Alber *et al.*, 2006; Arlotti *et al.*, 2004; Börner *et al.*, 2002; Bussemaker *et al.*, 1997; Dormann *et al.*, 2001), *Stem cell differentiation* (Knewitz and Mombach, 2006; Zhdanov and Kasemo, 2004a,b), *Dictyostelium discoideum* (Marée and Hogeweg, 2001, 2002; Marée *et al.*, 1999a,b; Savill and Hogeweg, 1997), *Simulated evolution* (Groenenboom and Hogeweg, 2002; Groenenboom *et al.*, 2005; Hogeweg, 2000; Johnston, 1998; Kesmir *et al.*, 2003; Pagie and Mochizuki, 2002), *General developmental patterning* (Honda and Mochizuki, 2002; Zhang *et al.*, 2011), *Convergent extension* (Zajac, 2002; Zajac *et al.*, 2002; Zajac *et al.*, 2003), *Epidermal formation* (Savill and Sherratt, 2003) *Hydra regeneration* (Mombach *et al.*, 2001; Rieu *et al.*, 2000), *Plant growth*, (Grieneisen *et al.*, 2007), *Retinal patterning* (Mochizuki, 2002; Takesue *et al.*, 1998), *Wound healing* (Dallon *et al.*, 2000; Maini *et al.*, 2002; Savill and Sherratt, 2003), *Biofilms* (Kreft *et al.*, 2001; Picioreanu *et al.*, 2001; Popławski *et al.*, 2008; Van Loosdrecht *et al.*, 2002), *Limb bud development* (Chaturvedi *et al.*, 2004; Popławski *et al.*, 2007), somite segmentation (Glazier *et al.*, 2008; Hester *et al.*, 2011), vascular system development (Merks and Glazier, 2006), choroidal neovascularization, lumen formation, cellular intercalation (Zajac *et al.*, 2000, 2003), *etc.*...

The GGH model represents a single region in space by multiple regular *lattices* (the *cell lattice* and optional *field lattices*). Most *GGH model objects* live on one of these lattices. The most fundamental GGH object, a *generalized cell,* may represent a biological cell, a subcellular compartment, a cluster of cells, or a piece of non-cellular material or surrounding *medium*. Each generalized cell is an extended domain of lattice pixels in the cell lattice that share a common index (referred to as the *cell index* σ). A biological cell can be composed of one or more generalized cells. In the latter case, the biological cell is defined as a cluster of generalized cells called *subcells*, which can describe cell compartments, complex cell shapes, cell polarity, *etc.*... For details on subcells, see Walther *et al.*, 2004; Börner *et al.*, 2002; Glazier *et al.*, 2007; Walther *et al.*, 2005.

Each generalized cell has an associated list of *attributes*, *e.g.*, *cell type*, *surface area* and *volume*, and more complex attributes describing its state, biochemical networks, *etc.*... *Interaction descriptions* and *dynamics* define how GGH objects behave.

The *effective energy* (H) Eq. (1) implements most cell properties, behaviors and interactions via constraint terms in H (Glazier *et al.*, 1998; Glazier and Graner, 1993; Glazier, 1993, 1996; Glazier *et al.*, 1995; Graner and Glazier, 1992; Mombach *et al.*, 1995; Mombach and Glazier, 1996). Since the terminology has led to some confusion in the past, we emphasize that the effective energy is simply a way to produce a desired set of cell behaviors and does **not** represent the physical energy of the cells.

In a typical GGH model, cells have defined volumes area, and interact via contact adhesion, so H is:

$$H = \sum_{\substack{\vec{i},\vec{j} \\ \text{neighbors}}} J(\tau(\sigma_{\vec{i}}), \tau(\sigma_{\vec{j}}))(1 - \delta(\sigma_{\vec{i}}, \sigma_{\vec{j}})) + \sum_{\sigma} [\lambda_{\text{vol}}(\sigma)(v(\sigma) - V_t(\sigma))^2]. \quad (1)$$

The first sum, over all pairs of neighboring lattice sites $\vec{i}$ and $\vec{j}$, calculates the *boundary* or *contactenergy* between neighboring cells to implement adhesive interactions. $J(\tau(\sigma_{\vec{i}}), \tau(\sigma_{\vec{j}}))$is the boundary energy per unit contact area for a pair of cells, with $\sigma_{\vec{i}}$ of type $\tau(\sigma_{\vec{i}})$ occupying cell-lattice site $\vec{i}$ and $\sigma_{\vec{j}}$ of type $\tau(\sigma_{\vec{j}})$ occupying cell-lattice site $\vec{j}$. The delta function restricts the contact-energy contribution to cell-cell interfaces. We specify $J(\tau(\sigma_{\vec{i}}), \tau(\sigma_{\vec{j}}))$ as a matrix indexed by the cell types. In practice, the range of cell types - $\tau(\sigma_{\vec{i}})$- is quite limited, whereas the range of cell indexes $\sigma_{\vec{i}}$ can be quite large, since σ enumerates all generalized cells in the simulation. Higher contact energies between cells result in greater repulsion between cells and lower contact energies result in greater adhesion between cells.

The second sum in (1), over all generalized cells, calculates the effective energy due to the volume constraint. Deviations of the volume area of cell σ from its target value ($V_t(\sigma)$), increase the effective energy, penalizing these deviations. On average, a cell will occupy a number of pixels slightly smaller than its target volume due to surface tension from the contact energies (J). The parameter λ_{vol} behaves like a Young's modulus, or *inverse compressibility*, with higher values reducing fluctuations of a cell's volume about its target value. The volume constraint defines $P = 2\lambda_{\text{vol}}(\sigma)(v(\sigma) - V_t(\sigma))$ as the *pressure* inside the cell. In similar fashion we can implement a constraint on cell's surface or membrane area.

Cell dynamics in the GGH model provide a simplified representation of cytoskeletally-driven cell motility using a stochastic modified Metropolis algorithm (Cipra, 1987) consisting of a series of index-copy attempts (see Figs. 1 and 2). Before each attempt, the algorithm randomly selects a *target site* in the cell lattice, $\vec{i}$, and a neighboring *source site* $\vec{i}'$. If different generalized cells occupy these sites, the algorithm sets $\sigma_{\vec{i}} = \sigma_{\vec{i}'}$ with probability $P(\sigma_{\vec{i}} \to \sigma_{\vec{i}'})$, given by the Boltzmann acceptance function (Metropolis *et al.*, 1953):

$$P(\sigma_{\vec{i}} \to \sigma_{\vec{i}'}) = \begin{cases} 1 & : \quad \Delta H \leq 0 \\ e^{-\frac{\Delta H}{T_m}} & : \quad \Delta H > 0 \end{cases}, \quad (2)$$

where ΔH is the change in the effective energy if the copy occurs and T_m is a parameter describing the amplitude of cell-membrane fluctuations. T_m can be specified globally or be cell specific or cell-type specific.

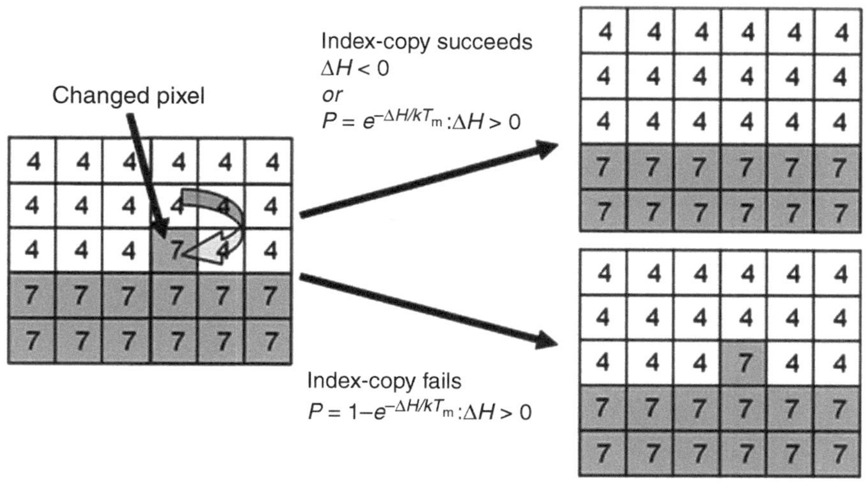

Fig. 1 GGH representation of an index-copy attempt for two cells on a 2D square cell lattice – The "white" pixel (source) attempts to replace the "grey" pixel (target). The probability of accepting the index copy is given by Eq. (2).

The average value of the ratio $\Delta H/T_m$ for a given generalized cell determines the amplitude of fluctuations of the cell boundaries. High $\Delta H/T_m$ results in rigid, barely- or non-motile cells and little cell rearrangement. For low $\Delta H/T_m$, large fluctuations allow a high degree of cell motility and rearrangement. For extremely low $\Delta H/T_m$, cells may fragment in the absence of a constraint sufficient to maintain the integrity of the borders between them. Because $\Delta H/T_m$ is a ratio, we can achieve appropriate cell motilities by varying either T_m or ΔH. Varying T_m allows us to explore the impact of global changes in cytoskeletal activity. Varying ΔH allows us to control the relative motility of the cell types or of individual cells by varying, for example, cells' inverse compressibility (λ_{vol}), the target volume (V_t) or the contact energies (J).

An index copy that increases the effective energy, *e.g.*, by increasing deviations from target values for cell volume or surface area or juxtaposing mutually repulsive cells, is improbable. Thus, the cell pattern evolves in a manner consistent with the biologically-relevant "guidelines" incorporated in the effective energy: cells maintain volumes close to their target values, mutually adhesive cells stick together, mutually repulsive cells separate, *etc.*.... The Metropolis algorithm evolves the cell-lattice configuration to simultaneously satisfy the constraints, to the extent to which they are compatible, with perfect damping (*i.e.*, average velocities are proportional to applied forces). Thus, the average time-evolution of the cell lattice corresponds to that achievable deterministically using finite-element or center-model methodologies with perfect damping.

A Monte Carlo Step (*MCS*) is defined as N index-copy attempts, where N is the number of sites in the cell lattice, and sets the natural unit of time in the model.

The conversion between MCS and experimental time depends on the average value of $\Delta H/T_m$. In biologically-meaningful situations, MCS and experimental time are proportional (Alber *et al.*, 2002, 2004; Novak *et al.*, 1999; Cickovski *et al.*, 2007).

In addition to generalized cells, a GGH model may contain other objects such as *chemical fields* and *biochemical networks* as well as *auxiliary equations* to describe behaviors like cell growth, division and rule-based differentiation. *Fields* evolve due to secretion, absorption, diffusion, reaction and decay according to appropriate PDEs. While complex coupled-PDEs are possible, most models require only secretion, absorption, diffusion and decay. Subcellular biochemical networks are usually described by *ordinary differential equations* (*ODEs*) inside individual generalized cells.

Extracellular chemical fields and subcellular networks affect generalized-cell behaviors by modifying the effective energy (*e.g.*, changes in cell target volume due to chemical absorption, chemotaxis in response to a field gradient or cell differentiation based on the internal state of a genetic network).

From a modeler's viewpoint the GGH technique has significant advantages compared to other methods. A single processor can run a GGH simulation of tens to hundreds of thousands of cells on lattices of up to 1024^3 sites. Because of the regular lattice, GGH simulations are often much faster than equivalent adaptive-mesh finite element simulations operating at the same spatial granularity and level of modeling detail. For smaller simulations, the speed of the GGH allows fine-grained sweeps to explore the effects of parameters, initial conditions, or details of biological models. Adding biological mechanisms to the GGH is as simple as adding new terms to the effective energy. GGH solutions are usually structurally stable, so accuracy degrades gracefully as resolution is reduced. The ability to model cells as deformable entities allows modelers to explore phenomena such as apical constriction leading to invagination, which are much harder to model using, for example, center models. However, the lattice-based representation of cells has also some drawbacks. The cell surface is pixelated, complicating measurements of surface area and curvature. The fixed discretization makes explicit modeling of fibers or membranes expensive, since the lattice constant must be set to the smallest scale to be explicitly represented. Cell membrane fluctuations are also caricatured as a result of the fixed spatial resolution. However, the latest versions of CC3D support a layer of finite-element links which have length but zero diameter. These can be used to represent fibers or membranes, allowing a simulation to combine the advantages of both methods at the cost of increased model complexity. In addition, the maximum speed with which cells can move on the cell lattice is approx. 0.1 pixel per MCS, which often fixes a finer time resolution than needed for other processes in a simulation. A more fundamental issue is that CC3D generalized cells move by destroying pixels and creating pixels, so rigid-body motion and advection are absent unless they are implemented explicitly. CC3D provides tools for both. The rigid-body simulators in CC3D are increasingly popular, but the advection solvers have so far been little used.

The canonical formulation of the GGH is derived from statistical physics. Consequently some of its terminology and concepts may initially seem unnatural to wet-lab biologist. To connect experimentally measured quantities to simulation parameters we employ a set of experimental and analysis techniques to extract parameter values. For example, even though the GGH intrinsic cell motility is not accessible in an experiment, the diffusion constant of cells in aggregates can be measured in both simulation and experiments. We can then adjust the GGH motility to make the diffusion constants match. Similarly, we can determine the effective form and strength of a cell's chemotaxis behavior from experimental dose response curves of net cell migration in response to net concentration gradients of particular chemoattractant. For example, if a cell of given type in a given gradient in a given environment moves with a given velocity, we can then fit the GGH chemotaxis parameters so the simulated cells reproduce that velocity. The GGH contact energies between cells can also be set to provide the experimentally accessible surface tensions between tissues (Glazier and Graner, 1992; Graner and Glazier, 1993; Glazier *et al.*, 2008; Steinberg, 2007). When experimental parameter values are not available, we perform a series of simulations varying the unknown parameter(s) and fit to match a macroscopic dynamic pattern which we can determine experimentally.

To speed execution, CompuCell3D models often reduce 3D simulations to their 2D analogs. While moving from 3D to 2D or *vice versa* is much easier in CC3D than in an adaptive mesh finite element simulation, the GGH formalism still requires rescaling of most model parameters. At the moment, such rescaling must be done by hand. *E.g.* in 2D, a pixel on a regular square lattice has 4 nearest neighbors, while in 3D it has 6 nearest neighbors. Therefore all parameters which involve areas surface (*e.g.* the surface area constraint, or contact energies) have to be rescaled. To simplify diffusion calculations, we often assume that diffusion takes place uniformly everywhere in space, with cells secreting or taking up chemicals at their centers of mass. This approach caricatures real diffusion, where chemicals are secreted through cell membranes and diffuse primarily in the extracellular space, which may itself have anisotropic or hindered diffusion. Since most CC3D simulations neglect intercellular spaces smaller than one or two microns, we connect to real extracellular diffusion by choosing the CC3D diffusion coefficient so that the effective diffusion length in the simulation corresponds to that measured in the experiment.

Overall, despite these issues, the mathematical elegance and simplicity of the GGH formalism has led to substantial popularity.

III. CompuCell3D

CC3D allows users to build sophisticated models more easily and quickly than does specialized custom code. It also facilitates model reuse and sharing.

A CC3D model consists of CC3DML scripts (an XML-based format), Python scripts, and files specifying the initial configurations of the cell lattice and of any

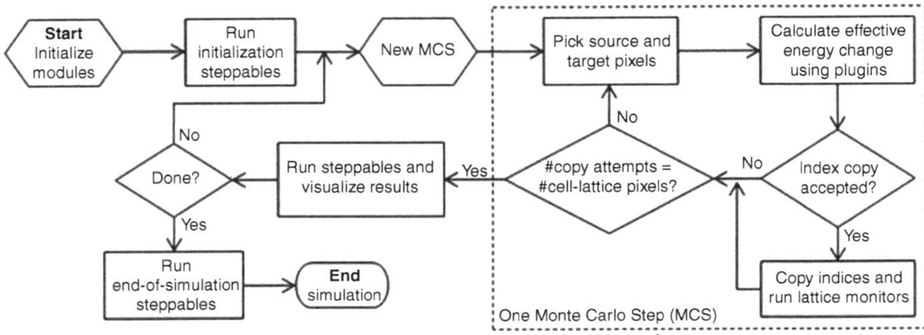

Fig. 2 Flow chart of the GGH algorithm as implemented in CompuCell3D.

fields. The CC3DML script specifies basic GGH parameters such as lattice dimensions, cell types, biological mechanisms, and auxiliary information, such as file paths. Python scripts primarily monitor the state of the simulation and implement changes in cell behaviors, for example, changing the type of a cell depending on the oxygen partial pressure in a simulated tumor.

CC3D is modular, loading only the modules needed for a particular model. Modules that calculate effective energy terms or monitor events on the cell lattice are called *plugins*. Effective-energy calculations are invoked every pixel-copy attempt, while cell-lattice monitoring plugins run whenever an index copy occurs. Because plugins are the most frequently called modules in CC3D, most are coded in C++ for speed.

Modules called *steppables* usually perform operations on cells, not on pixels. Steppables are called at fixed intervals measured in MCS. Steppables have three main uses: (1) to adjust cell parameters in response to simulation events,[1] (2) to solve PDEs, (3) to load simulation initial conditions or save simulation results. Most steppables are implemented in Python. Much of the flexibility of CC3D comes from user-defined Python steppables.

The CC3D kernel supports parallel computation in shared-memory architectures (via OpenMP), providing substantial speedups on multi-core computers.

Besides the computational kernel of CC3D, the main components of the CC3D environment are (1) Twedit++-CC3D – a model editor and code generator, (2) CellDraw – a graphical tool for configuring the initial cell lattice, (3) CC3D Player – a graphical tool for running, replaying, and analyzing simulations.

Twedit++-CC3D provides a Simulation Wizard that generates draft CC3D model code based on high-level specification of simulation objects such as cell types and their behaviors, fields and interactions. Currently, the user must adjust default parameters in the autogenerated draft code, but later versions will provide interfaces

[1] We will use the word *model* to describe the specification of a particular biological system and *simulation* to refer to a specific instance of the execution of such a model.

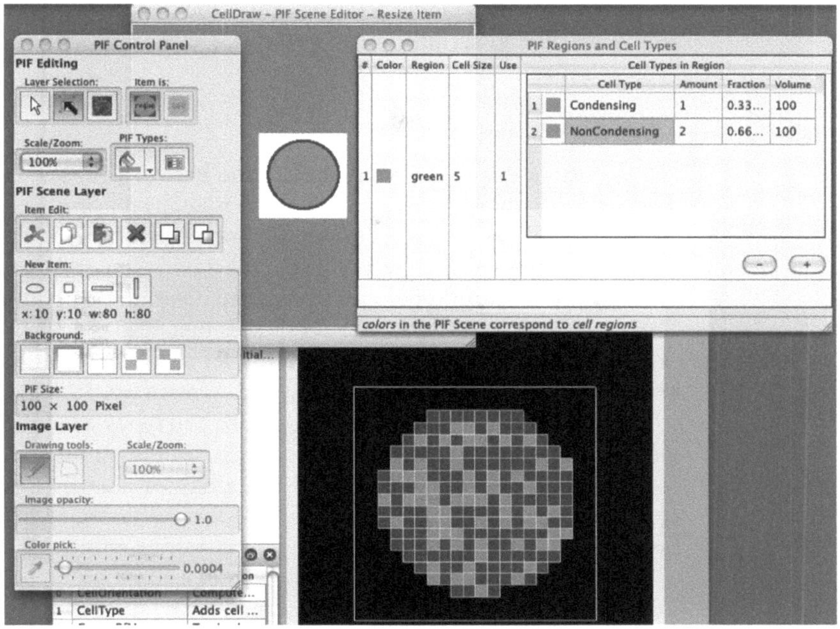

Fig. 3 CellDraw graphics tools and GUI. (For color version of this figure, the reader is referred to the web version of this book.)

for parameter specification. Twedit++-CC3D also provides a Python code-snippet generator, which simplifies coding Python CC3D modules.

CellDraw (Fig. 3) allows users to draw regions that it fills with cells of user-specified types. It also imports microscope images for manual segmentation, and automates the conversion of segmented regions – from TIFF sequences generated by 3rd party tools such as Fiji/ImageJ/TrakEM2 – for importing into CC3D.

CC3D Player is a graphical interface that loads and executes CC3D models. It allows users to change model parameters during execution (*steering*), define multiple 2D and 3D visualizations of the cell lattice and fields and conduct real-time simulation analysis. CC3D Player also supports batch mode execution on clusters.

IV. Building CC3D Models

This section presents some typical applications of GGH and CC3D. We use Twedit++-CC3D code generation and explain how to turn automatically generated draft code into executable models. All of the parameters appearing in the autogenerated simulation scripts are set to their default values.

A. Cell-Sorting Model

Cell sorting due to differential adhesion between cells of different types is one of the basic mechanisms creating tissue domains during development and wound healing and in maintaining domains in homeostasis. In a classic *in vitro* cell sorting experiment to determine relative cell adhesivities in embryonic tissues, mesenchymal cells of different types are dissociated, then randomly mixed and reaggregated. Their motility and differential adhesivities then lead them to rearrange to reestablish coherent homogenous domains with the most cohesive cell type surrounded by the less-cohesive cell types (Armstrong and Armstrong, 1984; Armstrong and Parenti, 1972). The simulation of the sorting of two cell types was the original motivation for the development of GGH methods. Such simple simulations show that the final configuration depends only on the hierarchy of adhesivities, whereas the sorting dynamics depends on the ratio of the adhesive energies to the amplitude of cell fluctuations.

To invoke the simulation wizard to create a simulation, we click `CC3DProject` → `New CC3D Project` in the Twedit++-CC3D menu bar (see Fig. 4). In the initial

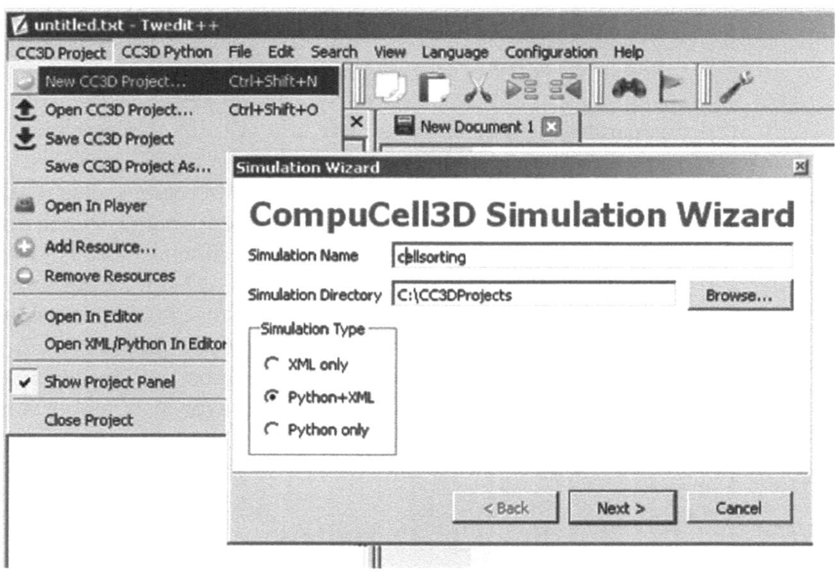

Fig. 4 Invoking the CompuCell3D Simulation Wizard from Twedit++. (For color version of this figure, the reader is referred to the web version of this book.)

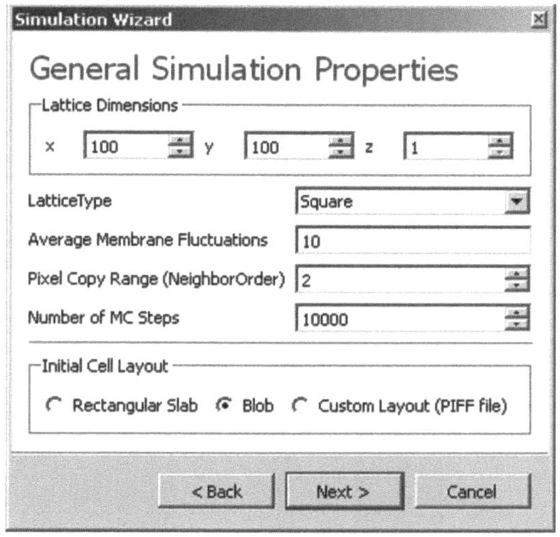

Fig. 5 Specification of basic cell-sorting properties in Simulation Wizard. (For color version of this figure, the reader is referred to the web version of this book.)

screen, we specify the name of the model (`cellsorting`), its storage directory (*C:\CC3DProjects*), and whether we will store the model as pure CC3DML, Python, and CC3DML or pure Python. This tutorial will use Python and CC3DML.

On the next page of the Wizard (see Fig. 5), we specify GGH global parameters, including cell-lattice dimensions, the cell-membrane fluctuation amplitude, the duration of the simulation in MCS and the initial cell-lattice configuration.

In this example, we specify a `100` × `100` × `1` cell lattice, that is, a 2D model, a fluctuation amplitude of 10, a simulation duration of 10,000 MCS, and a pixel-copy range of 2. `BlobInitializer` initializes the simulation with a disk of cells of specified size.

On the next Wizard page (see Fig. 6), we name the cell types in the model. We will use two cell types: `Condensing` (more cohesive) and `NonCondensing` (less cohesive). CC3D by default includes a special generalized cell type, `Medium`, with unconstrained volume that fills otherwise unspecified space in the cell lattice.

We skip the Chemical Field page of the Wizard and move to the Cell Behaviors and Properties page (see Fig. 7). Here, we select the biological behaviors we will include in our model. *Objects in CC3D (for example, cells) have no properties or behaviors unless we specify then explicitly.* Since cell sorting depends on differential adhesion between cells, we select the *Contact Adhesion* module from the Adhesion section (1) and give the cells a defined volume using the *Volume Flex* module from Constraints and Forces section.

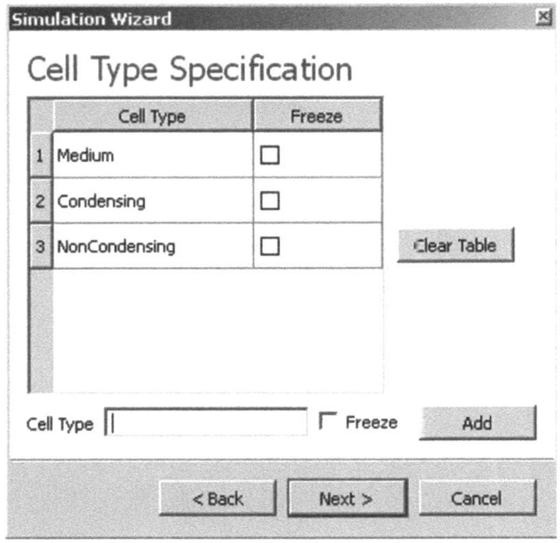

Fig. 6 Specification of cell-sorting cell types in Simulation Wizard. (For color version of this figure, the reader is referred to the web version of this book.)

We skip the next page related to Python scripting, after which Twedit++-CC3D generates the draft simulation code. Double-clicking on `cellsorting.cc3d` opens both the CC3DML (*cellsorting.xml*) and Python scripts for the model. Because the CC3DML file contains the complete model in this example, we postpone discussion of the Python script. A CC3DML file has three distinct sections. The first, the *Lattice Section* (lines 2–7) specifies global parameters like the cell-lattice size. The *Plugin Section* (lines 8–30) lists all the plugins used, for example, `CellType` and `Contact`. The *Steppable Section* (lines 32–39) lists all steppables; here we use only `BlobInitializer`.

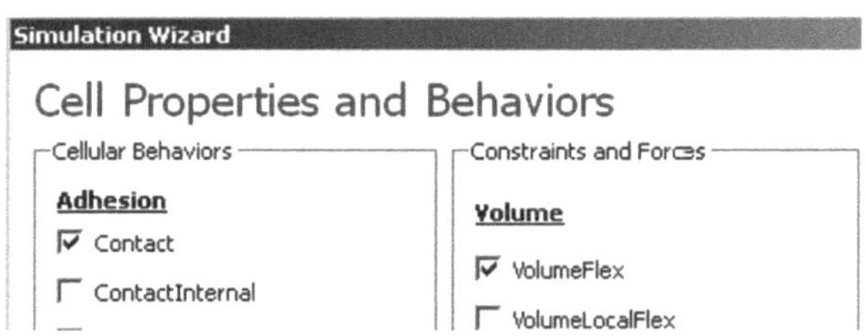

Fig. 7 Selection of cell-sorting cell behaviors in Simulation Wizard.[2] (For color version of this figure, the reader is referred to the web version of this book.)

[2] We have graphically edited the screenshots of Wizard pages to save space.

```
01  <CompuCell3D version="3.6.0">
02   <Potts>
03    <Dimensions x="100" y="100" z="1"/>
04    <Steps>10000</Steps>
05    <Temperature>10.0</Temperature>
06    <NeighborOrder>2</NeighborOrder>
07   </Potts>
08
09   <Plugin Name="CellType">
10    <CellType TypeId="0" TypeName="Medium"/>
11    <CellType TypeId="1" TypeName="Condensing"/>
12    <CellType TypeId="2" TypeName="NonCondensing"/>
13   </Plugin>
14
15   <Plugin Name="Volume">
16    <VolumeEnergyParameters CellType="Condensing"
       LambdaVolume="2.0" TargetVolume="25"/>
17    <VolumeEnergyParameters CellType="NonCondensing"
       LambdaVolume="2.0" TargetVolume="25"/>
18   </Plugin>
19
20   <Plugin Name="CenterOfMass"/>
21
22   <Plugin Name="Contact">
23    <Energy Type1="Medium" Type2="Medium">10</Energy>
24    <Energy Type1="Medium" Type2="Condensing">10</Energy>
25    <Energy Type1="Medium" Type2="NonCondensing">10</Energy>
26    <Energy Type1="Condensing"Type2="Condensing">10</Energy>
27    <Energy Type1="Condensing" Type2="NonCondensing">10</Energy>
28    <Energy Type1="NonCondensing" Type2="NonCondensing">10</Energy>
29    <NeighborOrder>2</NeighborOrder>
30   </Plugin>
31
32   <Steppable Type="BlobInitializer">
33    <Region>
34     <Center x="50" y="50" z="0"/>
35     <Radius>20</Radius>
36     <Width>5</Width>
37     <Types>Condensing,NonCondensing</Types>
38    </Region>
39   </Steppable>
40  </CompuCell3D>
```

Listing 1. Simulation-Wizard-generated draft CC3DML (XML) code for cell sorting.[3]

All parameters appearing in the autogenerated CC3DML script have default values inserted by Simulation Wizard. We must edit the parameters in the draft CC3DML script to build a functional cell-sorting model (Listing 1). The `CellType` plugin (lines 9–13) already provides three generalized cell types: `Condensing` (C), `NonCondensing` (N), and `Medium` (M), so we need not change it.

[3] We use indent each nested block by two spaces in all listings in this chapter to avoid distracting rollover of text at the end of the line. However, both Simulation Wizard and standard Python use an indentation of four spaces per block.

However, the boundary-energy (contact energy) matrix in the Contact plugin (lines 22–30) is initially filled with identical values, which prevents sorting. For cell sorting, Condensing cells must adhere strongly to each other (so we set J_{CC}=2), Condensing and NonCondensing cells must adhere more weakly (here, we set J_{CN}=11), and all other adhesions must be very weak (we set J_{NN}=J_{CM}=J_{NM}=16), as discussed in Section III. The value of J_{MM} = 0 is irrelevant, since the Medium generalized cell does not contact itself.

To reduce artifacts due to the anisotropy of the square cell lattice we increase the neighbor order range in the contact energy to 2 so the contact energy sum in Eq. (1) will include nearest and second-nearest neighbors (line 29).

In the Volume plugin, which calculates the volume-constraint energy given in Eq. (1) the attributes CellType, LambdaVolume, and TargetVolume inside the <VolumeEnergyParameters> tags specify $\lambda(\tau)$ and $V_t(\tau)$ for each cell type. In our simulations, we set $V_t(\tau) = 25$ and $\lambda(\tau) = 2.0$ for both cell types.

We initialize the cell lattice using the BlobInitializer, which creates one or more disks (solid spheres in 3D) of cells. Each disk (sphere) created is enclosed between <Region> tags. The <Center> tag with syntax <Center x= ''x_position'' y= ''y_position'' z= ''z_position''/> specifies the position of the center of the disk. The <Width> tag specifies the size of the initial square (cubical in 3D) generalized cells and the <Gap> tag creates space between neighboring cells. The <Types> tag lists the cell types to fill the disk. Here, we change the Radius in the draft BlobInitializer specification to 40. These few changes produce a working cell-sorting simulation.

To run the simulation, we right click cellsorting.cc3d in the left panel and choose the Open In Player option. We can also run the simulation by opening CompuCellPlayer and selecting cellsorting.cc3d from the File-> Open Simulation File dialog.

Fig. 8 shows snapshots of a simulation of the cell-sorting model. The less-cohesive NonCondensing cells engulf the more cohesive Condensing cells, which cluster and form a single central domain. By changing the boundary energies we can produce other cell-sorting patterns (Glazier and Graner, 1993; Graner and Glazier,

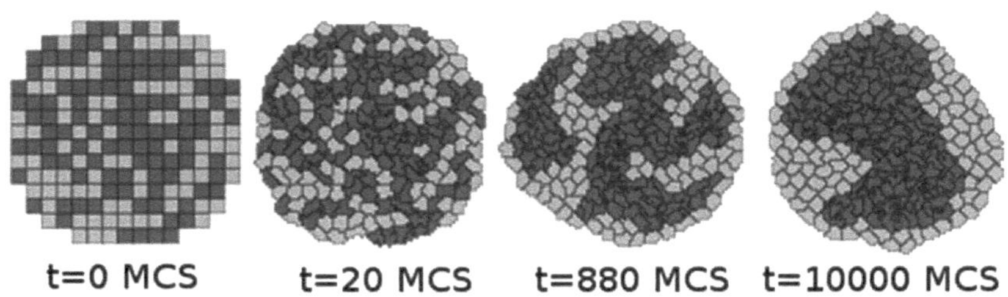

Fig. 8 Snapshots of the cell-lattice configurations for the cell-sorting simulation in Listing 1. The boundary-energy hierarchy drives NonCondensing (light grey) cells to surround Condensing (dark grey) cells. The white background denotes surrounding Medium.

1992). In particular, if we reduce the contact energy between the `Condensing` cell type and the `Medium`, we can force inverted cell sorting, where the `Condensing` cells surround the `NonCondensing` cells. If we set the heterotypic contact energy to be less than either of the homotypic contact energies, the cells of the two types will mix rather than sort. If we set the cell-medium contact energy to be very small for one cell type, the cells of that type will disperse into the medium, as in cancer invasion. With minor modifications, we can also simulate the scenarios for three or more cell types, for situations in which the cells of a given type vary in volume, motility or adhesivity, or in which the initial condition contains coherent clusters of cells rather than randomly mixed cells (engulfment).

B. Angiogenesis Model

Vascular development is central to both development and cancer progression. We present a simplified model of the earliest phases of capillary network assembly by endothelial cells based on cell adhesion and contact-inhibited chemotaxis. This model does a good job of reproducing the patterning and dynamics which occur if we culture human umbilical vein endothelial cells (HUVEC) on matrigel in a quasi-2D *in vitro* experiment (Merks and Glazier, 2006; Merks *et al.*, 2006, 2008). In addition to generalized cells modeling the HUVEC, we will need a diffusing chemical object, here, vascular endothelial growth factor (VEGF), cell secretion of VEGF, and cell-contact-inhibited chemotaxis to VEGF.

We will use a 3D voxel (pixel) with a side of 4 μm, that is, a volume of 64 μm^3. Since the experimental HUVEC speed is about 0.4 μm/min and cells in this simulation move at an average speed of 0.1 pixel/MCS, one MCS represents 1 min.

In the Simulation Wizard, we name the model `ANGIOGENESIS`, set the cell- and field-lattice dimensions to 50 × 50 × 50, the membrane fluctuation amplitude to 20, the pixel-copy range to 3, the number of MCS to 10,000, and select `BlobFieldInitializer` to produce the initial cell-lattice configuration. We have only one cell type – `Endothelial`.

In the `Chemical Fields` page (see Fig. 9), we create the `VEGF` field and select `FlexibleDiffusionSolverFE` from the `Solver` pull-down list.

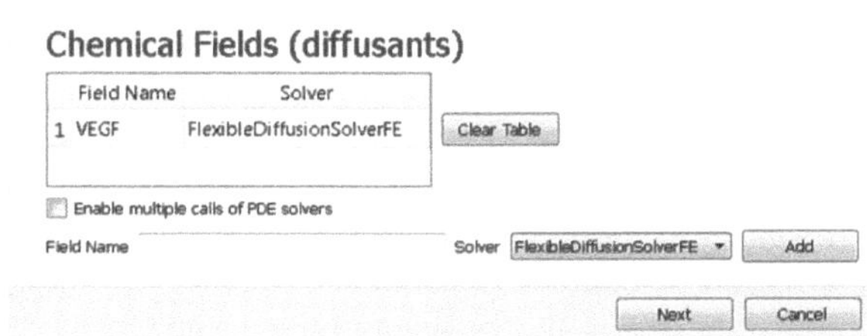

Fig. 9 Specification of the angiogenesis chemical field in Simulation Wizard. (For color version of this figure, the reader is referred to the web version of this book.)

Fig. 10 Specification of angiogenesis cell behaviors in Simulation Wizard. (For color version of this figure, the reader is referred to the web version of this book.)

Next, on the `Cell Properties and Behaviors` page (see Fig. 10), we select the `Contact` module from the `Adhesion-behavior` group and add `Secretion`, `Chemotaxis`, and `Volume-constraint` behaviors by checking the appropriate boxes.

Because we have invoked `Secretion` and `Chemotaxis`, the Simulation Wizard opens their configuration screens. On the `Secretion` page (see Fig. 11), from the pull-down list, we select the chemical to secrete by selecting `VEGF` in the `Field` pull-down menu and the cell type secreting the chemical (`Endothelial`), and enter the rate of 0.013 (50 pg/(cell h) = 0.013 pg/(voxel MCS), compare to Leith and Michelson, 1995). We leave the `Secretion Type` entry set to `Uniform`, so each pixel of an endothelial cell secretes the same amount of `VEGF` at the same rate. Uniform volumetric secretion or secretion at the cell's center of mass may be most appropriate in 2D simulations of planar geometries (e.g., cells on a petri dish or agar) where the biological cells are actually secreting up or down into a medium that carries the diffusant. CC3D also supplies a secrete-on-contact option to secrete outward from the cell boundaries and allows specification of which boundaries can secrete, which is more realistic in 3D. However, users are free to employ any of these methods in either 2D or 3D, depending on their interpretation of their specific biological situation. CC3D does not have intrinsic units for fields, so the amount of a chemical can be interpreted in units of moles, number of molecules, or grams. We click the `Add Entry` button to add the secretion information, then proceed to the next page to define the cells' chemotaxis properties.

On the `Chemotaxis` page, we select `VEGF` from the `Field` pull-down list and `Endothelial` for the cell type, entering a value for `Lambda` of 5000. When the `chemotaxis type` is `regular`, the cell's response to the field is linear; that is the effective strength of chemotaxis depends on the product of `Lambda` and the secretion rate of `VEGF`, for example, a `Lambda` of 5000 and a `secretion rate` of 0.013 has the same effective chemotactic strength as a `Lambda` of 500 and a

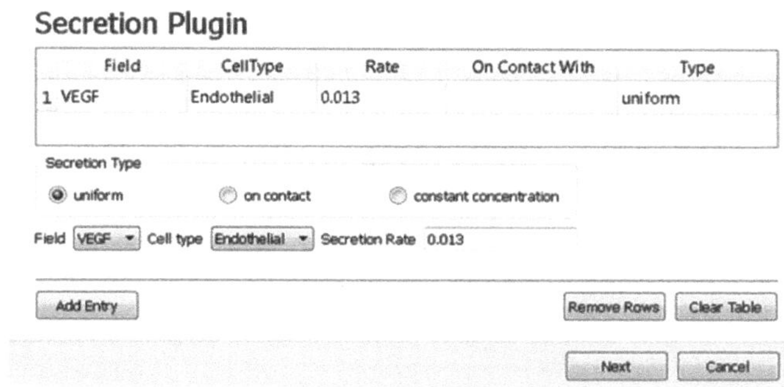

Fig. 11 Specification of angiogenesis secretion parameters in Simulation Wizard. (For color version of this figure, the reader is referred to the web version of this book.)

secretion rate of 0.13. Since endothelial cells do not chemotax at surfaces where they contact other endothelial cells (contact inhibition), we select Medium from the pull-down menu next to the Chemotax Towards button and click this button to add Medium to the list of generalized cell types whose interfaces with Endothelial cells support chemotaxis. We click the Add Entry button to add the chemotaxis information, then proceed to the final Simulation Wizard page Fig. 12.

Next, we adjust the parameters of the draft model. Pressure from chemotaxis to VEGF reduces the average endothelial cell volume by about 10 voxels from the target volume. So, in the Volume plugin, we set TargetVolume to 74 (64+10) and LambdaVolume to 20.0.

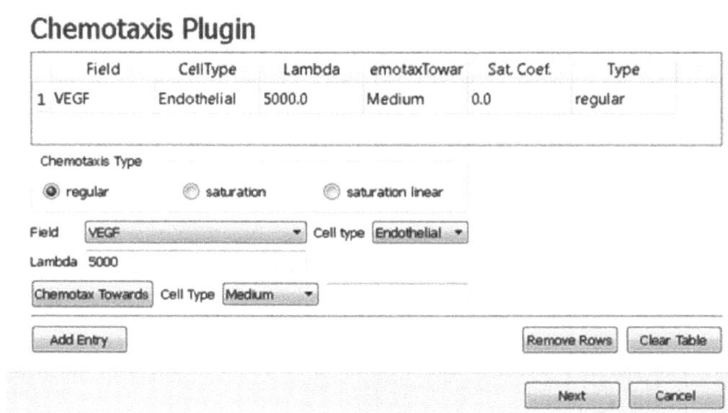

Fig. 12 Specification of angiogenesis chemotaxis properties in Simulation Wizard. (For color version of this figure, the reader is referred to the web version of this book.)

In experiments, in the absence of chemotaxis no capillary network forms and cells adhere to each other to form clusters. We therefore set J_{MM}=0, J_{EM}=12, and J_{EE}=5 in the `Contact` plugin (`M: Medium`, `E: Endothelial`). We also set the `NeighborOrder` for the `Contact` energy calculations to 4.

The diffusion equation that governs VEGF ($V(\vec{x})$) field evolution is

$$\frac{\partial V(\vec{x})}{\partial t} = D_{\text{VEGF}}^{\text{EC}} \nabla^2 V(\vec{x}) - \gamma_{\text{VEGF}} V(\vec{x}) \delta(\tau(\sigma(\vec{x})), M) + S^{\text{EC}} \delta(\tau(\sigma(\vec{x})), EC) \quad (3)$$

where $\delta(\tau(\sigma(\vec{x})), EC) = 1$ inside `Endothelial` cells and 0 elsewhere and $\delta(\tau(\sigma(\vec{x})), M) = 1$ inside `Medium` and 0 elsewhere. We set the diffusion constant $D_{\text{VEGF}} = 0.042\ \mu\text{m}^2/\text{s}$ (0.16 voxel^2/MCS, about two orders of magnitude smaller than experimental values),[4] the decay coefficient $\gamma_{\text{VEGF}} = 1\ \text{h}^{-1}$ [130,131] (0.016 MCS^{-1}) for `Medium` pixels and $\gamma_{\text{VEGF}} = 0$ inside `Endothelial` cells, and the secretion rate $S^{\text{EC}} = 0.013$ pg/(voxel MCS).

In the CC3DML script, describing `FlexibleDiffusionSolverFE` (Listing 2, lines 38–47) we set the values of the `<DiffusionConstant>` and `<DecayConstant>` tags to 0.16 and 0.016, respectively. To prevent chemical decay inside endothelial cells, we add the line `<DoNotDecayIn>Endothelial</DoNotDecayIn>` inside the `<DiffusionData>` tag pair.

Finally, we edit `BlobInitializer` (lines 49–56) to start with a solid sphere 10 pixels in radius centered at $x = 25$, $y = 25$, $z = 25$ with initial cell width 4, as in **Listing 2**.

```
01  <CompuCell3D version="3.6.0">
02
03  <Potts>
04   <Dimensions x="50" y="50" z="50"/>
05   <Steps>10000</Steps>
06   <Temperature>20.0</Temperature>
07   <NeighborOrder>3</NeighborOrder>
08  </Potts>
09
10  <Plugin Name="CellType">
11   <CellType TypeId="0" TypeName="Medium"/>
12   <CellType TypeId="1" TypeName="Endothelial"/>
13  </Plugin>
14
15  <Plugin Name="Volume">
16   <VolumeEnergyParameters CellType="Endothelial"
       LambdaVolume="20.0" TargetVolume="74"/>
17  </Plugin>
18
```

[4] `FlexibleDiffusionSolverFE` becomes unstable for values of $D_{\text{VEGF}} > 0.16$ voxel^2/MCS. For larger diffusion constants, we must call the algorithm multiple times per MCS (See the *Three-Dimensional Vascular Solid Tumor Growth* section).

```
19  <Plugin Name="Contact">
20   <Energy Type1="Medium" Type2="Medium">0</Energy>
21   <Energy Type1="Medium" Type2="Endothelial">12</Energy>
22   <Energy Type1="Endothelial" Type2="Endothelial">5</Energy>
23   <NeighborOrder>4</NeighborOrder>
24  </Plugin>
25
26  <Plugin Name="Chemotaxis">
27   <ChemicalField Name="VEGF" Source="FlexibleDiffusionSolverFE">
28    <ChemotaxisByType ChemotactTowards="Medium" Lambda="5000.0"
       Type="Endothelial"/>
29   </ChemicalField>
30  </Plugin>
31
32  <Plugin Name="Secretion">
33   <Field Name="VEGF">
34    <Secretion Type="Endothelial">0.013</Secretion>
37
35   </Field>
36  </Plugin>

38  <Steppable Type="FlexibleDiffusionSolverFE">
39   <DiffusionField>
40    <DiffusionData>
41     <FieldName>VEGF</FieldName>
42     <DiffusionConstant>0.16</DiffusionConstant>
43     <DecayConstant>0.016</DecayConstant>
44     <DoNotDecayIn> Endothelial</DoNotDecayIn>
45    </DiffusionData>
46   </DiffusionField>
47  </Steppable>
48
49  <Steppable Type="BlobInitializer">
50   <Region>
51    <Center x="25" y="25" z="25"/>
52    <Radius>10</Radius>
53    <Width>4</Width>
54    <Types>Endothelial</Types>
55   </Region>
56  </Steppable>
57
58  </CompuCell3D>
```

Listing 2. CC3DML code for the angiogenesis model.

The main behavior that drives vascular patterning is contact-inhibited chemotaxis (Listing 2, lines 26–30). VEGF diffuses away from cells and decays in `Medium`, creating a steep concentration gradient at the interface between `Endothelial` cells and `Medium`. Because `Endothelial` cells chemotax up the concentration gradient only at the interface with `Medium`, the `Endothelial` cells at the surface of the cluster compress the cluster of cells into vascular branches and maintain branch integrity.

We show screenshots of a simulation of the angiogenesis model in Fig. 13 (Merks *et al.*, 2008; Shirinifard *et al.*, 2009). We can reproduce either 2D or 3D primary capillary network formation and the rearrangements of the network agree with

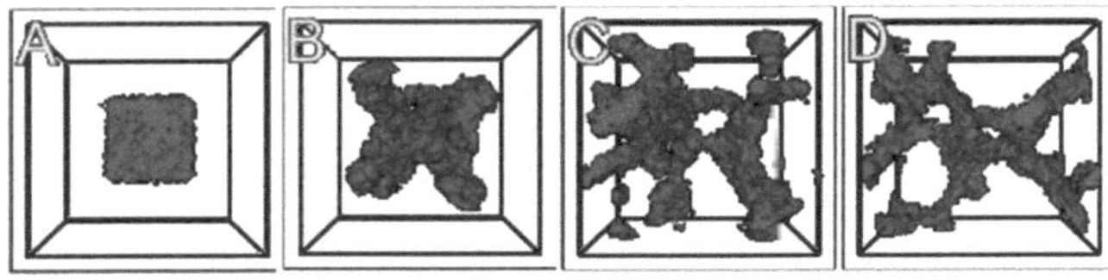

Fig. 13 An initial cluster of adhering endothelial cells forms a capillary-like network via sprouting angiogenesis. (A) 0 h (0 MCS); (B) ~2 h (100 MCS); (C) ~5 h (250 MCS); (D): ~18 h (1100 MCS). (For color version of this figure, the reader is referred to the web version of this book.)

experimentally observed dynamics. If we eliminate the contact inhibition, the cells do not form a branched structure (as observed in chick allantois experiments, Merks *et al.*, 2008). We can also study the effects of surface tension, external growth factors, and changes in motility and diffusion constants on the pattern and its dynamics. However, this simple model does not include the strong junctions HUVEC cells make with each other at their ends after a period of prolonged contact. It also does not attempt to model the vacuolation and linking of vacuoles that leads to a connected network of tubes.

Since real endothelial cells are elongated, we can include the `Cell-elongation` plugin in the Simulation Wizard to better reproduce individual cell morphology. However, excessive cell elongation causes cell fragmentation. Adding either the `Global` or `Fast Connectivity Constraint` plugin prevents cell fragmentation.

C. Overview of Python Scripting in CompuCell3D

In the models we presented above, all cells had parameter values fixed in time. To allow cell behaviors to change, we need to be able to adjust cell properties during a simulation. CC3D can execute Python scripts (CC3D supports Python versions 2.x) to modify the properties of cells in response to events occurring during a simulation, such as the concentration of a nutrient dropping below a threshold level, a cell reaching a doubling volume, or a cell changing its neighbors. Most such Python scripts have a simple structure based on `print` statements, `if-elif-else` statements, `for` loops, `lists`, and simple `classes` and do not require in-depth knowledge of Python to create.

This section briefly introduces the main features of Python in the CC3D context. For a more formal introduction to Python, see Lutz (2011) and *http://www.python.org*.

Python defines blocks of code, such as those appearing inside `if` statements or `for` loops (in general after "`:`"), by an increased level of indentation. This chapter uses two spaces per indentation level. For example, in Listing 3, we indent the body of the `if` statement by two spaces and the body of the inner `for` loop by additional two spaces. The `for` loop is executed inside the `if` statement, which checks if we are in the second MCS of the simulation. The command `pixelOffset=10` assigns to the variable `pixelOffset` a value of `10`. The `for` loop assigns to the variable `x` values ranging from `0` through `self.dim.x-1`, where `self.dim.x` is a CC3D internal variable containing the size of the cell lattice in the *x*-direction. When executed, Listing 3 prints consecutive integers from `10` to `10+self.dim.x-1`.

```
if mmcs==2m:
  pixelOffset = 10
  for x in range(self.dim.x):
    pixel = pixelOffset + x
    print pixel
```

Listing 3. Simple Python loop.

One of the advantages of Python compared to older languages like Fortran is that it can also iterate over members of a Python *list*, a *container* for grouping objects. Listing 4 executes a `for` loop over a list containing all cells in the simulation and prints the type of each cell.

```
for cell in self.cellList:
  print "cell type=", cell.type
```

Listing 4. Iterating over the inventory of CC3D cells in Python.

Lists can combine objects of any type, including integers, strings, complex numbers, lists, and, in this case, CC3D cells. CC3D uses lists extensively to keep track of cells, cell neighbors, cell pixels, etc.

CC3D allows users to construct custom Python code as independent modules called *steppables*, which are represented as classes. Listing 5 shows a typical CC3D Python steppable class. The first line declares the class name together with an argument (`SteppableBasePy`) inside the parenthesis, which makes the main CC3D objects, including cells, lattice properties, etc., available inside the class. The `def _init_(self,_simulator,_frequency=1):` declares the initializing function `_init_` which is called automatically during class object instantiation. After initializing the class and inheriting CC3D objects, we declare three main functions called at different times during the simulation: `start` is called before the simulation starts; `step` is called at specified intervals in MCS throughout the simulation; and `finish` is called at the end of the simulation. The `start` function iterates over all cells, setting their target volume and inverse compressibility to 25 and 5, respectively. Generically, we use the `start` function to define model initial conditions. The `step` function increases the target volumes of all cells by 0.001 after the tenth MCS, a typical way to implement cell growth in CC3D. The `finish` function prints the cell volumes at the end of the simulation.

```
class Example(SteppableBasePy):
  def __init__(self,_simulator,_frequency=1):
    SteppableBasePy.__init__(self,_simulator,_frequency)

  def start(self):
    print "Called at the beginning of the simulation"
    for cell in self.cellList:
      cell.targetVolume=25
      cell.lambdaVolume=5

  def step(self,mcs):
    print "Called every MCS"
    if mmcs>10:
      for cell in self.cellList:
        cell.targetVolume+=0.001

  def finish(self):
    print "Called at the end of the simulation"
    for cell in self.cellList:
      print "cell volume = ", cell.volume
```

Listing 5. Sample CC3D steppable class.

The `start`, `step`, and `finish` functions have default implementations in the base class `SteppableBasePy`. Therefore, we only need to provide definition of those functions that we want to override. In addition, we can add our own functions to the class.

The next section uses Python scripting to build a complex CC3D model.

D. Three-Dimensional Vascular Tumor Growth Model

The development of a primary solid tumor starts from a single cell that proliferates in an inappropriate manner, dividing repeatedly to form a cluster of tumor cells. Nutrient and waste diffusion limits the diameter of such *avascular tumor spheroids* to about 1 mm. The central region of the growing spheroid becomes necrotic, with a surrounding layer of cells whose hypoxia triggers VEGF-mediated signaling events that initiate tumor neovascularization by promoting growth and extension (*neoangiogenesis*) of nearby blood vessels. Vascularized tumors are able to grow much larger than avascular spheroids and are more likely to metastasize.

Here, we present a simplified 3D model of a generic vascular tumor that can be easily extended to describe specific vascular tumor types and host tissues. We begin with a cluster of proliferating tumor cells, `P`, and normal vasculature. Initially, tumor cells proliferate as they take up diffusing glucose from the field, *GLU*, which the preexisting vasculature supplies (in this model, we neglect possible changes in concentration along the blood vessels in the direction of flow and set the secretion parameters uniformly over all blood-vessel surfaces). We assume that the tumor cells (both in the initial cluster and later) are always hypoxic and secrete a long-diffusing isoform of VEGF-A, *L_VEGF*. When *GLU* drops below a threshold, tumor cells become necrotic, gradually shrink and finally disappear. The initial tumor cluster grows and reaches a maximum diameter characteristic of an avascular tumor spheroid. To reduce execution time in our demonstration, we choose our model parameters so that the maximum spheroid diameter will be about 10 times smaller than in experiments. A few preselected neovascular endothelial cells, `NV`, in the preexisting vasculature respond both by chemotaxing toward higher concentrations of pro-angiogenic factors and by forming new blood vessels via neoangiogenesis. The tumor-induced vasculature increases the growth rate of the resulting vascularized solid tumor compared to an avascular tumor, allowing the tumor to grow beyond the spheroid's maximum diameter. Despite our rescaling of the tumor size, the model produces a range of biologically reasonable morphologies that allow study of how tumor-induced angiogenesis affects the growth rate, size, and morphology of tumors.

We use the basic angiogenesis simulation from the previous section to simulate both preexisting vasculature and tumor-induced angiogenesis, adding a set of finite-element links between the endothelial cells to model the strong junctions that form between endothelial cells *in vivo*. We denote the short-diffusing isoform of VEGF-A, *S_VEGF*. Both endothelial cells and neovascular endothelial cells chemotax up gradients of `S_VEGF`, but only neovascular endothelial cells chemotax up gradients of `L_VEGF`.

In the Simulation Wizard, we name the model `TumorVascularization`, set the cell- and field-lattice dimensions to $50 \times 50 \times 80$, the membrane fluctuation amplitude to 20, the pixel-copy range to 3, the number of MCS to 10,000, and choose `UniformInitializer` to produce the initial tumor and vascular cells, since it automatically creates a mixture of cell types. We specify four cell types: `P`: proliferating tumor cells; `N`: necrotic cells; `EC`: endothelial cells; and `NV`: neovascular endothelial cells.

On the `Chemical Fields` page (see Fig. 14), we create the `S_VEGF` and `L_VEGF` fields and select `FlexibleDiffusionSolverFE` for both from the `Solver` pull-down list. We also check `Enable multiple calls of PDE solvers` to work around the numerical instabilities of the PDE solvers for large diffusion constants.

On the `Cell Behavior and Properties` page (see Fig. 15) we select both the `Contact` and `FocalPointPlasticity` modules from the `Adhesion` group, and add `Chemotaxis`, `Growth`, and `Mitosis`, `Volume Constraint`, and `Global Connectivity` by checking the appropriate boxes. We also track the `Center-of-Mass` (to access field concentrations) and `Cell Neighbors` (to implement contact-inhibited growth). Unlike in our angiogenesis simulation, we will implement secretion as a part of the `FlexibleDiffusionSolverFE` syntax.

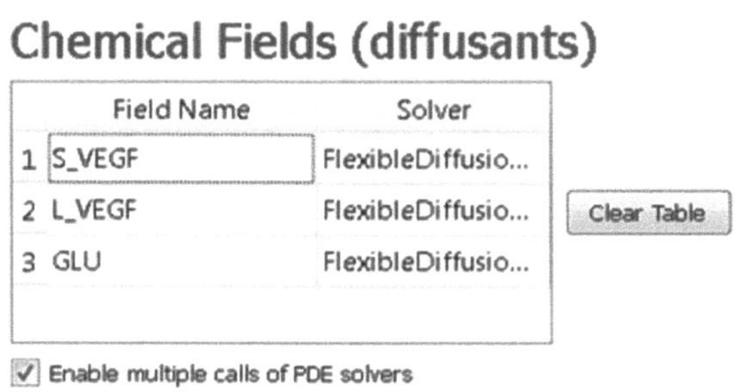

Fig. 14 Specification of vascular tumor chemical fields in the Simulation Wizard. (For color version of this figure, the reader is referred to the web version of this book.)

Fig. 15 Specification of vascular tumor cell behaviors in Simulation Wizard. (For color version of this figure, the reader is referred to the web version of this book.)

In the `Chemotaxis` page (see Fig. 16), for each cell-type/chemical-field pair we click the `Add Entry` button to add the relevant chemotaxis information, for example, we select `S_VEGF` from the `Field` pull-down list and `EC` and `NV` from the `cell-type` list and set `Lambda` to 5000. To enable contact inhibition of `EC` and `NV` chemotaxis, we select `Medium` from the pull-down menu next to the `Chemotax Towards` button and click the button to add `Medium` to the list. We repeat this process for the `T` and `N` cell types, so that `NV` cells chemotax up gradients of `L_VEGF`. We then proceed to the final `Simulation Wizard` page.

Twedit++ generates three simulation files – a CC3DML file specifying the energy terms, diffusion solvers, and initial cell layout, a main Python file that loads the

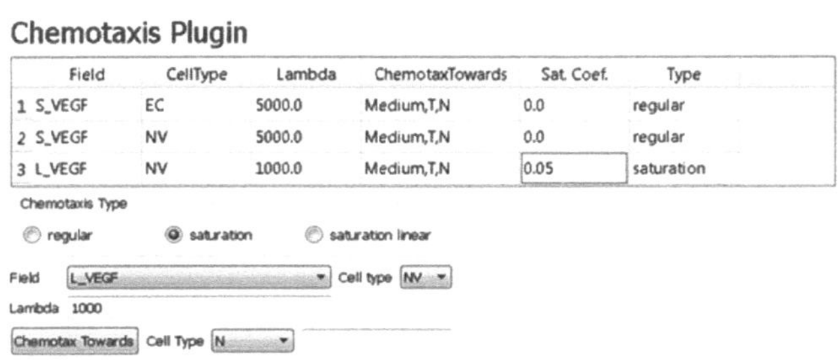

Fig. 16 Specification of vascular tumor chemotaxis properties in Simulation Wizard. (For color version of this figure, the reader is referred to the web version of this book.)

CC3DMLfile, sets up the CompuCell environment and executes the Python steppables and a Python steppables file. The main Python file is typically constructed by modifying the standard template in Listing 6. Lines 1–12 set up the CC3D simulation environment and load the simulation. Lines 14–20 create instances of two steppables – `MitosisSteppable` and `VolumeParamSteppable` – and register them with the CC3D kernel. Line 22 starts the main CC3D loop, which executes MCSs and periodically calls the steppables.

```
01   import sys
02   from os import environ
03   import string
04   sys.path.append(environ["PYTHON_MODULE_PATH"])
05
06   import CompuCellSetup
07   sim,simthread = CompuCellSetup.getCoreSimulationObjects()
08   CompuCellSetup.initializeSimulationObjects(sim,simthread)
09   import CompuCell
10
11   from PySteppables import SteppableRegistry
12   steppableRegistry=SteppableRegistry()
13
14   from VascularTumorSteppables import MitosisSteppable
15   mitosisSteppable=MitosisSteppable(sim,1)
16   steppableRegistry.registerSteppable(mitosisSteppable)
17
18   from VascularTumorSteppables import VolumeParamSteppable
19   volumeParamSteppable=VolumeParamSteppable(sim,1)
20   steppableRegistry.registerSteppable(volumeParamSteppable)
21
22   CompuCellSetup.mainLoop(sim,simthread,steppableRegistry)
```

Listing 6. The Main Python script initializes the vascular tumor simulation and runs the main simulation loop.

Next, we edit the draft autogenerated simulation CC3DML file in Listing 7.

```
<CompuCell3D>
<Potts>
 <Dimensions x="50" y="50" z="80"/>
 <Steps>100000</Steps>
 <Temperature>20</Temperature>
 <Boundary_x>Periodic</Boundary_x>
 <Boundary_y>Periodic</Boundary_y>
 <Boundary_z>Periodic</Boundary_z>
 <RandomSeed>313</RandomSeed>
 <NeighborOrder>3</NeighborOrder>
</Potts>

<Plugin Name="CellType">
 <CellType TypeName="Medium" TypeId="0"/>
 <CellType TypeName="P" TypeId="1"/>
 <CellType TypeName="N" TypeId="2"/>
 <CellType TypeName="EC" TypeId="3"/>
 <CellType TypeName="NV" TypeId="4"/>
```

```
</Plugin>

<Plugin Name="Chemotaxis">
 <ChemicalField Source="FlexibleDiffusionSolverFE" Name="S_VEGF">
  <ChemotaxisByType Type="NV" Lambda="5000" ChemotactTowards="Medium,P,N"/>
 </ChemicalField>
 <ChemicalField Source="FlexibleDiffusionSolverFE" Name="L_VEGF">
  <ChemotaxisByType Type="NV" Lambda="1000"
    ChemotactTowards="Medium,P,N" SaturationCoef="0.05"/>
 </ChemicalField>
 <ChemicalField Source="FlexibleDiffusionSolverFE" Name="S_VEGF">
  <ChemotaxisByType Type="EC" Lambda="5000" ChemotactTowards="Medium,P,N"/>
 </ChemicalField>
</Plugin>

<Plugin Name="CenterOfMass"/>
<Plugin Name="NeighborTracker"/>

<Plugin Name="Contact">
 <Energy Type1="Medium" Type2="Medium">0</Energy>
 <Energy Type1="P" Type2="Medium">10</Energy>
 <Energy Type1="P" Type2="P">8</Energy>
 <Energy Type1="N" Type2="Medium">15</Energy>
 <Energy Type1="N" Type2="P">8</Energy>
 <Energy Type1="N" Type2="N">3</Energy>
 <Energy Type1="EC" Type2="Medium">12</Energy>
 <Energy Type1="EC" Type2="P">30</Energy>
 <Energy Type1="EC" Type2="N">30</Energy>
 <Energy Type1="EC" Type2="EC">5</Energy>
 <Energy Type1="NV" Type2="Medium">12</Energy>
 <Energy Type1="NV" Type2="P">30</Energy>
 <Energy Type1="NV" Type2="N">30</Energy>
 <Energy Type1="NV" Type2="EC">5</Energy>
 <Energy Type1="NV" Type2="NV">5</Energy>
 <NeighborOrder>4</NeighborOrder>
</Plugin>

<Plugin Name="VolumeLocalFlex"/>

<Plugin Name="FocalPointPlasticity">
 <Parameters Type1="EC" Type2="NV">
  <Lambda>50.0</Lambda>
  <ActivationEnergy>-100.0</ActivationEnergy>
  <TargetDistance>5.0</TargetDistance>
  <MaxDistance>15.0</MaxDistance>
  <MaxNumberOfJunctions>2</MaxNumberOfJunctions>
 </Parameters>
 <Parameters Type1="EC" Type2="EC">
  <Lambda>400.0</Lambda>
  <ActivationEnergy>-100.0</ActivationEnergy>
  <TargetDistance>5.0</TargetDistance>
  <MaxDistance>15.0</MaxDistance>
  <MaxNumberOfJunctions>3</MaxNumberOfJunctions>
 </Parameters>
 <Parameters Type1="NV" Type2="NV">
  <Lambda>20.0</Lambda>
  <ActivationEnergy>-100.0</ActivationEnergy>
  <TargetDistance>5.0</TargetDistance>
  <MaxDistance>10.0</MaxDistance>
  <MaxNumberOfJunctions>2</MaxNumberOfJunctions>
 </Parameters>
 <NeighborOrder>1</NeighborOrder>
</Plugin>
```

```

<Plugin Name="ConnectivityGlobal">
 <Penalty Type="NV">10000</Penalty>
 <Penalty Type="EC">10000</Penalty>
</Plugin>

<Plugin Name="PDESolverCaller">
 <CallPDE PDESolverName="FlexibleDiffusionSolverFE" ExtraTimesPerMC="9"/>
</Plugin>

<Steppable Type="FlexibleDiffusionSolverFE">
 <!--endothelial-derived short diffusing VEGF isoform-->
 <DiffusionField>
  <DiffusionData>
   <FieldName>S_VEGF</FieldName>
   <ConcentrationFileName></ConcentrationFileName>
   <DiffusionConstant>0.016</DiffusionConstant>
   <DecayConstant>0.0016</DecayConstant>
   <DoNotDecayIn>EC</DoNotDecayIn>
   <DoNotDecayIn>NV</DoNotDecayIn>
   </DiffusionData>
   <SecretionData>
    <Secretion Type="NV">0.0013</Secretion>
    <Secretion Type="EC">0.0013</Secretion>
   </SecretionData>
  </DiffusionField>

 <!--tumor-derived long diffusing VEGF isoform-->
  <DiffusionField>
   <DiffusionData>
    <FieldName>L_VEGF</FieldName>
    <DiffusionConstant>0.16</DiffusionConstant>
    <DecayConstant>0.0016</DecayConstant>
   </DiffusionData>
   <SecretionData>
    <Secretion Type="P">0.001</Secretion>
    <Uptake Type="NV" MaxUptake="0.05" RelativeUptakeRate="0.5"/>
    <Uptake Type="EC" MaxUptake="0.05" RelativeUptakeRate="0.5"/>
   </SecretionData>
  </DiffusionField>

  <DiffusionField>
   <DiffusionData>
    <FieldName>GLU</FieldName>
    <ConcentrationFileName>GLU_300.dat</ConcentrationFileName>
    <DiffusionConstant>0.16</DiffusionConstant>
   </DiffusionData>
   <SecretionData>
    <Secretion Type="NV">0.4</Secretion>
    <Secretion Type="EC">0.8</Secretion>
    <Uptake Type="Medium" MaxUptake="0.0064" RelativeUptakeRate="0.1"/>
    <Uptake Type="P" MaxUptake="0.1" RelativeUptakeRate="0.1"/>
   </SecretionData>
  </DiffusionField>
 </Steppable>

 <Steppable Type="UniformInitializer">
  <Region>
   <BoxMin x="0" y="24" z="16"/>
   <BoxMax x="50" y="28" z="20"/>
   <Width>4</Width>
   <Types>EC</Types>
  </Region>
```

```
144    <Region>
145     <BoxMin y="0" x="24" z="16"/>
146     <BoxMax y="50" x="28" z="20"/>
147     <Width>4</Width>
148     <Types>EC</Types>
149    </Region>
150    <Region>
151     <BoxMin x="10" y="24" z="16"/>
152     <BoxMax x="50" y="28" z="20"/>
153     <Width>4</Width>
154     <Gap>25</Gap>
155     <Types>NV</Types>
156    </Region>
157    <Region>
158     <BoxMin y="8" x="24" z="16"/>
159     <BoxMax y="50" x="28" z="20"/>
160     <Width>4</Width>
161     <Gap>25</Gap>
162     <Types>NV</Types>
163    </Region>
164    <Region>
165     <BoxMin x="26" y="26" z="40"/>
166     <BoxMax x="34" y="34" z="48"/>
167     <Width>2</Width>
168     <Types>P</Types>
169    </Region>
170   </Steppable>
171
172   </CompuCell3D>
```

Listing 7. CC3DML specification of the vascular tumor model's initial cell layout, PDE solvers, and key cellular behaviors.

In Listing 7, in the `Contact` plugin (lines 36–53), we set $J_{MM}=0$, $J_{EM}=12$, and $J_{EE}=5$ (`M: Medium, E: EC`) and the `NeighborOrder` to 4. The `FocalPointPlasticity` plugin (lines 57–80) represents adhesion junctions by mechanically connecting the centers-of-mass of cells using a breakable linear spring (see Shirinifard *et al.*, 2009). `EC-EC` links are stronger than `EC-NV` links, which are, in turn, stronger than `NV-NV` links (see the CC3D manual for details). Since the Simulation Wizard creates code to implement links between all cell-type pairs in the model, we must delete most of them, keeping only the links between `EC-EC`, `EC-NV`, and `NV-NV` cell types.

We assume that `L_VEGF` diffuses 10 times faster than `S_VEGF`, so $D_{L\_VEGF}=0.42\ \mu m^2/s$ (1.6 $voxel^2/MCS$). This large diffusion constant would make the diffusion solver unstable. Therefore, in the CC3DML file (Listing 7, lines 108–114), we set the values of the `<DiffusionConstant>` and `<DecayConstant>` tags of the `L_VEGF` field to 0.16 and 0.0016, respectively, and use nine extra calls per MCS to achieve a diffusion constant equivalent to 1.6 (lines 87–89). We instruct `P` cells to secrete (line 116) into the L_VEGF field at a rate of 0.001 (3.85 pg/(cell h) = 0.001 pg/(voxel MCS)). Both `EC` and `NV` absorb `L_VEGF`. To simulate this uptake, we use the `<SecretionData>` tag pair (lines 117, 118).

Since the same diffusion solver will be called 10 times per MCS to solve `S_VEGF`, we must reduce the diffusion constant of `S_VEGF` by a factor of 10, setting the <DiffusionConstant> and <DecayConstant> tags of S.VEGF field to 0.016 and 0.0016, respectively. To prevent S_VEGF decay inside EC and NV cells, we add <DoNotDecayIn>EC</DoNotDecayIn> and <DoNotDecayIn>NV</DoNotDecayIn> inside the <DiffusionData> tag pair (lines 99, 100). We define S_VEGF to be secreted (lines 102–105) by both the EC and NV cells types at a rate of 0.013 per voxel per MCS (50 pg/(cell h) = 0.013 pg/(voxel MCS), compared to Leith and Michelson (1995).

The experimental glucose diffusion constant is about 600 $\mu m^2/s$. We convert the glucose diffusion constant by multiplying by appropriate spatial and temporal conversion factors: 600 $\mu m^2/s \times (\text{voxel}/4\ \mu m)^2 \times (60\ s/MCS)$=2250 $voxel^2/MCS$. To keep our simulation times short for this example, we use a simulated glucose diffusion constant 1500 smaller, resulting in much steeper glucose gradients and smaller maximum tumor diameters. We could use the steady-state diffusion solver for the glucose field to be more realistic.

Experimental GLU uptake by P cells is ~0.3 μmol/g/min. We assume that stromal cells (represented here without individual cell boundaries by Medium) take up GLU at a slower rate of 0.1 μmol/g/min. A gram of tumor tissue has about 10^8 tumor cells, so the glucose uptake per tumor cell is 0.003 pmol/MCS/cell or about 0.1 fmol/MCS/voxel. We assume that (at homeostasis) the preexisting vasculature supplies all the required GLU to Medium, which has a total mass of 1.28×10^{-5} grams and consumes GLU at a rate of 0.1 fmol/MCS/voxel, so the total GLU uptake (in the absence of a tumor) is 1.28 pmol/MCS. For this glucose to be supplied by 24 EC cells, their GLU secretion rate must be 0.8 fmol/MCS/voxel. We distribute the total GLU uptake (in the absence of a tumor) over all the Medium voxels, so the uptake rate is ~1.28 pmol/MCS/(~20,000 Medium voxels)=6.4×10^{-3} fmol/MCS/voxel.

We specify the uptake of GLU by Medium and P cells in lines 131 and 132 and instruct NV and EC cells to secrete GLU at the rate 0.4 and 0.8 pg/(voxel MCS), respectively (lines 129, 130).

We use UniformInitializer (lines 137–170) to initialize the tumor cell cluster and two crossing vascular cords. We also add two NV cells to each vascular cord, 25 pixels apart.

```
01   from PySteppables import *
02   from PySteppablesExamples import MitosisSteppableBase
03   import CompuCell
04   import sys
05   from random import uniform
06   import math
07
08   class VolumeParamSteppable(SteppableBasePy):
09     def __init__(self,_simulator,_frequency=1):
10       SteppableBasePy.__init__(self, _simulator,_frequency)
11       self.fieldL_VEGF = CompuCell.getConcentrationField('L_VEGF')
12       self.fieldGLU = CompuCell.getConcentrationField('GLU')
13
```

```
  def start(self):
    for cell in self.cellList:
      if (cell.type>=3):
        cell.targetVolume=64.0+10.0
        cell.lambdaVolume=20.0
      else:
        cell.targetVolume=32.0
        cell.lambdaVolume=20.0

  def step(self,mcs):
    pt=CompuCell.Point3D()
    for cell in self.cellList:
      if (cell.type==4): #Neovascular cells (NV)
        totalArea=0
        pt.x=int(round(cell.xCOM))
        pt.y=int(round(cell.yCOM))
        pt.z=int(round(cell.zCOM))
        VEGFconc=self.fieldL_VEGF.get(pt)
        cellNeighborList=self.getNeighborList(cell)
        for nsd in cellNeighborList:
          if (nsd.neighborAddress and nsd.neighborAddress.type>=3):
            totalArea+=nsd.commonSurfaceArea
        if (totalArea<45):
          cell.targetVolume+=2.0*VEGFconc/(0.01+VEGFconc)
      if (cell.type==1): #Proliferating Cells
        pt.x=int(round(cell.xCOM))
        pt.y=int(round(cell.yCOM))
        pt.z=int(round(cell.zCOM))
        gluConc=self.fieldGLU.get(pt)
        #Proliferating Cells become Necrotic when gluConc is low
        if (gluConc<0.001 and mcs>1000):
          cell.type=2
        else:
          cell.targetVolume+=0.022*gluConc/(0.05+gluConc)
      if cell.type==2: #Necrotic Cells
        cell.targetVolume-=0.1
        if cell.targetVolume<0.0:
          cell.targetVolume=0.0


class MitosisSteppable(MitosisSteppableBase):
  def __init__(self,_simulator,_frequency=1):
    MitosisSteppableBase.__init__(self,_simulator,_frequency)

  def step(self,mcs):
    cells_to_divide=[]
    for cell in self.cellList:
      if (cell.type==1 and cell.volume>64):
        cells_to_divide.append(cell)
      if (cell.type==4 and cell.volume>128):
        cells_to_divide.append(cell)
    for cell in cells_to_divide:
      self.divideCellRandomOrientation(cell)

  def updateAttributes(self):
    parentCell=self.mitosisSteppable.parentCell
    childCell=self.mitosisSteppable.childCell
    parentCell.targetVolume=parentCell.targetVolume/2
    parentCell.lambdaVolume=parentCell.lambdaVolume
    childCell.type=parentCell.type
    childCell.targetVolume=parentCell.targetVolume
    childCell.lambdaVolume=parentCell.lambdaVolume
```

Listing 8. Vascular tumor model Python steppables. The `VolumeParametersSteppable` adjusts the properties of the cells in response to simulation events and the `MitosisSteppable` implements cell division.

In the Python Steppable script in Listing 8, we set the initial target volume of both `EC` and `NV` cells to 74 (64 + 10) voxels and the initial target volume of tumor cells to 32 voxels (lines 14–21). All λ_{vol} are 20.0.

To model tumor cell growth, we increase the tumor cells' target volumes (lines 38–47) according to:

$$\frac{dV_t(\text{tumor})}{dt} = \frac{G_{\max} GLU(\vec{x})}{GLU(\vec{x}) + GLU_0} \tag{4}$$

where $GLU(\vec{x})$ is the `GLU` concentration at the cell's center-of-mass and GLU_0 is the concentration at which the growth rate is half its maximum. We assume that the fastest cell cycle time is 24 h, so $G_{\max}$ is 32 voxels/24 h = 0.022 voxel/MCS.

To account for contact-inhibited growth of `NV` cells, when their common surface area with other `EC` and `NV` cells is less than a threshold, we increase their target volume according to:

$$\frac{dV_t(NV)}{dt} = \frac{G_{\max} L\_VEGF(\vec{x})}{L\_VEGF(\vec{x}) + L\_VEGF_0} \tag{5}$$

where $L\_VEGF(\vec{x})$ is the concentration of `L_VEGF` at the cell's center-of-mass, L_VEGF_0 is the concentration at which the growth rate is half its maximum, and $G_{\max}$ is the maximum growth rate for `NV` cells. We calculate the common surface area between each `NV` cell and its neighboring `NV` or `EC` cells in lines 32–35. If the common surface area is smaller than 45, then we increase its target volume (lines 36, 37). When the volume of `NV` and `P` cells reaches a *doubling volume* (here, twice their initial target volumes), we divide them along a random axis, as shown in the `MitosisSteppable` (Listing 8, lines 54–75). The snapshots of the simulation are presented in Fig. 17

With this simple model we can easily explore the effects of changes in cell adhesion, nutrient availability, cell motility, sensitivity to starvation or dosing with chemotherapeutics or antiangiogenics on the growth and morphology of the simulated tumor.

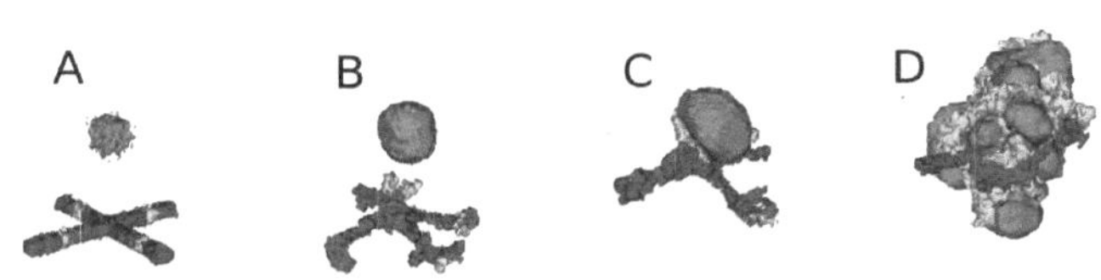

Fig. 17 Two-dimensional snapshots of the vascular tumor simulation taken at: (A) 0 MCS; (B) 500 MCS; (C) 2000 MCS; (D) 5000 MCS. Red and yellow cells represent endothelial cells and neovascular endothelial cells, respectively. (See color plate.)

E. Subcellular Simulations Using BionetSolver

While our vascular tumor model showed how to change cell-level parameters like target volume, we have not yet linked macroscopic cell behaviors to intracellular molecular concentrations. Signaling, regulatory, and metabolic pathways all steer the behaviors of biological cells by modulating their biochemical machinery. CC3D allows us to add and solve subcellular reaction-kinetic pathway models inside each generalized cell, specified using the SBML format (Hucka *et al.*, 2003), and to use such models (e.g., of their levels of gene expression) to control cell-level behaviors like adhesion or growth (Hester *et al.*, 2011).

We can use the same SBML framework to implement classic physics-based pharmacokinetic (PBPK) models of supercellular chemical flows between organs or tissues. The ability to explicitly model such subcellular and supercellular pathways adds greatly to the range of hypotheses CC3D models can represent and test. In addition, the original formulation of SBML primarily focused on the behaviors of biochemical networks within a single cell, whereas real signaling networks often involve the coupling of networks between cells. BionetSolver supports such coupling, allowing exploration of the very complex feedback resulting from intercell interactions linking intracellular networks, in an environment where the couplings change continuously due to cell growth, cell movement, and changes in cell properties.

As an example of such interaction between signaling networks and cell behaviors, we will develop a multi-cellular implementation of Delta–Notch mutual inhibitory coupling. In this juxtacrine signaling process, a cell's level of membrane-bound Delta depends on its intracellular level of activated Notch, which in turn depends on the average level of membrane-bound Delta of its neighbors. In such a situation, the Delta–Notch dynamics of the cells in a tissue sheet will depend on the rate of cell rearrangement and the fluctuations it induces. Although the example does not explore the wide variety of tissue properties due to the coupling of subcellular networks with intercellular networks and cell behaviors, it already shows how different such behaviors can be from those of their non-spatial simplifications. We begin with the ODE Delta–Notch patterning model of Collier *et al.* (1996) in which juxtacrine signaling controls the internal levels of the cells' Delta and Notch proteins. The base model neglects the complexity of the interaction due to changing spatial relationships in a real tissue:

$$\frac{dD}{dt} = v\left(\frac{1}{1 + bN^h} - D\right) \tag{6}$$

$$\frac{dN}{dt} = \frac{\overline{D}^k}{a + \overline{D}^k} - N \tag{7}$$

where D and N are the concentrations of activated Delta and Notch proteins inside a cell, $\overline{D}$ is the average concentration of activated Delta protein at the surface of the cell's neighbors, a and b are saturation constants, h and k are Hill coefficients, and v is a constant that gives the relative lifetimes of Delta and Notch proteins.

Notch activity increases with the levels of Delta in neighboring cells, whereas Delta activity decreases with increasing Notch activity inside a cell (Fig. 18). When

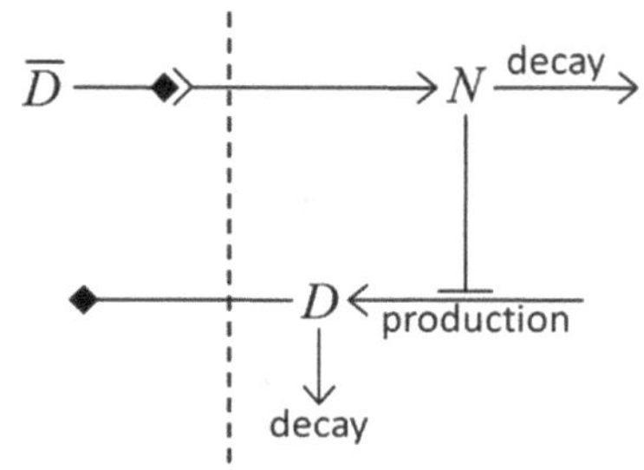

Fig. 18 Diagram of Delta–Notch feedback regulation between and within cells.

the parameters in the ODE model are chosen correctly, each cell assumes one of two exclusive states: a *primary fate*, in which the cell has a high level of Delta and a low level of Notch activity; and a *secondary fate*, in which the cell has a low level of Delta and a high level of Notch.

To build this model in CC3D, we assign a separate copy of the ODE model (6–7) to each cell and allow each cell to see the Delta concentrations of its neighbors. We use CC3D's BionetSolver library to manage and solve the ODEs, which are stored using the SBML standard.

The three files that specify the Delta–Notch model are included in the CC3D installation and can be found at *<CC3D-installation-dir>/DemosBionetSolver/DeltaNotch*: the main Python file (*DeltaNotch.py*) sets the parameters and initial conditions; the Python steppable file (*DeltaNotch_Step.py*) calls the subcellular models; and the SBML file (*DN_Collier.sbml*) contains the description of the ODE model. The first two files can be generated and edited using Twedit++, the last can be generated and edited using an SBML editor like Jarnac or JDesigner (both are open source). Listing 9 shows the SBML file viewed using Jarnac and can be downloaded from *http://sys-bio.org*.

```
p = defn cell
  vol compartment;
  var D, N;
  ext Davg, X;
  $X -> N; pow(Davg,k)/(a+pow(Davg,k))-N;
  $X -> D; v*(1/(1+b*pow(N,h))-D);
end;

p.compartment = 1;
p.Davg = 0.4;
p.X = 0;
p.D = 0.5;
p.N = 0.5;
p.k = 2;
p.a = 0.01;
p.v = 1;
p.b = 100;
p.h = 2;
```

Listing 9. Jarnac specification of the Delta–Notch coupling model in Fig. 17.

The main Python file (*DeltaNotch.py*) includes lines to define a steppable class (`DeltaNotchClass`) to include the ODE model and its interactions with the CC3D generalized cells (Listing 10).

```
from DeltaNotch_Step import DeltaNotchClass
deltaNotchClass=DeltaNotchClass(_simulator=sim,_frequency=1)
steppableRegistry.registerSteppable(deltaNotchClass)
```

Listing 10. Registering `DeltaNotchClass` in the main Python script, *DeltaNotch.py* in the Delta–Notch model.

The Python steppable file (Listing 11, *DeltaNotch_Step.py*) imports the BionetSolver library (line 1), then defines the class, and initializes the solver inside it (lines 2–5).

```
01  import bionetAPI
02  class DeltaNotchClass(SteppableBasePy):
03    def __init__(self,_simulator,_frequency):
04      SteppableBasePy.__init__(self,_simulator,_frequency)
05      bionetAPI.initializeBionetworkManager(self.simulator)
06
07    def start(self):
08      #Loading model
09      Name = "DeltaNotch"
10      Key  = "DN"
11      Path = os.getcwd()+"\DemosBionetSolver\DeltaNotch\DN_Collier.sbml"
12      IntegrationStep = 0.2
13      bionetAPI.loadSBMLModel(Name, Path, Key, IntegrationStep)
14
15      bionetAPI.addSBMLModelToTemplateLibrary(sbmlModelName,"TypeA")
16      bionetAPI.initializeBionetworks()
17
18      import random
19      for cell in self.cellList:
20        D = random.uniform(0.9,1.0)
21        N = random.uniform(0.9,1.0)
22        bionetAPI.setBionetworkValue("DN_D",D,cell.id)
23        bionetAPI.setBionetworkValue("DN_N",N,cell.id)
24        cellDict=CompuCell.getPyAttrib(cell)
25        cellDict["D"]=D
26        cellDict["N"]=N
```

Listing 11. Implementation of the `_init_` and `start` functions of the `DeltaNotchClass` in the Delta–Notch model.

The first lines in the `start` function (Listing 11, lines 9–12) specify the name of the model, its nickname (for easier reference), the path to the location where the SBML model is stored, and the time-step of the ODE integrator, which fixes the relation between MCS and the time units of the ODE model (here, 1 MCS corresponds to 0.2 ODE model time units). In line 13, we use the defined names, path and time-step parameter to load the SBML model.

In Listing 11, line 15 associates the subcellular model with the CC3D cells, creating an instance of the ODE solver (described by the SBML model) for each cell of type `TypeA`. Line 16 initializes the loaded subcellular models.

To set the initial levels of Delta (`D`) and Notch (`N`) in each cell, we visit all cells and assign random initial concentrations between 0.9 and 1.0 (Listing 11, lines 18–26). Line 18 imports the intrinsic Python random number generator. Lines 22 and 23 pass these values to the subcellular models in each cell. The first argument specifies the ODE model parameter to change with a string containing the nickname of the model, here `DN`, followed by an underscore and the name of the parameter as defined in the SBML file. The second argument specifies the value to assign to the parameter, and the last argument specifies the cell id. For visualization purposes, we also store the values of `D` and `N` in a dictionary attached to each cell (lines 25, 26).

Listing 12 defines a `step` function of the class, which is called by every MCS, to read the Delta concentrations of each cell's neighbors to determine the value of $\overline{D}$ (the average Delta concentration around the cell). The first three lines in Listing 12 iterate over all cells. Inside the loop, we first set the variables `D` and `nn` to zero. They will store the total Delta concentration of the cell's neighbors and the number of neighbors, respectively. Next, we get a list of the cell's neighbors and iterate over them. Line 9 reads the Delta concentration of each neighbor (the first argument is the name of the parameter and the second is the id of the neighboring cell) summing the total Delta and counting the number of neighbors. Note the `+=` syntax (e.g., `nn+=1` is equivalent to `nn=nn+1`). Lines 3 and 7 skip `Medium` (`Medium` has a value 0, so `if (Medium)` is false).

```
01  def step(self,mcs):
02    for cell in self.cellList:
03      if cell:
04        D=0.0; nn=0
05        cellNeighborList=self.getCellNeighbors(cell)
06        for nsd in cellNeighborList:
07          if nsd:
08            nn+=1
09            D+=bionetAPI.getBionetworkValue("DN_D",nsd.neighborAddress.id)
10        if (nn>0):
11          D=D/nn
12        bionetAPI.setBionetworkValue("DN_Davg",D,cell.id)
13        cellDict=CompuCell.getPyAttrib(cell)
14        cellDict["D"]=D
15        cellDict["N"]=bionetAPI.getBionetworkValue("DN_N",cell.id)
16    bionetAPI.timestepBionetworks()
```

Listing 12. Implementation of a `step` function (continuation of the code from Listing 11) to calculate $\overline{D}$ in the `DeltaNotchClass` in the Delta–Notch model.

After looping over the cell's neighbors, we update the variable $\overline{D}$, which in the SBML code has the name `Davg`, to the average neighboring Delta (`D`) concentration, ensuing that the denominator, `nn`, is not zero (Listing 12, lines 10–12).

The remaining lines (Listing 12, lines 13–15) access the cell dictionary and store the cell's current Delta and Notch concentrations. Line 16 then calls BionetSolver and tells it to integrate the ODE model with the new parameters for one integration step (0.2 time units in this case).

Fig. 19 shows a typical cell configurations and states for the simulation. The random initial values gradually converge to a pattern with cells with low levels of Notch (primary fate) surrounded by cells with high levels of Notch (secondary fate).

In Listing 13, lines 2–4 define two new visualization fields in the main Python file (*DeltaNotch.py*) to visualize the Delta and Notch concentrations in CompuCell Player. To fill the fields with the Delta and Notch concentrations, we call the steppable class, `ExtraFields` (Listing 13, lines 6–9). This code is very similar to our previous steppable calls, with the exception of line 8, which uses the function `setScalarFields()` to reference the visualization `Fields`.

```
01   #Create extra player fields here or add attributes
02   dim=sim.getPotts().getCellFieldG().getDim()
03   DeltaField=simthread.createScalarFieldCellLevelPy("Delta")
04   NotchField=simthread.createScalarFieldCellLevelPy("Notch")
05
06   from DeltaNotch_Step import ExtraFields
07   extraFields=ExtraFields(_simulator=sim,_frequency=5)
08   extraFields.setScalarFields(DeltaField,NotchField)
09   steppableRegistry.registerSteppable(extraFields)
```

Listing 13. Adding extra visualization fields in the main Python script *DeltaNotch.py* in the Delta–Notch model.

In the steppable file (Listing 14, *DeltaNotch_Step.py*) we use `setScalarFields()` to set the variables `self.scalarField1` and `self.scalarField2` to point to the fields `DeltaField` and `NotchField`, respectively. Lines 10 and 11 of the `step` function clear the two fields using `clearScalarValueCellLevel()`. Line 12 loops over all cells, line 13 accesses a cell's dictionary, and lines 14 and 15 use the `D` and `N` entries to fill in the respective visualization fields, where the first argument specifies the visualization field, the second the cell to be filled, and the third the value to use.

```
01   class ExtraFields(SteppableBasePy):
02     def __init__(self,_simulator,_frequency=1):
03       SteppableBasePy.__init__(self,_simulator,_frequency)
04
05     def setScalarFields(self,_field1,_field2):
06       self.scalarField1=_field1
07       self.scalarField2=_field2
08
09     def step(self,mcs):
10       clearScalarValueCellLevel(self.scalarField1)
11       clearScalarValueCellLevel(self.scalarField2)
12       for cell in self.cellList:
13         cellDict=CompuCell.getPyAttrib(cell)
14         fillScalarValueCellLevel(self.scalarField1,cell,cellDict["D"])
15         fillScalarValueCellLevel(self.scalarField2,cell,cellDict["N"])
```

Listing 14. Steppable to visualize the concentrations of Delta and Notch in each cell in the Delta–Notch model.

The two fields can be visualized in CompuCell Player using the `Field-selector` button of the `Main Graphics Window` menu (second-to-last button, Fig. 19).

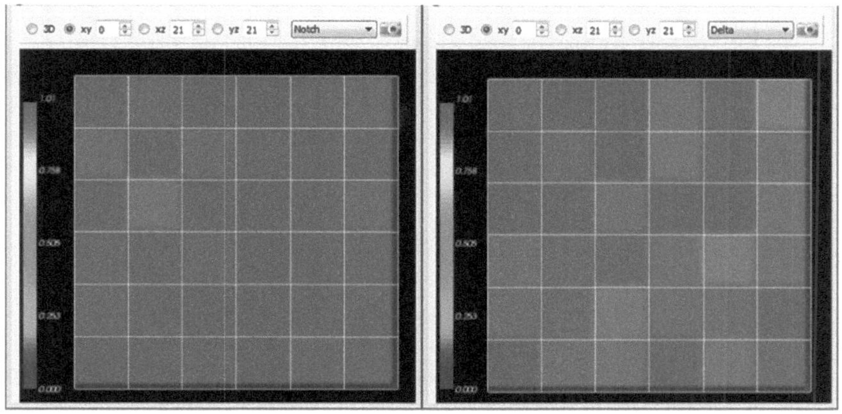

Fig. 19 Initial Notch (left) and Delta (right) concentrations in the Delta–Notch model. (For color version of this figure, the reader is referred to the web version of this book.)

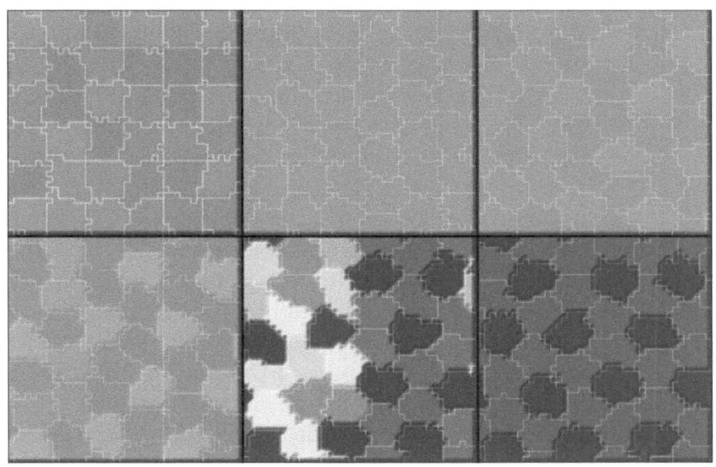

Fig. 20 Dynamics of the Notch concentrations of cells in the Delta–Notch model. Snapshots taken at 10, 100, 300, 400, 450, and 600 MCS. (See color plate.)

As we illustrate in Fig. 20, the result is a roughly hexagonal pattern of activity with one cell of low-Notch activity for every two cells with high Notch activity. In the presence of a high level of cell motility, the identity of high and low Notch cells can change when the pattern rearranges. We could easily explore the effects of Delta–Notch signaling on tissue structure by linking the Delta–Notch pathway to one of its known downstream targets. For example, if we wished to simulate embryonic feather-bud primordial in chicken skin or the formation of colonic crypts, we could start with an epithelial sheet of cells in 3D on a rigid support, and couple the growth of the cells to their level of Notch activity by having Notch inhibit cell growth. The result would be clusters of cell growth around the initial low-Notch cells, leading to a patterned 3D buckling of the epithelial tissue. Such mechanisms are capable of extremely complex and subtle patterning, as observed *in vivo*.

V. Conclusion

Multi-cell modeling, especially when combined with subcell (or supercell) modeling of biochemical networks, allows the creation and testing of hypotheses concerning many key aspects of embryonic development, homeostasis, and developmental disease. Until now, such modeling has been out of reach to all but experienced software developers. CC3D makes the development of such models much easier, though it still does involve a minimal level of hand editing. We hope the examples we have shown will convince readers to evaluate the suitability of CC3D for their research.

Furthermore, CC3D directly addresses the current difficulty researchers face in reusing, testing, or adapting both their own and published models. Most published multi-cell, multi-scale models exist in the form of Fortran/C/C++ code, which is often of little practical value to other potential users. Reusing such code involves digging into large code bases, inferring their function, extracting the relevant code, and trying to paste it into a new context. CC3D improves this status quo in at least three ways: (1) it is fully open source; (2) CC3D models can be executed cross-platform and do not require compilation; (3) CC3D models are modular, compact, and shareable. Because Python-based CC3D models require much less effort to develop than does custom code programming: simulations are fast and easy to develop and refine. Even with these convenience features, CC3D 3.6 often runs as fast or faster than custom code solving the same model. Current CC3D development focuses on adding GPU-based PDE solvers, MPI parallelization, and additional cell behaviors. We are also developing a high-level cell-behavior model description language that will compile into executable Python, removing the last need for model builders to learn programming techniques.

All examples presented in this chapter are included in the CC3D binary distribution and will be curated to ensure their correctness and compatibility with future versions of CC3D.

Acknowledgments

We gratefully acknowledge support from the National Institutes of Health, National Institute of General Medical Sciences grants R01 GM077138 and R01 GM076692, the Environmental Protection Agency, and the Office of Vice President for Research, the College of Arts and Sciences, the Pervasive Technologies Laboratories, and the Biocomplexity Institute at Indiana University. GLT acknowledges support from the Brazilian agencies Conselho Nacional de Pesquisa e Desenvolvimento (CNPq) and

Fundação de Amparo à Pesquisa do Estado do Rio Grande do Sul (FAPERGS) under the grant PRONEX-10/0008-0. Indiana University's University Information Technology Services provided time on their BigRed cluster for simulation execution. Early versions of CompuCell and CompuCell3D were developed at the University of Notre Dame by JAG, Dr. Mark Alber, and Dr. Jesus Izaguirre, and collaborators with the support of National Science Foundation, Division of Integrative Biology, grant IBN-00836563. Since the primary home of CompuCell3D moved to Indiana University in 2004, the Notre Dame team have continued to provide important support for its development. We especially would like to thank our current collaborators, Herbert Sauro and Ryan Roper, from University of Washington for developing the subcellular reaction kinetics model simulator BionetSolver.

References

Alber, M. S., Jiang, Y., and Kiskowski, M. A. (2004). Lattice gas cellular automation model for rippling and aggregation in myxobacteria. *Physica D* **191**, 343–358.

Alber, M. S., Kiskowski, M. A., Glazier, J. A., and Jiang, Y. (2002). On cellular automation approaches to modeling biological cells. In "*Mathematical Systems Theory in Biology, Communication and Finance*," (J. Rosenthal, and D. S. Gilliam, eds.), pp. 1–40. Springer-Verlag, New York.

Alber, M., Chen, N., Glimm, T., and Lushnikov, P. (2006). Multiscale dynamics of biological cells with chemotactic interactions: from a discrete stochastic model to a continuous description. *Phys. Rev. E* **73**, 051901 (PMID 16802961).

Armstrong, P. B., and Armstrong, M. T. (1984). A role for fibronectin in cell sorting out. *J. Cell. Sci.* **69**, 179–197.

Armstrong, P. B., and Parenti, D. (1972). Cell sorting in the presence of cytochalasin B. *J. Cell. Biol.* **55**, 542–553.

Chaturvedi, R., Huang, C., Izaguirre, J. A., Newman, S. A., Glazier, J. A., and Alber, M. S. (2004). A hybrid discrete-continuum model for 3-D skeletogenesis of the vertebrate limb. *Lect. Notes Comput. Sci.* **3305**, 543–552.

Cickovski, T., Aras, K., Alber, M. S., Izaguirre, J. A., Swat, M., Glazier, J. A., Merks, R. M. H., Glimm, T., Hentschel, H. G. E., and Newman, S. A. (2007). From genes to organisms via the cell: a problem-solving environment for multicellular development. *Comput. Sci. Eng.* **9**, 50.

Cipra, B. A. (1987). An introduction to the Ising-model. *Amer. Math. Monthly* **94**, 937–959.

Collier, J. R., Monk, N. A. M., Maini, P. K., and Lewis, J. H. (1996). Pattern formation by lateral inhibition with feedback: a mathematical model of Delta–Notch intercellular signaling. *J. Theor. Biol.* **183**, 429–446.

Dallon, J., Sherratt, J., Maini, P. K., and Ferguson, M. (2000). Biological implications of a discrete mathematical model for collagen deposition and alignment in dermal wound repair. *IMA J. Math. Appl. Med. Biol.* **17**, 379–393.

Drasdo, D., Kree, R., and McCaskill, J. S. (1995). Monte-Carlo approach to tissue-cell populations. *Phys. Rev. E* **52**, 6635–6657.

Glazier, J. A. (1993). Cellular patterns. *Bussei Kenkyu* **58**, 608–612.

Glazier, J. A. (1996). Thermodynamics of cell sorting. *Bussei Kenkyu* **65**, 691–700.

Glazier, J. A., and Graner, F. (1992). Simulation of biological cell sorting using a two-dimensional extended Potts model. *Phys. Rev. Lett.* **69**, 2013–2016.

Glazier, J. A., and Graner, F. (1993). Simulation of the differential adhesion driven rearrangement of biological cells. *Phys. Rev. E* **47**, 2128–2154.

Glazier, J. A., Raphael, R. C., Graner, F., and Sawada, Y. (1995). The energetics of cell sorting in three dimensions. In "*Interplay of Genetic and Physical Processes in the Development of Biological Form*," (D. Beysens, G. Forgacs, F. Gaill, eds.), pp. 54–66. World Scientific Publishing Company, Singapore.

Glazier, J. A., Balter, A., and Poplawski, N. (2007). Magnetization to morphogenesis: a brief history of the Glazier–Graner–Hogeweg model. In "*Single-Cell-Based Models in Biology and Medicine*," (A. R. A. Anderson, M. A. J. Chaplain, K. A. Rejniak, eds.), pp. 79–106. Birkhauser Verlag Basel, Switzerland.

Glazier, J. A., Zhang, Y., Swat, M., Zaitlen, B., and Schnell, S. (2008). Coordinated action of N-CAM, N-cadherin, EphA4, and ephrinB2 translates genetic prepatterns into structure during somitogenesis in chick. *Curr. Top. Dev. Biol.* **81**, 205–247.

Graner, F., and Glazier, J. A. (1992). Simulation of biological cell sorting using a 2-dimensional extended Potts model. *Phys. Rev. Lett.* **69**, 2013–2016.

Grieneisen, V. A., Xu, J., Maree, A. F. M., Hogeweg, P., and Schere, B. (2007). Auxin transport is sufficient to generate a maximum and gradient guiding root growth. *Nature* **449**, 1008–1013.

Groenenboom, M. A., and Hogeweg, P. (2002). Space and the persistence of male-killing endosymbionts in insect populations. *Proc. Biol. Sci.* **269**, 2509–2518.

Groenenboom, M. A., Maree, A. F. M., and Hogeweg, P. (2005). The RNA silencing pathway: the bits and pieces that matter. *PLoS Comput. Biol.* **1**, 55–165.

Hester, S. D., Belmonte, J. M., Gens, J. S., Clendenon, S. G., and Glazier, J. A. (2011). A Multi-cell, Multi-scale Model of Vertebrate Segmentation and Somite Formation. *PLoS Comput. Biol* **7**, e1002155.

Hogeweg, P. (2000). Evolving mechanisms of morphogenesis: on the interplay between differential adhesion and cell differentiation. *J. Theor. Biol.* **203**, 317–333.

Holm, E. A., Glazier, J. A., Srolovitz, D. J., and Grest, G. S. (1991). Effects of lattice anisotropy and temperature on domain growth in the two-dimensional Potts model. *Phys. Rev. A* **43**, 2662–2669.

Honda, H., and Mochizuki, A. (2002). Formation and maintenance of distinctive cell patterns by coexpression of membrane-bound ligands and their receptors. *Dev. Dynamics* **223**, 180–192.

Hucka, M., Finney, A., Sauro, H. M., Bolouri, H., Doyle, J. C., Kitano, H., Arkin, A. P., Bornstein, B. J., Bray, D., Cornish-Bowden, A., Cuellar, A. A., Dronov, S., Gilles, E. D., Ginkel, M., Gor, V., Goryanin, I. I., Hedley, W. J., Hodgman, T. C., Hofmeyr, J. -H., Hunter, P. J., Juty, N. S., Kasberger, J. L., Kremling, A., Kummer, U., Le Novère, N., Loew, L. M., Lucio, D., Mendes, P., Minch, E., Mjolsness, E. D., Nakayama, Y., Nelson, M. R., Nielsen, P. F., Sakurada, T., Schaff, J. C., Shapiro, B. E., Shimizu, T. S., Spence, H. D., Stelling, J., Takahashi, K., Tomita, M., Wagner, J., and Wang, J. (2003). The Systems biology markup language (SBML): a medium for representation and exchange of biochemical network models. *Bioinformatics* **19**, 524–531.

Johnston, D. A. (1998). Thin animals. *J. Phys. A* **31**, 9405–9417.

Kesmir, C., and de Boer, R. J. (2003). A spatial model of germinal center reactions: cellular adhesion based sorting of B cells results in efficient affinity maturation. *J. Theor. Biol.* **222**, 9–22.

Kesmir, C., van Noort, V., de Boer, R. J., and Hogeweg, P. (2003). Bioinformatic analysis of functional differences between the immunoproteasome and the constitutive proteasome. *Immunogenetics* **55**, 437–449.

Knewitz, M. A., and Mombach, J. C. M. (2006). Computer simulation of the influence of cellular adhesion on the morphology of the interface between tissues of proliferating and quiescent cells. *Comput. Biol. Med.* **36**, 59–69.

Leith, J. T., and Michelson, S. (1995). Secretion rates and levels of vascular endothelial growth factor in clone A or HCT-8 human colon tumour cells as a function of oxygen concentration. *Cell Prolif.* **28**, 415–430.

Longo, D., Peirce, S. M., Skalak, T. C., Davidson, L., Marsden, M., and Dzamba, B. (2004). Multicellular computer simulation of morphogenesis: blastocoel roof thinning and matrix assembly in *Xenopus laevis*. *Dev. Biol.* **271**, 210–222.

Lutz, M. (2011). *Programming Python.* O'Reilly & Associates, Inc, Sebastopol, CA.

Maini, P. K., Olsen, L., and Sherratt, J. A. (2002). Mathematical models for cell-matrix interactions during dermal wound healing. *Int. J. Bifurcation Chaos* **12**, 2021–2029.

Marée, A. F. M., and Hogeweg, P. (2001). How amoeboids self-organize into a fruiting body: multicellular coordination in *Dictyostelium discoideum*. *Proc. Natl. Acad. Sci. USA* **98**, 3879–3883.

Marée, A. F. M., and Hogeweg, P. (2002). Modelling *Dictyostelium discoideum* morphogenesis: the culmination. *Bull. Math. Biol.* **64**, 327–353.

Marée, A. F. M., Panfilov, A. V., and Hogeweg, P. (1999a). Migration and thermotaxis of *Dictyostelium discoideum* slugs, a model study. *J. Theor. Biol.* **199**, 297–309.

Marée, A. F. M., Panfilov, A. V., and Hogeweg, P. (1999b). Phototaxis during the slug stage of *Dictyostelium discoideum*: a model study. *Proc. Royal Soc. Lond. Ser. B* **266**, 1351–1360.
Merks, R. M., Brodsky, S. V., Goligorksy, M. S., Newman, S. A., and Glazier, J. A. (2006). Cell elongation is key to *in silico* replication of *in vitro* vasculogenesis and subsequent remodeling. *Dev. Biol.* **289**, 44–54.
Merks, R. M., and Glazier, J. A. (2006). Dynamic mechanisms of blood vessel growth. *Nonlinearity* **19**, C1–C10.
Merks, R. M., Perryn, E. D., Shirinifard, A., and Glazier, J. A. (2008). Contact-inhibited chemotactic motility can drive both vasculogenesis and sprouting angiogenesis. *PLoS Comput. Biol.* **4**, e1000163.
Metropolis, N., Rosenbluth, A., Rosenbluth, M. N., Teller, A. H., and Teller, E. (1953). Equation of state calculations by fast computing machines. *J. Chem. Phys.* **21**, 1087–1092.
Meyer-Hermann, M., Deutsch, A., and Or-Guil, M. (2001). Recycling probability and dynamical properties of germinal center reactions. *J. Theor. Biol.* **210**, 265–285.
Mochizuki, A. (2002). Pattern formation of the cone mosaic in the zebrafish retina: A cell rearrangement model. *J. Theor. Biol.* **215**, 345–361.
Mombach, J. C. M., and Glazier, J. A. (1996). Single cell motion in aggregates of embryonic cells. *Phys. Rev. Lett.* **76**, 3032–3035.
Mombach, J. C. M., de Almeida, R. M. C., Thomas, G. L., Upadhyaya, A., and Glazier, J. A. (2001). Bursts and cavity formation in Hydra cells aggregates: experiments and simulations. *Physica A* **297**, 495–508.
Mombach, J. C. M., Glazier, J. A., Raphael, R. C., and Zajac, M. (1995). Quantitative comparison between differential adhesion models and cell sorting in the presence and absence of fluctuations. *Phys. Rev. Lett.* **75**, 2244–2247.
Nguyen, B., Upadhyaya, A., van Oudenaarden, A., and Brenner, M. P. (2004). Elastic instability in growing yeast colonies. *Biophys. J.* **86**, 2740–2747.
Novak, B., Toth, A., Csikasz-Nagy, A., Gyorffy, B., Tyson, J. A., and Nasmyth, K. (1999). Finishing the cell cycle. *J. Theor. Biol.* **199**, 223–233.
Popławski, N. J., Shirinifard, A., Swat, M., and Glazier, J. A. (2008). Simulation of single-species bacterial-biofilm growth using the Glazier–Graner–Hogeweg model and the CompuCell3D modeling environment. *Math. Biosci. Eng.* **5**, 355–388.
Popławski, N. J., Swat, M., Gens, J. S., and Glazier, J. A. (2007). Adhesion between cells diffusion of growth factors and elasticity of the AER produce the paddle shape of the chick limb. *Physica A* **373**, C521–C532.
Rieu, J. P., Upadhyaya, A., Glazier, J. A., Ouchi, N. B., and Sawada, Y. (2000). Diffusion and deformations of single hydra cells in cellular aggregates. *Biophys. J.* **79**, 1903–1914.
Savill, N. J., and Hogeweg, P. (1997). Modelling morphogenesis: from single cells to crawling slugs. *J. Theor. Biol.* **184**, 229–235.
Savill, N. J., and Sherratt, J. A. (2003). Control of epidermal stem cell clusters by Notch-mediated lateral induction. *Dev. Biol.* **258**, 141–153.
Steinberg, M. S. (2007). Differential adhesion in morphogenesis: a modern view. *Curr. Opin. Genet. Dev.* **17**(4), 281–286.
Shirinifard, A., Gens, J. S., Zaitlen, B. L., Popławski, N. J., Swat, M. H., and Glazier, J. A. (2009). 3D multi-cell simulation of tumor growth and angiogenesis. *PLoS ONE* **4**, e7190.
Takesue, A., Mochizuki, A., and Iwasa, Y. (1998). Cell-differentiation rules that generate regular mosaic patterns: modelling motivated by cone mosaic formation in fish retina. *J. Theor. Biol.* **194**, 575–586.
Turner, S., and Sherratt, J. A. (2002). Intercellular adhesion and cancer invasion: a discrete simulation using the extended Potts model. *J. Theor. Biol.* **216**, 85–100.
Walther, T., Reinsch, H., Grosse, A., Ostermann, K., Deutsch, A., and Bley, T. (2004). Mathematical modeling of regulatory mechanisms in yeast colony development. *J. Theor. Biol.* **229**, 327–338.
Walther, T., Reinsch, H., Ostermann, K., Deutsch, A., and Bley, T. (2005). Coordinated growth of yeast colonies: experimental and mathematical analysis of possible regulatory mechanisms. *Eng. Life Sci.* **5**, 115–133.

Wearing, H. J., Owen, M. R., and Sherratt, J. A. (2000). Mathematical modelling of juxtacrine patterning. *Bull. Math. Biol.* **62**, 293–320.

Zajac, M. (2002). Modeling convergent extension by way of anisotropic differential adhesion. Ph.D. thesis, University of Notre Dame.

Zajac, M., Jones, G. L., and Glazier, J. A. (2000). Model of convergent extension in animal morphogenesis. *Phys. Rev. Lett.* **85**, 2022–2025.

Zajac, M., Jones, G. L., and Glazier, J. A. (2003). Simulating convergent extension by way of anisotropic differential adhesion. *J. Theor. Biol.* **222**, 247–259.

Zhang, Y., Thomas, G. L., Swat, M., Shirinifard, A., and Glazier, J. A. (2011). Computer imulations of Cell Sorting Due to Differential Adhesion. *PLoS ONE* **6**(10), e24999.

Zhdanov, V. P., and Kasemo, B. (2004a). Simulation of the growth and differentiation of stem cells on a heterogeneous scaffold. *Phys. Chem. Chem. Phys.* **6**, 4347–4350.

Zhdanov, V. P., and Kasemo, B. (2004b). Simulation of the growth of neurospheres. *Europhys. Lett.* **68**, 134–140.

CHAPTER 14

Multiscale Model of Fibrin Accumulation on the Blood Clot Surface and Platelet Dynamics

Zhiliang Xu[*], Scott Christley[†], Joshua Lioi[‡], Oleg Kim[*], Cameron Harvey[§], Wenzhao Sun[*], Elliot D. Rosen[¶] and Mark Alber[*,||]

[*]Department of Applied and Computational Mathematics, University of Notre Dame, Notre Dame, Indiana, USA

[†]Department of Surgery, University of Chicago, Chicago, Illinois, USA

[‡]Department of Applied and Computational Mathematics and Statistics, University of Notre Dame, Notre Dame, Indiana, USA

[§]Department of Physics, University of Notre Dame, Notre Dame, Indiana, USA

[¶]Department of Medical and Molecular Genetics, Indiana University School of Medicine, Indianapolis, Indiana, USA

[||]Department of Medicine, Indiana University School of Medicine, Indianapolis, Indiana, USA

0091-679X/10 $35.00
DOI 10.1016/B978-0-12-388403-9.00014-X

Abstract

A multiscale computational model of thrombus (blood clot) development is extended by incorporating a submodel describing formation of fibrin network through "fibrin elements" representing regions occupied by polymerized fibrin. Simulations demonstrate that fibrin accumulates on the surface of the thrombus and that fibrin network limits growth by reducing thrombin concentrations on the thrombus surface and decreasing adhesivity of resting platelets in blood near thrombus surface. These results suggest that fibrin accumulation may not only increase the structural integrity of the thrombus but also considerably contribute toward limiting its growth. Also, a fast Graphics Processing Unit implementation is described for a multiscale computational model of the platelet–blood flow interaction.

I. Introduction

To restrict the loss of blood following rupture of blood vessels, the human body rapidly forms a clot consisting of platelets and fibrin. Although hemostasis (blood clotting) is essential to prevent hemorrhage, inappropriate clotting initiated by vessel wall damage or dysfunction of endothelial cells lining the lumen of the vessel wall can lead to intravascular clots (thrombi) that may disrupt flow causing damage to tissues and organs in the flow field. Venous thromboembolic disease is a significant biomedical problem with the annual incidence in the United States being estimated as high as 900,000 cases per year leading to 300,000 deaths (Wakefield *et al.*, 2009).

The assembly of a blood thrombus requires complex interactions among multiple molecular and cellular components in the blood and occurs under fluid flow. Under normal conditions, clotting is started by initiation events requiring components in flowing blood to form complexes with components in the wall at the injury site. These include platelet adhesion to matrix proteins in the vessel wall leading to platelet activation, recruitment of more platelets in the flowing blood, and formation of coagulation initiation complexes. This results in the activation of a network of coagulation reactions that generate thrombin. Thrombin converts fibrinogen in blood to fibrin, which self-polymerizes forming a fibrin network that is a major structural component of the clot. Thrombin is also a potent platelet activator. Although most of the components of the hemostatic system and interactions among them have been identified in biochemical and genetic research during the last several decades, the regulation of the interactions to provide proper clotting to limit bleeding but block pathological thrombosis remains to be elucidated. Current biological models of thrombus formation (Diamond, 2009; Esmon, 1993, 2001, 2009; Mann *et al.*, 2006) suggest that the surface of a growing clot recruits resting platelets in flowing blood to adhere and to become activated on the thrombus. The activated platelets support coagulation reactions that produce thrombin leading to activation of more platelets and the formation of a fibrin network. Thus, the growing thrombus continuously generates a new prothrombotic surface to support continued growth.

We have found that as a thrombus develops, there is a significant increase in the surface fraction composed of fibrin and a decrease in the fraction composed of platelets (Fig. 1) (Kamocka *et al.*, 2010). We propose that this fibrin network on the surface of the developing thrombus limits further development of the thrombus, and that it does so using two mechanisms: (1) by impeding the transport of clot-promoting and clot-building materials between the interior of the clot and the clot surface; (2) by binding less stably to platelets, allowing them to be easily washed off by blood flow. As fibrin is a major structural component of the thrombus contributing to structural stability and that individuals deficient for fibrinogen suffer bleeding disorders, fibrin(ogen) is considered a prothrombotic component (Lord, 2007). In addition, fibrin(ogen) is also considered as an anti-thrombotic agent (Lishko, 2007 and Yermolenko *et al.*, 2010). We use an extended multiscale model to test our hypothesis that fibrinogen also functions to halt thrombus growth, which may be considered antithrombotic, suggests that classifying components as pro- or antithrombotic is an oversimplification. The model is extended by incorporating a phenomenological fibrin cell submodel to simulate the dynamically formed fibrin network.

Implementation of a multiscale model on a Graphics Processing Unit (GPU) is described in the second part of this chapter for speeding up latest 3D simulations of

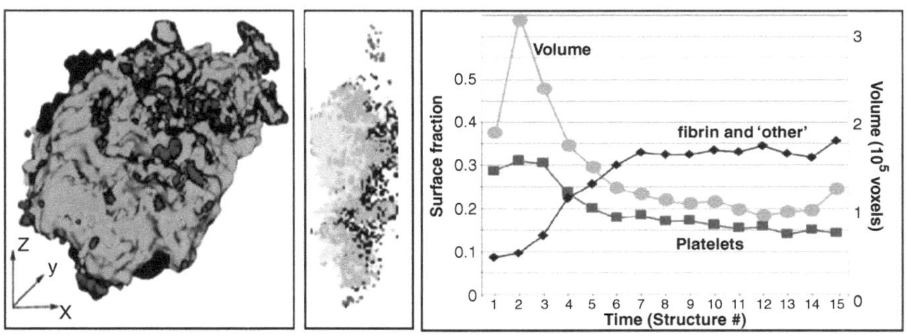

Fig. 1 Surface composition of developing thrombi. Vertical stacks of images were collected by multi-photon microscopy of laser-induced injuries in mesenteric veins of a mice. 3D image reconstruction of a late stage thrombus is shown in (Left) (luminal view) and (Middle) (cross-section in *yz* plane (wall is on the left, lumen is on the right). Regions composed primarily of platelets are red, primarily of fibrin—green, composed of platelets and fibrin—yellow; and regions excluding plasma, fibrin and platelets (other material, cells)—black. (Right) Shows changing surface composition of the thrombus as it stabilizes. Stabilization is associated with decreasing amounts of platelets and increasing amounts of fibrin on the surface (A and B are reprinted from Mu *et al.* (2009)), with permission from the Royal Society of Chemistry. Mu *et al.* (2009) is available online at http://pubs.rsc.org/en/content/articlepdf/2011/sm/c0sm01528h). (See color plate.)

thrombus formation (Sweet *et al.*, 2011). *In vivo* blood clot takes minutes to develop; whereas most computational models that include detailed description of blood clot components at various spatial and temporal scales are limited to simulating clotting events only for seconds. The subcellular element model (SCEM) is used in (Sweet *et al.*, 2011) to represent platelets, whereas the Lattice Boltzmann (LB) equation is utilized for simulating plasma flow. The coupling between SCEM and LB equation is implemented through the Langevin method. Using this GPU implementation, we are capable of simulating dynamics of 100 platelets moving in a blood flow over a period of 0.5 s using about 2000 s of computer time. This provides a significant advantage in terms of simulation efficiency because it normally takes days of computer time to run simulations using CPU-based implementation (without parallelization).

II. Biological Background

Thrombus formation is the result of two interrelated processes, platelet interactions, and activation of the coagulation pathway. Immediately after vessel damage, platelets adhere to the site of vessel injury forming a single cell layer (Cranmer *et al.*, 1999; Gruner *et al.*, 2003; Jackson *et al.*, 2000; Kuijpers *et al.*, 2004; Nieswandt and Watson, 2003). Following adhesion, platelets form multicellular aggregates mediated by the binding of the platelet receptor on different platelets binding to the same bridging molecule: fibrin(ogen), von Willebrand factor (vWF), fibrinectin. In addition, platelets undergo activation involving significant morphological changes, the exposure of new proteins on the platelet surface and the extracellular release of contents of alpha and dense granules found in resting platelets. These contents include a variety of hemostatic proteins and effector molecules that activate resting platelets.

The recruitment of free-flowing platelets to sites of injury is a key step in the formation of blood thrombus. Targeting of platelets to these sites is a multistep process with the sequential involvement of distinct adhesion molecules on platelets and subendothelial cell surfaces. There are multiple mechanisms ranging from the molecular to cellular level that affect cell rolling on subendothelial cells. The process is initiated by the binding of GPIba with the subendothelial collagen-bound vWF. This interaction maintains platelets in close contact with the surface, even though with platelet rolling, until other receptors and ligands mediate a stable attachment after activation. When vWF is bound to collagen, the transition from rolling to stable adhesion occurs in seconds. Moreover, this cascade of highly regulated molecular events is dictated by local circulatory hemodynamics and the mechanical and kinetic properties of participating adhesion molecules and the cellular material properties such as cell deformation and microvillus viscoelasticity, which may critically affect the dynamics of platelet – injury wall interaction (Doggett *et al.*, 2002). Finally, the type and spatial distribution of the receptors play a key role in cell rolling. However, their relative roles have yet to be categorized quantitatively.

In addition to platelet interactions, coagulation factor VII (FVII) in the blood is exposed to tissue factor (TF), expressed on cells in the vessel wall, initiating the network of coagulation reactions. These reactions lead to the generation of thrombin (FIIa), which converts fibrinogen to fibrin, the major matrix protein in a thrombus, and activates FXIII to cross-link fibrin. Thrombin triggers a positive feedback loop of propagation reactions. Thrombin is also a potent activator of resting platelets further promoting thrombus development. The activated platelets provide a procoagulant surface that promotes the surface-dependent coagulation reactions at the site of injury.

Thrombus development is a complex process with a lot to be explored. For instance, current biological models of thrombogenesis (Diamond, 2009; Esmon, 1993, 2001, 2009; Mann *et al.*, 2006) suggest that the surface of a growing clot recruits resting platelets in flowing blood to adhere and to become activated on the thrombus. The activated platelets support coagulation reactions that produce thrombin leading to activation of more platelets and the formation of a fibrin network. Thus, the growing thrombus continuously generates a new prothrombotic surface to support continued growth. However, subocclusive thrombi generated by vascular injury stop growing and stabilize within minutes after injury. This suggests that negative feedback mechanisms limit continuous growth of developing thrombi.

Several processes activated after thrombus initiation and platelet components that have been identified may provide the negative feedback function to limit thrombus growth. The protein C (PC) anticoagulant pathway is activated by the thrombin generated in the developing thrombus. However, the spatial separation between the sites where thrombin and PC are generated may prevent aPC from limiting thrombus growth (Fogelson and Tania, 2005; Xu *et al.*, 2010). Activation of PC would require either thrombin generated in the thrombus to reach upstream endothelial cells or aPC generated on downstream endothelial cells to migrate upstream to the thrombus (Fogelson and Tania, 2005). Moreover, previous simulations (Xu *et al.*, 2010) showed that while the generation of aPC reduces the thrombin concentration, it was still high enough so that fibrin production is almost not affected.

Platelet ESAM is released from alpha granules after platelet activation and functions to destabilize the thrombus and interfere with late events in thrombus development (Stalker *et al.*, 2009). Additionally, platelet Pecam1 inhibits thrombus growth in murine arterioles following laser injury (Falati *et al.*, 2003). Although expressed on resting platelets, Pecam1-mediated outside-in signaling is initiated after the initiation of thrombus growth and may provide a negative feedback mechanism to inhibit continued platelet accumulation. However, contradictory results have been reported (Rosen *et al.*, 2001). The work described in this chapter explores a novel mechanism in which the accumulation of fibrin on the developing thrombus surface limits further growth. In what follows, we give an overview of a multiscale model (Kamocka *et al.*, 2010; Kim *et al.*, 2011; Mu *et al.*, 2009, 2010; Sweet *et al.*, 2011; Xu *et al.*, 2009a, 2010, 2011) of thrombogenesis in concert with laboratory experiments to test this mechanism. The details of the computational method are given in the next section.

III. Overview of the Modeling Approach

Several models have been developed to study various aspects of thrombosis but only few simplified models are available for studying the effects of fibrin on blood clot formation. (See recent review chapters (Diamond, 2009; Xu *et al.*, 2011) for detailed discussions of these models.) A unique feature of the earlier introduced multiscale model of thrombogenesis (Kamocka *et al.*, 2010; Kim *et al.*, 2011; Mu *et al.*, 2009, 2010; Sweet *et al.*, 2011; Xu *et al.*, 2009a, 2010, 2011) is its representation of the movement, adhesivity, and activity of each individual platelet and blood cell as objects with volumes bounded by fluctuating membranes. The model involves components that operate at different scales: platelets, cells, plasma, vessel wall, injury, coagulation factors, and platelet activators. It couples processes and interactions among components including platelet–platelet adhesion, activation states of platelets, blood plasma flow, as well as coagulation and anticoagulation reactions that take into account plasma-phase and membrane-phase reactions using the general approach introduced in (Fogelson and Tania, 2005; Kuharsky and Fogelson, 2001). Submodels at specific scales are as follows:

(a) Biochemical reactions submodel: systems of ODEs and PDEs are used to describe the coagulation cascade;
(b) Cell submodel: discrete stochastic Cellular Potts Model (Graner and Glazier, 1992; Glazier and Graner, 1993) represents different types of cells as well as describes cell–cell and platelet-injury adhesion, platelet activation, cell movements, cell state changes, and platelet aggregation;
(c) Fibrin network submodel: extended discrete stochastic CPM includes "fibrin elements," which represent a small region occupied by polymerized fibrin.
(d) Flow submodel: incompressible Navier–Stokes (NS) equations and Darcy's law describe dynamics of viscous blood plasma.

The CPM for simulating blood clot formation consists of a list of generalized cells on a lattice, a set of chemical diffusants and local rules based on experimental observations describing cellular biological and physical behavior. Each platelet or other blood cell is represented in the CPM by a cluster of lattice sites. Distribution of multidimensional indices associated with lattice sites determines current system configuration. The Metropolis algorithm, based on the Monte-Carlo Boltzmann acceptance rule, is used to determine dynamics of the CPM. The effective CPM energy mixes true energies, like platelet–platelet adhesion, and terms that mimic energies, for example, the response of a cell to the chemical fields or to the flow as well as area and volume constraints. Biochemical reactions as well as chemical production are modeled on the membrane of each cell. (The details can be found in (Xu *et al.*, 2008, 2009a, 2009b, 2010).) Using this model blood clot development was simulated within a 2D rectangular channel representing blood vessel (Kamocka *et al.*, 2010; Kim *et al.*, 2011; Mu *et al.*, 2009, 2010; Sweet *et al.*, 2011; Xu *et al.*, 2009a, 2010, 2011). (See also Fig. 5.)

IV. Study of the Role of the Fibrin Network

A. Experimental Materials and Methods

1. Injury Protocol and Visualization

Fluorescently labeled probes detecting platelets, fibrin, and plasma were injected into anesthetized mice. Direct laser induced,injuries were made to the luminal surface of the top of exposed mesenteric venules. Image data were acquired by the sequential collection of Z-stacks enabling one to generate and analyze 3D reconstructions of the developing thrombus. (The injury and imaging protocols are described in detail in Kamocka *et al*. (2010).)

2. Image Analysis

Z-stacks obtained by multiphoton microscopy were analyzed using algorithms described in (Mu *et al*., 2009, 2010). Briefly, algorithms determined thresholds of each fluorescent signal and each pixel was categorized as either containing (a) fibrin, (b) platelets, (c) platelets and fibrin, or (d) "other material" (if signals for all probes, including plasma, were below threshold). Algorithms grouped like contiguous voxels in the Z-stacks enabling one to define surfaces of subdomains as well as the entire thrombus.

3. Results

Following laser-induced injury of mesenteric venules, thrombi were monitored by multiphoton microscopy (Kamocka *et al*., 2010). Using image processing algorithms (Mu *et al*., 2009, 2010), 3D reconstructions of the developing thrombus were generated. The volume tracing in Fig. 1 indicates that the thrombus grows rapidly in the first two to three structures, and then decreases in size reaching a stable volume by the sixth to eighth structure (8–11 min). As the thrombus stabilizes, the composition of the thrombus surface changes from one primarily composed of platelets to one composed mainly of fibrinogen. (See Fig. 1A.) These results suggested the hypothesis that the accumulation of fibrin on the thrombus surface might provide a self-limiting mechanism for thrombus growth.

B. Simulations Under Normal Conditions

To examine how fibrin elements affect the outcome of a simulation, we have used different threshold thrombin concentrations required for fibrin polymerization. Using a thrombin threshold (1 nM/mL) corresponding to thrombin concentrations (Weisel and Nagaswami, 1992) that promote fibrin accumulation in experimental thrombi and a probability (>80%), we simulated fibrin networks formation in the thrombus. Use of low probability (<1%) resulted in thrombogenesis with only few fibrin elements forming. (See Section 5 for detailed description of the multiscale model.)

Figure 2 compares the growth of simulated thrombi using normal and very high thrombin concentrations required for fibrin generation. Using both conditions, resting platelets adhere to the injury site are activated and can support surface-mediated coagulation reactions. In proximity to TF in the vessel wall, coagulation is initiated and maintained on activated platelets in the developing thrombus generating thrombin. Resting platelets in blood contacting the thrombus may adhere and become activated in response to released platelet activators and thrombin. However, under conditions in which fibrinogen can readily be converted to fibrin, the simulation predicts rapid generation of thrombin between 50 and 200 s, leading to the formation of fibrin elements which cover the surface (Fig. 2). Although the concentration of thrombin and platelet activators on the thrombus surface is sufficient for platelet activation, it takes too long before activated platelets bind stably to fibrin to incorporate platelets under flow. Eventually, the polymerized fibrin network is thick enough to impede the diffusion of thrombin generated on platelets within the thrombus to reach the surface

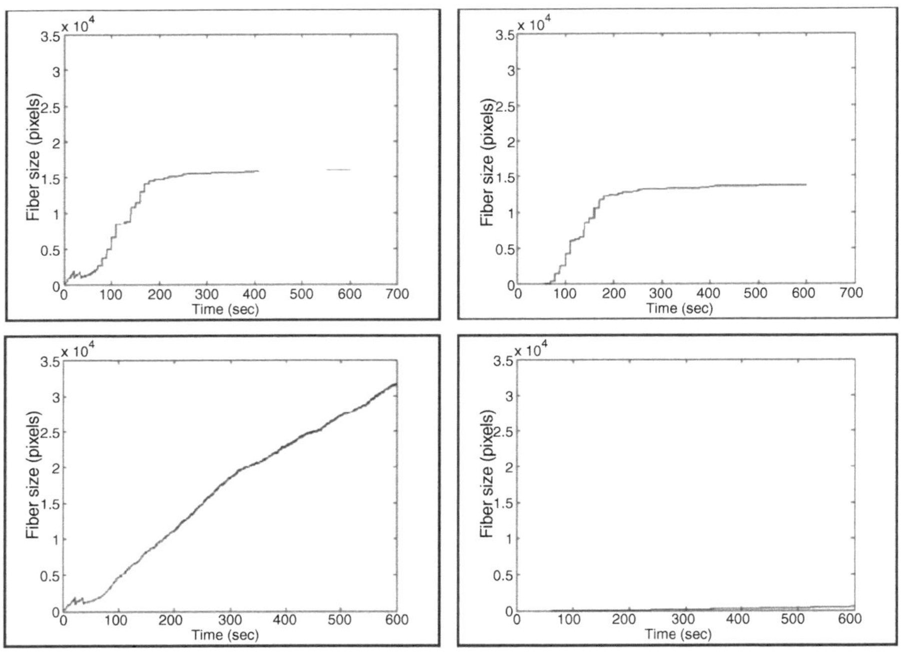

Fig. 2 Simulated thrombus growth. The top panel represents the growth of the thrombus (left) and accumulation of fibrin elements (right) in a simulation of thrombus growth in which the thrombin concentration required for the generation of fibrin corresponds to values observed in wild-type animals permitting normal formation of fibrin elements. The bottom panel represents the growth of the thrombus (left) and accumulation of fibrin elements (right) in a simulation of thrombus growth in which fibrin elements are generated with a very low (<1%) probability in the normal thrombin concentration. In the absence of fibrin generation, few fibrin elements form but thrombus growth continues for 600 s (lower left) compared to the case with normal fibrin generation (upper left) where growth stops at 200 s.

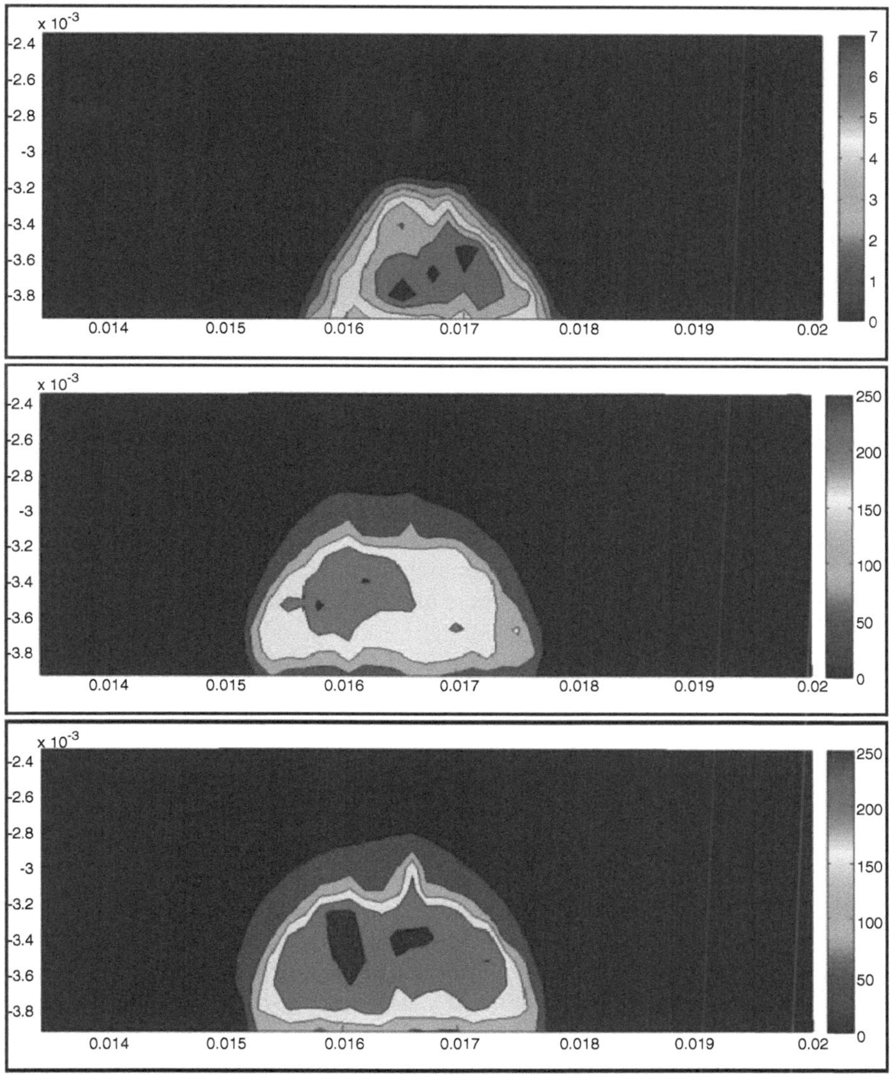

Fig. 3 Thrombin distribution in a simulation of a developing thrombus. The color map shows concentration of thrombin in and around the thrombus at 60 s (top), 220 s (middle), and 600 s (bottom) after injury. The thrombin concentration required for fibrin generation used in the simulation is the same as in Fig. 2; top panel. The dashed blank line represents the boundary of the thrombus. The simulation includes platelets but not other blood cells. The simulation indicates low thrombin concentration on the thrombus surface after the thrombus stabilizes. (For color version of this figure, the reader is referred to the web version of this book.)

Fig. 4 Time sequence of the simulated wild-type clot structure with fibrin elements formation. Colors: inactivated platelets are red, activated platelets are blue and fibrin elements are black. (A) time = 60 s; (B) time = 220 s; (C) time = 600 s. (For interpretation of the references to color in this figure legend, the reader is referred to the web version of this book.)

and the surface thrombin concentration falls (Fig. 3). By 500–600 s the thrombus is stabilized as the surface thrombin concentration is too low to promote platelet recruitment and to generate new fibrin elements (Fig. 4).

C. Challenges

In this chapter we explore a novel mechanism involving accumulation of fibrin on the developing thrombus surface and limiting further growth. The chapter describes an extension of the computational model of thrombogenesis (Kamocka *et al.*, 2010; Kim *et al.*, 2011; Mu *et al.*, 2009, 2010; Sweet *et al.*, 2011; Xu *et al.*, 2009a, 2010, 2011), which incorporates formation of "fibrin elements" to represent conversion of fibrinogen into polymerized fibrin. This was motivated by experimental observations that thrombus stabilization was associated with increased fibrin accumulation on the surface. Accumulation of the fibrin also increases the structural integrity of the thrombus. However, simulations also show that it creates a barrier to blood borne zymogens impeding the generation of thrombin. Additionally, fibrin elements reduce diffusion of platelet activators and prevent thrombin generated on activated platelets from reaching the surface limiting further platelet recruitment. Furthermore, the adhesivity of resting platelets in blood to the thrombus surface decreases as the percentage of the surface covered with fibrin increases.

The simulations are also consistent with observations of thrombus development in fibrinogen-deficient mice. Unexpectedly, thrombi develop in fibrinogen knockouts following ferric chloride, laser-induced, or Rose Bengal experiential injury models. Thrombi are as large as in wild-type mice but are less stable undergoing initial growth, embolization, and regrowth. The results are consistent with the hypothesis that in the absence of fibrin, thrombus stability is reduced but that thrombi retain the ability to regrow for an extended time. However, platelet adhesion and signaling events mediated by fibrinogen-GPIIbIIIa interactions are absent in fibrinogen-deficient mice. Therefore, usage of fibrinogen mutants that retain normal GPIIbIIIa binding but do not polymerize will enable one (Bowley *et al.*, 2008; Bowley and Lord, 2009) to more directly test the hypothesis that polymerized fibrin contributes to the cessation of thrombus growth.

Mosesson *et al.* (2009) have generated transgenic mice enabling one to test the effects of fibrin sequestration of thrombin. A naturally occurring splice variant of human gamma fibrinogen (γ'-fib) generates an isoform containing a COOH-terminal amino acid sequence sequestering thrombin. The analogous murine splice variant does not contain the thrombin-binding sequence. Mosesson *et al.* (2009) show impaired thrombus development in mice expressing human γ'-fibrinogen compared to thrombus development in control mice. These results suggest that ability of fibrin to sequester and reduce the diffusion of thrombin inhibits thrombus growth consistent with the results presented in this manuscript.

The multiscale computational model describing formation of the fibrin network in the form of fibrin elements requires only thrombin concentration on the computational grid to remain at a level over threshold for a time required for fibrin element generation (10 s for wild-type case) (Weisel and Nagaswami, 1992), which is a highly simplification of real scenario. Adhesivity and porosity of the fibrin elements are chosen using estimates from experimental data in the literature. At this point, we do not model molecular dynamics of fibrin polymerization or the structure of the fiber network within the fibrin element. Similarly, the model does not include intracellular signaling pathways regulating the behavior of platelets in response to platelet activators (Purvis *et al.*, 2008). Modeling polymerization in detail and integrating polymerization into a multiscale model to simulate thrombus development is a considerable challenge because it requires modeling events (or components) from molecular to micron structure scales.

V. Multiscale Model of Thrombus Development with Fibrin Network Formation

A. Blood Flow Submodel

Blood clot is treated in this chapter as a porous medium to account for the transport property of blood flow. In the case of vein thrombus formation studied in the present chapter, the Reynolds number of the blood flow is $\leq O(1)$. Therefore, we describe the blood flow within the clot by Darcy's law and the flow elsewhere by Stokes equation. Complete flow sub-model consists of the combination of the unsteady Stokes equations (1) and Darcy's law (2). Namely, we solve Stokes equations outside the clot, use Darcy's law within the clot and couple them using domain decomposition approach

$$\begin{aligned} \frac{\partial \boldsymbol{u}}{\partial t} &= -\frac{1}{\rho}\nabla p + \nu\nabla^2\boldsymbol{u} \\ \nabla\cdot\boldsymbol{u} &= 0 \end{aligned} \tag{1}$$

$$\boldsymbol{u} = -\frac{k}{\nu}\nabla p \tag{2}$$

where ρ is the fluid density, $\boldsymbol{u}$ and p are volume-averaged velocity and pressure, respectively, and ν is the shear viscosity. In our computations, the permeability k

varies from 10^{-8} to 10^{-11} cm^2 to examine the effects of permeability on the clot size. Up to date, no specific permeability data are available for actual platelet-contracted, flow-compacted blood clots, which contain layers of fibrin, platelets, and red blood cells (Diamond and Anand, 1993). In coarse fibrin gels of clotted plasma, the permeability k is about 10^{-8} cm^2; whereas permeability of fine fibrin gels of clotted plasma is about 10^{-10} cm^2. The effective pore diameter is on the order of a few microns in coarse gels of clotted plasma; whereas the pore diameter is on the order of 10^{-1} μm in fine gels. In flow-compacted coarse fibrin (porosity $\sim$0.75), the specific permeability may be as low as 10^{-11} cm^2 or less. See (Diamond and Anand, 1993) for a summary of these type measurements.

B. Coagulation Pathway Submodel

We utilize the coagulation model introduced in (Xu *et al.*, 2010) which includes both solution-phase and membrane-phase reactions with concentrations of membrane-binding sites being limited and treated as control variables using method introduced in (Kuharsky and Fogelson, 2001). The PDEs describe rates of change of the concentration of each solution-phase factor or complex, and the ODEs describe rates of change of the concentration of each membrane-phase factor or complex. Activated platelets provide membrane-binding sites where surface-binding zymogens and enzymes react.

C. "Fibrin Element" Submodel

We expand the multiscale model (Xu *et al.*, 2008, 2009a, 2009b, 2010) by introducing "fibrin elements," which are CPM lattice nodes representing polymerized fibrin (see Xu *et al.* (2010), Xu *et al.* (2009a), Xu *et al.* (2008), Xu *et al.* (2009b)) for details about CPM biological cell representation). A fibrin element can adhere to other fibrin elements, vessel wall, and activated or resting platelets. At each simulation time step we compute thrombin production on the surface of each individual activated platelet and its distribution over space and time. Based on the thrombin concentration and distribution, we compute fibrin generation. When fibrin concentration remains higher than a threshold value at a lattice cite for a period of time corresponding to the time required to promote fibrin polymerization *in vitro*, we introduce with a certain probability a special type of CPM "cell," which we call fibrin element? The porosity of the fibrin elements affects advection and diffusion of coagulation factors and diffusants released by platelets. The model does not include molecular details of fibrin polymerization and network formation because such computational tools are unavailable at this time. The internal structure of the fibrin network is neglected as well. Nevertheless, by dynamically introducing "fibrin elements," one can simulate general role played by fibrin network in thrombogenesis. The algorithm for employing CPM to model formation of fibrin network consists of the following steps:

(1) Compute concentration of thrombin using coagulation pathway submodel;
(2) Estimate concentration of polymerized fibrinogen converted by thrombin from.

$$\frac{\partial \phi}{\partial t} + \boldsymbol{u} \cdot \nabla \phi = D_\phi \nabla^2 \phi - \kappa_\phi [II]_a \phi \tag{3}$$

$$\frac{d\Phi}{dt} = \kappa_\phi [II]_a \phi \tag{4}$$

where $[\boldsymbol{II}]_a$ is the concentration of the thrombin; ϕ is the concentration of fibrinogen; Φ is the concentration of fibrin and κ_ϕ is the fibrin production (polymerization) rate. Note that Eq. (4) does not contain an advection term due to the fact that fibrin grows within the platelet aggregate where the flow velocity is almost zero;

(3) Introduce with a probability p_f "fibrin elements" to represent space filled with the fibrin network. When Φ around platelet (or already created "fibrin elements") is higher than a specified threshold value for a period of time representing time it takes fibrinogen to polymerize, a new fibrin cell is created near the platelet or the existing "fibrin cell". For the wild-type case, the probability is set to $p_f > 0.8$. We use a probability ($p_f < 0.01$) to simulate thrombogenesis where few fibrin elements form.

Other simplifications used in the model are as follows:

- While resting platelets are discoid, initially after activation they become spherical with irregular protrusions. With the extension of filopodia, which attaches to the matrix components of the thrombus, platelets spread and flatten. Additionally, entrapped platelets can retract the fibrin to about a tenth of the original volume by squeezing plasma from the clot. We neglect in the current model these effects. We do take into account the permeability and porosity of the clot as we study the role of the fiber network.
- Platelet–platelet connection through fibrin or fibrinogen is modeled by changing adhesion between platelets. We do not explicitly model the fiber that bridges two platelets.
- Fibrinolysis is not taken into consideration. This is justified by the fact that (a) we study the initial clot formation process on a time scale of several minutes; (b) for moderately sized clots, substantial fibrinolysis does not occur on a time scale of seconds or a few minutes (Diamond and Anand, 1993).

D. Numerical Schemes Used for Running Simulations

The model system is solved in a 2D rectangular domain representing a small section of the blood vessel (see also Fig. 5). New inactivated platelets are introduced in the simulation domain from the inlet on one side of the vessel

Fig. 5 Snapshot of a simulated clot. Colors: fibrin elements are black, blood cells are grey, platelets connected by fibrin (ogen) or vWF are blue, and inactivated platelets are red. (For interpretation of the references to color in this figure legend, the reader is referred to the web version of this book.)

lumen domain at a specified rate and are carried by the flow to the outlet on the other side.

An injury site is set in the middle of the lower boundary. In our model the injury site is represented by CPM cells with specific positions in space to which platelets can adhere and get activated. The adhesive interactions between the injury cells and resting platelets in our model correspond to the adhesion of resting platelets to the subendothelial matrix that is exposed to blood following injury. In addition we assume the injury cells expose TF that can complex with FVIIa in the flowing blood to initiate the extrinsic pathway of coagulation.

At each simulation time step the following elements of the algorithm are performed:

- Navier–Stokes equations and Darcy's law equation for the subdomain identified as a blood clot are solved resulting in an update of the blood flow field in the whole computational domain.
- Then convection-reaction-diffusion equations and ODEs, which model the coagulation reactions, are solved to evolve the biochemical coagulation species in space and time.
- The soluble fibrinogen concentration Eq. (3) is solved to update fibrinogen distribution in space.
- Next, the CPM is used to update positions and status of the blood cells. For example, a resting platelet can be activated by thrombin or by touching activated platelet.
- Then fibrin polymerization Eq. (4) is solved to simulate production of fibrin monomers.
- Fibrin monomer concentration is checked at each lattice cite. If the concentration at a lattice cite adjacent to the developing clot stays above the threshold for a time period that it takes fibrin monomers to form cross links resulting in fiber network, a "fiber element" is created with certain probability at the lattice cite to represent the polymerized fiber network.

VI. GPU Implementation of the Simulation of the Platelet–Blood Flow Interaction

In what follows we describe an implementation of 3D simulations of the platelet dynamics at low Reynolds numbers in linear flow (Sweet *et al.*, 2011) on the GPUs. We review in subsections 6.1 and 6.2 the multiscale platelet-flow interaction model (Sweet *et al.*, 2011) consisting of two submodels: (1) a subcellular element model (Newman, 2005, 2007, 2008) (SCEM) for representing platelets; and (2) a LB submodel for describing fluid dynamics of plasma in a vessel. The Langevin equation approach developed in (Sweet *et al.*, 2011) is used to couple SCEM with the LB. Subsection 6.3 describes the GPU algorithm.

A. Subcellular Element Model

The SCEM of a platelet uses $N = 53$ elements to represent a cell. These elements (called subcellular elements (SCEs) or nodes) are connected by bonds, which form the structure of the cell. SCEs are arranged in five layers of ten SCEs each, as well top and bottom layers made of one SCE each. One SCE is in the center of the cell. We assume that individual platelet has spherical shape with initially equal bond lengths between the center node and all others. (See Fig. 6.) All SCEs are bonded to the center node as well as to their nearest neighbors, seven bonds per element, except for the central node which bonds with all other elements with a total of 52 bonds. Each

Fig. 6 Schematic diagram of the SCEM representation of an inactivated platelet. (For color version of this figure, the reader is referred to the web version of this book.)

bond is a spring of target length (determined by the equilibrium length l_{ij}) and bond strength (determined by the spring constant k_{ij}). Thus, the associated potential energy function for nodes i and j is as follows

$$U_{ij} = \frac{k_{ij}}{2}(||\boldsymbol{r}_{ij}|| - l_{ij})^2 \tag{5}$$

where $\boldsymbol{r}_{ij} = \mathbf{x}_i - \mathbf{x}_j$ is the position vector difference between nodes i and j. The force vector corresponding to U^e_{ij} and acting on nodes i and j is

$$\boldsymbol{F}^e_{ij} = -\nabla_x U_{ij}(\mathbf{x}) = -k_{ij}(||\boldsymbol{r}_{ij}|| - l_{ij})\hat{\boldsymbol{r}_{ij}} \tag{6}$$

Here $\boldsymbol{r}_{ij}$ is a unit vector and $\hat{\boldsymbol{r}_{ij}} = (\boldsymbol{r}_{ij}/||\boldsymbol{r}_{ij}||)$. The platelet's shape and movement is determined entirely by the forces acting on the elements. Bonds deviating from their target length create forces acting on the two adjacent elements. Similarly, force created by fluid acts on the elements of a cell.

B. Coupled LB Equation and SCEM

Coupling between the flow and the SCEM is implemented by using one-way Langevin method (Sweet *et al.*, 2011). We assume that movements of cells do not disturb the flow field. The LB equation is used to model the fluid flow and is solved by using the D3Q19 method. We simulate the flow in a rectangular channel with dimensions 50 μm × 50 μm × 600 μm. The lattice spacing is 1 μm. The platelet dimensions are about 2–4 μm. The simulation is initialized with the platelets positioned near the influx of the channel. The initial flow condition is set to satisfy a Pouiselle flow. The no-slip condition is enforced along the boundary of the channel. A flux boundary condition is maintained at the inlet and outlet boundary.

Motion of a cell is modeled as movement of a particle in the fluid flow due to collisions with the molecules of the fluid. Here we briefly summarize the method developed in (Sweet *et al.*, 2011). Given a system of N particles represented by the position vector $\mathbf{x}$, the following modified Langevin equation describes the motion of a cell

$$\boldsymbol{M}\ddot{\mathbf{x}} = \boldsymbol{F}(\mathbf{x}) - \gamma(\dot{\mathbf{x}} - \delta\mathbf{v}^f) + \sqrt{2\gamma NkT}\boldsymbol{Z} \tag{7}$$

Here $\boldsymbol{Z}$ is a vector of normally distributed random variables and $\boldsymbol{M}$ is a diagonal matrix of particle masses. Force vector is given by $\boldsymbol{F}(\mathbf{x}) = -\nabla U(\mathbf{x})$ (see Eq. (6)). k is the Boltzmann coefficient, T is the temperature and γ is the friction coefficient. $\delta\mathbf{v}^f$, the flow perturbation vector, is defined by $\mathbf{v}^f = \mathbf{v}^f - \langle \mathbf{v}^f \rangle$, where $\mathbf{v}^f$ and its average $\langle \mathbf{v}^f \rangle$ is defined as follows. The flow velocity v^f_n at the nth SCE ($1 \leq n \leq N$) is determined by taking the weighted average (based on distance) of the surrounding LB nodes for each of the x, y, and z directions. The vector of the flow velocity in the Langevin equation is $\mathbf{v}^f = \left[\mathbf{v}^f_1, \ldots, \mathbf{v}^f_N\right]^T$. Therefore, observed velocity $\mathbf{v}_o$ of the SCEs is defined by $\mathrm{v}_o = \dot{\mathrm{x}} + \langle \mathbf{v}^f \rangle$. A Langevin Leapfrog method (Sweet *et al.*, 2011) is used to solve Eq. (7).

C. GPU Implementation of the Multiscale Model

Software toolkit CUDA (Nvidia CUDA, in press) provided by Nvidia has been used for implementing multiscale model of thrombus formation on the GPU as follows. Simulation code combines C-language with the CUDA computing language. The CPU or host-side of the program handles the initialization and input/output for the simulation, whereas the GPU or device-side of the program carries out all calculations. EVGA GeForce GTX 480(Fermi) video card is used for running simulations. We describe below single-GPU implementation limited by the available GPU memory.

In what follows, we describe details of implementing the SCEM and LB equation model on GPUs as well as Langevin coupling. Each submodel has independent data arrays that store the necessary information. For the SCEM, positions, velocities, and forces acting on each SCE are stored as 2D arrays similar to (Christley *et al.*, 2010), whereas for the LB equation, three components of the fluid-velocity (u_x, u_y, u_z), density (ρ), and obstacle location are each stored in 3D arrays.

Current implementation of the 3DQ19 method of solving LB equation on GPU uses an approach similar to the "ping-pong" memory allocation scheme from (Myre *et al.*, 2011), which uses two separate fluid-packet matrices for managing the data between the collision and streaming steps. During the collision step, fluid-input-matrix (fIN) is used for collision algorithm and assigns the results to the fluid-output-matrix (fOUT). Then, on the streaming step, the fOUT is used as input for the streaming algorithm that assigns the results back to the fIN for the next step. There are two sets of 19 fluid-vector 3D arrays that manage the fluid packet collision and streaming processes. Various 3D arrays for the LB are determined by the LB grid size. The LB equation submodel requires approximately 1000 times the amount of memory needed for the SCEM. The 43 arrays (19 fIN, 19 fOUT, ρ, u_x, u_y, u_z, and obstacle) for $50 \times 50 \times 600$ LB nodes at 4 bytes (float) or 8 bytes (double) results in 246 MB or 492 MB of memory use, respectively. For the SCEM, the 12 arrays (force, velocity, and position in 3D) for 100 platelets containing 53 SCE giving 248 kB in floating point precision or 496 kB in double precision. Even 1000 platelets in double precision would only need $\sim$5 MB of memory. The available memory on the GPUs used for our code development is about 1.5 GB. Although LB requires considerable amount of memory, the algorithm is highly parallel and fits well to the GPU architecture. In the future if a larger LB system is required that exceeds the memory available on a single GPU, then a straightforward partitioning of the algorithm across multiple GPUs can be devised.

During a single simulation time-step, we first solve LB equation and then the Langevin Leapfrog integration scheme is executed. For each SCE, the half-time step velocity is updated based on the force calculated for that element from previous step. Second, the position of the element is updated based on the half-time step velocity. Third, the force is calculated based on the updated positions of the elements. Finally, the half-time step velocity is updated with the new force values and the velocity for the next time step is determined.

For a serial code to implement the SCEM, a number of functions sequentially step through every element to update the values of velocity, force, and position. Serial codes must access every element, one after another, in order to read either read the velocity and write a new position or read the force and write a new velocity value. When writing new force values from the current position of SCEs, one element must read multiple positions of other SCEs. Most schemes take advantage of Newton's third law that states the force acting on one element by a second element is equal in magnitude and opposite in direction to the force acting on the second element by the first element. Therefore, force calculation functions only need to calculate the distance of separation between two elements once in order to determine the force acting on each element due to the pair-wise interaction. In order to take advantage of GPU parallelization, which requires independent accessing and writing to memory by the parallel threads, the interaction between elements is handled in a different manner. We assign GPU-threads (or independent tasks) to each SCE during the various kernels (or GPU functions) responsible for updating position, velocity, and force. Groups of GPU-threads are processed simultaneously. To update a SCE's velocity (position), individual GPU-threads will use the current force (velocity) values held in one particular SCE's memory location within a force (velocity) data array. The GPU-threads then write the updated velocity (position) to the SCE's memory location in the velocity (position) data array. These updates can be done in parallel because velocity and position updates for individual SCEs are completely independent of the velocity and position of other elements. (The number of simultaneous GPU-threads is determined by the hardware.)

Force update for each element requires positions of multiple SCEs that results in multiple GPU-threads needing to access the same memory locations for the position data. This can lead to some fraction of memory reads to be done in serial, but does not lead to errors in calculations. However, our code is written so that each GPU-thread can only write updated force values to a single SCE. The force acting on each SCE for a given simulation step is the summation of multiple pair-wise interactions. If multiple GPU-threads were allowed to update this sum, then it would be possible, in theory, for a GPU-thread to use an incorrect value for the sum. This could happen if another GPU-thread wrote a new force value to a particular SCE in between the time another GPU-thread read the force value and attempted to update that value. This limitation prevents taking advantage of Newton's third law and leads to additional calculations for each SCE. For instance, the distance between two SCE would have to be calculated twice, once for each GPU-thread responsible for the two connected SCEs. However, the additional speed up gain through the GPU parallelization justifies the additional calculations.

D. Simulation Results

We simulated a channel with two million LB nodes as a $50 \times 50 \times 600$ lattice for a single-component single-phase (scsp) fluid using GPU card described above. The code, which could be further optimized, is able to perform $\sim$120 millions lattice-node

Table I
Execution time of coupled LB equation and SCEM by cell numbers

Number of cells	Real time (s)	User time (s)	System time (s)
1	609.971	600.981	8.990
25	703.968	689.070	10.770
50	1045.715	1029.590	9.770
100	1772.029	1586.370	11.660

updates per second (MLUPS). Code performance of 418 MLUPS has been achieved in (Myre *et al.*, 2011) for the "ping-pong" implementation. Therefore, further optimization of our code is possible. Additional increase in the performance of our existing code can be achieved by optimizing memory access through the arrangement of lattice-nodes into memory blocks on the GPU. We expect to achieve an approximately two- to three-fold speed up by assigning lattice nodes to thread blocks in a manner that allows for a better coalesced memory access. This is done in (Myre *et al.*, 2011) by arranging lattice-nodes into single-node columns on thread blocks rather than in cubic or rectangular volumes as is implemented in our current code. We ran simulations to test the performance of coupled LB equation and SCEM simulation code over a period of 0.5 s with 50,000 time steps (Table I). The second column shows the wall-clock time spent on individual simulation. The third column shows the time spent on GPU calculations; while the last column shows the time spent on operating system tasks such as memory transfers, file I/O and CPU-GPU context switching. Note that the simulation must periodically copy data from the GPU memory to the CPU memory and then save that data into output files so the results can be analyzed and visualized. Column 4 of Table I indicates that this noncomputational data transfer is only a small fraction of the total time ($\leq$2%), especially for large 3D arrays.

VII. Concluding Remarks

A multiscale modeling framework is extended and used in this chapter for studying fibrin accumulation on the surface of a forming thrombus. Namely, previously introduced multiscale model (Kamocka *et al.*, 2010; Kim *et al.*, 2011; Mu *et al.*, 2009, 2010; Sweet *et al.*, 2011; Xu *et al.*, 2009a, 2010, 2011) is refined to include a phenomenological submodel of a polymerized fibrin network formation represented by a set of "fibrin elements." Simulations predict that the development of a polymerized fibrin network coating the surface of a thrombus can inhibit continued thrombus growth, which is consistent with the hypothesis suggested by our experiments. This study suggests a possible and previously neglected mechanism regulating the size of a developed thrombus. Further combined experimental and simulation study is needed to confirm current simulation findings. For instance, studying how resting-state and activated platelets adhere to fibrin network, how coagulation factors transport by diffusion and blood flow convection through fibrin network will provide new insight into different roles of the fibrin network.

We also present a GPU-based implementation of a platelet-flow interaction model introduced in (Sweet *et al.*, 2011). Preliminary simulations using this GPU-based implementation show a significant performance increase. However, current platelet-flow interaction model cannot be used at this time to simulate all processes involved in blood clot formation. To make the model more biologically relevant we are currently working on adding coagulation pathways, platelet activation and adhesion submodels to the multiscale modeling environment.

Although simulations of thrombogenesis involve many simplifications, they are starting to provide quantitative descriptions of thrombus development and are beginning to predict results of experimental manipulations that were not obvious solely from earlier experimental data. Continued development of multiscale approaches combining experiments with biologically relevant simulations will provide biologists with an import tool to perform simulation *in silico* to generate novel hypotheses that can be tested experimentally. Finally, the ability to predict the outcomes of simultaneous variation at multiple hemostatic components will have significant biomedical value possibly enabling one to more accurately evaluate hemorrhagic or thrombotic risk for individual patients

Acknowledgments

This research was partially supported by the NSF grant DMS-0800612 and DMS-1115887, and NIH grant 1 R01 GM100470-01 and Gerber Foundation for M. Alber and O. Kim.

References

Bowley, S. R., and Lord, S. T. (2009). Fibrinogen variant BbetaD432A has normal polymerization but does not bind knob "B". *Blood* **113**(18), 4425–4430.

Bowley, S. R., Merenbloom, B. K., Okumura, N., Betts, L., Heroux, A., Gorkun, O. V., and Lord, S. T. (2008). Polymerization-defective fibrinogen variant gammaD364A binds knob "A" peptide mimic. *Biochemistry* **47**(33), 8607–8613.

Christley, S., Lee, B., Dai, X., and Nie, Q. (2010). Integrative multicellular biological modeling: a case study of 3D epidermal development using GPU algorithms. *BMC Syst. Biol.* **4**, 107.

Cranmer, S. L., Ulsemer, P., Cooke, B. M., Salem, H. H., de la Salle, C., Lanza, F., and Jackson, S. P. (1999). Glycoprotein (GP) ib-IX-transfected cells roll on a von Willebrand factor matrix under flow. importance of the GPib/actin-binding protein (ABP-280) interaction in maintaining adhesion under high shear. *J. Biol. Chem.* **274**(10), 6097–6106.

Diamond, S. L. (2009). Systems biology to predict blood function. *J. Thromb. Haemost.* **Suppl. 1**, 177–180.

Diamond, S. L., and Anand, S. (1993). Inner clot diffusion and permeation during fibrinolysis. *Biophys. J.* **65**, 2622–2643.

Doggett, T. A., Girdhar, G., Lawshé, A., Schmidtke, D. W., Laurenzi, I. J., Diamond, S. L., and Diacovo, T. G. (2002). Selectin-like kinetics and biomechanics promote rapid platelet adhesion in flow: the GPIb$-vWF tether bond. *Biophys. J.* **83**, 194–205.

Esmon, C. T. (1993). Cell mediated events that control blood coagulation and vascular injury. *Annu. Rev. Cell. Biol.* **9**, 1–26.

Esmon, C. T. (2001). Role of coagulation inhibitors in inflammation. *Thromb. Haemost.* **86**(1), 51–56.

Esmon, C. T. (2009). Basic mechanisms and pathogenesis of venous thrombosis. *Blood Rev.* **23**(5), 225–229.

Falati, S., Liu, Q., Gross, P., Merrill-Skoloff, G., Chou, J., Vandendries, E., Celi, A., Croce, K., Furie, B. C., and Furie, B. (2003). Accumulation of tissue factor into developing thrombi in vivo is dependent upon microparticle P-selectin glycoprotein ligand 1 and platelet P-selectin. *J. Exp. Med.* **197**(11), 1585–1598.

Fogelson, A. L., and Tania, N. (2005). Coagulation under flow: the influence of flow-mediated transport on the initiation and inhibition of coagulation. *Pathophysiol. Haemost. Thromb.* **34**, 91–108.

Glazier, J. A., and Graner, F. M. C. (1993). Simulation of the differential adhesion driven rearrangement of biological cells. *Phys. Rev. E* **47**(3), 2128–2154.

Graner, F. M. C., and Glazier, J. A. (1992). Simulation of biological cell sorting using a two-dimensional extended potts model. *Phys. Rev. Lett.* **69**, 2013–2016.

Gruner, S., Prostredna, M., Schulte, V., Krieg, T., Eckes, B., Brakebusch, C., and Nieswandt, B. (2003). Multiple integrin-ligand interactions synergize in shear-resistant platelet adhesion at sites of arterial injury in vivo. *Blood* **102**, 4021–4027.

Jackson, S. P., Mistry, N., and Yuan, Y. (2000). Platelets and the injured vessel wall—"rolling into action": Focus on glycoprotein Ib/V/IX and the platelet cytoskeleton. *Trends Cardiovasc. Med.* **10**(5), 192–197.

Kamocka, M. M., Mu, J., Liu, X., Chen, N., Zollman, A., Sturonas-Brown, B., Dunn, K., Xu, Z., Chen, D. Z., and Alber, M. S., *et al.* (2010). 2-Photon intravital imaging of thrombus development in vivo. *J. Biomed. Optics* **15**(1), 016020.

Kim, E., Kim, O., Machlus, K. R., Liu, X., Kupaev, T., Lioi, J., Wolberg, A. S., Chen, D. Z., Rosen, E. D., and Xu, Z. L., *et al.* (2011). Correlation between fibrin network structure and mechanical properties: an experimental and computational analysis. *Soft Matter* **7**(10), 4983–4992.

Kuharsky, A. L., and Fogelson, A. L. (2001). Surface-mediated control of blood coagulation: the role of binding site densities and platelet deposition. *Biophys. J.* **80**, 1050–1074.

Kuijpers, M. J., Schulte, V., Oury, C., Lindhout, T., Broers, J., Hoylaerts, M. F., Nieswandt, B., and Heemskerk, J. W. (2004). Facilitating roles of murine platelet glycoprotein ib and alphaIIbbeta3 in phosphatidylserine exposure during vWF-collagen-induced thrombus formation. *J. Physiol.* **558**, 403–415.

Lishko, Valeryi K., Timothy Burke and Tatiana Ugarova. (2007). Antiadhesive effect of fibrinogen: a safeguard for thrombus stability. *Blood*, **109**:1541-1549.

Lord, S. T. (2007). Fibrinogen and fibrin: scaffold proteins in hemostasis. *Curr. Opin. Hematol.* **14**(3), 236–241.

Mann, K. G., Brummel-Ziedins, K., Orfeo, T., and Butenas, S. (2006). Models of blood coagulation. *Blood Cells Mol. Dis.* **36**(2), 108–117.

Mosesson, M. W., Cooley, B. C., Hernandez, I., Diorio, J. P., and Weiler, H. (2009). Thrombosis risk modification in transgenic mice containing the human fibrinogen thrombin-binding gamma' chain sequence. *J. Thromb. Haemost.* **7**(1), 102–110.

Mu, J., Liu, X., Kamocka, M. M., Xu, Z., Alber, M. S., Rosen, E. D., and Chen, D. Z. (2010). Segmentation, reconstruction, and analysis of blood thrombi in 2-photon microscopy images. *EURASIP J. Adv. Signal Process.* **2010**, 1.

Mu, J., Liu, X., Kamocka, M. M., Xu, Z., Alber, M. S., Rosen, E. D., and Chen, D. Z. (2009). Segmentation, reconstruction, and analysis of blood thrombi in 2-photon microscopy images. Albuquerque, New Mexico.

Myre, J., Walsh, S. D. C., Lilja, D., and Saar, M. O. (2011). Performance analysis of single-phase, multiphase, and multicomponent lattice—Boltzmann fluid flow simulations on GPU clusters. *Concurrency Computat: Pract. Exper.* **23**, 332–350.

Newman, T. (2005). Modeling multicellular systems using subcellular elements. *Math. Biosci. Eng.* **2**, 613–624.

Newman, T. (2007). Modeling multicellular structures using the subcellular element model. *Phys. Biol.*

Newman, T. (2008). Grid-free models of multicellular systems, with an application to large-scale vortices accompanying primitive streak formation. *Curr. Topics Dev. Biol.* **81**, 157–182.

Nieswandt, B., and Watson, S. P. (2003). Platelet-collagen interaction: is GPVI the central receptor? *Blood* **102**, 449–461.

Nvidia CUDA [http://www.nvidia.com/object/cuda_home.html] [Internet].

Purvis, J. E., Chatterjee, M. S., Brass, L. F., and Diamond, S. L. (2008). A molecular signaling model of platelet phosphoinositide and calcium regulation during homeostasis and P2Y1 activation. *Blood* **112**, 4069–4079.

Rosen, E. D., Raymond, S., Zollman, A., Noria, F., Sandoval-Cooper, M., Shulman, A., Merz, J. L., and Castellino, F. J. (2001). Laser-induced noninvasive vascular injury models in mice generate platelet- and coagulation-dependent thrombi. *Am. J. Pathol.* **158**(5), 1613–1622.

Sweet, C. R., Chatterjee, S. C., Xu, Z., Bisordi, K., Rosen, E. D., and Alber, M. (2011). Modeling platelet blood flow interaction using subcellular element langevin method. *J. R. Soc. Interf.* **8**(65), 1760–1771.

Stalker, T. J., Wu, J., Morgans, A., Traxler, E. A., Wang, L., Chatterjee, M. S., Lee, D., Quertermous, T., Hall, R. A., and Hammer, D. A., *et al.* (2009). Endothelial cell specific adhesion molecule (ESAM) localizes to platelet–platelet contacts and regulates thrombus formation in vivo. *J. Thromb. Haemost.* **7**, 1886–1896.

Wakefield, T. W., McLaerty, R. B., Lohr, J. M., Caprini, J. A., Gillespie, D. L., and Passman, M. A. (2009). Call to action to prevent venous thromboembolism. *J. Vasc. Surg.* **49**(6), 1620–1623.

Weisel, J. W., and Nagaswami, C. (1992). Computer modeling of fibrin polymerization kinetics correlated with electron microscope and turbidity observations: clot structure and assembly are kinetically controlled. *Biophys. J.* **63**, 111–128.

Xu, Z. L., Chen, N., Shadden, S., Marsden, J. E., Kamocka, M. M., Rosen, E. D., and Alber, M. S. (2009a). Study of blood flow impact on growth of thrombi using a multiscale model. *Soft Matter* **5**, 769–779.

Xu, Z. L., Komocka, M. M., Alber, M., and Rosen, E. D. (2011). Computational approaches to studying thrombus development. *Arterioscler. Thromb. Vasc. Biol.* **31**, 500–505.

Xu, Z., Lioi, J., Mu, J., Liu, X., Kamocka, M. M., Rosen, E. D., Chen, D. Z., and Alber, M. S. (2010). A multiscale model of venous thrombus formation with surface-mediated control of blood coagulation cascade. *Biophys. J.* **98**(9), 1723–1732.

Xu, Z., Chen, N., Kamocka, M. M., Rosen, E. D., and Alber, M. S. (2008). Multiscale model of thrombus development. *J. R. Soc. Interf.* **5**, 705–722.

Xu, Z., Mu, J., Liu, X., Kamocka, M. M., Rosen, E. D., Chen, D. Z., and Alber, M. S. (2009b). Combined experimental and simulation study of blood clot formation. Toronto, Canada357–362.

Yermolenko, Ivan S., Fuhrmann, Alexander., Magonov, Sergei N., Lishko, Valeryi K., Oshkadyerov, Stanislav P., Ros, Robert., and Ugarova, Tatiana P. (2010). Origin of the Nonadhesive Properties of Fibrinogen Matrices Probed by Force Spectroscopy. *Langmuir* **26**(22), 17269–17277.

INDEX

T

U

V

VOLUMES IN SERIES

Founding Series Editor
DAVID M. PRESCOTT

Volume 1 (1964)
Methods in Cell Physiology
Edited by David M. Prescott

Volume 2 (1966)
Methods in Cell Physiology
Edited by David M. Prescott

Volume 3 (1968)
Methods in Cell Physiology
Edited by David M. Prescott

Volume 4 (1970)
Methods in Cell Physiology
Edited by David M. Prescott

Volume 5 (1972)
Methods in Cell Physiology
Edited by David M. Prescott

Volume 6 (1973)
Methods in Cell Physiology
Edited by David M. Prescott

Volume 7 (1973)
Methods in Cell Biology
Edited by David M. Prescott

Volume 8 (1974)
Methods in Cell Biology
Edited by David M. Prescott

Volume 9 (1975)
Methods in Cell Biology
Edited by David M. Prescott

Volume 10 (1975)
Methods in Cell Biology
Edited by David M. Prescott

Volume 11 (1975)
Yeast Cells
Edited by David M. Prescott

Volume 12 (1975)
Yeast Cells
Edited by David M. Prescott

Volume 13 (1976)
Methods in Cell Biology
Edited by David M. Prescott

Volume 14 (1976)
Methods in Cell Biology
Edited by David M. Prescott

Volume 15 (1977)
Methods in Cell Biology
Edited by David M. Prescott

Volume 16 (1977)
Chromatin and Chromosomal Protein Research I
Edited by Gary Stein, Janet Stein, and Lewis J. Kleinsmith

Volume 17 (1978)
Chromatin and Chromosomal Protein Research II
Edited by Gary Stein, Janet Stein, and Lewis J. Kleinsmith

Volume 18 (1978)
Chromatin and Chromosomal Protein Research III
Edited by Gary Stein, Janet Stein, and Lewis J. Kleinsmith

Volume 19 (1978)
Chromatin and Chromosomal Protein Research IV
Edited by Gary Stein, Janet Stein, and Lewis J. Kleinsmith

Volume 20 (1978)
Methods in Cell Biology
Edited by David M. Prescott

Advisory Board Chairman
KEITH R. PORTER

Volume 21A (1980)
Normal Human Tissue and Cell Culture, Part A:
Respiratory, Cardiovascular, and Integumentary Systems
Edited by Curtis C. Harris, Benjamin F. Trump, and Gary D. Stoner

Series Editor
LESLIE WILSON

Volume 31 (1989)
Vesicular Transport, Part A
Edited by Alan M. Tartakoff

Volume 32 (1989)
Vesicular Transport, Part B
Edited by Alan M. Tartakoff

Volume 33 (1990)
Flow Cytometry
Edited by Zbigniew Darzynkiewicz and Harry A. Crissman

Volume 34 (1991)
Vectorial Transport of Proteins into and across Membranes
Edited by Alan M. Tartakoff

Selected from Volumes 31, 32, and 34 (1991)
Laboratory Methods for Vesicular and Vectorial Transport
Edited by Alan M. Tartakoff

Volume 35 (1991)
Functional Organization of the Nucleus: A Laboratory Guide
Edited by Barbara A. Hamkalo and Sarah C. R. Elgin

Volume 36 (1991)
***Xenopus laevis:* Practical Uses in Cell and Molecular Biology**
Edited by Brian K. Kay and H. Benjamin Peng

Series Editors
LESLIE WILSON AND PAUL MATSUDAIRA

Volume 37 (1993)
Antibodies in Cell Biology
Edited by David J. Asai

Volume 38 (1993)
Cell Biological Applications of Confocal Microscopy
Edited by Brian Matsumoto

Volume 39 (1993)
Motility Assays for Motor Proteins
Edited by Jonathan M. Scholey

Volume 40 (1994)
A Practical Guide to the Study of Calcium in Living Cells
Edited by Richard Nuccitelli

Volume 41 (1994)
Flow Cytometry, Second Edition, Part A
Edited by Zbigniew Darzynkiewicz, J. Paul Robinson, and Harry A. Crissman

Volume 91 (2009)
Cilia: Structure and Motility
Edited by Stephen M. King and Gregory J. Pazour

Volume 92 (2009)
Cilia: Motors and Regulation
Edited by Stephen M. King and Gregory J. Pazour

Volume 93 (2009)
Cilia: Model Organisms and Intraflagellar Transport
Edited by Stephen M. King and Gregory J. Pazour

Volume 94 (2009)
Primary Cilia
Edited by Roger D. Sloboda

Volume 95 (2010)
Microtubules, in vitro
Edited by Leslie Wilson and John J. Correia

Volume 96 (2010)
Electron Microscopy of Model Systems
Edited by Thomas Müeller-Reichert

Volume 97 (2010)
Microtubules: In Vivo
Edited by Lynne Cassimeris and Phong Tran

Volume 98 (2010)
Nuclear Mechanics & Genome Regulation
Edited by G.V. Shivashankar

Volume 99 (2010)
Calcium in Living Cells
Edited by Michael Whitaker

Volume 100 (2010)
The Zebrafish: Cellular and Developmental Biology, Part A
Edited by: H. William Detrich III, Monte Westerfield and Leonard I. Zon

Volume 101 (2011)
The Zebrafish: Cellular and Developmental Biology, Part B
Edited by: H. William Detrich III, Monte Westerfield and Leonard I. Zon

Volume 102 (2011)
Recent Advances in Cytometry, Part A: Instrumentation, Methods
Edited by Zbigniew Darzynkiewicz, Elena Holden, Alberto Orfao, William Telford and Donald Wlodkowic

Volume 103 (2011)
Recent Advances in Cytometry, Part B: Advances in Applications
Edited by Zbigniew Darzynkiewicz, Elena Holden, Alberto Orfao, Alberto Orfao and Donald Wlodkowic

Volume 104 (2011)
The Zebrafish: Genetics, Genomics and Informatics 3rd Edition
Edited by H. William Detrich III, Monte Westerfield, and Leonard I. Zon

Volume 105 (2011)
The Zebrafish: Disease Models and Chemical Screens 3rd Edition
Edited by H. William Detrich III, Monte Westerfield, and Leonard I. Zon

Volume 106 (2011)
Caenorhabditis elegans: Molecular Genetics and Development 2nd Edition
Edited by Joel H. Rothman and Andrew Singson

Volume 107 (2011)
Caenorhabditis elegans: Cell Biology and Physiology 2nd Edition
Edited by Joel H. Rothman and Andrew Singson

Volume 108 (2012)
Lipids
Edited by Gilbert Di Paolo and Markus R Wenk

Volume 109 (2012)
Tetrahymena thermophila
Edited by Kathleen Collins

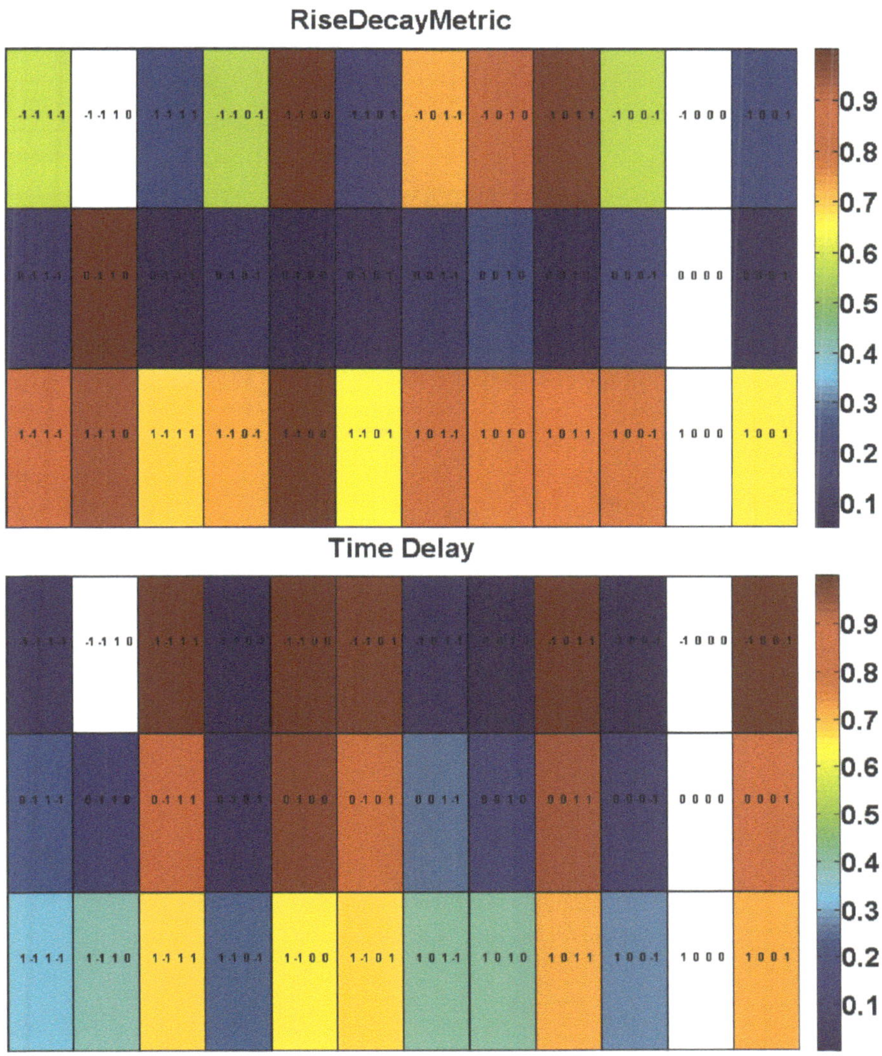

Plate 1 (Figure 1.4 on page 11 of this volume).

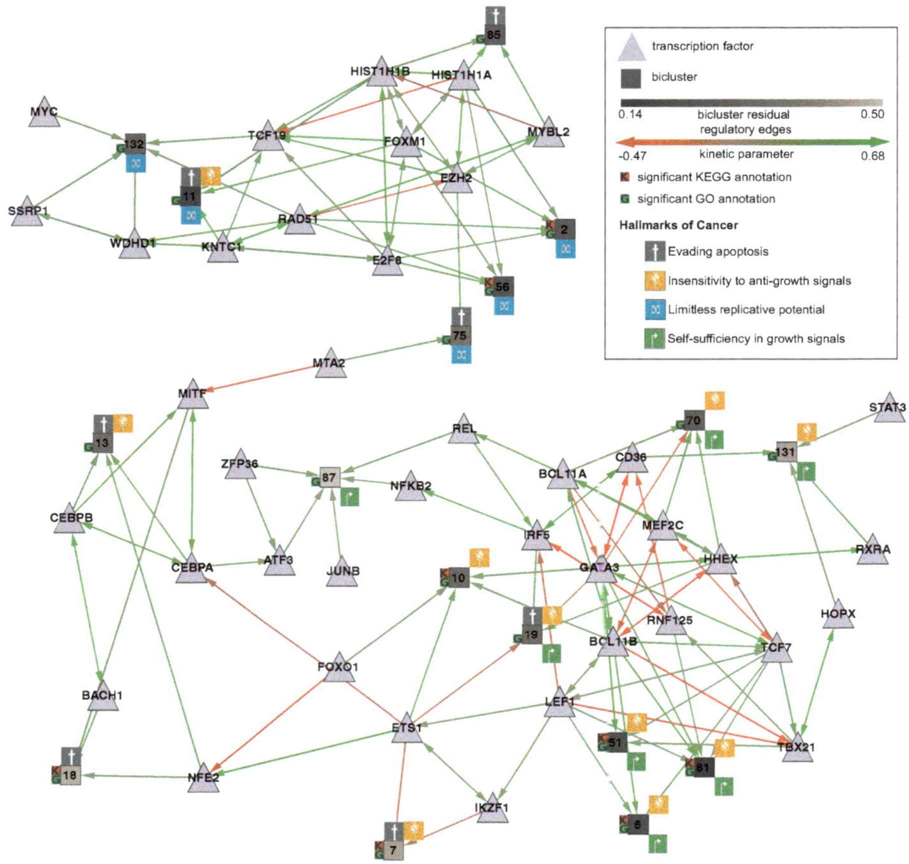

Plate 2 (Figure 2.3 on page 34 of this volume).

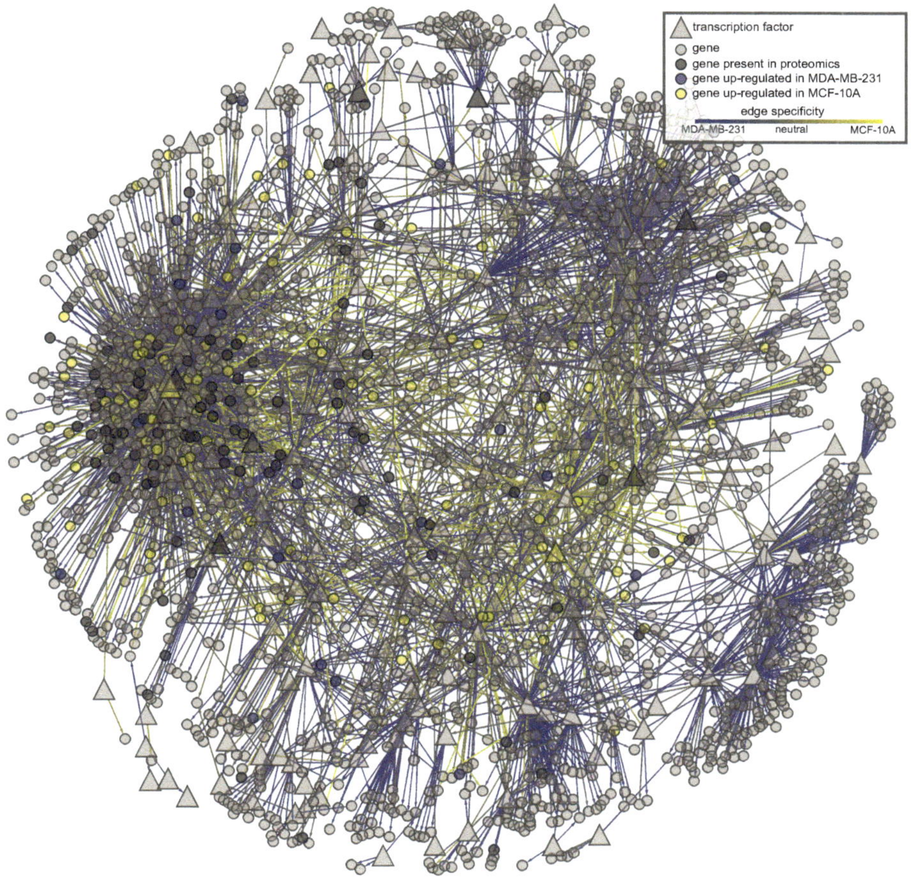

Plate 3 (Figure 2.4 on page 35 of this volume).

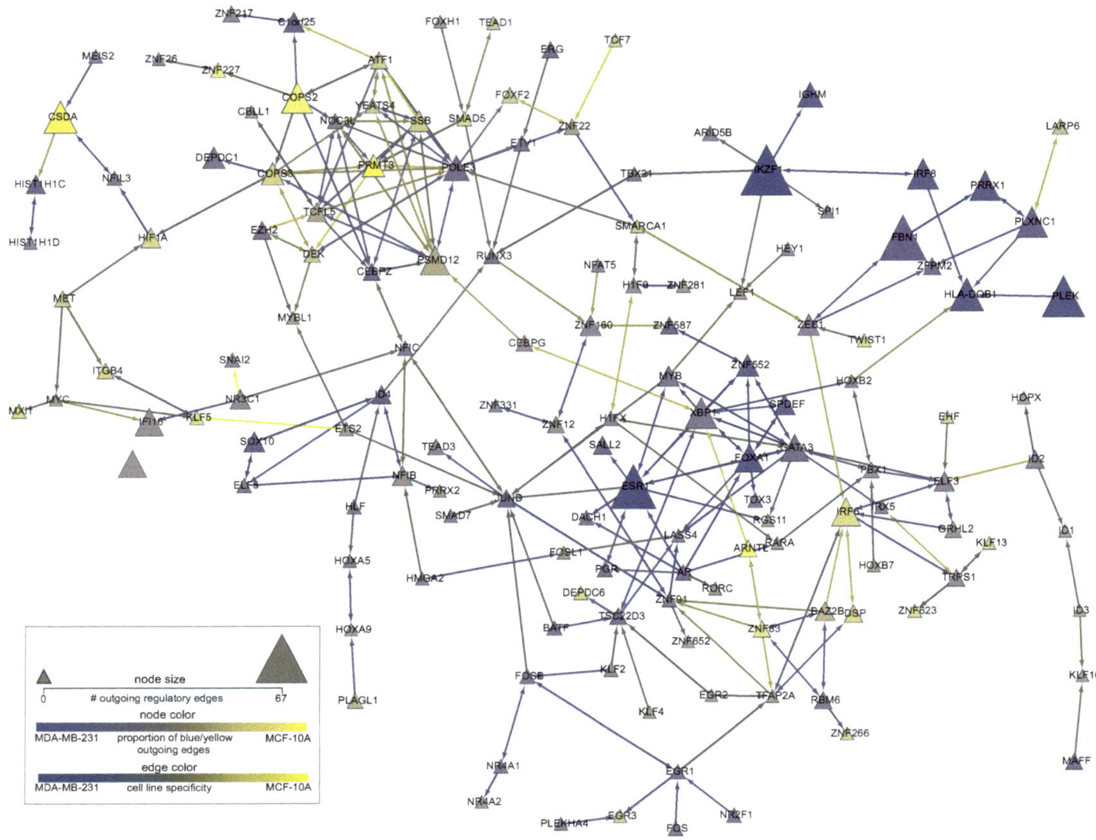

Plate 4 (Figure 2.5 on page 36 of this volume).

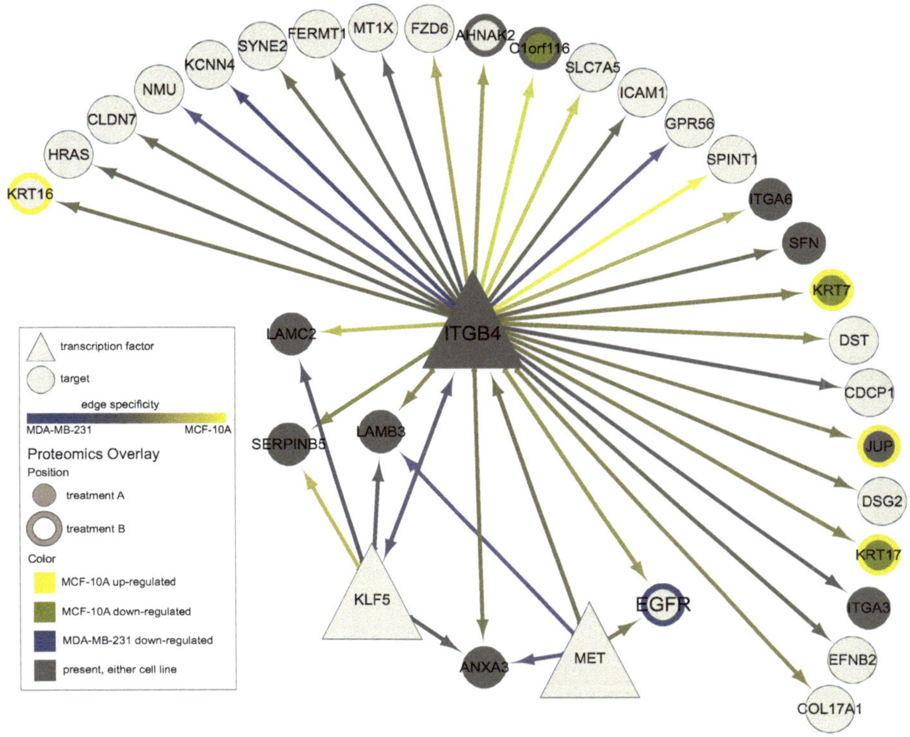

Plate 5 (Figure 2.6 on page 37 of this volume).

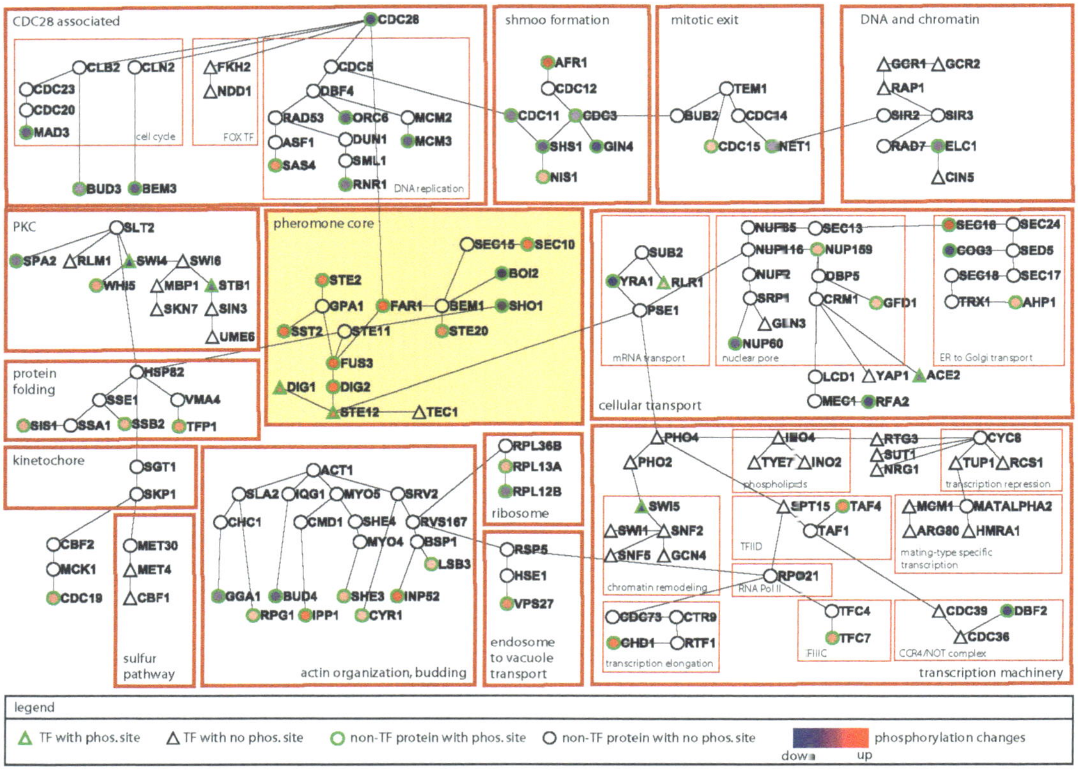

Plate 6 (Figure 3.2 on page 63 of this volume).

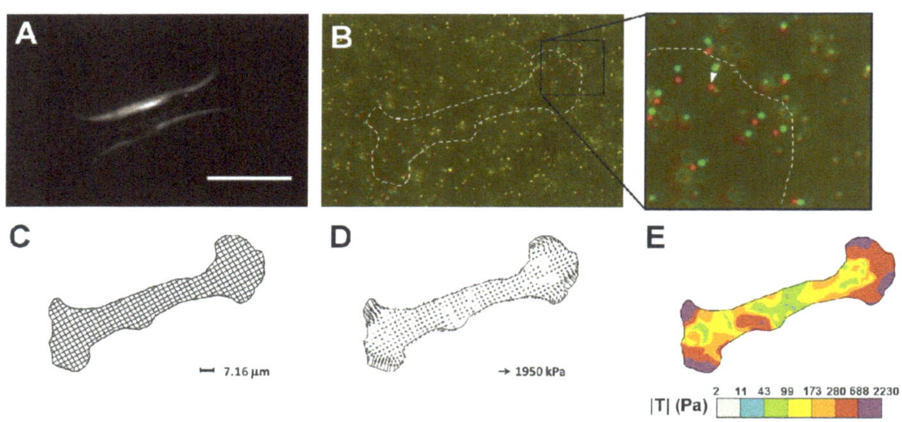

Plate 7 (Figure 6.1 on page 143 of this volume).

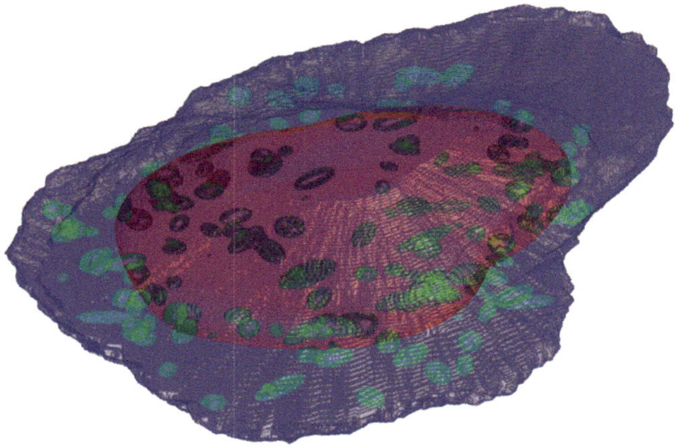

Plate 8 (Figure 7.7 on page 189 of this volume).

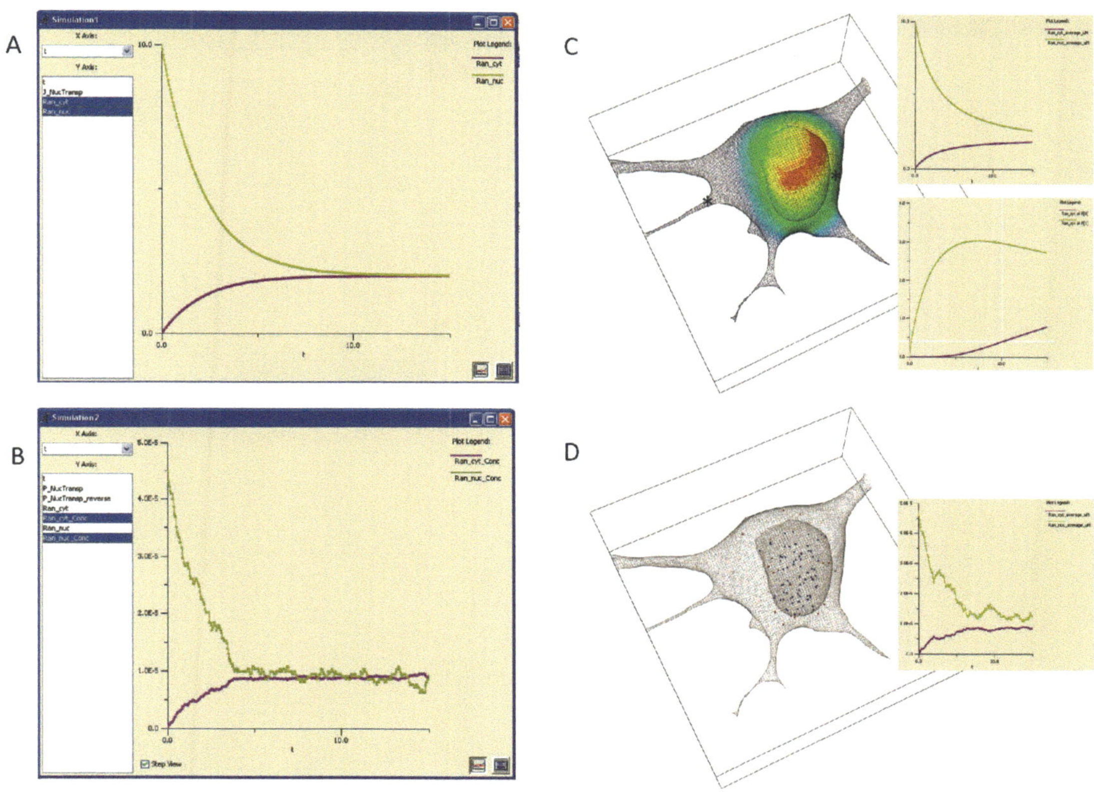

Plate 9 (Figure 8.1 on page 199 of this volume).

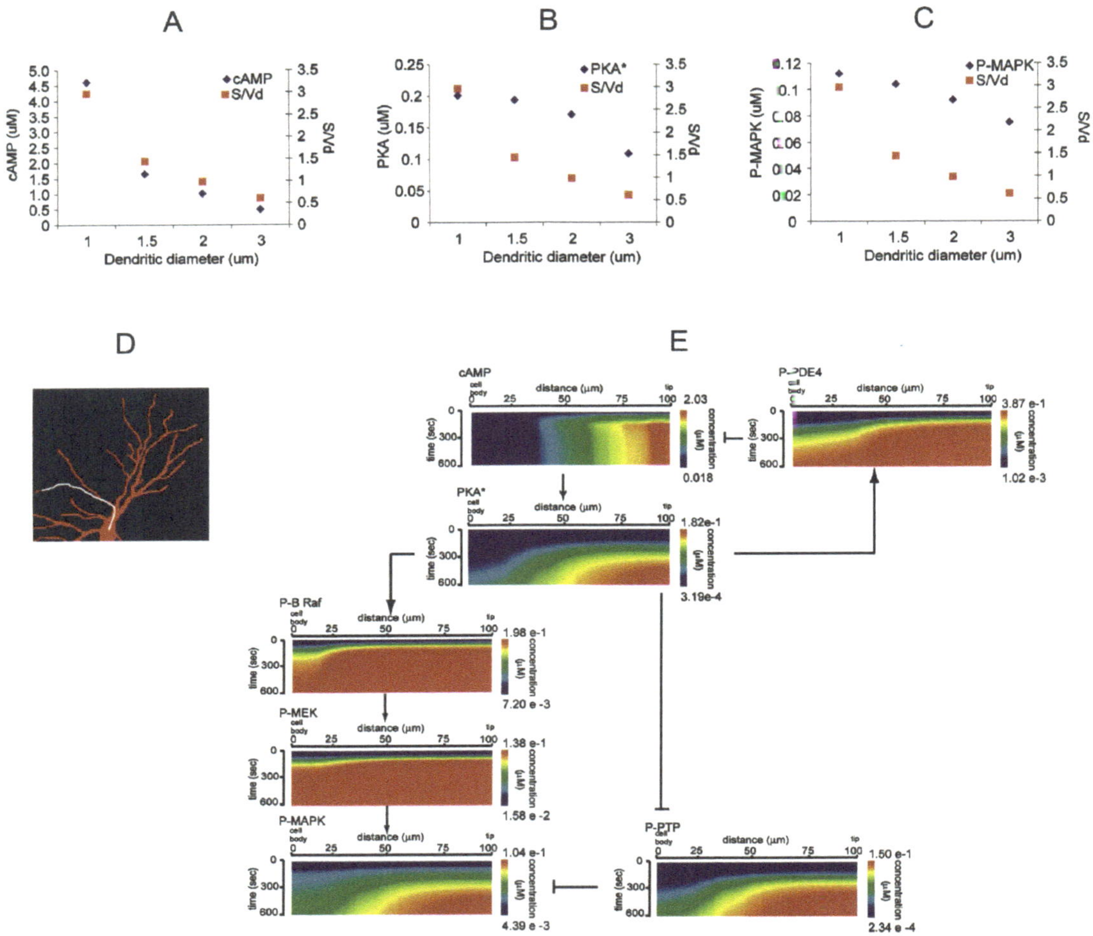

Plate 10 (Figure 8.8 on page 215 of this volume).

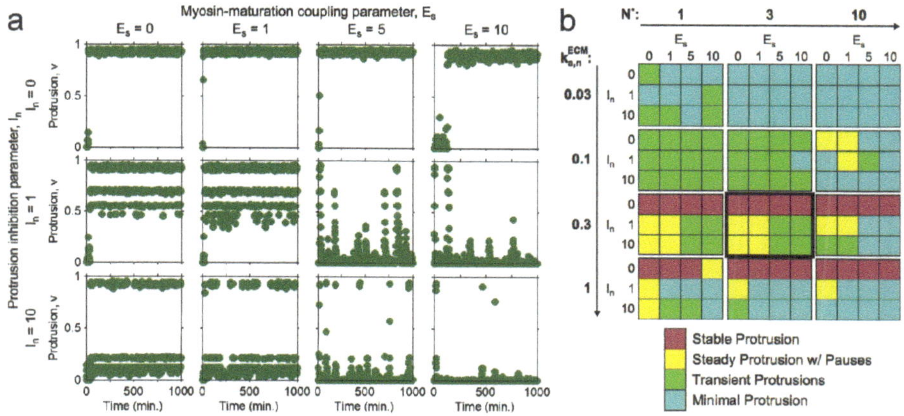

Plate 11 (Figure 9.3 on page 230 of this volume).

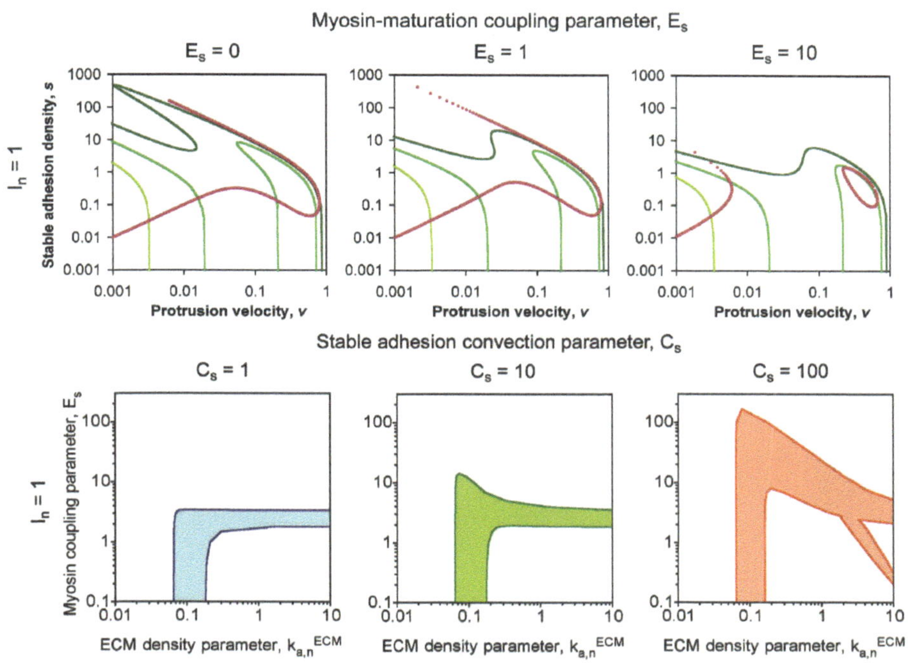

Plate 12 (Figure 9.6 on page 232 of this volume).

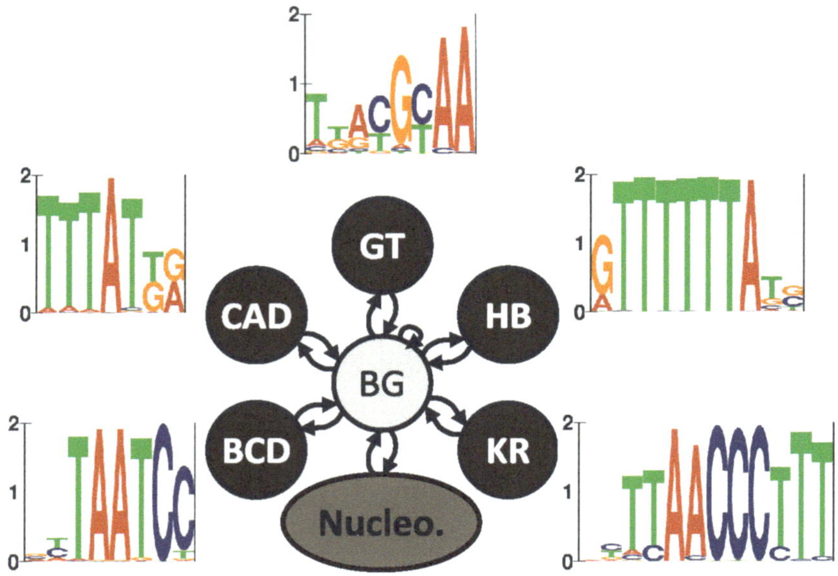

Plate 13 (Figure 11.1 on page 267 of this volume).

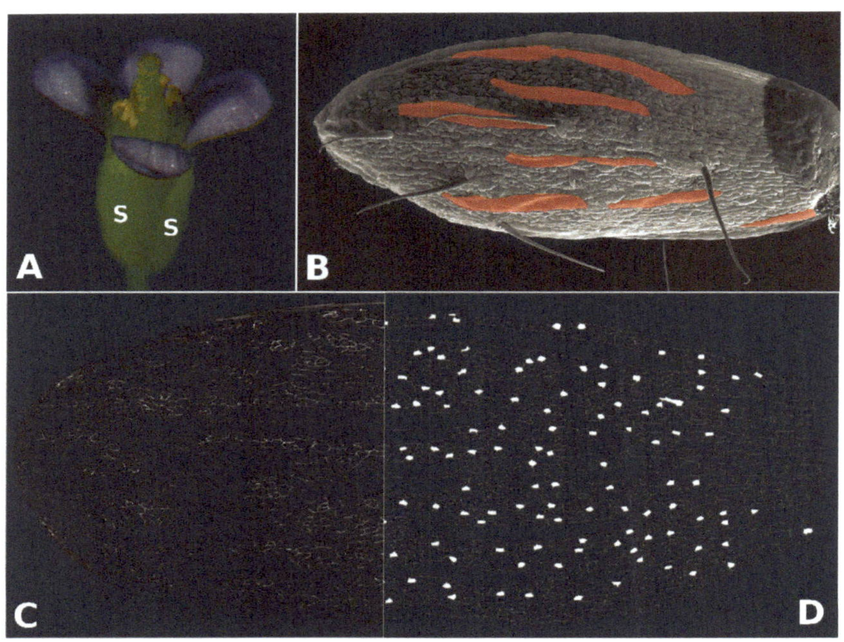

Plate 14 (Figure 12.1 on page 289 of this volume).

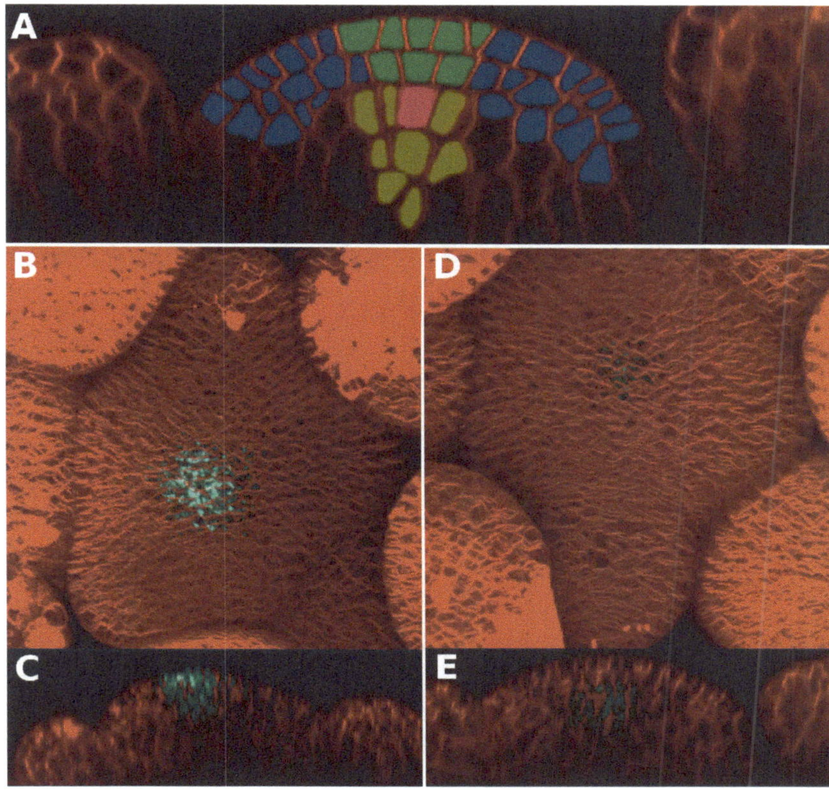

Plate 15 (Figure 12.2 on page 290 of this volume).

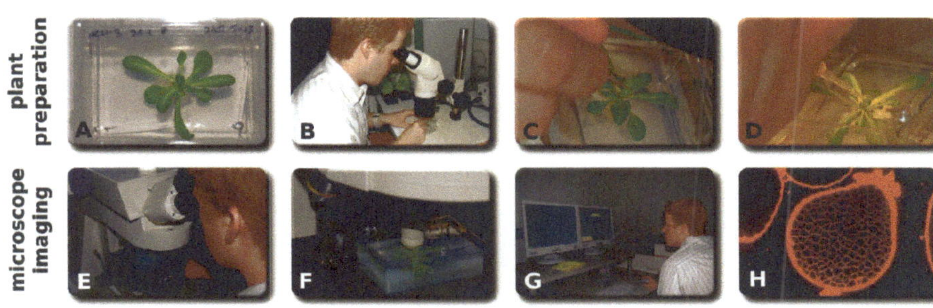

Plate 16 (Figure 12.4 on page 302 of this volume).

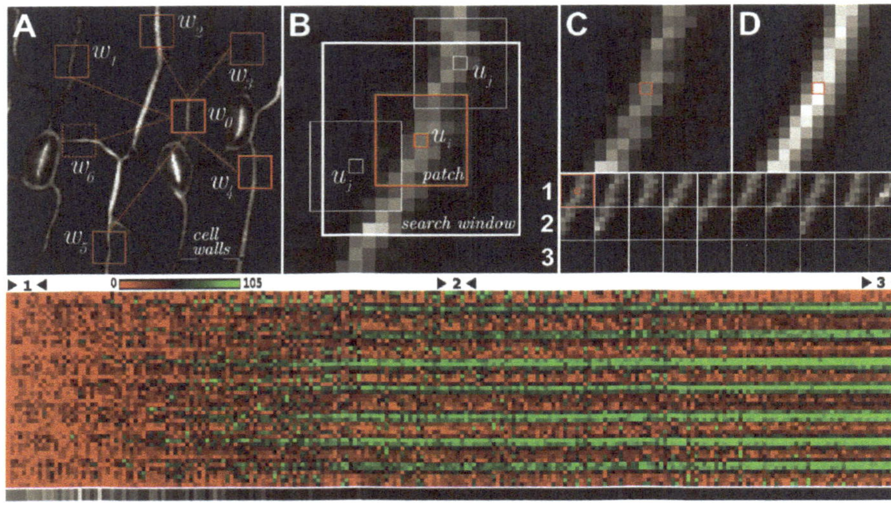

Plate 17 (Figure 12.5 on page 310 of this volume).

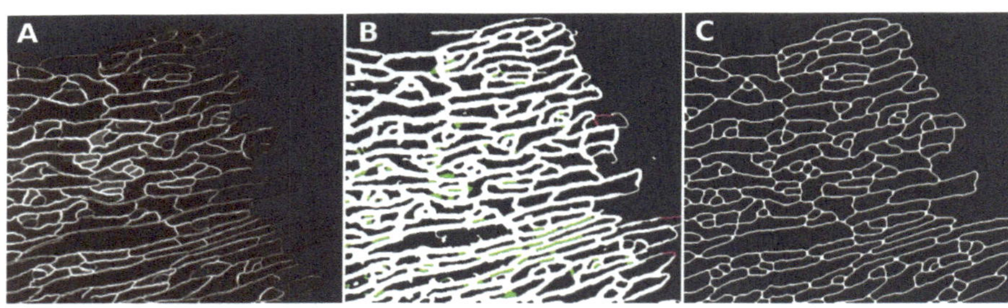

Plate 18 (Figure 12.7 on page 316 of this volume).

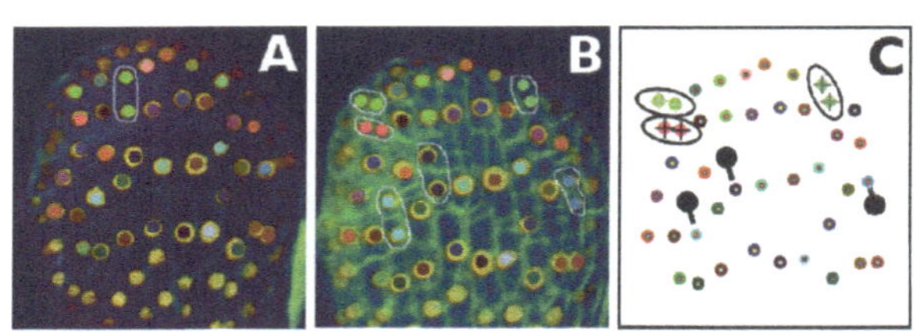

Plate 19 (Figure 12.11 on page 319 of this volume).

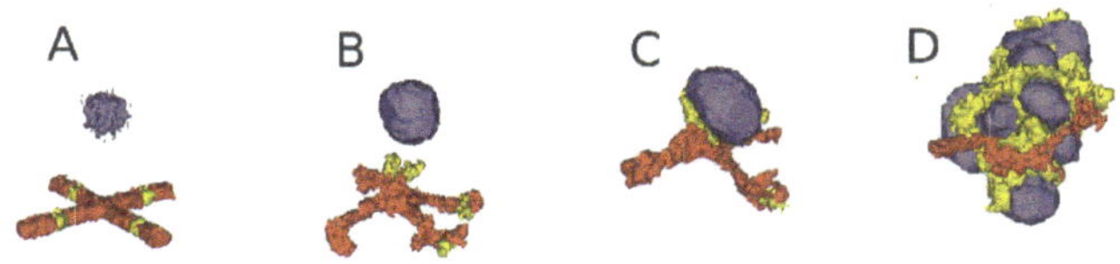

Plate 20 (Figure 13.17 on page 355 of this volume).

Plate 21 (Figure 13.20 on page 361 of this volume).

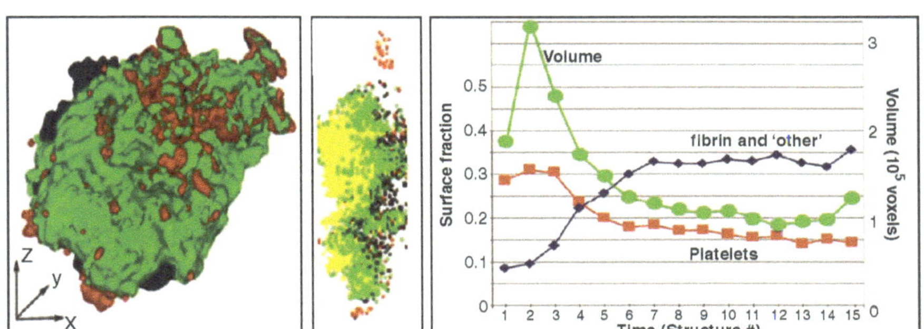

Plate 22 (Figure 14.1 on page 369 of this volume).

Kolibri 360 GmbH
Ferdinand-Jühlke-Straße 7
99095 Erfurt, Deutschland
produktsicherheit@kolibri360.de